Solid Electrolytes for Advanced Applications

Ramaswamy Murugan · Werner Weppner
Editors

Solid Electrolytes for Advanced Applications

Garnets and Competitors

Springer

Editors
Ramaswamy Murugan
Department of Physics
Pondicherry University
Puducherry, India

Werner Weppner
LS Sensorik und Festkörperionik
Universität zu Kiel
Kiel, Germany

ISBN 978-3-030-31580-1   ISBN 978-3-030-31581-8 (eBook)
https://doi.org/10.1007/978-3-030-31581-8

This Springer imprint is published by the registered company Springer Nature Switzerland AG
The registered company address is: Gewerbestrasse 11, 6330 Cham, Switzerland

Preface

Epochal changes are presently going on in the automotive industry. Mobility is a key need of our life and it has become obvious that we cannot continue for much longer burning oil in view of pollution and abundance. Regenerative sources have to be used for our increasing consumption of energy. Fortunately, solar cells and wind turbines provide high efficiencies in the meantime, but the energy needs to be stored both for mobile and stationary applications. Storage of generated electrical energy directly as electrical energy in batteries without changing the energy form is the most efficient way. However, the performance of conventional batteries is limited mainly with regard to not only energy density but also other shortcomings. This has changed with the development of lithium-ion batteries in view of the high reaction energy of lithium. It is clear that this technology is very important, but is a first step only. It is necessary to improve this technology in view of safety, cost, energy density, and lifetime. The main problems come from the liquid organic electrolyte. There is hardly any other solution visible than the use of chemically stable solid electrolytes, where the newly discovered lithium garnets have presently the highest priority. All components are readily available, inflammable, nonpoisonous, environmentally benign, safe, and will provide long lifetimes of the batteries. These improvements will become important not only for mobility but also for stationary energy storage. In the future, every home will store their own generated energy, e.g., by solar cells on the roof, in safe small batteries. We will largely become independent of utility companies and power grids. The new solid electrolyte based batteries will therefore have an enormous economical effect besides solving the problems of pollution and limited energy density as well as safety of the present day liquid electrolyte batteries.

Besides the new generation of all-solid-state batteries, garnet-type solid electrolytes are also important for a variety of other applications, such as supercapacitors, all-solid-state electrochromic displays, and chemical sensors for environmental and process gases with direct conversion of concentrations into electrical signals.

The present book is based on the "1st World Conference on Solid Electrolytes for Advanced Applications: Garnets and Competitors" held at Puducherry, India, where the most recent findings were discussed among the most actively involved scientists in the field. The outcome is the present description of the scientific results of garnet-based solid electrolytes and electrochemistry as well as material aspects and the state of practical applications. The reader can take advantage of this compilation of all scientific and practical aspects of recent developments in the field of garnet-type solid electrolytes.

Puducherry, India Ramaswamy Murugan
Kiel, Germany Werner Weppner

Contents

Part I
Solid Electrolyte

Chapter 1
Solid-State Electrolytes: Structural Approach

Suresh Mulmi and Venkataraman Thangadurai

Abstract The chapter systematically describes how the structural framework dictates the pathways for ion mobility (e.g., 1D, 2D and 3D) in solid-state electrolytes. In lithium-stuffed garnets, for example, Li^+-ion shows three-dimensional nature of ion transport; whereas, the motion of same Li^+-ion occurs in one- and two-dimensions in β-eucryptite ($LiAlSiO_4$) and Li_3N, respectively. In addition to Li^+-ion, Na^+, H^+ and O^{2-} ion-conducting solid-state electrolytes are also introduced in the chapter recognizing their greater importance on developing novel materials for renewable energy applications.

1.1 Introduction

Mass transport in solids was known since Faraday's period. Solid electrolytes exhibiting proton, lithium, sodium, potassium, silver, fluoride, and oxide ion conduction have shown promising technological scope for the development of a wide variety of all-solid-state electrochemical devices such as batteries, fuel cells, gas sensors, capacitors and smart windows. Several solid electrolytes exhibit ionic conductivity in the order of 10^{-3} to 10^{-1} S cm^{-1} at room temperature; which is comparable to liquid electrolytes. This chapter explains the transportation of ions in their crystalline framework and conduction phenomenon. The ion mobility in the structural lattice of ceramics contributing to higher conductivity depends upon various parameters, including concentration and size of active mobile ions, lattice structure and defect formations, conduction mechanisms and ions' mobility.

The term 'solid state ionics' became popular in the 1960s when interest in solid electrolytes was immense due to its real applications in batteries. For example, a fast Na-ion transport phenomenon discovered in layered (2D) Na-β-alumina ($Na_2O.11Al_2O_3$) for Na-ion batteries [1, 2]. Subsequently, Goodenough demonstrated Na super ion conductor (NASICON) with chemical formula of

S. Mulmi · V. Thangadurai (✉)
Department of Chemistry, University of Calgary, 2500 University Drive NW, Calgary T2N 1N4, Canada
e-mail: vthangad@ucalgary.ca

R. Murugan and W. Weppner (eds.), *Solid Electrolytes for Advanced Applications*,
https://doi.org/10.1007/978-3-030-31581-8_1

$Na_{1+x}Zr_2Si_xP_{3-x}O_{12}$; $(0 \leq x \leq 3)$ that exhibited excellent room-temperature conductivities (~10^{-4} S cm^{-1}) and showed anisotropic Na-ion conductivity unlike 2D Na-β-alumina [3–5]. Figure 1.1 shows brief historical timeline of the development of fast-ion conducting solid electrolytes.

In an era of battery research that progressed on liquid electrolytes, the discoveries of solid-state fast-ion conductors certainly helped to move forward to overcome the safety and stability issues associated with organic and aqueous liquid electrolytes. Similar to aqueous/liquid electrolyte, commonly used in batteries, solid-state electrolyte involves mass and charge transport—unlike electron transport in metals and semiconductors. In contrast to liquid electrolytes, solid electrolytes exhibit predominately anion or cation conductivity, and it is very uncommon to have both these charges equally mobile in solid-state system like in aqueous electrolytes. Therefore, intensive research on ion-conducting solid electrolytes—exhibiting fast ionic conductivity at room temperature in the range close to aqueous/liquid electrolytes—could be alternatives for fabricating all-solid-state electrochemical devices. The additional reasons for developing solid-state electrolytes versus its liquid counterpart are electrode corrosion, leakage and evolution of toxic gases (safety). These issues could easily be fixed by utilizing solid electrolytes with suitable ion conduction at their operating temperature.

The first attempt on fabricating all-solid-state battery was made using alkali/silver halides. However, the insulating behaviour of these halides at room temperature due to high activation energy made them poor electrolyte choices. Phase transition of AgI from β-AgI (space group *P6₃mc*)—a phase that is stable at room temperature—to highly conducting α-AgI (space group *Im-3m*, stable at 147–555 °C) offered more than the equivalent sites for the Ag^+-ions mobility (Fig. 1.2). Introducing impurity ions via aliovalent doping is one of the common strategies to improve the ionic conductivity in solid electrolytes [7]. Oxygen-ion conduction in yttria-stabilized zirconia (YSZ) is one of the famous examples of solid electrolytes that show excellent oxide ion conductivity at high temperature due to chemical substitution (doping). Millions of oxygen sensors are manufactured each year using YSZ to check the air/fuel ratio in automobiles. This phenomenon was first practically implied as broadband light source almost 120 years ago [8–10]. The gas sensor device operated under the principle of motion of defects in oxide-ion lattice of ZrO_2 [11]. However, it took

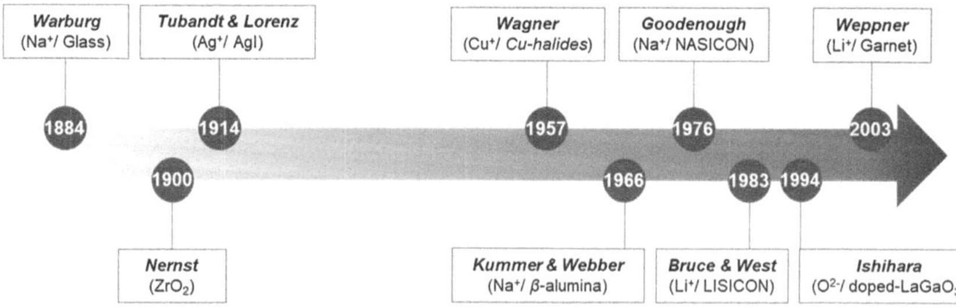

Fig. 1.1 A brief historical outline of the development of solid-state electrolytes

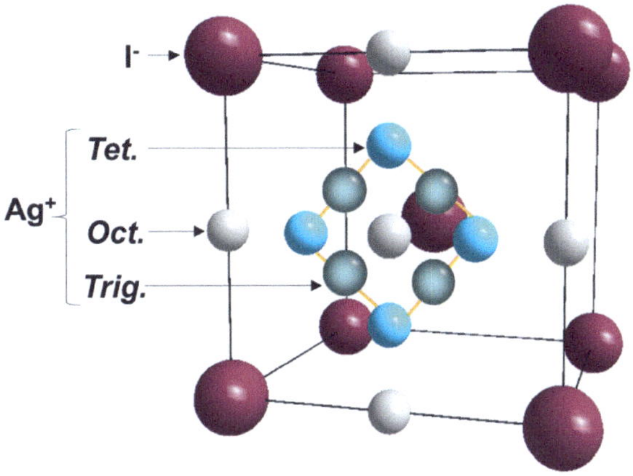

Fig. 1.2 Crystal structure of α-AgI, showing the bcc anion (I^-) sublattice and the locations of Ag^+ cations in octahedral (*Oct*), tetrahedral (*Tet*) and trigonal (*Trig*) positions (structure drawn using crystallographic information file (CIF) from [6])

several years to find the proper solid electrolyte with high ionic conductivity at room temperature. MAg_4I_5 (M = K, Rb) [12, 13] and Na-β-alumina [14] are typical solid electrolytes that show ionic conduction of ~10^{-1} S cm^{-1} of Ag^+ and Na^+, respectively at 25°C (Table 1.1). Since these discoveries, several solid electrolytes exhibiting H^+, Li^+, Cu^+, F^- and O^{2-} have been intensively studied. It has been more than about six decades—these fast-ion conducting materials have been widely referred as 'Solid Electrolytes' or 'Fast Ion Conductors' or 'Ceramic Electrolytes' [15–17].

Solid-state electrolytes are generally designed with the following requirements for application in solid-state ionic devices: (i) highly ionically conducting and electronically insulating, (ii) chemically stable at high temperatures under oxidizing and reducing environments, (iii) dense and free from porosity, (iv) thinner and uniform (lowering ohmic losses) and (v) thermal expansion coefficient close to electrodes. There are merely few solid electrolytes which suffice all these requirements. As a result, this brings challenges and opportunities for further improvements. The crystalline structured framework generally determines the nature of conduction (fast or slow) in solid-state devices under their operating conditions. Therefore, it is worth understanding the fundamental aspects of ions migrating through the solid-state electrolytes. Thus, this chapter is devoted to describing the ion transportation phenomenon in crystalline network in solid-state system.

1.2 Ion Conduction in Solids

In this chapter, ionic conduction in crystalline/polycrystalline electrolytes will be discussed very briefly, and description on amorphous and solid polymer electrolytes are not considered in this chapter. Crystalline solids show ordered arrangement of atoms in three dimensions. In order for an ion to move, ions are either partially occupied in their lattice and/or structure should exhibit interstitial vacant sites for ion migration.

Table 1.1 Selected examples of 1D, 2D and 3D solid-state electrolytes

Compound	σ (S cm^{-1})	T (°C)	Ea (eV)	Dimension	Lattice constant (Å)	Space group	References
Ag^+-ion conductors							
α-AgI	1.6	200	0.10	3D	a = 5.106	*Im-3m* (229)	[18, 19]
Ag-β-alumina	6.7 × 10^{-3}	25	0.16	2D	a = 5.106 c = 22.5131	*P63/mmc* (194)	[14]
$RbAg_4I_5$	2.5 × 10^{-1}	25	0.07	3D	a = 11.2393(4)	*P4$_1$32* (213)	[20, 21]
KAg_4I_5	2.1 × 10^{-1}	25	–	3D	a = 11.1582(7)	*P4$_1$32* (213)	[22, 23]
Li^+-ion conductors							
LiI	1.0 × 10^{-7}	25	0.43	3D	a = 6.000(7)	*Fm-3m* (225)	[24, 25]
Li-β-alumina	1.3 × 10^{-4}	25	0.19	2D	a = 7.9080	*P4$_3$32* (212)	[26–28]
Li_3N	4.0 × 10^{-4}	25	0.30	2D	a = 3.652(3) c = 3.866(2)	*P6/mmm* (191)	[29, 30]
$LiAlSiO_4$	4.7 × 10^{-5}	25	0.95	1D	a = 10.4818 c = 11.1750	*P6$_4$22* (181)	[31]
LISICON ($Li_{14}Zn(GeO_4)_4$)	1.3 × 10^{-1}	300	0.50	3D	a = 10.828(2) b = 6.251(1) c = 5. 140(1)	*Pnma* (62)	[32–34]
$La_{0.55}Li_{0.36}TiO_3$	1.5 × 10^{-3}	30	0.33	3D	a = 3.8717(1)	*Pm-3m* (221)	[35–37]
$Li_7La_3Zr_2O_{12}$	7.7 × 10^{-4}	25	0.30	3D	a = 12.9682(6)	*Ia-3d* (230)	[38]
Cu^+-ion conductors							
α-CuBr	5.0 × 10^{-1}	469	–	3D	–	–	[39]

(continued)

Table 1.1 (continued)

Compound	σ (S cm^{-1})	T (°C)	Ea (eV)	Dimension	Lattice constant (Å)	Space group	References
α-CuI	9.0×10^{-2}	400	–	3D	–	–	[39, 40]
KCu_4I_5	6.0×10^{-1}	25	–	3D	–	–	[21, 22]
Na^+-ion conductors							
Na-β-alumina	1.4×10^{-2}	25	0.15	2D	a = 5.5840 c = 22.4500	*P63/mmc* (194)	[41]
NASICON ($Na_3Zr_2PSi_2O_{12}$)	4.5×10^{-1}	300	0.07	3D	a = 8.8043 c = 22.7585	*R-3c* (167)	[4, 42, 43]
O^{2-}-ion conductors							
$Ce_{0.8}Gd_{0.2}O_{1.9}$	5.0×10^{-2}	727	variable	3D	a = 5.425	*Fm-3m* (225)	[44, 45]
$Zr_{0.85}Y_{0.15}O_{2-x}$ (YSZ)	1.2×10^{-1}	1000	0.80	3D	a = 5.132	*Fm-3m* (225)	[46–50]
$La_{0.9}Sr_{0.1}Ga_{0.8}Mg_{0.2}O_{2.85}$	1.0×10^{-1}	800	0.70	3D	a = 5.5740(2) c = 13.6187(1)	*R3c* (161)	[51–54]
H^+ conductors							
$BaCe_{0.85}Yb_{0.15}O_{3-\delta}$	7.0×10^{-4}	300	0.54	3D	–	–	[55]
$BaZr_{0.8}Y_{0.2}O_{3-\delta}$	1.1×10^{-1}	500	0.63	3D	a = 6.0605 c = 8.6855	*Imma* (74)	[56–58]
$Ba_3Ca_{1.18}Nb_{1.82}O_{8.73}$	5.5×10^{-4}	600	0.52	3D	a = 8.3928	*Fm-3m* (225)	[59, 60]

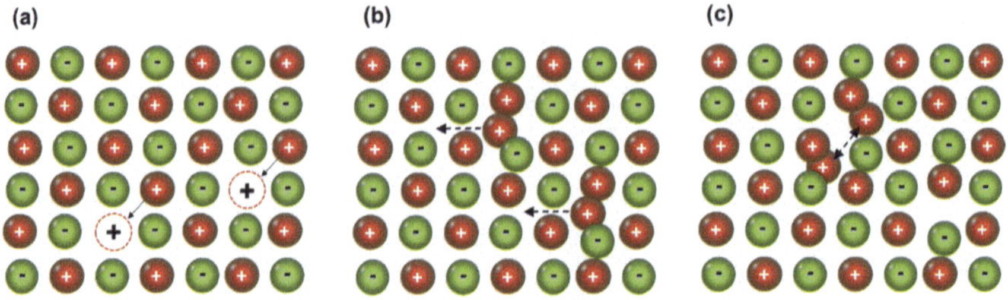

Fig. 1.3 Schematic illustration of ion transportation in solid electrolytes with Schottky/Frenkel defects corresponding: **a** vacancy mechanism, **b** interstitial mechanism and **c** interstitialcy mechanism

In crystalline fast-ion conducting solid electrolyte, ions are facilitated through disordered sublattice that occurs within an enclosed rigid cage-like framework; where, concentration gradients and electrical fields are the typical driving forces. The ion transportation in molten sublattice can be considered as a liquid-like movement.

The ion conduction (σ) in solid-state electrolytes is expressed by the equation [61]:

$$\sigma = n_i q \mu_i \tag{1.1}$$

where, n_i, q and μ_i are the mobile ion 'i' concentration, elementary charge and ionic mobility, respectively. Parameters σ, n, and μ are all temperature dependant factors that follow the Arrhenius-type behaviour. The electrical conductivity is related to the diffusivity of mobile ions—which is given by the following Nernst–Einstein equation [61]:

$$D_i = \frac{\sigma_i k_B T}{n_i z_i^2 q^2} \tag{1.2}$$

where k_B and T are Boltzmann's constant and temperature, respectively.

In general, ion conduction in materials can be explained using Schottky and Frenkel point defects—where the transportation of ions is mainly caused by either vacancy/interstitial jump or interstitially. Schematic diagram of this mechanism is shown in Fig. 1.3.

In solid electrolytes, the ionic conductivity is enhanced by creating highly disordered defects for ions to move; and defects are created by aliovalent doping, which has been successfully applied to synthesize several fast-ion conducting solid electrolytes. Ions in solid electrolytes move through the low energy path (low activation energy). The ionic transport in solid electrolytes follows Arrhenius-type equation [62], i.e.

$$\sigma T = A \exp\left[\frac{-E_a}{k_B T}\right] \tag{1.3}$$

where, A is the pre-exponential factor and E_a is the activation energy for ion transport in solid electrolyte.

For practical purpose, solid electrolytes are supposed to possess ion transference close to unity under the operating temperatures and atmopsheres. The ionic transference number is defined as ratio of ionic conductivity to that of total conductivity. There are several other ways to evaluate the transference number. The determination of transference number becomes easier if only one type of ionic species is predominantly conducting; for example, an oxide-ion conducting membrane (e.g. YSZ)—where the open circuit voltage (OCV) is related to pO_2 at each electrode compartment. For mixed ionic and electron conductors (MIECs), the experimental OCV (E_{exp}) could be related to ion transference number (t_i) by [63]:

$$t_{\text{ion}} = \frac{E_{\text{exp}}}{E_{\text{th}}} \tag{1.4}$$

where E_{th} is the theoretical cell voltage (generally observed when negligible electric conduction occurs in solid electrolytes during cell operation). The electronic conductivity due to free electrons and holes generally show higher conductivity over ionic conductivity in electrode materials.

Figure 1.4 shows the electrical conductivity plot of typical solid-state electrolytes exhibiting cation and/or anion conduction [4, 20, 41, 64–66]. Among the various cation conductors, garnet-type $Li_7La_3Zr_2O_{12}$ with high Li-ion conductivity has drawn a lot of interest regarding its use as an electrolyte in all-solid-state Li-ion batteries [38, 67, 68]. In the case of anion conductors, yttria-stabilized zirconia (YSZ) has already made its mark as a commercial high-temperature oxygen sensor since 1970s. YSZ exhibits high conductivity at elevated temperature (800–1000 °C), where

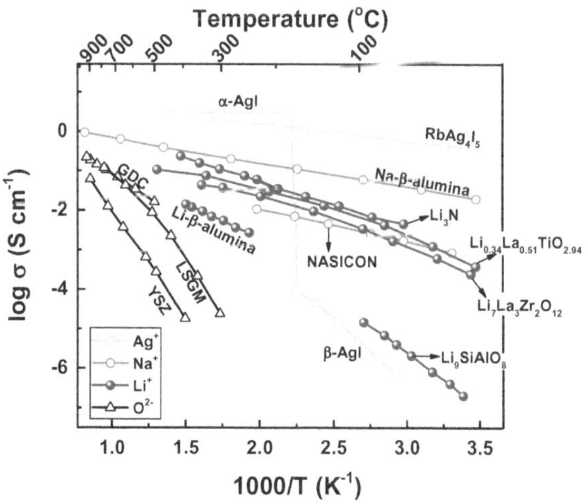

Fig. 1.4 Arrhenius plot of fast-ion conducting materials with conducting species Ag^+, Na^+, Li^+ and O^{2-} ions

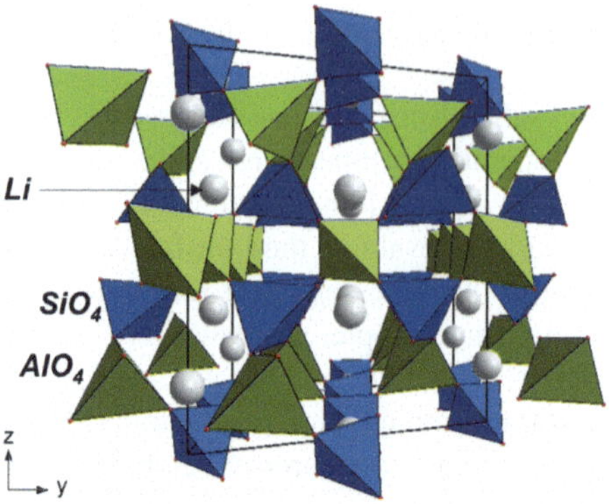

Fig. 1.5 Crystal structure of β-eucryptite (structure drawn using CIF file from [70])

oxide-ion transference number (t_i) is close to unity. The oxide-ion conduction mechanism in YSZ is explained by migration of oxide-ion vacancies. In addition, YSZ possess high chemical and mechanical stability.

Table 1.1 shows the ionic conductivity, activation energy and crystal structure of some known solid electrolytes that show 1D, 2D and 3D pathways for ion migration. The framework structure of crystalline lattice determines the type of paths for ion migration. Generally, layered structure with 2D migration shows lower activation energy compared to 3D for same ion migration, as evident from Na-ion migration in 2D Na-β-alumina versus 3D NASICON structure. In the following section, we describe the ionic conduction in various 1D, 2D and 3D solid-state structures.

1.3 Ion Conduction in 1D Solid-State Structure

1.3.1 Lithium Aluminosilicate

Typical example of 1D ionic conductor is Li-ion conducting $LiAlSiO_4$, where ionic conduction along the *c*-axis is remarkably enhanced. $LiAlSiO_4$ is commonly known as β-eucryptite. The solid solution of lithium aluminosilicate consists of a structural framework of core shared SiO_4 and AlO_4, this results in 1D 'quartz channel' framework along the *c*-axis [31]. The crystalline structure consists of an alternating layer of corner-shared SiO_4 and AlO_4 tetrahedrons. Li ions occupy the 1D channel (Fig. 1.5) [69]. The parent silicate Li_4SiO_4 provides moderate Li-ion conduction. However, substituting Si with Al provides interstitial Li sites, which further increases the mobile Li-ion concentration in the structure.

As shown in Fig. 1.6, the 1D quartz channel in $LiAlSiO_4$ spreads along the *c*-axis. The Li^+ ions alternately occupy the site in quartz channel and exhibit strong anisotropic ionic conduction that runs parallel along *c*-axis [71–73]. In β-eucryptite, the unidirectional (*c*-axis) Li-ion transference measurements were carried out using Li-metal as electrode [74].

1.3.2 Apatites

Traditional fluorite and perovskite-type oxides are considered to be benchmark materials for oxide-ion conduction—where the transportation of ions involves 3D motion. Rare earth oxy-apaties also offer high oxide-ion conductivity at an intermediate temperatures [75–82] and ions in oxy-apatites facilitate through 1D channel. In general, the crystallographic formula for hexagonal apatite structure can be written as $R_{10}(MO_4)O_{2\pm\delta}$, where R and M represent rare-earth or alkaline-earth cation (R = La^{3+}, Mg^{2+}, Ca^{2+}) and *p*-block element (M = Si^{4+}, Ge^{4+}, P^{5+}), respectively. Si-based apatite typically is stabilized in hexagonal symmetry (a = 9.7–9.9 Å; c = 7 Å) with $P6_3/m$ space group [76]. Researchers are interested to explore apatite silicates as an electrolyte in intermediate temperature solid oxide fuel cells (IT-SOFCs) because of their high oxide-ion conduction along the *c*-axis. In apatite silicates, SiO_4 tetrahedra are aligned in such a way that it constructs a framework with 2 channels running along the *c*-axis; where first channel is comprised of rows of La ions in ring formation with oxide ion at the centre and second one is with the rows of only La ions. The interstitial oxide ion moves along the *c*-axis in a sinusoidal-like pathway (Fig. 1.7). In addition to apatite silicates, apatite-type germanates also show significant oxide-ion conduction. Unlike 1D ion migration in silicates, the conduction mechanism in germanates is more isotropic [78, 81].

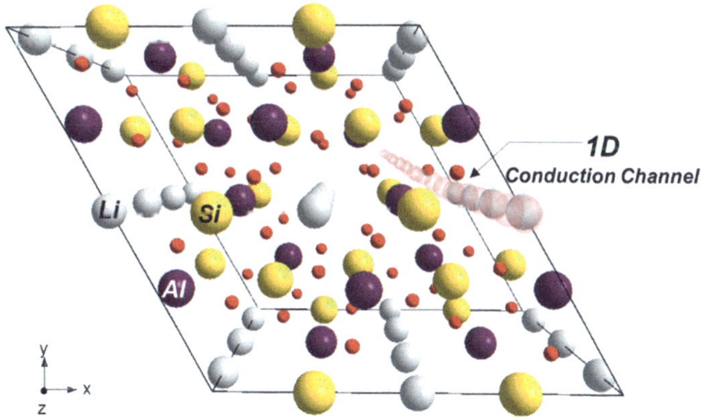

Fig. 1.6 Idealized structure showing 1D conduction pathway of Li^+ in $LiAlSiO_4$ (structure drawn using CIF file from [70])

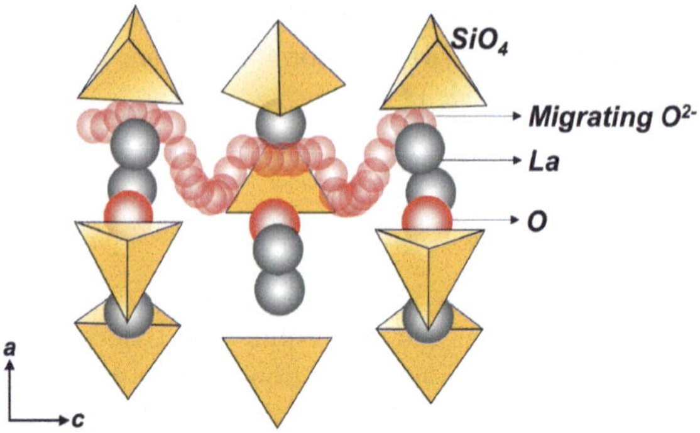

Fig. 1.7 Hexagonal apatite structure ($La_{9.33}(SiO_4)_6O_2$) showing a proposed phenomenon of interstitial oxide-ion migration in a non-linear 1D pathway [78, 80]

1.4 Ion Conduction in 2D Solid-State Structure

1.4.1 Lithium Nitride (Li_3N)

In 1937, Zintl and Brauer first characterized Li_3N and it is crystallized in *P6/mmm* space group [83]. Figure 1.8 shows the crystal structure of alpha polymorph of Li_3N. Li_3N is composed of [Li_2N] layers, where Li ions occupy the space between these layers. This graphite-like 2D structure is comprised of hexagonal layers of planar lithium-nitrogen hexagons. Such NLi_6 layer is connected above and below in the *ab* plane by further lithium ions which links the plane towards the third dimension (along the *c*-axis). There are two more phases of Li_3N, beta and gamma (not discussed here), where nitrogen atoms are packed in a different manner. Boukamp and Huggins showed that polycrystalline

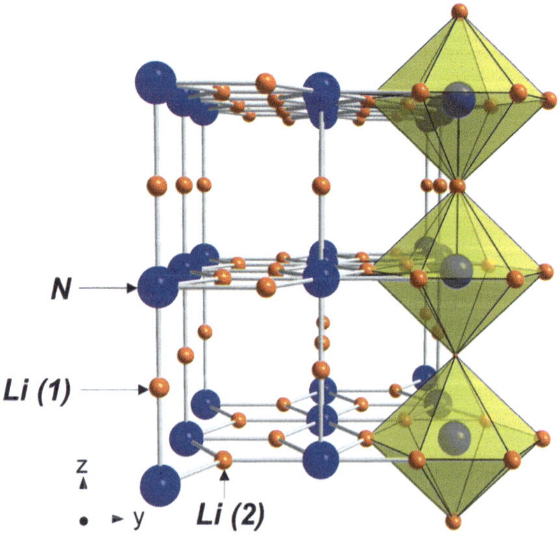

Fig. 1.8 Hexagonal structure adopted by α-Li_3N (structure drawn using CIF file from [84])

Li_3N exhibit remarkable Li-ion conductivity (4×10^{-4} S cm^{-1} at room temperature) [30]. This fast-ion conduction in Li_3N is due to the Li vacancies within the Li_2N layers. At room temperature, α-phase of polycrystalline Li_3N exhibits 3% of Li-ion vacancies [29].

1.4.2 Na-β-Alumina

β-aluminas are layered (2D) structures, where the ions conduct through the loosely packed alternative layers that are stacked between spinel blocks of alumina. In this classic Na^+-ion conductor, β-alumina structures exist mainly in β ($NaAl_{11}O_{17}$) and β" ($NaAl_5O_8$). They differ in stacking arrangement of the spinel blocks and the conduction planes (Fig. 1.9) [85]. Na^+ ions are transported via interstitial mechanism (Fig. 1.3), where excess Na^+ ions occupy interstitial sites. When these interstitial sites are empty (less Na^+ ions on conduction planes)—as is found commonly for β-alumina—the conductivity is highly reduced. In contrast, β"-alumina ($NaAl_5O_8$) conducts Na^+ ions via the vacancy process (Fig. 1.3a)—which seems to have higher Na^+ migration rate compared to β-alumina.

In Na-β-aluminas, Na^+ ions migrate through the 2D channel because the spinel blocks act as a barrier for Na^+ ions to move in 3D. The 2D channels are schematically illustrated in Fig. 1.9. The conductivity of Na-β-alumina is shown in Fig. 1.4 and compared against other superionic conductors, where the activation energy is found to be as low as 0.15 eV for Na-β-alumina. β-alumina structural framework was utilized to study the various monovalent ions' transportation behaviour other than Na^+ [1, 28, 87]. Dunn and co-workers further explored the conductivity of divalent and trivalent cations (e.g. Sr^{2+}, Ca^{2+}, Gd^{3+}, Nd^{3+}) in β-alumina structure [88]. Figure 1.10 compares the ion conductivities of such monovalent, divalent and trivalent cations in β-alumina. The activation energy seems to gradually increase with increasing valency of cations in β-alumina. Similarly, the faster mobility of monovalent cations (except Li^+) in host β-alumina is translated into the higher conductivity [89–91].

1.5 Ion Conduction in 3D Solid-State Structure

1.5.1 NASICON ($Na_3Zr_2PSi_2O_{12}$)

NASICON is primarily a non-stoichiometric solid solution of zirconium phosphate and silicate—which creates a 3D framework and make Na^+ ions to facilitate easily [42]. The framework is built in such a way that Na^+ ions in NASICON show low

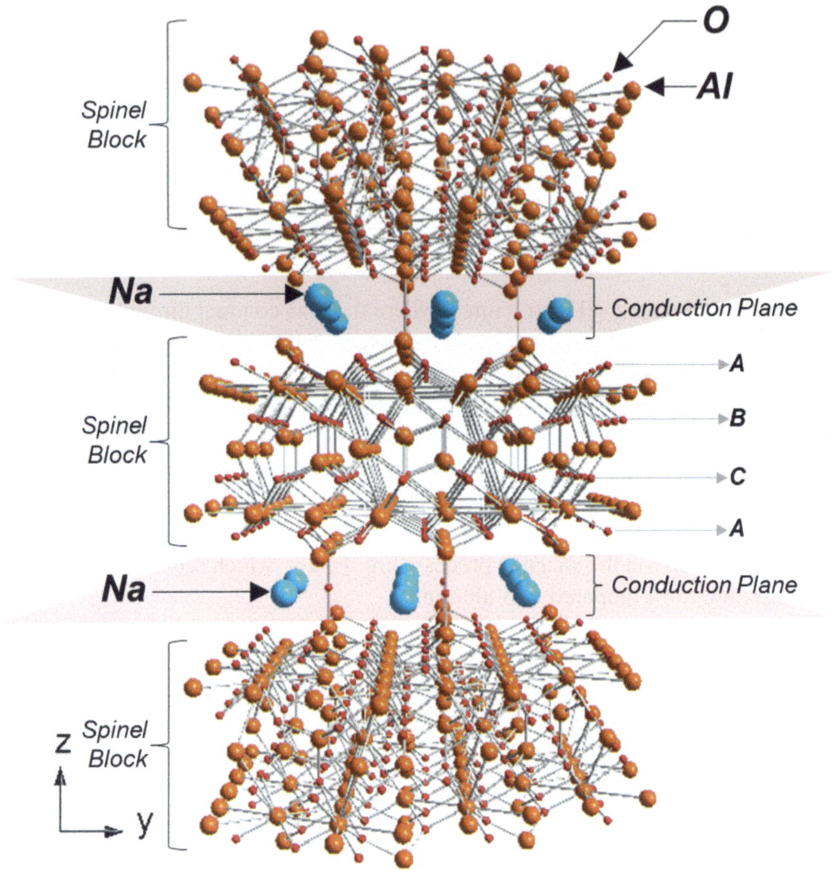

Fig. 1.9 Schematic diagram of Na-β-alumina with Na ions in conduction planes located between densely packed spinel-like blocks of aluminium oxides. The stacking sequence of spinel-like blocks consist of four thick oxide layers in an ABCA sequence (structure drawn using CIF file from [86])

activation energy for conduction. NASICON framework consists of (Si, P)O_4 tetrahedra and ZrO_6 octahedra, which are interconnected to provide 3D transportation of Na^+ through Na(1) and Na(2) sites (Fig. 1.11). Na(1) and Na(2) occupy six- and eight-coordinate sites, respectively.

In NASICON electrolyte, the diffusion path of the mobile ions is not restricted in dimensionality by the structural framework. The stability and flexibility of NASICON framework has led researchers to foster further into NASICON-related 3D framework structures with the inclusion of various transition metals for battery applications [93–98]. The substitution of Na by Li ions in NASICON led to the development of $Li_{1.3}Al_{0.3}Ti_{1.7}(PO_4)_3$ which has shown the highest bulk conductivity for Li^+ ion at room temperature (7×10^{-4} S cm^{-1}) [99, 100].

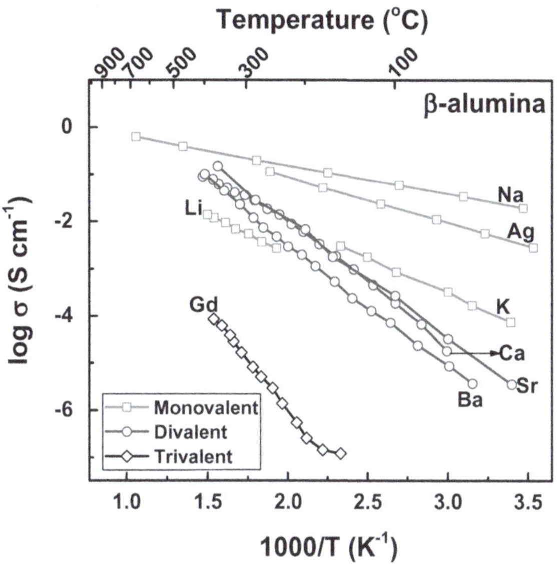

Fig. 1.10 Arrhenius plot with conductivities of mono- (Li^+, Na^+, K^+ and Ag^+), di- (Ca^{2+}, Sr^{2+}, Ba^{2+}) and tri- (Gd^{3+}) valent cations in β-alumina structural framework [88]

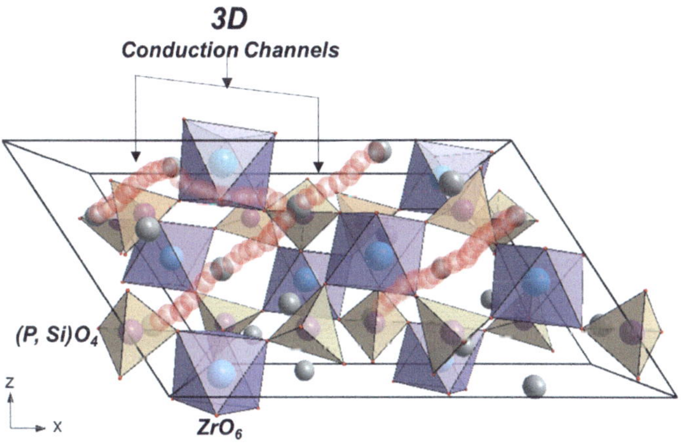

Fig. 1.11 NASICON structure showing possible Na^+-ion migration pathways (structure drawn using the CIF file from [92])

1.5.2 *Oxide-Ion Conductors*

1.5.2.1 Fluorite-Type Oxide-Ion Conductors

Fulorite-structured conductors are commonly utilized as SOFC electrolyte. The idealized structure of fluorite-type oxide is denoted by AO_2, where A corresponds to tetravalent cations such as Zr and Ce. As mentioned above, doping is one of the

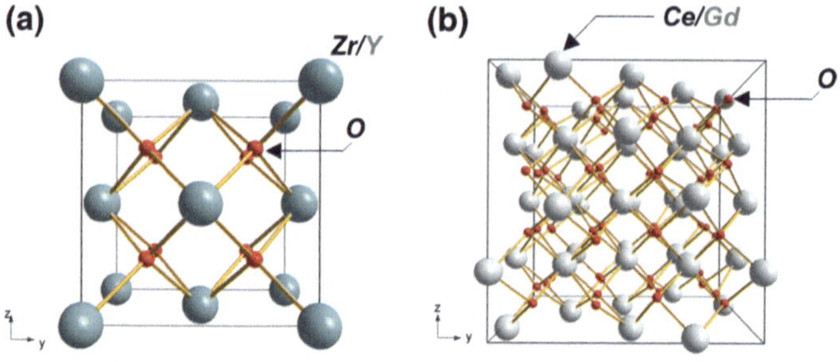

Fig. 1.12 **a** Zirconia- and **b** ceria-based solid electrolytes showing their framework for conducting oxygen ion via doping strategy. High oxide-ion conductivity is obtained by doping (**a**) and (**b**) using Y and Gd as dopant(s), respectively (structures drawn using CIF files from [113, 114])

main strategies in increasing the concentration of oxygen vacancies. For doping, acceptor dopants, typically trivalent cations, are introduced into the cation sublattice of parent fluorite-type oxide to stabilize its cubic structure [101–104]—that is why the best known fluorite-type oxide-ion conductor so-far is yttria-stabilized zirconia (YSZ). The oxide-ion conduction in fluorite-type oxides occurs via vacancy hopping. Another most studied fluorite-type oxide-ion conductor is doped-CeO_2 [105–111]. Typical dopants include gadolinium-doped ceria (GDC) and samarium-doped ceria (SDC). Both GDC and SDC have higher oxide-ion conductivities than YSZ at lower temperatures (<600 °C). In contrast, their use at higher temperature (>600 °C) is not recommended since the t_i for oxide ion decreases due to the significant presence of n-type electronic conductivity at elevated temperatures. The t_i at 500–700 °C for GDC is found to be >0.9 [44, 112]. The structural arrangement of YSZ and GDC are shown in Fig. 1.12. Furthermore, some oxide-ion conducting electrolytes for SOFCs with their conductivity at 800 °C is listed in Table 1.2.

1.5.2.2 Perovskite-Type Oxide-Ion Conductors

In an ideal perovskite ABO_3 large A cation occupies a twelve-coordination-site, whereas B cation with six-coordination site forms a corner-sharing BO_6 octahedral network. Similar to GDC in fluorite-type, LSGM ($La_{1-x}Sr_xGa_{1-y}Mg_yO_{3-\delta}$)—based upon doped-$LaGaO_3$—is widely studied perovskite-type oxide-ion electrolytes. High oxide-ion conduction in perovskite-type doped-$LaGaO_3$ was first reported by Ishihara et al. [123]. LSGM exhibit unit t_i for oxide ions at wide range of pO_2 ($10^{-20} < pO_2 < 1$) [116, 128–132].

The oxide ion in LSGM is facilitated by vacancy hopping between the oxygen sites along a GaO_6 octahedral edge [135]. Both experimental [136] and computational techniques [133, 134] have been used to propose the possible oxide-ion migration mechanism occurring in LSGM. Oxygen-ion migration process in $LaGaO_3$

Table 1.2 Major oxide-ion conducting electrolytes in solid oxide cells (SOCs) with their conductivity at 800 °C

Electrolytes		Composition	$\sigma_{800\,°C}$ (S cm^{-1})	References
Zironica-based	YSZ	$(ZrO_2)_{1-x}(Y_2O_3)_x$ ($x \approx 0.08$–0.1)	2.00×10^{-2}	[115, 116]
	SSZ	$(ZrO_2)_{1-x}(Sc_2O_3)_x$ ($x \approx 0.8$)	1.00×10^{-1}	[117–119]
	CaSZ	$Zr_{0.85}Ca_{0.15}O_{1.85}$	3.16×10^{-3}	[120]
Ceria-based	GDC	$Ce_xGd_{1-x}O_y$ ($x \approx 0.8$, $y \approx 1.8$)	1.58×10^{-1}	[108, 121]
	SDC	$Ce_xSm_{1-x}O_y$ ($x \approx 0.8$, $y \approx 1.9$)	3.55×10^{-2}	[122]
	YDC	$Ce_xY_{1-x}O_y$ ($x \approx 0.8$, $y \approx 1.96$)	2.82×10^{-2}	[105]
Lanthanum based	LSGM	$La_xSr_{1-x}Ga_yMg_{1-y}O_3$ ($x \approx 0.9$, $y \approx 0.8$)	1.58×10^{-1}	[123, 124]
	LSGMC	$La_xSr_{1-x}Ga_yMg_{1-y-z}Co_zO_3$ ($x \approx 0.9$, $y \approx 0.8$)	2.24×10^{-1}	[125]
Other electrolytes	YSB	$(Bi_2O_3)_{1-x}(Y_2O_3)_x$ ($x \approx 0.08$–0.25)	3.16×10^{-1}	[126]
	YSTh	$(ThO_2)_{1-x}(Y_2O_3)x$ ($x \approx 0.08$–0.1)	1.26×10^{-3}	[127]

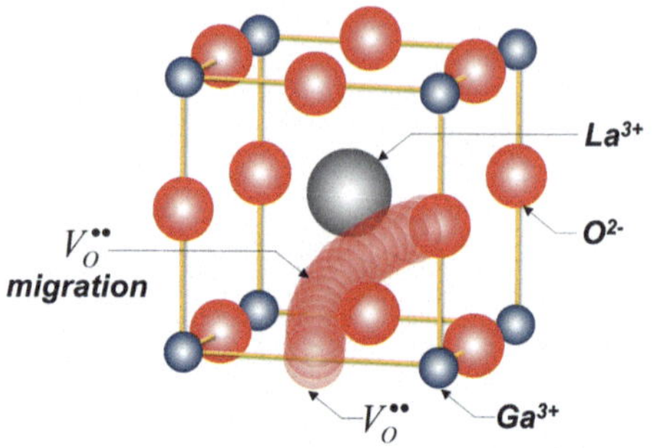

Fig. 1.13 Possible oxygen vacancy migration pathway in $LaGaO_3$ perovskite [133, 134]

perovskite taking a curved path is schematically illustrated in Fig. 1.13. In addition to LSGM, few other perovskites that match the ion conductivities to that of YSZ are: $NdGa_{0.9}Mg_{0.1}O_{2.95}$ [137, 138] and $Gd_{0.85}Ca_{0.15}AlO_{2.925}$ [139].

1.5.3 Li-Ion Conducting Garnets

These garnet oxides are mesosilicates with the common formula $A_3B_2(SiO_4)_3$, where A- and B-sites usually occupy divalent and trivalent cations, respectively. The trade-off between high ion conductivity and high stability for safety purposes is well balanced in Li-ion conducting garnets. In lithium-conducting garnets, lithium generally occupies tetrahedral sites as in an ideal garnet $Li_3Nd_3Te_2O_{12}$. Additional lithium ions can be introduced because the structural framework supports the introduction of more Li ions—few examples include $Li_5La_3M_2O_{12}$ (M = Nb, Ta) [67], $Li_6ALa_2Ta_2O_{12}$ (A = Sr, Ba) and $Li_7La_3Zr_2O_{12}$ [38] by adjusting the valance of the A and B cations in garnets; which reflects a significant structural flexibility of garnet-type oxides. These additional lithium ions are distributed over tetrahedral (*24d*) and distorted octahedral (*48g*/*96h*) sites. From the prospect of conductivity, the total ion conductivity in $Li_7La_3Zr_2O_{12}$ [38] is × 9 orders of magnitude higher (7.7×10^{-4} S cm^{-1} at 25 °C) compared to $Li_3Ln_3Te_2O_{12}$ [68]. Mobile Li ions (grey spheres) in a unit cell of a cubic garnet is depicted in Fig. 1.14.

Perovskite-type oxide is another good framework to host Li ions and facilitate its migration. $La_{2/3}TiO_3$ is one of those perovskites, where one-third of A-site cations is deficient [141]. For mobility of Li, some of the La is partially substituted by lithium. $La_{0.5}Li_{0.34}TiO_{2.94}$ (LLT) has lithium-ion conductivity of 1×10^{-3} S cm^{-1} at room temperature [142], with the activation energy reaching ~0.30 eV [143]. A schematic diagram showing the possible paths for mobile Li ions is drawn in Fig. 1.15. Table 1.3 shows examples of some solid Li-ion electrolytes.

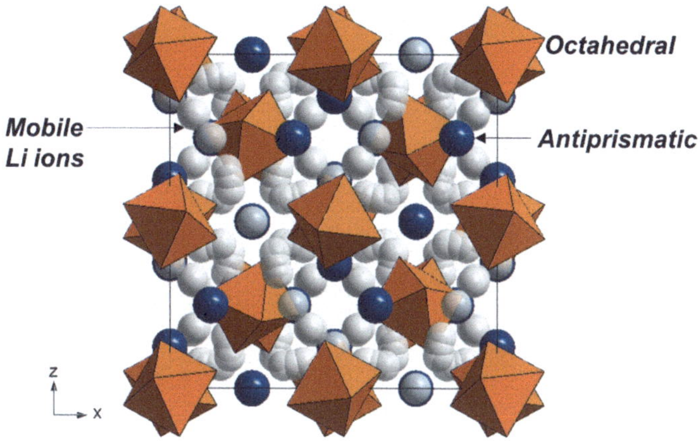

Fig. 1.14 A cubic unit cell of ideal garnet with space group = *Ia-3d* (230). Mobile Li ions are shown in grey spheres (structure drawn using CIF file from [140])

Fig. 1.15 Lithium-ion conduction path via A-site (La-site) of perovskite-type $La_{0.55}Li_{0.36}TiO_3$ (vacancies in La-sites are not shown for clarity) (structure drawn using CIF file from [144])

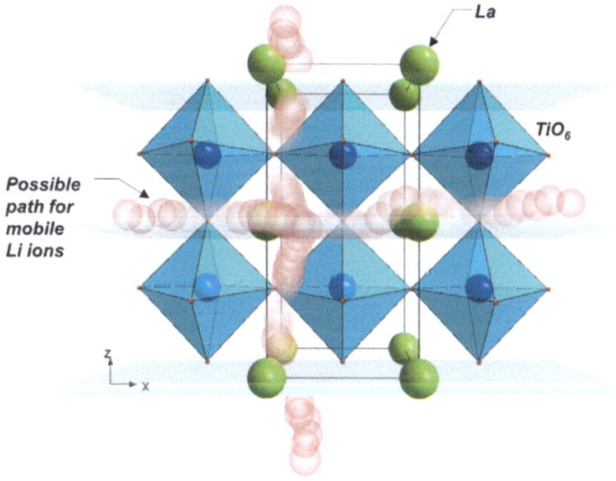

1.5.4 Proton Conductors

Compared to oxide-ion conduction in ceramic oxides, proton conduction generally has lower activation energy (E_a) (Tables 1.1 and 1.2). Therefore, proton-conducting metal oxides have huge potential for high electrochemical performance, especially at relatively lower temperatures versus oxide-ion conducting materials [145–147]. The proton conduction in ceramic oxides—filling oxygen-ion vacancies with hydroxyl ions at high temperature—was originally proposed by Stotz and Wagner in 1966 [148]. Iwahara further expanded the field by performing systematic investigations in 1980 [149–151]. The transportation of protons in ceramics is unique in itself as they are incorporated into the structure from water vapour or hydrogen gas at high temperatures.

Table 1.3 Some solid Li-ion conductors with high conductivity and low activation energy

Li-ion conductors	Lattice constant (Å)	Space group	E_a (eV)	σ (S cm^{-1})	References
$LiZr_2(PO_4)_2$	$a = 8.85$ $c = 22.24$	*R-3c* (167)	0.43 (300–400°C)	5.0×10^{-3} (350 °C)	[152]
$LiTi_2(PO_4)_2$	$a = 8.512$ $c = 20.858$	*R-3c* (167)	0.30 (30–200°C)	6.3×10^{-3} (350 °C)	[153]
$Li_{6.4}La_3Zr_{1.4}Ta_{0.6}O_{12}$	$a = 12.923$	*Ia-3d* (230)	0.35 (25–157°C)	1.0×10^{-3} (25 °C)	[154]
$Li_{6.5}La_{2.5}Ba_{0.5}ZrTaO_{12}$	$a = 12.783(4)$	*Ia-3d* (230)	0.31 (25–350°C)	6.0×10^{-3} (100 °C)	[155]
$Li_{6.5}La_3Nb_{1.25}Y_{0.75}O_{12}$	$a = 12.9488(11)$	*Ia-3d* (230)	0.36 (25–325°C)	1.2×10^{-3} (75 °C)	[156]
$La_{0.55}Li_{0.36}Ti_{0.995}Al_{0.005}O_3$	$a = 3.870(1)$	*Pm-3m* (221)	0.28 (25–100°C)	1.1×10^{-3} (25 °C)	[37]

Typical solid-state electrolytes for proton conduction are zirconates or cerates of alkaline earth elements [157]; where partial substitution with trivalent cations are preferred. In general, these electrolytes exhibit conductivity of 10^{-3}–10^{-2} S cm^{-1} at 600–900 °C under hydrogen gas; i.e. electrochemical transport of hydrogen across the ceramic membrane [150, 158, 159]. $SrCe_{0.95}Yb_{0.05}O_{3-\delta}$ is one of such membranes that shows 1.0×10^{-2} S cm^{-1} proton conductivity at 900 °C under hydrogen-containing atmosphere.

1.6 Conclusions

The high ionic conductivity of solid-state ionic electrolytes can be utilized in a wide variety of applications. All-solid-state battery is one such example where solid electrolyte is employed to avoid the safety issues, which is mainly caused due to the use of liquid/aqueous electrolytes. The control of the transportation of either cations and/or anions within the framework of solid-state electrolyte determines the potential applications. For example, cations such as Li^+ and Na^+ ions are facilitated through the solid-state electrolyte in battery applications; whereas, O^{2-} conducting ceramic membranes are used in several energy sectors including solid oxide fuel cells, oxygen separation and oxygen sensors. In addition, solid electrolytes' application areas has further been expanded to displays and solar cells depending upon the nature of ion and the operation conditions (e.g. temperature). The basic strategy with superionic materials is to take their advantages to overcome the barriers that persist with their liquid/aqueous counterpart—which is possible by embracing the structural approach as one of the tools to solve the structure–property relationships in solid electrolytes; particularly, electrochemical properties.

References

1. Kummer JT (1972) Prog Solid State Chem 7:141
2. Kummer JT, Weber N (1968) SAE Trans 76:1003
3. Takahashi T, Kuwabara K, Shibata M (1980) Solid State Ion 1:163
4. Goodenough JB, Hong HYP, Kafalas JA (1976) Mater Res Bull 11.203
5. Goodenough JB, Hong HYP, Kafalas JA, Dwight K (1976) Massachusetts Institute of Technology, Lexington (USA). Lincoln Lab
6. Wright AF, Fender BEF (1977) J Phys C: Solid State Phys 10:2261
7. Kammampata SP, Thangadurai V (2018) Ionics 24:639
8. Nernst W (1901) United States Patent Office, 1
9. Eddy DS, Trans IEEE (1974) Veh Technol 23:125
10. Cutler RA, Reynolds JR, Jones A (1992) J Am Ceram Soc 75:2173
11. Reginald LA, George WW (1943) United States Patent Office, 1
12. Huber H, Mali M, Roos J, Brinkmann D (1984) J. Phys. Colloques 45:75
13. Awano T, Nanba T, Ikezawa M (1992) Solid State Ionics 53–56:1269
14. Whittingham MS, Huggins RA (1971) J Electrochem Soc 118:1
15. Jonghe LC (1979) J Am Ceram Soc 62:289

16. West AR (1989) Ber Bunsenges Phys Chem 93:1235
17. Steele BCH (1989) High conductivity solid ionic conductors, recent trends and applications, Takahashi T (ed). World Scientific, Singapore
18. Wilsey RB (1923) Lond Edinb Dublin Philos Mag J Sci 46, 487
19. Tubandt C, Lorenz F (1914) Z Phys Chem 87U:543
20. Bradley JN, Greene PD (1967) Trans Faraday Soc 63:424
21. Hull S, Keen D, Sivia D, Berastegui P (2002) J Solid State Chem 165:363
22. Chandra S, Mohabey VK (1979) Phys Status Solidi A 53:63
23. Funke K, Banhatti RD, Wilmer D, Dinnebier R, Fitch A, Jansen M (2006) J Phys Chem A 110:3010
24. Liang CC (1973) J Electrochem Soc 120:1289
25. Poulsen FW (1981) Solid State Ion 2:53
26. Kaneda T, Bates JB, Wang JC (1978) Solid State Commun 28:469
27. Briant JL, Farrington GC (1981) J Electrochem Soc 128:1830
28. Roth WL, Farrington GC (1977) Science 196:1332
29. Gregory DH, O'Meara PM, Gordon AG, Hodges JP, Short S, Jorgensen JD (2002) Chem Mater 14:2063
30. Boukamp BA, Huggins RA (1978) Mater Res Bull 13:23
31. Nagel W, Böhm H (1982) Solid State Commun 42:625
32. Kanno R, Murayama M (2001) J Electrochem Soc 148:A742
33. Bruce PG, West AR (1983) J Electrochem Soc 130:662
34. Hong HYP (1978) Mater Res Bull 13:117
35. Stramare S, Thangadurai V, Weppner W (2003) Chem Mater 15:3974
36. Thangadurai V, Weppner W (2006) Ionics 12:81
37. Thangadurai V, Weppner W (2000) Ionics 6:70
38. Murugan R, Thangadurai V, Weppner W (2007) Angew Chem Int Ed 46:7778
39. Wagner JB, Wagner C (1957) J Chem Phys 26:1597
40. Bührer W, Hälg W (1977) International symposium on solid ionic and ionic-electronic conductors, p 701
41. Whittingham MS, Huggins RA (1971) J Chem Phys 54:414
42. Kreuer KD, Kohler H, Maier J (1989) World Scientific, 242
43. Kang HB, Cho NH (1999) J Mater Sci 34:5005
44. Kudo T, Obayashi H (1976) J Electrochem Soc 123:415
45. Kossoy A, Frenkel AI, Wang Q, Wachtel E, Lubomirsky I (2010) Adv Mater 22:1659
46. Schoch B, Weppner W (1989) Ber Bunsenges Phys Chem 93:1212
47. Filal M, Petot C, Mokchah M, Chateau C, Carpentier J (1995) Solid State Ionics 80:27
48. Kilner JA (2008) Nat Mater 7:838
49. Goodenough JB (2000) Nature 404:821
50. Eichler A (2001) Phys Rev B 64:174103
51. Huang YH, Dass RI, Xing ZL, Goodenough JB (2006) Science 312:254
52. Ishihara T, Akbay T, Furutani H, Takita Y (1998) Solid State Ion 113:585
53. Ishihara T, Furutani H, Honda M, Yamada T, Shibayama T, Akbay T, Sakai N, Yokokawa H, Takita Y (1999) Chem Mater 11:2081
54. Slater PR, Irvine JTS, Ishihara T, Takita Y (1998) J Solid State Chem 139:135
55. Stevenson DA, Jiang N, Buchanan RM, Henn FEG (1993) Solid State Ion 62:279
56. Haile S, Staneff G, Ryu K (2001) J Mater Sci 36:1149
57. Fabbri E, Pergolesi D, Traversa E (2010) Chem Soc Rev 39:4355
58. Pagnier T, Charrier-Cougoulic I, Ritter C, Lucazeau G (2000) Eur Phys J Appl Phys 9:1
59. Malavasi L, Fisher CAJ, Islam MS (2010) Chem Soc Rev 39:4370
60. Iwahara H (1996) Solid State Ion 86–88:9
61. Tsagarakis ED, Weppner W (2005) Ionics 11:240
62. Murugan R, Thangadurai V, Weppner W (2008) Appl Phys A 91:615
63. Weppner W, Huggins RA (1978) Annu Rev Mater Sci 8:269
64. Etsell TH, Flengas SN (1970) Chem Rev 70:339

65. Hu YW, Raistrick ID, Huggins RA (1976) Mater Res Bull 11:1227
66. Steele BCH (1984) Solid State Ion 12:391
67. Thangadurai V, Kaack H, Weppner WJF (2003) J Am Ceram Soc 86:437
68. Thangadurai V, Narayanan S, Pinzaru D (2014) Chem Soc Rev 43:4714
69. Jochum T, Reimanis I (2010) J Am Ceram Soc 93:1591
70. Pillars WW, Peacor DR (1973) Am Miner 58:681
71. Schulz H, Tscherry V (1972) Acta Crystallogr Sect B: Struct Sci 28:2168
72. Schulz H, Tscherry V (1972) Acta Crystallogr Sect B: Struct Sci 28:2174
73. Johnson RT, Morosin B, Knotek ML, Biefeld RM (1975) Phys Lett A 54:403
74. Alpen UV, Schulz H, Talat GH, Böhm H (1977) Solid State Commun 23:911
75. Susumu N, Hiromichi A, Yoshihiko S (1995) Chem Lett 24:431
76. Nakayama S, Kageyama T, Aono H, Sadaoka Y (1995) J Mater Chem 5:1801
77. Sansom JEH, Richings D, Slater PR (2001) Solid State Ion 139:205
78. Tolchard JR, Islam MS, Slater PR (2003) J Mater Chem 13:1956
79. Najib A, Sansom JEH, Tolchard JR, Slater PR, Islam MS (2004) J Phys C Solid State Phys Dalton Trans 3106
80. Sansom JEH, Kendrick E, Tolchard JR, Islam MS, Slater PR (2006) J Solid State Electrochem 10:562
81. Kendrick E, Islam MS, Slater PR (2007) J Mater Chem 17:3104
82. Slater PR, Francesconi MG (1995) Annu Rep Sect A Inorg Chem 92:433
83. Brauer G, Zintl E (1937) Z Phys Chem 37B:323
84. Kawada I, Isobe M, Okamura FP, Watanabe H, Ohsumi K, Horiuchi H, Sato T, Ishii T (1986) Mineral J 13:28
85. West AR (2014) Wiley
86. Beevers C, Ross MA (1937) Zeitschrift für Kristallographie-Crystalline Materials, 97, 59
87. Schäfer GW, Weppner W (1992) Solid State Ion 53–56:559
88. Dunn B, Farrington GC, Thomas JO (1989) MRS Bull 14:22
89. Ni J, Tsai TT, Whitmore DH (1981) Solid State Ion 5:199
90. Farrington GC, Dunn B (1982) Solid State Ion 7:267
91. Dunn B, Farrington GC (1983) Solid State Ion 9–10:223
92. Baur W, Dygas J, Whitmore D, Faber J (1986) Solid State Ion 18:935
93. Padhi AK, Manivannan V, Goodenough JB (1998) J Electrochem Soc 145:1518
94. Thangadurai V, Shukla AK, Gopalakrishnan J (1999) J Mater Chem 9:739
95. Nanjundaswamy KS, Padhi AK, Goodenough JB, Okada S, Ohtsuka H, Arai H, Yamaki J (1996) Solid State Ion 92:1
96. Masquelier C, Padhi AK, Nanjundaswamy KS, Goodenough JB (1998) J Solid State Chem 135:228
97. Ortiz GF, López MC, Lavela P, Vidal-Abarca C, Tirado JL (2014) Solid State Ion 262:573
98. Norhaniza R, Subban RHY, Mohamed NS (2013) J Power Sources 244:300
99. Aono H, Sugimoto E, Sadaoka Y, Imanaka N, Adachi GY (1990) J Electrochem Soc 137:1023
100. Aono H, Sugimoto E, Sadaoka Y, Imanaka N, Adachi GY (1990) Chem Lett 19:1825
101. Smith AW, Meszaros FW, Amata CD (1966) J Am Ceram Soc 49:240
102. Burke LD, Rickert H, Steiner R (1971) Elektrochemische Untersuchungen zur Teilleitfähigkeit, Beweglichkeit und Konzentration der Elektronen und Defektelektronen in dotiertem Zirkondioxid und Thoriumdioxid+. Z Phys Chem 74:146
103. Kitazawa K, Coble RL (1974) J Am Ceram Soc 57:360
104. Hui S, Roller J, Yick S, Zhang X, Decès-Petit C, Xie Y, Maric R, Ghosh D (2007) J Power Sources 172:493
105. Eguchi K, Setoguchi T, Inoue T, Arai H (1992) Solid State Ion 52:165
106. Tuller HL, Nowick AS (1975) J Electrochem Soc 122:255
107. Kilner JA (2008) Chem Lett 37:1012
108. Mogensen M, Sammes NM, Tompsett GA (2000) Solid State Ion 129:63
109. Mogensen M, Lybye D, Bonanos N, Hendriksen PV, Poulsen FW (2004) Solid State Ion 174:279

110. Steele BCH (2000) Solid State Ion 129:95
111. Inaba H, Tagawa H (1996) Solid State Ion 83:1
112. Milliken C, Guruswamy S, Khandkar A (2002) J Am Ceram Soc 85:2479
113. Lamas D, De Reca NW (2000) J Mater Sci 35:5563
114. Kümmerle EA, Heger G (1999) J Solid State Chem 147:485
115. Casselton REW (1974) J Appl Electrochem 4:25
116. Steele BCH, Heinzel A (2001) Nature 414:345
117. Politova TI, Irvine JTS (2004) Solid State Ion 168:153
118. Ishii T, Iwata T, Tajima Y, Yamaji A (1992) Solid State Ion 57:153
119. Badwal SPS, Ciacchi FT, Milosevic D (2000) Solid State Ion 136–137:91
120. Kingery WD, Pappis J, Doty ME, Hill DC (1959) J Am Ceram Soc 42:393
121. Balazs GB, Glass RS (1995) Solid State Ion 76:155
122. Lu C, Worrell W, Gorte R, Vohs J (2003) J Electrochem Soc 150:A354
123. Ishihara T, Matsuda H, Takita Y (1994) J Am Chem Soc 116:3801
124. Feng M, Goodenough J (1995) ChemInform, 26
125. Stevenson J, Hasinska K, Canfield N, Armstrong T (2000) J Electrochem Soc 147:3213
126. Battle P, Catlow C, Heap J, Moroney L (1986) J Solid State Chem 63:8
127. Subbarao EC, Sutter PH, Hrizo J (1965) J Am Ceram Soc 48:443
128. Goodenough JB (2003) Annu Rev Mater Res 33:91
129. Haile SM (2003) Acta Mater 51:5981
130. Brett DJ, Atkinson A, Brandon NP, Skinner SJ (2008) Chem Soc Rev 37:1568
131. Orera A, Slater P (2009) Chem Mater 22:675
132. Kilner J (2007) Faraday Discuss 134:9
133. Khan MS, Islam MS, Bates DR (1998) J Phys Chem B 102:3099
134. Islam MS, Davies RA (2004) J Mater Chem 14:86
135. Kajitani M, Matsuda M, Hoshikawa A, Harjo S, Kamiyama T, Ishigaki T, Izumi F, Miyake M (2005) Chem Mater 17:4235
136. Yashima M, Nomura K, Kageyama H, Miyazaki Y, Chitose N, Adachi K (2003) Chem Phys Lett 380:391
137. Ishihara T, Matsuda H, Bustam MA, Takita Y (1996) Solid State Ionics 86:197
138. Petric A, Huang P (1996) Solid State Ion 92:113
139. Sinha A, Näfe H, Sharma BP, Gopalan P (2008) J Electrochem Soc 155:B309
140. Cussen EJ (2006) Chem Commun 412
141. Latie L, Villeneuve G, Conte D, Le Flem G (1984) J Solid State Chem 51:293
142. Inaguma Y, Liquan C, Itoh M, Nakamura T, Uchida T, Ikuta H, Wakihara M (1993) Solid State Commun 86:689
143. Inaguma Y, Itoh M (1996) Solid State Ion 86–88:257
144. Ibarra J, Varez A, León C, Santamarıa J, Torres-Martınez L, Sanz J (2000) Solid State Ion 134:219
145. Yamazaki Y, Hernandez-Sanchez R, Haile SM (2009) Chem Mater 21:2755
146. Tong J, Clark D, Hoban M, O'Hayre R (2010) Solid State Ion 181:496
147. Pergolesi D, Fabbri E, D'Epifanio A, Di Bartolomeo E, Tebano A, Sanna S, Licoccia S, Balestrino G, Traversa E (2010) Nat Mater 9:846
148. Stotz S, Wagner C (1966) Ber Bunsenges Phys Chem 70:781
149. Takahashi T, Iwahara H (1971) Energy Convers 11:105
150. Iwahara H, Esaka T, Uchida H, Maeda N (1981) Solid State Ion 3:359
151. Takahashi T, Iwahara H (1980) Rev Chim Minér 17:243
152. Aono H, Sugimoto E, Sadaoka Y, Imanaka N, Adachi GY (1991) Solid State Ion 47:257
153. Aono H, Sugimoto E, Sadaoka Y, Imanaka N, Adachi GY (1993) J Electrochem Soc 140:1827
154. Li Y, Han JT, Wang CA, Xie H, Goodenough JB (2012) J Mater Chem 22:15357
155. Ramzy A, Thangadurai V, Appl ACS (2010) Mater. Interfaces 2:385
156. Narayanan S, Ramezanipour F, Thangadurai V (2012) J Phys Chem C 116:20154
157. Iwahara H, Uchida H, Ono K, Ogaki K (1988) J Electrochem Soc 135:529
158. Iwahara H, Uchida H, Maeda N (1982) J Power Sources 7:293
159. Iwahara H (1999) Solid State Ion 125:271

Chapter 2
Synthesis of Nanostructured Garnets

J. M. Weller and Candace K. Chan

Abstract The garnet-type Li^+ ion conductor $Li_7La_3Zr_2O_{12}$ (LLZO) is a promising solid electrolyte for all-solid-state lithium batteries. Numerous synthesis methods have been developed as alternatives to the conventional solid-state reaction first used to prepare LLZO. This chapter provides a comprehensive review of the research progress that has been made on synthetic methods to obtain nanostructured LLZO, with a focus on the synthesis of materials prepared using sol–gel/combustion, electrospinning, cellulose templating, spray pyrolysis, co-precipitation, and molten salt methods, as well as to highlight some unique properties of fine-grained and nanostructured LLZO.

2.1 Introduction

The application of nanostructures in lithium-ion batteries has focused mainly on improving the charge storage capacity, power rates, and phase transformations in the active materials of the anode and cathode. There has been particular emphasis on utilizing the features of 1D nanostructures such as nanorods and nanowires, for example, to direct electron transport along the long axis of the nanostructure while using the nanometer radial dimension to exploit the large interfacial area with the liquid electrolyte and facilitate phase and volume transformations [1–4]. There has been less attention given to studying how 1D nanostructured materials can play a role in improving the properties of solid electrolytes for lithium-ion batteries. However, recent advances in synthesis methods for Li-containing materials have enabled the preparation of a variety of nanostructured, Li^+ conducting solid electrolytes.

Among the solid-state ceramic Li^+ conductors, the garnet family of materials has received a great deal of interest since the first report of the synthesis and characterization of garnet-type $Li_7La_3Zr_2O_7$ (LLZO) in 2007 by Murugan et al. [5], due to its high ionic conductivity and chemical stability against both air and metallic lithium [6–8].

J. M. Weller · C. K. Chan (✉)
Materials Science and Engineering, School for Engineering of Matter, Transport and Energy, Arizona State University, Tempe, AZ 85041, USA
e-mail: candace.chan@asu.edu

R. Murugan and W. Weppner (eds.), *Solid Electrolytes for Advanced Applications*,
https://doi.org/10.1007/978-3-030-31581-8_2

Early studies on LLZO focused on understanding the structure and crystal chemistry of materials prepared by conventional solid-state reaction (SSR) and consolidated into pellets. The cubic phase of LLZO (c-LLZO, space group $Ia\bar{3}d$) generally displays high Li^+ conductivity on the order of $0.1–1 \times 10^{-3}$ S/cm, while the tetragonal phase (t-LLZO, space group $I4_1/acd$) has poorer ionic conductivity, usually around $10^{-6}–10^{-7}$ S/cm. Typically, extrinsic dopants are added to LLZO in order to stabilize the thermodynamically unfavorable cubic phase at room temperature and avoid the formation of t-LLZO.

It soon became evident that the high temperatures and long sintering times needed to prepare LLZO using conventional ceramics methods would not only be prohibitive for scale-up of the synthesis and incorporation of the materials into robust (i.e., non-brittle) all-solid-state batteries, but could also have deleterious effects on the properties of LLZO. One particularly important issue was Li^+ loss during the high-temperature heat treatments, which results in the decomposition of LLZO into the lanthanum zirconate pyrochlore phase, $La_2Zr_2O_7$, which has no Li^+ conductivity [9, 10]. Hence, other synthesis methods aside from SSR have received much attention, particularly those that can enable the formation of the desired structure and composition of LLZO at lower temperatures. As the ionic conductivity of LLZO is highly dependent on its Li content [10–12] as well as crystal structure, many low-temperature methods, which often also provide smaller particle sizes that facilitate processing into thin films, have been developed for LLZO and its doped variants. Even then, it should be noted that many syntheses for LLZO include anywhere from 10–50% excess of the Li source to compensate for Li^+ volatilization.

Oxide systems, in general, have the benefit of a large range of synthesis and processing procedures available for various nanostructured morphologies. This makes LLZO an ideal candidate for studying size-dependent properties compared to other inorganic solid electrolytes that are deposited using sputtering (e.g., lithium phosphate oxynitride, "LiPON") [13], or display high air and water reactivity (e.g., sulfide electrolytes) [14]. The benefits of shrinking LLZO to the nanoscale are still under investigation. While the advantageous effects of nanocrystalline grain sizes on ionic conductivity have been demonstrated in other systems, (e.g., for O^{2-} conducting CeO_2 electrolytes [15, 16] and in NASICON-type Li^+ conductors [17–19]) the relationship between the nanometric grain size and LLZO performance in an all-solid-state battery is still unclear.

Several recent studies have shown advantages of smaller grain sizes compared to bulk LLZO, such as higher ionic conductivity [20], densification rate [21], cycling performance and current density [22], and fracture strength [23]. Hence, improved understanding of size-dependent materials properties may be important for the development of better LLZO-based electrolytes for safer Li-ion batteries. For example, Cheng et al. [22, 24] demonstrated that there is a grain size effect on the air reactivity (i.e., formation of surface Li_2CO_3 upon exposure to H_2O and CO_2), interfacial impedance, and critical current density (CCD), the current density above which Li dendrites propagate and cause a short circuit. In all cases, a finer grained microstructure (20–40 μm) was correlated with better properties—lower interfacial resistance correlated with reduced formation of Li_2CO_3 upon exposure to air, and a higher CCD before which Li dendrites caused a short circuit—compared to the large-grained (100–200 μm) LLZO. This was explained by the more distributed Li^+ ionic current density through LLZO containing a larger number of grain boundaries (i.e., from smaller grain sizes) compared to the more focused current at fewer failure points for LLZO with larger grain sizes. However, the CCD in this study was still only 0.3 mA/cm^2, about 3× lower than the minimum value needed for practical battery applications [25]. Also, a later study by Xia et al. [26] showed improved stability of large-grained LLZO to the deleterious surface reactions with air, which was explained in terms of a reduction in grain boundary area and high interfacial energy regions where H_2O and CO_2 can react.

A highly systematic study by Sharafi et al. [27] to investigate the effect of grain size on the mechanical and electrochemical properties of LLZO pellets prepared by hot-pressing demonstrated an opposite trend to that of Cheng et al. [22, 24], wherein very large grain sizes were correlated with improved CCD—the CCD increased from 0.3 to 0.6 mA/cm^2, one of the highest reported to date, when the LLZO grain size increased from 5 to 600 μm. This seemingly contradictory result was explained by separating the grain size effect on the interfacial impedance from its effect on Li dendrite propagation—in the work by Cheng et al. [22] the interfacial impedance of the small and large-grained samples was not constant. It also remains to be seen if the nature of the grain boundaries in LLZO pellets prepared by hot-pressing is different from those prepared in conventional sintering. Sharafi et al. also found that the fracture toughness of LLZO was found to decrease when the grain size increased from 5 to 40 μm and remained constant for grain sizes ≥40 μm; the CCD, however, did not appear correlated to the mechanical properties of the LLZO.

The grain sizes of the materials investigated in the aforementioned studies were still on the micron-scale and similar studies extrapolated to nanoscaled grains have yet to be performed in a systematic way. Nonetheless, the properties of nanostructured ceramics in general, such as densification at milder sintering conditions (i.e., reduced sintering temperatures and times) [28, 29], decreased impurity segregation to grain boundaries [30], and superplasticity [31] may lead to improved properties in pellets derived from nanostructured LLZO. Further, for practical applications, the solid electrolyte thickness in an all-solid-state battery must be mechanically robust, yet thin enough for the battery to maintain high-energy density. To remain competitive with the energy density in a conventional Li-ion battery using liquid electrolyte, the solid electrolyte thickness must be <100 μm to maintain the same bulk resistance if the solid electrolyte has an ionic conductivity of ~5×10^{-3} S/cm [32]. Hence, moving to smaller particles that can be easily densified into thin films will be important for the application of LLZO into batteries of practical and commercial relevance.

Despite the as-of-yet unresolved relationships between nanocrystalline grain size and LLZO performance in pellet samples, fine-grained and nanostructured LLZO has already found application as ceramic fillers in polymer composite electrolytes. The application of nanosized particles as ceramic fillers was already demonstrated some time ago to be effective for enhancing the mechanical stability and ionic conductivity of polymer-based solid electrolytes [33–36], but these fillers mostly comprised of spherical particles of inert or "passive" components without intrinsic Li^+ conductivity

[35, 37]. Several recent studies using various LLZO nanostructured morphologies embedded into polymer films have revealed a number of advantages, such as low filler contents required for maximum ionic conductivity and fast Li^+ conduction along interfacial regions [38–42]. Therefore, preparation of nanosized LLZO is well motivated for applications in composite polymer electrolytes.

Another interesting, yet still not well-understood observation is the formation of c-LLZO without the presence of extrinsic dopants, which is frequently observed in low-temperature syntheses (e.g., sol–gel methods [43–46], but also in molten salt reaction [47]) and/or in nanostructured LLZO. The mechanism for the phase stability of c-LLZO at room temperature in nanostructured LLZO is still not clear. Other metal oxides with multiple polymorphs (e.g., titania, zirconia) also display similar size-dependent phase stabilization due to differences in surface energy and crossovers in polymorphic stability as the surface area of the materials increases [48–50]. It is plausible that a similar phenomenon could account for cubic stabilization of LLZO below a critical size parameter. Other factors such as concentration of Li vacancies [11, 51, 52], CO_2 adsorption [53, 54], and H_2O doping [55] might also play a role. As using dopants to stabilize c-LLZO requires careful control of dopant concentrations due to the narrow composition range for peak conductivity (e.g., for Ta and Nb substitution of Zr) [12, 56–58] and also to avoid formation of nonconducting compounds that may segregate to grain boundaries (e.g., $LaAlO_3$ for Al doping) [22, 59], nanostructuring may be an attractive alternative route for obtaining c-LLZO.

In this chapter, the synthesis of nanostructured LLZO powders using a variety of different methods is summarized. The preparation of LLZO thin films (e.g., using chemical vapor deposition or high vacuum deposition techniques) is omitted from this review. Solid-state reaction is first discussed, as it is the conventional method for preparing LLZO, and it is possible to use milling techniques for particle reduction to the nanoscale. Next, low-temperature sol–gel methods such as Pechini-type and combustion-based approaches are discussed, followed by co-precipitation methods. Finally, nanocrystal synthesis approaches such as electrospinning, nanocellulose templating, flame spray pyrolysis, and molten salt synthesis, are described.

2.2 Synthesis by Solid-State Reaction Methods

The majority of research investigating LLZO use solid-state reaction (SSR) as the preparation method of the materials, as it is a well-established technique for the synthesis of complex oxides. Briefly, solid-state reactions rely on intimate mixing to provide physical contact (preferably, with high interfacial surface area) between powdered reagents followed by heating to high temperatures (often between 1000–1500 °C) to provide the thermodynamic driving force for diffusion of atomic species and formation of the desired phase [60]. In many cases, SSR is performed by pressing pellets of the reagents to maximize the contact area, followed by one or potentially several heat treatments with repeated grinding/milling and pelletizing between each treatment.

For LLZO, the typical precursors are LiOH or Li_2CO_3, La_2O_3, and ZrO_2, which are calcined between 800–1000 °C prior to the final pellet sintering temperatures, which usually exceed 1000 °C. Since the desired cubic structure is a non-quenchable, high-temperature phase [9, 53, 59] heterovalent dopants to stabilize the cubic phase over the tetragonal one at room temperature are also added as the oxide, e.g., Al_2O_3, Ga_2O_3, Ta_2O_5, Nb_2O_5. Figure 2.1 shows example SSR synthesis procedures for preparing Al-doped LLZO (ALLZO) [61] and Ta-doped LLZO (LLZTO) [12] and corresponding measured room-temperature ionic conductivities.

Due to the hygroscopic nature of La_2O_3, this precursor is often heated at 900 °C for 10–12 h prior to use. It should also be noted that the reaction conditions required to prepare LLZO by SSR are often changed with the introduction of dopants, which suggests a change in the thermodynamics or kinetics of the reaction. For example, ALLZO prepared following the procedures illustrated in Fig. 2.1a required sintering at 1200 °C for 40 h to obtain the maximum ionic conductivity of 3.41×10^{-4} S/cm [61], but Ta-doped LLZO only required sintering at 1150 °C for 4 h using the same powder processing and calcination steps to reach 7.1×10^{-4} S/cm [62]. This has also been observed using in situ neutron diffraction heating studies to investigate the formation process of LLZO from the ball-milled precursors (it should be mentioned that ball-milling forms an amorphous precursor phase) [63]. For undoped LLZO, it was found that the ball-milled precursors (Li_2CO_3, La_2O_3, and ZrO_2) start to crystallize and form intermediate phases $La_2O_2CO_3$ and $La_2Zr_2O_7$

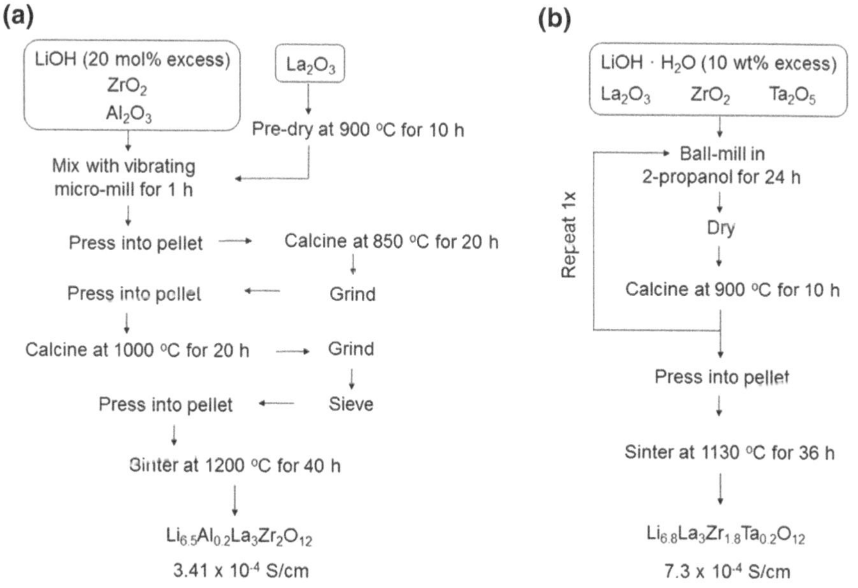

Fig. 2.1 Conventional solid-state reaction (SSR) method for synthesizing **a** Al-doped LLZO (from procedures described by Tsai et al. [61]) and **b** Ta-doped LLZO (from procedures described by Wang et al. [12])

at 540 °C, with LLZO formation beginning above 800 °C once Li_2CO_3 had melted (melting point: 720 °C) and accelerating above ~850 °C; LLZO is the dominant phase at 950 °C and the Li_2CO_3 melt begins to evaporate at 980 °C. In contrast, for Nb-doped LLZO, only 335 °C was sufficient to crystallize the amorphous precursor and melting was observed at lower temperatures compared to the undoped LLZO; Nb-LLZO formation was observed to begin at 750 °C, ~50° lower than the undoped case.

The reaction time needed to obtain phase-pure LLZO is important, as incompletely reacted intermediates or impurity phases such as $Li_6Zr_2O_7$ and Li_2ZrO_3, which have significantly lower ionic conductivity than LLZO [64–66], are thought to accumulate on grain boundaries, block Li^+ transport, and create large grain boundary resistances. On the other hand, calcination times that are too long can lead to Li^+ loss and the formation of $La_2Zr_2O_7$ [10], which has no Li^+ conductivity and poor sinterability [67, 68] or tetragonal LLZO, which can form even with Al^{3+} doping [10]. It should also be noted that the long calcination times at high temperatures can also promote interdiffusion of Al^{3+} from alumina crucibles into LLZO, which was identified as the source of adventitious doping in the beginning stages of LLZO research [59]. This can be prevented by using Pt crucibles [26, 69]. MgO crucibles are a cost-effective alternative to Pt crucibles, and placing pellets on a bed of mother powder can also prevent reaction or sticking of the pellets to the crucibles [61].

In addition to the large amount of time, repetitive processing steps, and energy required for preparing LLZO using SSR, the morphology of the as-synthesized material tends to be irregular and the particle size not monodisperse. As this can affect the ability of the materials to be processed into dense pellets [70], post-synthesis processing using milling techniques may be used to improve the sinterability and density of the pellets. Although there is a practical lower limit when high-energy milling techniques are used for particle size reduction [71], there have been a number of examples where particles in the nanoscale can be obtained using these techniques. Lee et al. [72] found that using increased rotation speeds in a planetary mill during the mixing of the precursors led to smaller and more homogenous grains, which correlated with higher pellet densities and ionic conductivities (Fig. 2.2). Yoshima et al. [73] reported average particle sizes of 200–400 nm could be obtained by bead milling LLZO prepared by SSR; however, no electron microscopy images or ionic conductivity values of these materials were reported, as the powders were incorporated into gel-polymer electrolytes.

LLZTO particles with a starting particle size of ~10 μm were reduced down to ~40 nm in the work by Zhang et al. [41] using first planetary ball-milling to reduce the particles to ~3 μm, followed by high-energy ball-milling under argon using 0.1 mm diameter zirconia balls. As shown in Fig. 2.3, measurements obtained using a laser particle size analyzer showed that average particle sizes of ~400, ~200, and ~40 nm were obtained (the precise milling times were not disclosed), although the scanning electron microscopy (SEM) images showed the particle morphologies are not homogenous. The authors found that polymer composite membranes containing

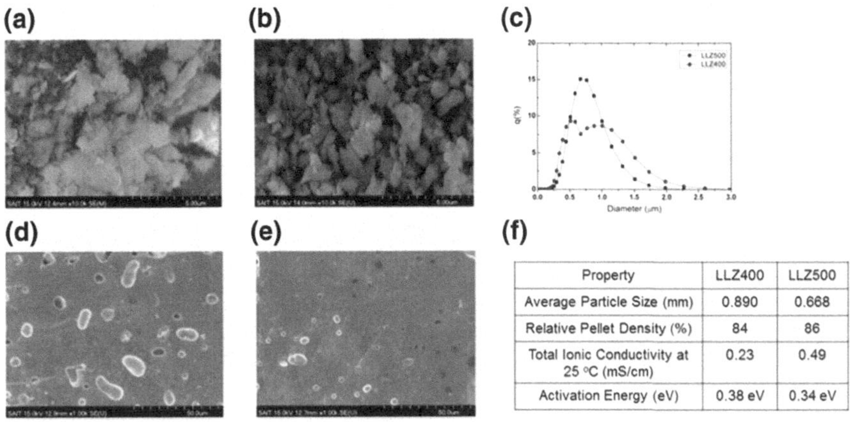

Property	LLZ400	LLZ500
Average Particle Size (mm)	0.890	0.668
Relative Pellet Density (%)	84	86
Total Ionic Conductivity at 25 °C (mS/cm)	0.23	0.49
Activation Energy (eV)	0.38 eV	0.34 eV

Fig. 2.2 **a**, **b** SEM images of powders, **c** particle size distribution plots, and **d**, **e** cross-section SEM images of pellets (sintered at 1125 °C for 20 h) of LLZO after milling at rotation speeds of 400 and 500 rpm, respectively, for 12 h. Adapted from Lee et al. [72] with permission from Elsevier

12.7 vol% of the ~40 nm LLZTO particles as ceramic fillers displayed ionic conductivities of 2.1×10^{-4} S/cm, which was nearly 2 orders of magnitude higher than those measured in the membranes containing the larger particles.

One outstanding case of using SSR for preparation of LLZO is that of Huang et al. [69]. In this work, LLZTO (5–10 μm particles) was synthesized using conventional SSR in large quantities and subsequently attrition milled to result in primary particles of ~300 nm diameter. In this case, 300 nm is apparently the smallest particle size achievable by attrition milling. That said, in order to mill the as-synthesized powders to such a small size requires substantial amounts of solvent—in this case, a 4:1 mixture of water to ethanol was used, resulting in a relatively inexpensive solvent system compared to typical systems used for attrition milling like acetone or 2-propanol. Still, to obtain the attrition milled powders, spray drying was performed, resulting in particle aggregates on the order of several microns, and requiring handling of a large amount of solvent waste. Despite the larger number of processing steps, the higher surface area of these powders combined with the relative simplicity of SSR make this an attractive approach (Fig. 2.4).

Finally, modification of SSR with wet-chemical approaches has also been demonstrated to be effective for obtaining LLZO nanoparticles. Kumar et al. [74] designed a two-step reaction scheme as shown in Fig. 2.5, where ZrO_2 nanoparticles were ball-milled with $La(OH)_3$ and AlOOH, followed by impregnation of the powder with lithium acetate. Pure phase cubic ALLZO with average primary particle size around 100 nm (agglomerate sizes around 1–2 μm) was obtained after calcining at 750 °C for 3 h, and dense pellets (relative density of 90.5%) obtained after sintering at 1100 °C for 3 h displayed ionic conductivity of 1.6×10^{-4} S/cm at room temperature, which is comparable to conventional SSR. These results show that this

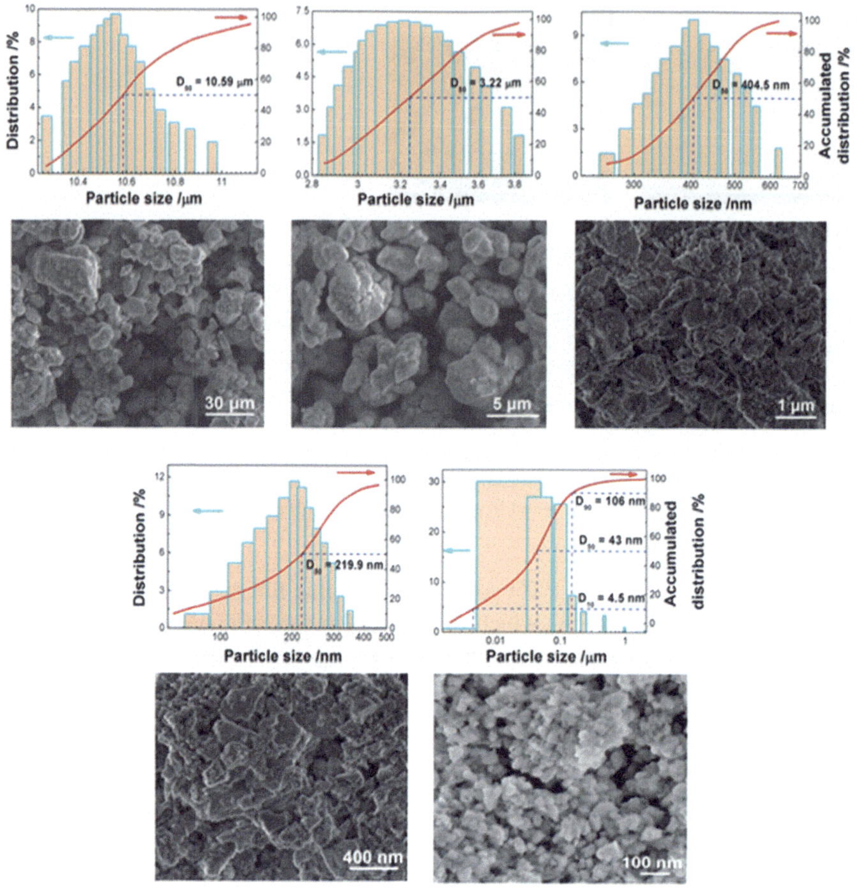

Fig. 2.3 Size distribution of LLZTO obtained with ball-milling process as determined with a laser particle size analyzer and corresponding SEM images. Adapted from Zhang et al. [41] with permission from Elsevier

so-called "low-temperature solid-state route" can not only greatly reduce the calcination temperature and times required to obtain LLZO, but is also effective for preparing materials with the desired ionic conductivity.

In summary, despite the simplicity of SSR and the high ionic conductivity of the resultant materials, the large time and energy costs and poor morphology of the products (i.e., large particle size, large size distribution, irregular shaped particles, etc.) has motivated the use of alternative synthesis methods for LLZO, which will be reviewed in the following sections.

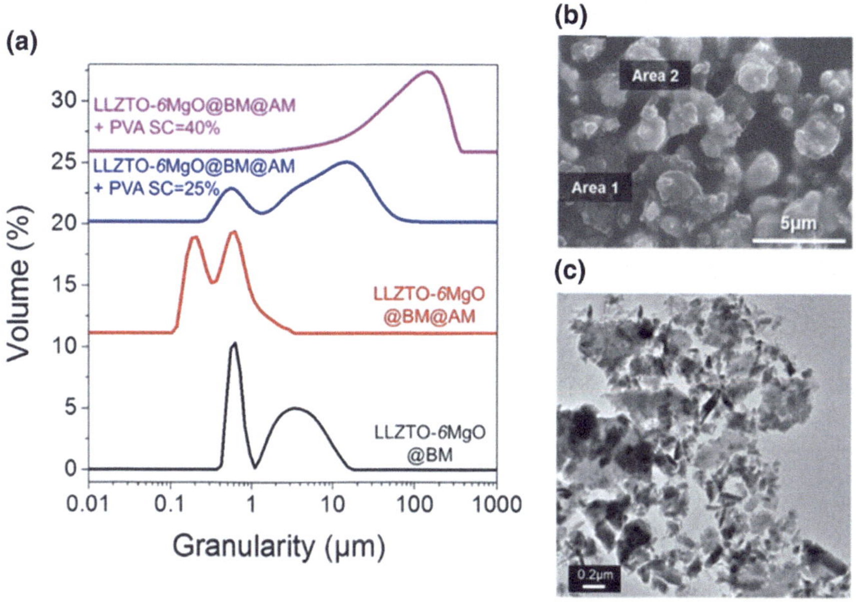

Fig. 2.4 **a** Distribution of particle sizes in the slurry of LLZTO-MgO (6 wt%) after ball-milling (black), after ball-milling and attrition milling (red), and after addition of polyvinyl alcohol binder (PVA) at solid loadings of 25 and 40% (blue and pink, respectively). **b** SEM image of LLZTO/MgO (6 wt%) composite particles after ball-milling. **c** TEM image of LLZO/MgO (6 wt%) composite particles after attrition milling. Adapted from Huang et al. [69] with permission from the American Chemical Society

2.3 Sol–Gel Synthesis Methods

An extremely common method for production of complex oxides, and particularly nanosized oxides, is that of sol–gel synthesis [75, 76]. Sol–gel chemistry enables the formation of inorganic materials from solution through the use of liquid precursors that first form a colloidal solution or "sol" that eventually transforms into a network structure known as a "gel" [75]. Commonly, metal organic precursors (e.g., metal alkoxides) are used, although metal salts may also be employed with the correct chelating agents in the solution. Depending on the conditions, sol–gel techniques can be used to deposit thin films, precipitate particles, form aerogels/xerogels, and are otherwise versatile in the synthesis of complex oxides [8, 42, 75, 77]. Sol–gel methods may also yield materials with better compositional homogeneity than those obtained with SSR due to molecular mixing of the precursors, as well as lower processing temperatures, shorter synthesis times, and particles with greater uniformity in terms of particle morphology and size [76]. In general, the sol–gel synthesis of materials can be realized, as schematically illustrated in Fig. 2.6, (a) via connection of colloidal particles through van der Waals or hydrogen bonding, (b) through hydrolysis and condensation of metal alkoxides, (c) from aqueous metal-complex (e.g., citrate) solutions, (d) from in situ formation of a coordinating polymer

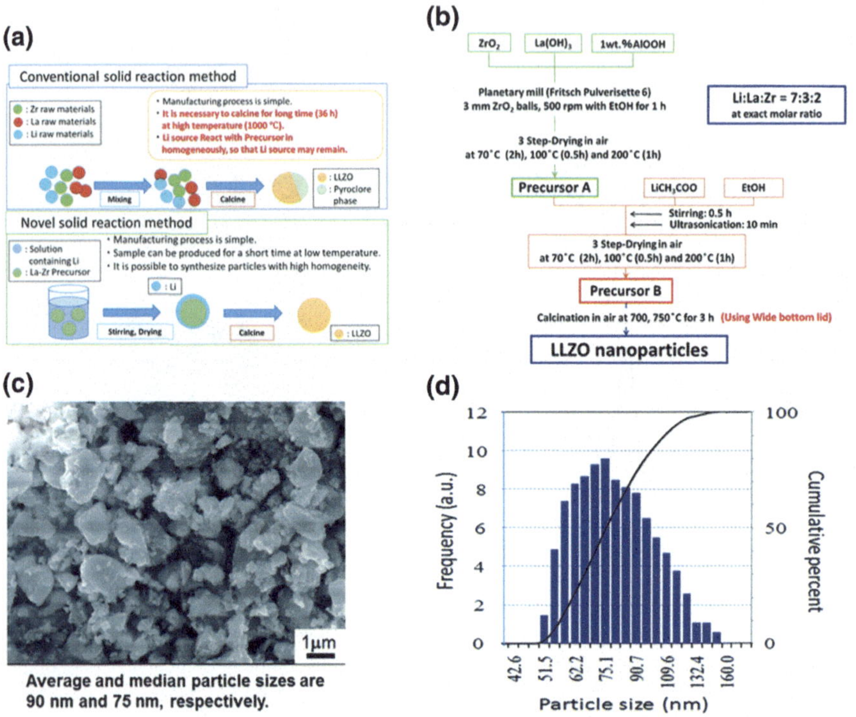

Fig. 2.5 **a** Schematic of modified SSR compared to conventional SSR; **b** Synthetic procedures for modified SSR; **c** SEM image and **d** particle size distribution of particles prepared by modified SSR. Adapted from Kumar et al. [74] with permission from the Royal Society of Chemistry

via polyesterification between a polyhydroxy alcohol (e.g., ethylene glycol) and carboxylic acid with metal complex (e.g., metal-citrate) (also known as the Pechini method), or (e) from a coordinating polymer and metal-salt solution [76], followed by calcination to remove the organic components and obtain the crystalline ceramic.

Table 2.1 lists sol–gel syntheses reported for LLZO powders and thin films, along with the reagents and calcination conditions used, and the reported ionic conductivities (in studies where a series of samples at different conditions were studied, only the highest conductivity is displayed). Unless otherwise noted, the LLZO obtained after calcination adopted the cubic phase. Note that in cases where the precise composition of the obtained Zr precursor is not known (e.g., for $ZrO(NO_3)\cdot xH_2O$ or the zirconium carbonate basic hydrate $[ZrO_2]_2\cdot CO_2\cdot xH_2O$), thermogravimetric analysis can be performed to obtain the degree of hydration [51, 78, 79], which is particularly important for obtaining the precise doping level on the Zr site. Utilization of La_2O_3 as a lanthanum source also often requires preheating before use due to its hygroscopic nature, e.g., 900 °C for 12 h [79, 80]. Some oxide precursors (e.g., Ga_2O_3) can be used but must first be dissolved/digested in nitric acid [81].

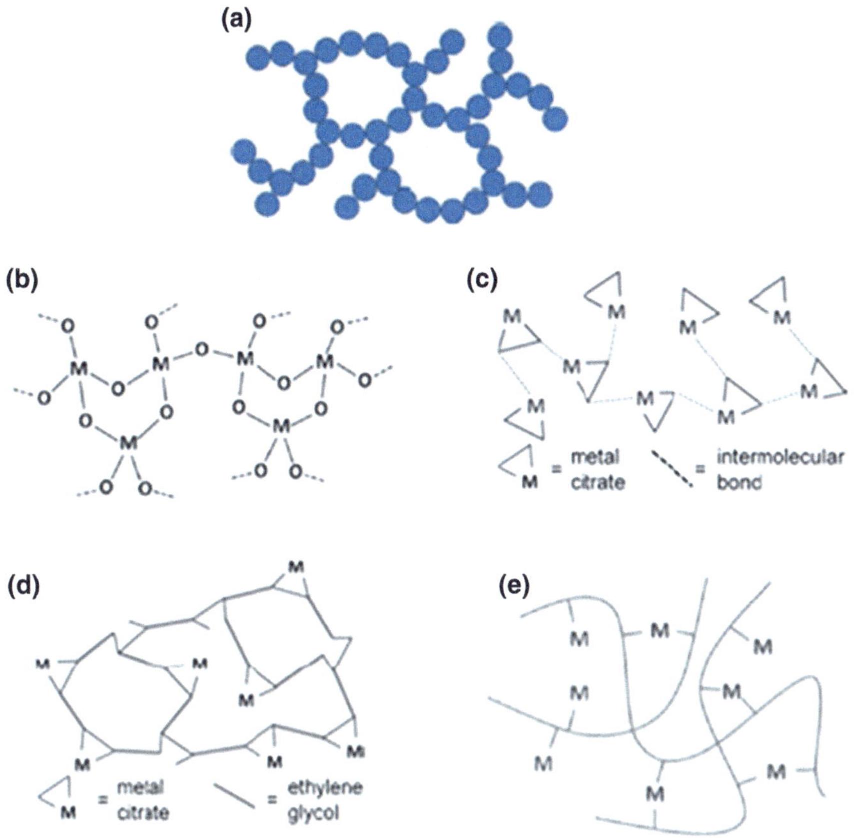

Fig. 2.6 Schematics of different types of gels used in the sol–gel synthesis of materials (see text for explanation). Adapted from Danks et al. [76] with permission from the Royal Society of Chemistry

A particularly notable example of the synthesis of nanostructured ALLZO using a sol–gel method is that of Sakamoto et al. [23], wherein 50 nm Al_2O_3 nanocrystals and soluble Li, La, and Zr precursors were employed to result in a gel network with an Al_2O_3 "backbone", as demonstrated by supercritical drying of the gelled precursor to preserve the morphology and chemistry in the dried state. The resulting material displayed an aerogel morphology consisting of a highly interconnected solid network with ~10 nm wide nanowire-like cross-links, surface area of 292 m^2/g, and large fraction of microporosity (33%). Unfortunately, this supercritically dried gel was not calcined or studied further, but conventional drying of the gel revealed that cubic ALLZO could be obtained after heat treatment in an argon environment. This is attributed to the metal-oxygen bonds formed during the gelation period, in contrast to Pechini based methods, where heating in air is required to provide sufficient oxygen to form the crystallized ceramic. Sakomoto et al. also found that ALLZO could be formed with calcination temperature as low as 600 °C in air, and induction

Table 2.1 Details of reagents, heat treatment, and total Li^+ ionic conductivity (unless otherwise indicated) at room temperature for sol–gel approaches to prepare LLZO

Composition	Reagents	Calcination conditions	Ionic conductivity	References
Tetragonal LLZO	Li_2CO_3 (dissolved in dil. HNO_3) La_2O_3 (dissolved in dil. HNO_3) $Zr(OC_2H_5)_4$ (in ethanol) Citric acid Ethylene glycol	Gel dried at 200 °C for 24 h Powders calcined at 800 °C for 5 h Pellets sintered at 900 °C for 4 h	3.12×10^{-7} S/cm	Kokal et al. [43]
	Li_2CO_3 La_2O_3 $ZrO(NO_3)_2$ Citric acid Dilute HNO_3	Gel dried and pyrolyzed at 200 °C; Powder calcined at 900 °C for 3 h Pellets sintered at 1100 °C for 1 h	5.19×10^{-6} S/cm	Il'ina et al. [82]
LLZO (undoped, cubic phase)	Li_2CO_3 $La(NO_3)_3{\cdot}6H_2O$ $Zr(CH_3COO)_4$ Acetic acid	Powder calcined at 750 °C for 4 h	Not reported	Xie et al. [83]
	$LiCH_3COO{\cdot}6H_2O$ (30 wt% excess) $La(CH_3COO)_3{\cdot}1.5H_2O$ $Zr(CH_3COO)_4$ Acetic acid	Powders heated to 400 °C; Pellets sintered at 750 °C for 8 h	2.85×10^{-6} S/cm	Xie et al. [44]
	Li_2CO_3 (10 mol% excess) La_2O_3 (dissolved in dil. HNO_3) $ZrO(NO_3){\cdot}6H_2O$ (dissolved in dil. HNO_3) Citric acid	Powders heated at 800 °C for 24 h Pellets sintered at 1230 °C for 8 h	1.67×10^{-4} S/cm	Tan et al. [84]

(continued)

Table 2.1 (continued)

Composition	Reagents	Calcination conditions	Ionic conductivity	References
	$LiCH_3COO$ $La(CH_3COO)_3 \cdot 1.5H_2O$ $Zr(C_4H_9O)_4$ Acetic acid Ethanol	Gel heated at 700 °C for 4 h Pellets sintered at 1100 °C for 6 h	3.5×10^{-6} S/cm	Kotobuki et al. [85]
Al-doped LLZO	$LiNO_3$ $La(NO_3)_3 \cdot 6H_2O$ $ZrO(NO_3) \cdot xH_2O$ Citric acid Ethylene glycol (Al from adventitious doping from Al_2O_3 crucible)	Powder calcined at 350 °C for 5 h Pellets sintered at 800 °C for 20 h, then 1180 °C for 35 h in air	5.69×10^{-4} S/cm (bulk); 1.39×10^{-4} S/cm (grain boundary)	Shimonishi et al. [86]
	$LiNO_3$ (10% excess) $La(NO_3)_3 \cdot 6H_2O$ $ZrOCl_2 \cdot 3H_2O$ $Al(NO_3)_3 \cdot 9H_2O$ Citric acid Ethylene Glycol	Mixture heated at 130 °C to form foam, Foam calcined at 900 °C for 6 h Pellet sintered at 1200 °C for 6 h	2×10^{-4} S/cm (for LLZO with 1.2 wt% Al)	Jin et al. [87]
	$LiNO_3$ (10 wt% excess) $La(NO_3)_3 \cdot 6H_2O$ $ZrO(NO_3) \cdot xH_2O$ $Al(NO_3)_3 \cdot 9H_2O$ Citric acid (as organic complexing agent) Butan-1-ol or isopropyl alcohol (as surface active agent)	Gel heated at 250 °C for 1 h Powder calcined at 500 °C for 2 h, then 700 °C for 12 h	Not reported	Janani et al. [78]

(continued)

Table 2.1 (continued)

Composition	Reagents	Calcination conditions	Ionic conductivity	References
	$LiNO_3$ (10 wt% excess) $La(NO_3)_3·6H_2O$ $ZrO(NO_3)·xH_2O$ $Al(NO_3)_3·9H_2O$ Citric acid Propan-2-ol (as polymerizing agent)	Gel calcined at 500 °C for 2 h, then 700 °C for 12 h Pellets sintered at 1100 °C for 1 h	2.12×10^{-5} S/cm	Hubaud et al. [88]
	$Li(NO_3)_3·xH_2O$ $La(NO_3)_3·6H_2O$ $Zr(OC_3H_7)$ 1-Propanol Acetic acid (as chelating agent) Al_2O_3	Gels calcined at 450 °C for 4 h in air; Pellets sintered with hot-pressing at 1000 °C for 4 h under 40 MPa	~4×10^{-4} S/cm	Sakamoto et al. [23]
	$LiNO_3$ $La(NO_3)_3·6H_2O$ $Zr(O\text{-}n\text{-}C_3H_7)_4$ $Al(O\text{-}sec\text{-}C_4H_9)_3$ Ethyl acetoacetate (as stabilizing agent)	Dried gel calcined at 700 °C for 5 h Pellets sintered at 900 °C for 10 h	1.6×10^{-6} (without additives); 1.9×10^{-5} S/cm (with Li_3BO_3)	Takano et al. [89]
	$LiNO_3·xH_2O$ $La(NO_3)_3·6H_2O$ $Zr(OC_3H_7)_4$ 1-Proponal Acetic acid (as chelating agent) 2-Methoxyethanol Al_2O_3	Dried gel was heated at 450 °C; Pellets sintered at 900 °C for 4 h	Not reported	Yoo et al. [90]

(continued)

Table 2.1 (continued)

Composition	Reagents	Calcination conditions	Ionic conductivity	References
	Li_2CO_3 La_2O_3 $ZrO(NO_3)_2$ $Al(NO_3)_3{\cdot}9H_2O$ Citric acid Nitric acid	Gel dried and pyrolyzed at 200 °C; Pellets sintered at 1150 °C for 1 h	3.4×10^{-4} S/cm for $Li_7La_3Zr_{1.89}Al_{0.15}O_{12}$	Raskovalov et al. [91]
	$LiNO_3$ (10 wt% excess) $La(NO_3)_3{\cdot}6H_2O$ $ZrO(NO_3){\cdot}xH_2O$ $Al(NO_3)_3{\cdot}9H_2O$ Citric acid Butan-1-ol	Gel heated at 250 °C for 1 h, then 900 °C for 12 h Pellets sintered at 1200 °C for 12 h	2.1×10^{-4} S/cm w/o additive; 6.1×10^{-4} S/cm with 1 wt% Li_4SiO_4 additive	Janani et al. [92]
	$LiNO_3$ (10 wt% excess) $La(NO_3)_3{\cdot}6H_2O$ Zr acetylacetonate $Al(NO_3)_3{\cdot}9H_2O$ Ethanol	Gel calcined at 650 °C for 15 h Pellets sintered at 1070 °C for 10 h	5×10^{-4} S/cm (bulk)	van den Broek et al. [93, 94]
	Li acetate La_2O_3 $ZrO(NO_3){\cdot}xH_2O$ Ammonium hydroxide EDTA Acetic acid Amorphous aluminum oxide (from aluminum acetylacetonate)	Gel calcined at 550–600 °C, then 900 °C for 1 h Powder mixed with aluminum oxide precursor and sintered at 1100 °C for 3 h	~3×10^{-4} S/cm	El-Shinawi et al. [95]

(continued)

Table 2.1 (continued)

Composition	Reagents	Calcination conditions	Ionic conductivity	References
	$LiNO_3$ $La(NO_3)_3 \cdot 6H_2O$ $Zr(NO_3)_2 \cdot 6H_2O$ $Al(NO_3)_3 \cdot 9H_2O$ Polyvinyl alcohol Glutaraldehyde (as cross-linker) Cetyltrimethylammonium bromide (as surfactant)	Dried gel calcined at 800 °C for 2 h	Not reported	Bae et al. [96]
	$LiCH_3COO \cdot 2H_2O$ (10% excess) $La(CH_3COO)_3 \cdot 1.5H_2O$ $Zr(OCH_2CH_2CH_3)_4 \cdot 2H_2O$ $AlCl_3 \cdot 6H_2O$ Acetylacetone Ethanol	Gel calcined at 1000 °C for 7 h Pellet sintered at 800 °C for 5 h	1.5×10^{-5} S/cm	Rosenkiewitz et al. [97]
Al-doped LLZO (Thin Films)	Lithium tert-butoxide (10% excess) $La(NO_3)_3 \cdot 6H_2O$ Zirconium n-propoxide 2-Methoxyethanol	Thin-film layers spin coated and then dried at 400 °C for 2 min, then calcined at 500 °C for 2 min in air; After repeating with multiple layers, film was calcined at 600 °C for 10 min	1.67×10^{-6} S/cm	Chen et al. [98]

(continued)

Table 2.1 (continued)

Composition	Reagents	Calcination conditions	Ionic conductivity	References
	$LiNO_3$ $La(NO_3)_3{\cdot}6H_2O$ $Zr(O\text{-}n\text{-}C_3H_7)_4$ $Al(O\text{-}sec\text{-}C_4H_9)_3$ 1-Propanol Ethyl acetoacetate Lithium dodecyl sulfate (LDS) (0.1 wt% as surfactant)	Thin films dip coated onto MgO and calcined at 450 °C for 15 min, then 900 °C (w/Li_2CO_3 to avoid Li loss)	2.8×10^{-7} S/cm (w/o LDS); 2.4×10^{-6} S/cm (w/0.1 wt% LDS)	Tadanaga et al. [99]
Ga-doped LLZO	Lithium acetate La_2O_3 $ZrO(NO_3){\cdot}xH_2O$ Gallium acetylacetonate Dilute HNO_3 EDTA (dissolved in NH_3)	Gel heated at 250 °C; Powder calcined at 550 °C for 18 h Pellets calcined at 1085 °C for 6 h	5.4×10^{-4} S/cm	El-Shinawi et al. [79]
	$LiNO_3$ $La(NO_3)_3{\cdot}6H_2O$ Zirconium (IV) 2,4-pentanedionate Ga_2O_3 Ethanol	Gel powder calcined at 600–650 °C for 1–10 h Powder mixed with Ga_2O_3 (0.2 mol%) and calcined at 950 °C for 6 h as pellets	~4.0×10^{-4} S/cm $Li_{6.4}Ga_{0.2}La_3Zr_2O_{12}$	Afyon et al. [100]
Mo-doped LLZO	$LiCO_3$ La_2O_3 H_2MoO_4 Citric acid Nitric acid	Powder calcined at 750 °C for 6 h Pellets sintered at 1100 °C for 3 h	5.09×10^{-4} S/cm for $Li_{6.6}La_3Zr_{1.8}Mo_{0.2}O_{12}$	Liu et al. [101]

(continued)

Table 2.1 (continued)

Composition	Reagents	Calcination conditions	Ionic conductivity	References
Nb-doped LLZO	$LiNO_3$ (10% excess) $La(NO_3)_3 \cdot 6H_2O$ $Zr(OC_3H_7)_4$ $Nb(OC_2H_5)_5$ Ethyl acetoacetate 2-Propanol Ethanol	Dried gels calcined at 700 °C for 5 h Pellets sintered at 900 °C for 10 h with Li_3BO_3	7×10^{-5} S/cm when sintered with 6.5 wt% Li_3BO_3	Rosero-Navarro et al. [102]
Ta-doped LLZO	$LiNO_3 \cdot xH_2O$ $La(NO_3)_3 \cdot xH_2O$ $Zr(OC_3H_7)_4$ $Ta(OC_2H_5)_5$ 1-Propanol or 2-Methoxyethanol Acetic acid (as chelating agent)	Dried gel calcined at 450 °C for 4 h, then 950 °C for 4 h Pellets sintered at 1100 °C for 4 h	1.4×10^{-4} S/cm	Yoon et al. [103]
Ga and Sc co-doped LLZO	$LiNO_3$ (10% excess) $La(NO_3)_3 \cdot 6H_2O$ $Zr(C_5H_7O_2)_4$ Sc_2O_3 Ga_2O_3 (digested in HNO_3) Citric acid	Gel calcined at 600 °C for 12 h and then 800 °C for 12 h Pellets sintered at 1200 °C for 6 h	1.8×10^{-3} S/cm for $Li_{6.65}Ga_{0.15}La_3Zr_{1.90}Sc_{0.10}O_{12}$	Buannic et al. [81]
Ca and Nb co-doped LLZO	$LiNO_3$ (10% excess) $La(NO_3)_3$ $ZrO(NO_3)_2$ $Ca(NO_3)_2$ $NbCl_5$ Citric acid Ethylene glycol	Gel calcined at 400 °C for 10 h Pellets sintered at 800 °C for 10 h, then 1050 °C for 12 h	2.2×10^{-4} S/cm for $Li_7La_{2.75}Ca_{0.25}Zr_{1.75}Nb_{0.25}O_{12}$	Han et al. [104]

(continued)

Table 2.1 (continued)

Composition	Reagents	Calcination conditions	Ionic conductivity	References
Al or Nb-doped LLZO/carbon foam	LiOH $La(NO_3)_3{\cdot}6H_2O$ $ZrO(NO_3){\cdot}xH_2O$ $Al(NO_3)_3{\cdot}9H_2O$ or $NbCl_5$	Sol dried at 250 °C for 3 h to obtain gel; Gel carbonized at 850 °C for 2 h under Ar to form LLZO/carbon foam, or 850 °C for 2 h in air to obtain LLZO nanoparticles	Not reported	Tao et al. [105]
Y-doped LLZO	$LiNO_3$ (10 wt% excess) $La(NO_3)_3{\cdot}6H_2O$ $ZrO(NO_3){\cdot}xH_2O$ $Y(NO_3)_3{\cdot}6H_2O$ Citric acid Butan-1-ol	Gel heated at 250 °C for 1 h Powder calcined at 700 °C for 12 h	Not reported	Deviannapoorani et al. [106]
Y-doped LLZO (Thin Film)	$LiNO_3{\cdot}3H_2O$ (20% excess) $La(NO_3)_3{\cdot}6H_2O$ $(CH_3CO_2)_xZr(OH)_y$ $(x + y \sim 4)$ $Y(NO_3)_3{\cdot}6H_2O$ Ethanol Methoxyethanol	Thin films dip coated and spin coated, then calcined at 800 °C for 3 h Pellets sintered at 1150 °C for 3 h	3.4×10^{-8} at 45 °C for thin film; 2.1×10^{-5} S/cm for pellet prepared from sol–gel	Bitzer et al. [107]
Ta and Ca co-doped LLZO	Li_2CO_3 (30 wt% excess) La_2O_3 $Zr(CH_3COO)_4$ Ta_2O_5 $CaCO_3$ Acetic acid	Powders calcined at 400 °C for 2 h, then 750 °C for 8 h Pellets sintered at 1125 °C for 6 h	2.84×10^{-4} S/cm for $Li_{6.45}Ca_{0.05}La_{2.95}Ta_{0.6}Zr_{1.4}O_{12}$	Chen et al. [108]

hot-pressing resulted in sintered pellets with densities up to 96% of the theoretical density of ALLZO with conductivities up to 4×10^{-4} S/cm [23]. Critically, the use of hot-pressing and nanocrystalline starting material allowed a final ceramic with an average grain size of only 260 nm, more than 10× smaller than using conventional SSR-derived powder [23].

A modification of sol–gel type synthesis is the Pechini method (Fig. 2.6d) [76]. This method, named after M. P. Pechini from his 1967 patent [109] essentially utilizes a chelating agent such as citric acid to form complexes with dissolved metal nitrates, followed by addition of a glycol (e.g., ethylene glycol) and subsequent in situ polymerization (i.e., transesterification between citrate and glycol as depicted schematically in Fig. 2.7a) to form a network of metal chelates entrapped in a citrate-glycol polymer backbone. This precursor is heated to mild temperatures resulting in the decomposition of nitrate anions and partial combustion of the material, yielding a reactive precursor which, upon calcination, generally results in ultrafine crystalline material. A scheme demonstrating the process is shown in Fig. 2.7b.

The literature is rife with variations on the Pechini method, and many researchers have used similar methods to synthesize LLZO. Shimonishi et al. [86] synthesized LLZO starting from Li_2CO_3, La_2O_3, and $Zr(EtO)_4$, all of which were dissolved in HNO_3, mixed with citric acid and ethylene glycol and subsequently treated at elevated temperature. Once the precursor was gelled, it was dried at elevated temperatures to form a partially combusted black powder, which could be converted to LLZO via calcination at 800 °C. Similarly, Kokal et al. [43] used essentially the same so-called "modified-Pechini" method to synthesize apparently undoped c-LLZO and t-LLZO depending on the processing conditions. This method yielded no LLZO when a calcination temperature of 650 °C was used, but rather the $La_2Zr_2O_7$ pyrochlore phase, with phase-pure c-LLZO obtained at 700 °C. Heating to higher temperatures yielded t-LLZO. SEM images of these powders can be observed in Fig. 2.8. The

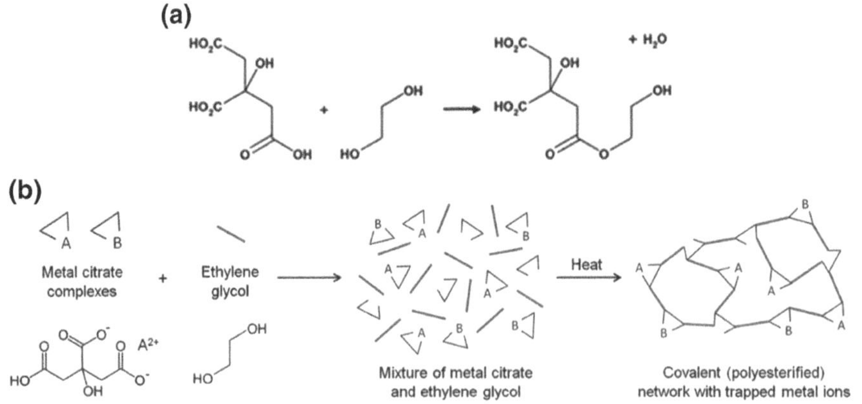

Fig. 2.7 Schematics of **a** transesterification and **b** metal-complex formation in Pechini sol–gel method (see text for more explanation). Adapted from Danks et al. [76] with permission from the Royal Society of Chemistry

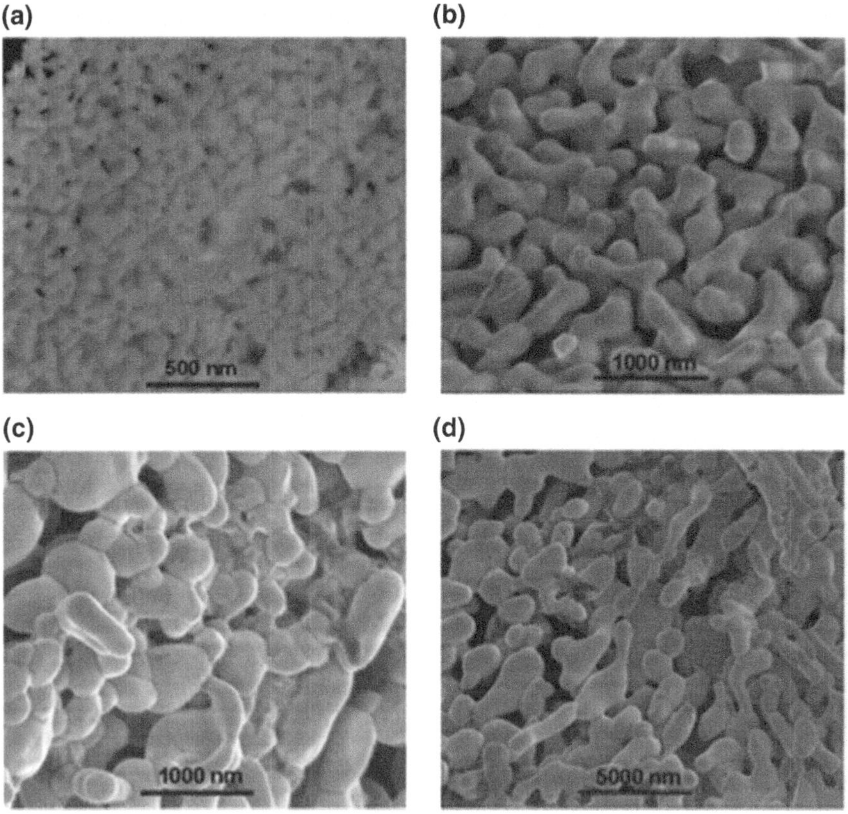

Fig. 2.8 SEM micrographs of the LLZO powders prepared by modified Pechini sol–gel method after calcination at **a** 650, **b** 700, **c** 800, and **d** 900 °C in air for 5 h. Reproduced from Kokal et al. [43] with permission from Elsevier

coral-like ligament morphology observed is quite characteristic of LLZO prepared by sol–gel based methods. The conductivity of these materials was quite poor, as the powder converted to t-LLZO upon pellet sintering.

Janani et al. [78] also used a similar method to prepare LLZO, albeit without ethylene glycol. In this case, both undoped c-LLZO and ALLZO were synthesized between 700 and 800 °C. Ionic conductivity measurements were not performed, but it was noted that dopants were evidently not required to stabilize the cubic phase in some cases. It was also observed that the use of isopropyl alcohol or butanol as surfactants affected the size of the LLZO products, with the former resulting in larger average grain sizes. Il'ina et al. [82] compared the properties of tetragonal LLZO prepared using conventional SSR and the citrate-nitrate method. The particles from the latter approach were agglomerated but three times smaller than the materials prepared by SSR; further, the ionic conductivity at 20 °C was higher for the sol–gel method (5.2×10^{-6} S/cm vs. 1.3×10^{-6} S/cm). Preparation of cubic ALLZO was also accomplished using the citrate-nitrate method, with ionic conductivity of 3.4 × 10^{-4} S/cm at 25 °C obtained for the optimized composition [91]. Chen et al. reported a modified, nitrate-free method that eliminated NO_x emissions during the synthesis and was also effective for co-doping of Ta and Ca into cubic LLZO, which was found to have a synergistic effect on the ionic conductivity [108].

Tao et al. [105] used a one-step Pechini method to prepare composites consisting of LLZO nanoparticles loaded onto carbon foam. The gel prepared from the LLZO nitrate precursors and citrate was carbonized at 850 °C for 2 h under argon. The resulting material displayed a cross-linked, porous hierarchical structure from the pyrolysis of the citric acid and nitrates and consisted of LLZO nanoparticles 30–200 nm in size decorated on an amorphous carbon matrix. The composites were investigated as a 3D scaffold for the cathode of a lithium-sulfur battery and showed excellent cycling stability. The unsupported LLZO nanoparticles could be obtained by calcining the gel in air instead of argon; composite polymer electrolytes using these nanoparticles as ceramic fillers (15 wt%) could achieve ionic conductivity of 9.5 × 10^{-6} S/cm at 20 °C.

Another notable example is from Afyon et al. [100], in which Ga-doped LLZO (GLLZO) was prepared using a sol–gel combustion route that avoided the use of carboxylic acids and ethylene glycol so that lower temperatures and shorter calcination times could be used. Uniquely, fine powders of t-LLZO (300–1 μm in size) could be obtained at only 650 °C, which were then combined with Ga_2O_3 after synthesis (Fig. 2.9). Evidently, Ga^{3+} was incorporated into the LLZO lattice to form c-LLZO at temperatures as low as 100 °C to form GLLZO particles in the range of 200–300 nm. Subsequently, these powders were densified at a quite low temperature of only 950 °C for 6 h, and pellets exhibited a conductivity of 4 × 10^{-4} S/cm at a relative density of only 76% [100].

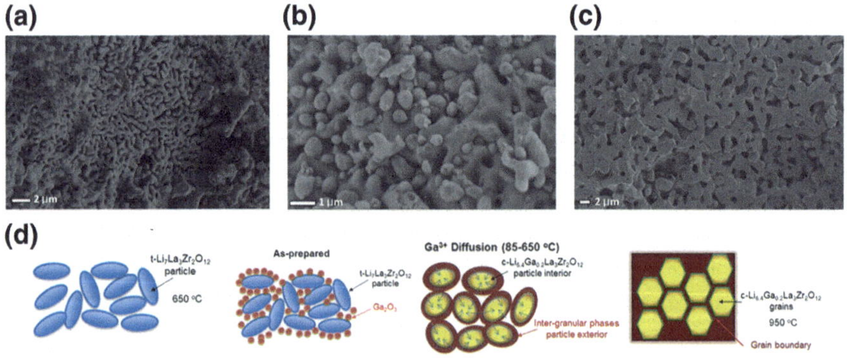

Fig. 2.9 SEM images of **a** t-LLZO powder obtained after synthesis at 650 °C using modified sol–gel combustion method; c-$Li_{6.4}Ga_{0.2}La_3Zr_2O_{12}$ **b** powder and **c** pellet cross-section after sintering at 950 °C for 6 h. **d** Schematic showing a proposed model for Ga^{3+} diffusion into t-LLZO to form cubic GLLZO, followed by sintering to form dense pellets. Adapted from Afyon et al. [100] with permission from the Royal Society of Chemistry

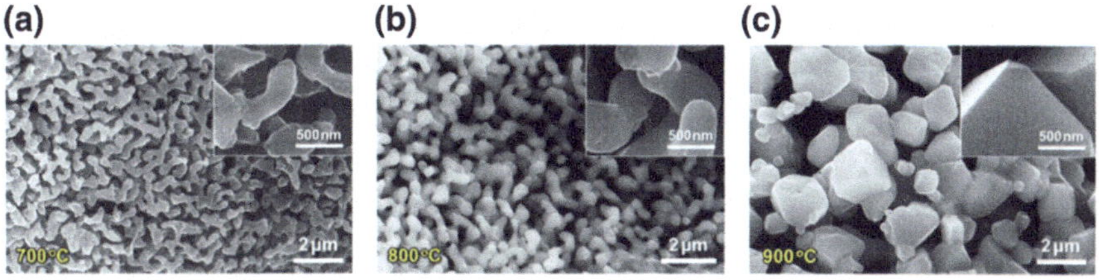

Fig. 2.10 SEM images of ALLZO prepared from polyvinyl alcohol-based hydrogel after calcination at **a** 700, **b** 800, **c** 900 °C for 2 h Adapted from Bae et al. [96] with permission from Elsevier

The so-called "polymer sol–gel method" (Fig. 2.6e) shares some similarities with the Pechini-type synthesis in that an organic material is chosen to form complexes with metal ions followed by heat treatment and combustion to generate fine powders. However, it differs in that a preexisting, well-characterized polymer is used as the organic template, rather than the somewhat ill-defined polymer that forms in situ from transesterification between citric acid and ethylene glycol in the Pechini synthesis. The polymer sol–gel method was reported for the synthesis of ALLZO by Li et al. [110], wherein hydroxypropyl cellulose (molecular weight: 100 k) was used as the polymer; however, no morphology or particle size information was reported. Bae et al. [96] used polyvinyl alcohol, glutaraldehyde as cross-linker, and nitrate precursors for ALLZO to form hydrogels which were calcined for 2 h to form different morphologies, as shown in Fig. 2.10, where the similar coral or ligament morphology seen in other sol–gel methods is observed when calcined at 700 or 800 °C.

Another particularly notable example is the use of sol–gel processing to obtain uniform thin films of LLZO. In the report by Bitzer et al., a combination of dip-coating followed by spin-coating was used to prepare Y-doped (0.2 mol%) LLZO [107]. The sol particle size was controlled by varying the aging time (Fig. 2.11a) and using a complex-forming solvent consisting of 25% 2-methoxyethanol and 75% ethanol. As shown in Fig. 2.11b, the approach was quite effective for obtaining transparent sols with an average particle size of 10 nm. Thin films 50 nm thick could be obtained with good homogeneity using dip-coating (Fig. 2.11c–d) and LLZO could be obtained after calcination at 800 °C in both air and argon atmospheres. However, the low ionic conductivity (3.4×10^{-8} S/cm for the thin film, 2.1×10^{-5} S/cm for a pellet made from the sol–gel) indicates that further optimization is still needed.

2.4 Electrospinning Synthesis of LLZO Nanowires/Nanofibers

Electrospinning has become a versatile method for synthesizing a variety of ceramic nanowires and nanofibers through the incorporation of inorganic sols into a polymer matrix and use of an electric field to form fibers from the precursor solution;

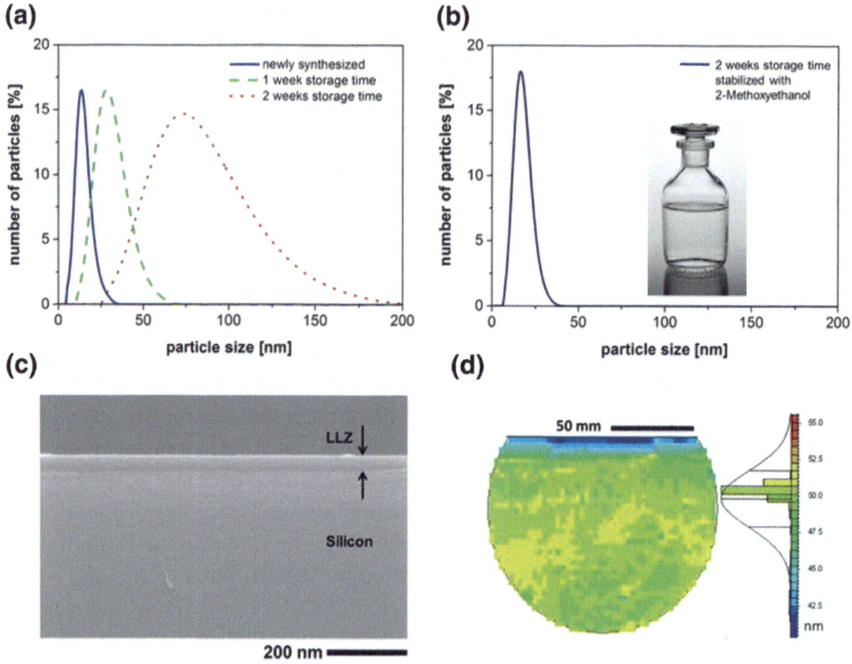

Fig. 2.11 Particle size distribution of Y-doped LLZO sol after **a** varying storage times in ethanol solution; **b** after stabilization with 2-methoxyethanol, inset shows a photograph of the transparent sol; LLZO thin film obtained by dip-coating sol onto Si wafer **c** SEM images and **d** thickness map from laser ellipsometry. Adapted from Bitzer et al. [107] with permission from Elsevier

calcination is then used to remove the organic compounds and form the polycrystalline ceramic nanowires [111, 112]. Similar to polymer sol–gel methods, the functional groups on the polymer are used to bind dissolved metal ions to form a gel and constrain particle nucleation and growth [76]. However, during crystallization, coalescence is further minimized due to the separation of the nanowires, with the final grain size limited by the dimensions of the nanowires. Therefore, the electrospinning synthesis of ceramics can be considered a modification of the sol–gel and combustion-related synthesis techniques described previously. The preparation of most common binary metal oxides (e.g., TiO_2, ZnO, SiO_2 Al_2O_3) using electrospinning has already been well-established, while recently, electrospinning has also been used to prepare complex oxides more relevant for battery applications [113] such as $LiCoO_2$ [114], $LiMn_2O_4$ [115], $Li_4Ti_5O_{12}$ [116], and Li, La(TiO_3) [117, 118].

The advantage of electrospinning compared to other chemical methods for preparing 1D nanostructures is that nanowire networks are formed as porous mats at the collector plate during the synthesis process, eliminating the need for an additional film assembly step. Furthermore, these networks display interconnected continuous pore channels (i.e., open pores), which have been shown to inhibit grain growth in nanocrystalline ceramics via grain boundary pinning [119]. Yang et al. showed that Li, $La(TiO_3)$ (LLTO) nanowires prepared by electrospinning could crystallize at much shorter calcination times than conventional LLTO prepared by SSR; further, pellets derived from LLTO nanowires displayed higher relative densities and total ionic conductivity compared to pellets prepared from LLTO powders made using conventional sol–gel synthesis [118]. These results show that electrospinning can be a promising route toward solid electrolyte nanowires with advantageous properties over materials prepared by SSR and even conventional sol–gel methods.

A typical lab-scale electrospinning setup [111, 112] consists of a spinneret (e.g., a metallic needle tip), a metallic collector and a high voltage power supply, as depicted schematically in Fig. 2.12a and shown in the photograph in Fig. 2.12b. A syringe (or optionally, multiple syringes, Fig. 2.12c) is filled with a viscous solution containing the ceramic sol–gel/polymer precursor mixture, and the needle of the syringe and the grounded metal collector are connected to the power supply. During the electrospinning process, the precursor solution is pushed out of the syringe tip using a syringe pump, forming a droplet. Since the droplet is electrified, it will experience dragging forces toward the collector exerted by the electric field, but is still balanced by the surface tension of the liquid at this stage. When the applied voltage exceeds a critical value, surface tension can no longer hold the liquid in place. The droplet will then shape into a cone (known as a Taylor cone) and a jet will shoot out from the apex. The jet is accelerated through the electric field and elongated due to the high viscosity, forming a continuous fiber that is deposited onto the collector as a mat, which can be removed as a free-standing membrane (Fig. 2.13a) and then calcined to obtain the crystalline ceramic nanowires.

Yang et al. [45] demonstrated the synthesis of LLZO nanowires using electrospinning for the first time by utilizing a similar sol preparation method as that used by

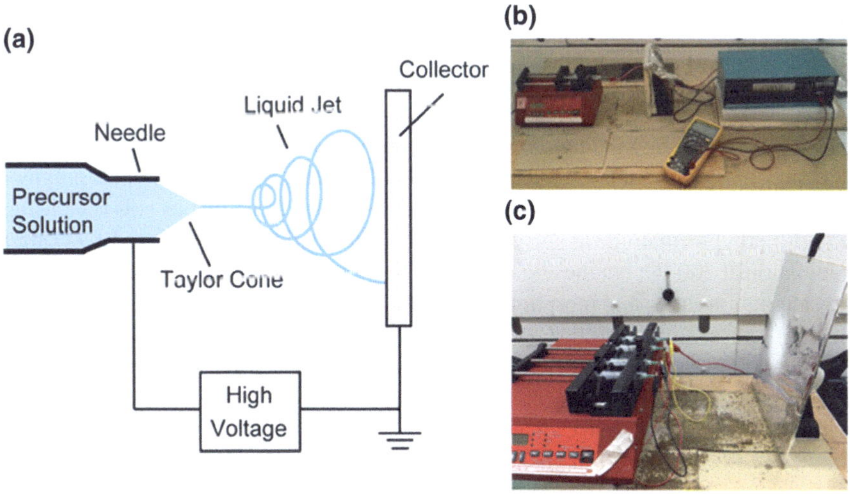

Fig. 2.12 **a** Schematic setup for electrospinning; **b** single-syringe and **c** multi-syringe electrospinning setup. Adapted from Chan et al. [42] with permission from Elsevier

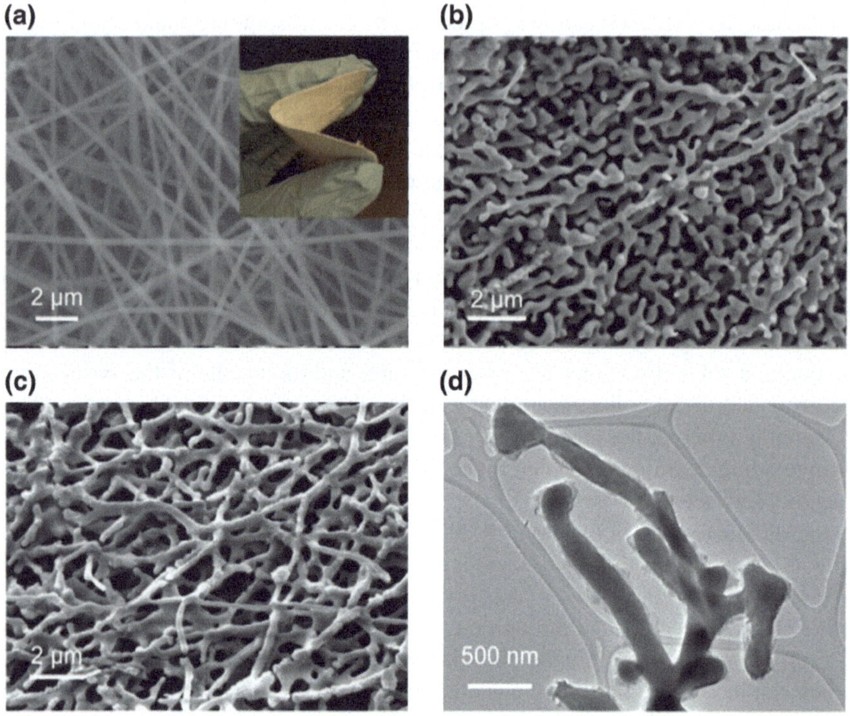

Fig. 2.13 LLZO nanowires prepared by electrospinning **a** SEM images of as-spun nanofibers composed of poly(vinylpyrrolidone) and LLZO sol–gel precursors; inset shows a photograph of a free-standing, as-spun nanowire mat. **b** SEM image of cubic LLZO nanowires from aqueous precursor obtained after calcination at 700 °C for 3 h. Adapted from Yang et al. [45] with permission from the American Chemical Society. **c** SEM and **d** TEM image of cubic LLZO nanowires from nonaqueous precursor after calcination at 700 °C for 1 h. Adapted from Yang et al. [39] with permission from the American Chemical Society

Janani et al. [78], in conjunction with a solution of a 15 wt% polyvinylpyrrolidone (PVP, molecular weight ~1,300,000) solution in acetic acid. The as-spun polymer fibers (Fig. 2.13a) prepared from both aqueous sols and nonaqueous sols (i.e., using N,N-dimethyl formamide as solvent) could be converted into phase-pure LLZO after calcination at 700 °C, but the choice of electrospinning precursor was found to affect the calcination times and morphologies of the resulting nanowires [42]. Using a calcination temperature of 700 °C, 3 h were required to fully crystallize the nanowires from the aqueous precursor [45], but the c-LLZO that was formed adopted a morphology that can be described as interconnected, coral-like ligaments with dimensions of 100–200 nm (Fig. 2.13b). On the other hand, c-LLZO could be obtained from the nanowires electrospun from the nonaqueous precursor (Fig. 2.13c–d) using only 1 h calcination at 700 °C [39]. Due to the shorter calcination time, there is also less shape change in the nanowires, although there is still a tendency to form interconnected morphologies. These morphologies arise if the nanowires are in contact with each other, since they can form junctions and merge during calcination.

Further, Yang et al. showed that the cubic phase of LLZO was formed even when there were no extrinsic dopants added [39, 45]. As the calcination time increased, the nanowires became thicker and underwent coalescence to form larger ligaments. These morphology changes are also accompanied by a phase transformation from c-LLZO to t-LLZO, particularly after longer heating times that promote the coalescence of the ligaments to form micron-sized particles [45]. Varying the electrospinning conditions to synthesize larger diameter LLZO nanowires also revealed that the tetragonal distortion in the cubic structure could be seen in these materials [42]. These results point to a diameter dependence on the phase stability in electrospun LLZO nanowires, with c-LLZO found in smaller diameter nanowires and t-LLZO present in larger diameter ones.

A possible explanation for this phenomenon is that, similar to many other metal oxide systems (such as TiO_2 or ZrO_2), the reduction of crystallite size results in an otherwise non-thermodynamically stable crystalline phase, most likely due to surface energy considerations [50]. To confirm this assertion, size reduction of bulk t-LLZO by high-energy milling to yield submicron LLZO particles was shown to result in conversion from t-LLZO to c-LLZO, further corroborating the size-stabilization effect [45]. Nonetheless, dopant-stabilized c-LLZO nanowires were also reported by adding aluminum nitrate [38, 39] or tantalum ethoxide [39] into the electrospinning precursors. LLZO nanowires prepared from electrospinning have been investigated as ceramic fillers in polyethylene oxide [38] or polyacrylontitrile [39] matrices with ionic conductivities in the 1–2.5 $\times$ 10^{-4} S/cm range for the composite electrolytes.

2.5 Cellulose Templating

Cellulose templating is another approach that can be used to prepare a variety of porous materials [120–122]. In this method, the formation process of the ceramic is similar to the case with electrospinning, except that preexisting polymer fibers (the cellulose in this case) are utilized as sacrificial templates and to absorb the sol–gel precursors rather than the polymer/sol fibers formed during electrospinning. Different from the polymer-assisted sol–gel combustion processes, in cellulose templating, it is possible to obtain nanostructured replicates with good correspondence to the original template [120, 123].

Gordon et al. [46] demonstrated that templating various cellulose substrates (commercial Whatman 42 filter paper, Kimwipes laboratory tissue paper, and nanocellulose fibers) with the same LLZO sol used in Yang's electrospinning synthesis [45] could be used to obtain nanostructured LLZO. Nanocellulose templates (Fig. 2.14a) were found to yield undoped c-LLZO particles (Fig. 2.14b) and the other paper templates resulted in t-LLZO with larger ligament feature sizes in the micron range. In this case as well, it was supposed that the nanosized pores of the nanocellulose allowed size restriction of the LLZO precursors, which in turn resulted in

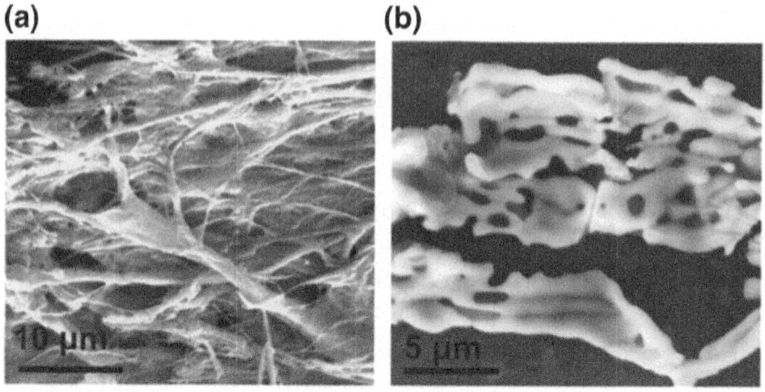

Fig. 2.14 SEM images of **a** nanocellulose fibers and **b** undoped c-LLZO templated from nanocellulose Adapted from Gordon et al. [46] with permission from the American Chemical Society

undoped c-LLZO with diameters ~500 nm. Xie et al. [124] adopted a similar concept, except using cellulose produced by bacteria from the genus *Acetobacter* as template (Fig. 2.14c) and with aluminum as dopant to obtain c-LLZO. The diameter of the bacteria cellulose nanofibers was around 30–100 nm, while the obtained LLZO structures after calcination had diameters around 500–1 μm and were similar in morphology to the LLZO nanostructures reported by Gordon et al. Despite this, the LLZO formed an interconnected and 3D porous network that could be infiltrated with a solution containing a Li salt and polyethylene oxide, forming a flexible composite polymer electrolyte with ionic conductivity 1.12×10^{-4} S/cm [124].

One of the observations that can be made from the aforementioned nanocellulose templating studies is that the LLZO synthesized from this method does not actually show a good replication of the original template. One of the reasons why could be because the nanocellulose forms agglomerated fibers, which can allow the impregnated LLZO sol to be distributed over the bundles and cause the templated LLZO to eventually coalesce into larger ligament structures during the calcination. Another reason is that the temperatures (>700 °C) required to crystallize LLZO sol–gel precursors are quite high compared to other metal oxides that can be prepared by cellulose templating, which thus enhances the solid-state diffusion of the species, grain coalescence, and morphology changes. For example, the nanoscale features and hierarchical morphology of filter paper can be replicated in templated titania, which only requires calcination at 450 °C to form crystalline anatase [120].

Gong et al. [125] also used the cellulose templating approach to prepare LLZO, but instead of using nanocellulose, the templates were cellulose textiles consisting of microfibers ~10 μm in diameter bundled into ~200 μm yarns and arranged in a woven pattern with interfiber pores of ~10–20 μm. After pretreatment (annealing in air at 270 °C for 10 h and washing with ethanol), the cellulose textiles were soaked in the LLZO sol–gel solution (with Al-dopant precursor) for 24 h and then calcined at 800 °C. As shown in Fig. 2.15a, the morphology of the cellulose template was well-preserved after templating and calcination, with the interfiber pores increasing after

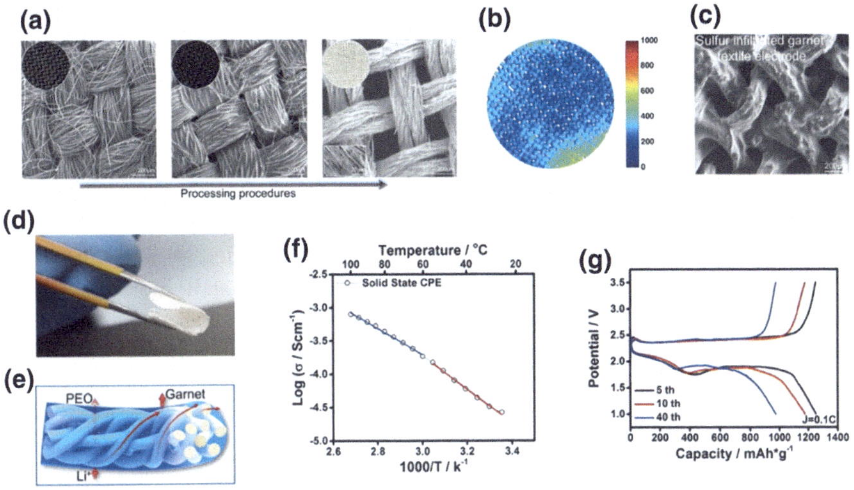

Fig. 2.15 SEM images of LLZO ceramic textile **a** after pretreatment (left), impregnation with LLZO precursor solution (center), and calcination to LLZO (right); **b** Reconstructed model of flatness uniformity of LLZO textile as generated by 3D laser scanning; **c** SEM image of composite cathode after loading 10.8 mg/cm^2 of sulfur into LLZO textile; **d** Photograph, **e** schematic of Li^+ diffusion mechanism, and **f** Arrhenius plot of ceramic textile-based composite polymer electrolyte; **g** Voltage profiles of solid-state lithium-sulfur battery containing 3D, ceramic textile supported sulfur cathode Adapted from Gong et al. [125] with permission from Elsevier

the high-temperature treatment as well. Further, 3D laser scanning to characterize the topography of the LLZO textiles showed good uniformity in the height profile (Fig. 2.15b) and the textiles displayed good flexibility. The LLZO textiles were investigated as ceramic fillers in composite polymer electrolytes, which displayed a Li-ion conductivity of 7×10^{-5} S/cm at 25 °C (Fig. 2.15d–f), slightly lower than other composite electrolytes using electrospun LLZO nanowires as fillers [38, 39]. The porous structure of the LLZO textiles was further exploited as a 3D scaffold for the sulfur cathode in a lithium/sulfur battery. The LLZO textile was sintered onto a dense LLZO pellet, and then infiltrated with a slurry containing sulfur powder and carbon nanotube conducting additives (Fig. 2.15c). A small amount of liquid electrolyte was still needed to reduce the interfacial impedance at the sulfur/LLZO interface. The charge/discharge behavior of the battery (Fig. 2.15g) showed characteristics typical of the Li/S system, but a large polarization (~0.6 V) and capacity decay was observed, which was attributed to the still non-optimized LLZO thicknesses and volume utilization of the 3D electrode architecture. Nonetheless, these results show that such cellulose-templated LLZO materials can have great potential for composite electrolytes as well as composite electrodes.

2.6 Spray Pyrolysis

Spray pyrolysis is another technique commonly used to prepare ceramic materials that have been adapted to the synthesis of LLZO. Solution-based aerosol processes are attractive in that unlike other solution-based syntheses, the precipitation, calcination, and sintering stages of the powder synthesis occur in an integrated process, which can offer improved control over the nanoparticle formation and morphology [126]. A spray pyrolysis method for preparing ALLZO was reported by Djenadic et al. [127] using $Zr(C_5H_7O_2)_4$ and Li, La, and Al nitrate salts (Li at 30 mol% excess) in water/methanol mixtures. This precursor solution was nebulized with an ultrasonic generator and the mist was transported into a reactor under flowing oxygen with a syringe pump. The nanoparticles, synthesized at 900 °C, displayed a broad particle size distribution and were described as hollow structures with shell comprised of nanometer-sized particles (crystallite size 7–9 nm). Pellets obtained after sintering of the nanoparticles at 1000 °C for 1 h had rather low relative densities (~50%) and porous morphologies, with ionic conductivity only in the ~10^{-6} S/cm range.

A different spray-pyrolysis-based synthetic procedure for LLZO was demonstrated by Yi et al. based on liquid-feed flame spray pyrolysis (LF-FSP) [71, 128]. Flame spray pyrolysis is a versatile method that is already widely used in industry for the ton-scale manufacturing of nanomaterials [129]. LF-FSP has been used by Laine et al. to prepare a variety of ultrafine ceramic particles [130–134], whereby flame combustion of metal organic precursors occurs within an aerosol of a combustible solvent (typically ethanol) and oxygen, which serves as the atomizing gas and co-reactant. The organic ligands are oxidatively decomposed, resulting in an inorganic vapor containing oxo and hydroxyl species from which the particles can grow via nucleation, vapor deposition, and cluster collision [131, 134]. The schematic in Fig. 2.16a shows the LF-FSP reactor apparatus, where the precursor solution is aerosolized with oxygen in a quartz chamber, followed by combustion with a methane/pilot torch; the nanoparticles are collected downstream in an electrostatic precipitator operating at 10 kV [128].

A major advantage of LF-FSP is that un-aggregated and uniform spherical nanoparticles <100 nm in size can be prepared with this method, which can greatly facilitate the use of thin film processing routes (e.g., tape casting) for these materials. Further, the LF-FSP particles can be suspended directly into slurries for tape casting after the synthesis, unlike sol–gel prepared materials, which must be subjected to high-energy milling before tape casting to reduce agglomerates that form during calcination. As a result, LF-FSP is claimed to be the only technique that can provide nanopowders that can be directly used for tape casting of thin films with thickness <100 μm [135]. A disadvantage of the method, however, is that metal organic compounds, which often must be synthesized due to commercial unavailability, need to be used as the precursors. For instance, for LLZO, lithium propionate (used at 50 wt% excess), lanthanum isobutyrate, zirconium isobutyrate, and alumatrane [$Al(OCH_2CH_2)_3N$] or gallium-atrane [$Ga(OCH_2CH_2)_3N$] for dopants were used as precursors [71, 128]. Lower-cost precursors, such as nitrates, are unsuitable

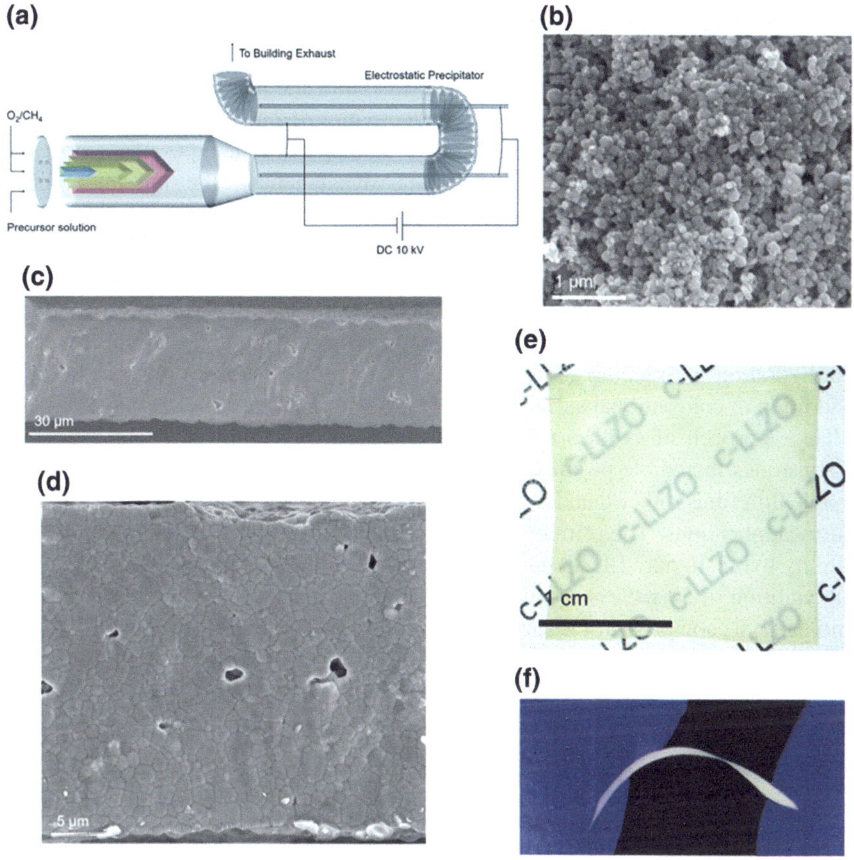

Fig. 2.16 **a** Schematic of reactor apparatus for the synthesis of LLZO using liquid-feed flame spray pyrolysis (LF-FSP); **b** SEM image of as-produced Ga-doped "decomposed" LLZO nanoparticles; **c** SEM fracture surface and **d** thermally etched fracture surface of GLLZO film sintered at 1130 °C for 0.3 h followed by 900 °C for 2 h; **e**–**f** Photographs of translucent and flexible ALLZO thin film with dimensions of 2 cm × 2 cm × ~30 μm. (**a**), (**e**), and (**f**) adapted from Yi et al. [128] with permission from the Royal Society of Chemistry; (**b**)–(**d**) adapted from Yi et al. [71] with permission from Elsevier

since they can melt easily and decompose thermally to NOx without combusting; the covalent linkages from tri- and bidentate organic ligands to the metal was determined to be important for creating the combustion processes needed to obtain small, uniform particles [134].

The particles prepared using LF-FSP of the LLZO precursors were quite uniform, with an average size of 90 nm (Fig. 2.16b). However, the obtained particles were not LLZO, but rather "decomposed LLZO" consisting of the $La_2Zr_2O_{12}$ pyrochlore phase, an unknown but most likely noncrystalline La-containing phase, and Li_2CO_3. The authors attribute the formation of these "decomposed LLZO" nanoparticles to the high flame temperatures >1500 °C and presence of H_2O and CO_2 as combustion byproducts.

However, when these powders were tape-cast, thin and phase-pure dense (>94%) LLZO membranes were obtained, with conductivities of 2×10^{-4} S/cm for ALLZO [128] and 1.3×10^{-3} S/cm for GLLZO [71]. Examples of the types of membranes obtained in this work are seen in Fig. 2.16c–f. Advantageously, the sintering process used for these tape-cast films required lower temperatures and times than many other heating schemes used by other researchers for pellets—1090 °C for 1 h for ALLZO [128] and 1130 °C for 0.3 h for GLLZO [71]. Further, it was proposed that Li^+/H^+ exchange on the surface of the LLZO nanoparticles upon air exposure, which typically is deleterious to sintering, may have enhanced the densification process, with hydration introducing additional mechanisms for densification which operate in addition to conventional surface and bulk diffusion during sintering [71].

This approach of synthesizing nanosized precursors followed by tape casting demonstrates incredible promise for application of LLZO as a solid electrolyte, as flexible thin films rather than large dense pellets are an ideal geometry for practical battery applications. The primary drawback to the LF-FSP synthesis method is the inability to obtain pure LLZO nanoparticles rather than "decomposed LLZO," even with addition of 50 wt% excess Li precursor. Nonetheless, these studies show that nanomaterials comprised of intermediate phases or constituent oxides of LLZO can also provide a promising route toward obtaining LLZO solid electrolytes.

2.7 Co-precipitation Methods

Co-precipitation methods are alternative wet chemistry routes that avoid some of the drawbacks of sol–gel and combustion-based syntheses, such as the formation of nitrous gases or use of expensive precursors. Several reports of co-precipitation methods for LLZO have been reported. Langer et al. [136] dissolved the La, Al, and Zr (oxo)nitrate precursors for LLZO into water and precipitated the metal ions as hydroxides at pH 9.5 using ammonium hydroxide. After drying and mixing with $LiNO_3$, the powder was calcined at 850 °C to form crystalline LLZO. While this is a much lower calcination temperature than that needed for SSR and similar to those used in sol–gel methods, the obtained particles were agglomerated and sintered pellets had low density (57%) and room temperature ionic conductivity of only 3.2×10^{-6} S/cm.

A similar wet co-precipitation method reported by Kim et al. [137] (using NaOH and NH_4OH to modify the pH, with LiOH hydrate as the Li source), resulted in LLZO particles with a plate-like morphology (lengths of 300–600 nm and widths of 50–100 nm) after calcination at 700 °C; the particles became more spherical and increased in size as the calcination temperature increased, with t-LLZO (and other impurity phases) observed after calcination at 900 °C. Shao et al. ball-milled the co-precipitated hydroxides with Li_2CO_3 for 8 h prior to pellet sintering and found that this was effective for preparing LLZO with primary particle size 0.5 to

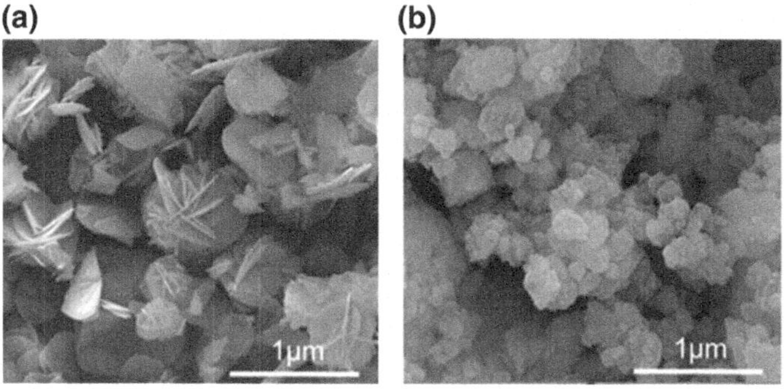

Fig. 2.17 SEM images for the **a** $La_3Zr_2Al_{0.25}CO_3/OH$ particles formed after the LLZO co-precipitation reaction; **b** precursor powders after coating and crystallization of $Li_2C_2O_4$ shell on the co-precipitates. Reproduced from Zhang et al. [140] with permission from Elsevier

~1 μm and ionic conductivity of 2×10^{-4} S/cm [138]. Other reports showed that co-precipitation could also be used to obtain Ga, Ta, Al, Nb, and co-doped LLZO, with pellets displaying ionic conductivities in the ~10^{-4} S/cm range [51, 57, 139], demonstrating that materials with comparable conductivity as those prepared by SSR are possible using co-precipitation.

Another co-precipitation-based approach to prepare LLZO nanoparticles [140] used ammonium carbonate as a precipitant to make flaky $La_3Zr_2Al_{0.25}CO_3/OH$ particles <1 micron in size; this precipitate was subsequently coated with a lithium oxalate ($Li_2C_2O_4$) shell <100 nm thick (Fig. 2.17). This core-shell structure was calcined and c-LLZO was formed starting at 700 °C, but with coexistence of $La_2Zr_2O_7$, via $LaCO_3(OH)$ and $Li_2C_2O_4$ phases that were observed at 500 °C. Pure phase, Al-doped c-LLZO was obtained after calcination at 900 °C. The particle sizes of the obtained c-LLZO were not provided, but the grain sizes of the $LaCO_3(OH)$ and $Li_2C_2O_5$ intermediate phases were calculated to be 66.8 nm and 51 nm, respectively, using the Scherrer equation.

2.8 Molten Salt Methods

While the various sol–gel, combustion, and co-precipitation wet-chemical methods are successful for preparing nanostructured LLZO, in general they do not confer an easily scalable method to form the materials, given the generally high cost for sol–gel precursors (e.g., metal alkoxides) and large amounts of "sacrificial" reagents such as excess solvents (alcohols, glycols, etc.), and combustible material (e.g., polymers, nanocellulose); further, agglomeration of particles after calcination is often an issue. Therefore, solution-based methods of fine particles would be more attractive for scale-up of the synthesis of nanostructured LLZO. While to our knowledge, no successful hydrothermal or solvothermal syntheses of LLZO have yet to be reported, synthesis of LLZO in a liquid salt medium has shown great promise for the scalable production of LLZO fine particles.

Flux-based and molten salt synthetic methods have long been an attractive approach to synthesize metal oxide crystals, including complex ternary or quarternary oxides [141, 142]. In molten salt synthesis (MSS), precursors usually consisting of metal oxides or metal salts are mixed intimately with an inorganic compound (often metal oxides, hydroxides, halides, and oxosalts, or mixtures thereof), followed by heat treatment above the melting point of the compound(s) to provide the thermodynamic driving force for either dissolution of the precursors or topochemical reaction to promote the formation of the desired crystalline phase, typically at temperatures well below the normal melting point of the crystal being synthesized [143]. A type of solution-based synthesis, the crystal growth process in flux solvents is similar as to what occurs in aqueous or organic solvents: (1) dissolution of the reactants, (2) diffusion of the reactants through the molten flux, (3) nucleation of crystallites, and (4) further crystal growth from the nuclei [142, 143]. MSS is quite versatile and has been used to synthesize large crystals (several mm) [141] as well as nanomaterials [144], often as well-faceted crystals due to the preference for the formation of the low free energy faces (i.e., those with low Miller indices) [143]. Further, MSS often results in high-quality nanocrystalline products at temperatures lower than those required by SSR for the same target material owing to the greatly improved diffusional kinetics of dissolved species in the flux compared to the extremely limited mobility of atoms diffusing in the solid-state [144].

Although a complex oxide with many components (five or more if considering dopants), LLZO could be well-suited for flux-based syntheses since the Li_2CO_3 or LiOH precursors are molten during the conventional solid-state reactions anyways (the melting points are 720 °C and 462 °C, respectively). This was speculated to facilitate Al^{3+} dissolution from the Al_2O_3 crucibles in adventitious doping [59]; the dissolution of the LLZO oxide precursors in a molten Li_2CO_3 flux during the SSR synthesis has since been confirmed through detailed thermogravimetric and in situ neutron diffraction experiments [63].

The first report of a flux-based synthesis for LLZO was by Awaka et al. [145] in 2009, where tetragonal phase LLZO single crystals were synthesized by taking polycrystalline LLZO powder prepared by SSR (Li_2CO_3, La_2O_3, and ZrO_2 heated at 900 °C for 17 h, followed by 980 °C for 5 h) and dissolving it in a Li_2CO_3 melt at a 1:1 weight ratio at 1040 °C for 48 h. The single crystals had maximum diameter of 50 μm (Fig. 2.18) and sintered pellets prepared from the samples had bulk Li-ion conductivity of 1.63×10^{-6} S/cm and grain boundary conductivity of 5.59×10^{-7} S/cm at 27 °C, which is much lower than the typical values obtained for cubic LLZO. It should be pointed out that in this reaction, LLZO itself was used as the starting material for the flux synthesis, not the LLZO precursors, and that the flux synthesis was also performed at a higher temperature than the SSR itself.

Subsequently, in 2011 Awaka et al. [146] reported the flux synthesis of cubic LLZO from La_2O_3, ZrO_2, and a number of Li sources (i.e., LiCl, LiOH, $LiNO_3$, and Li_2CO_3) in excess to serve as self-fluxes. The components were mixed as a powder and heated in an alumina crucible at 1150 °C for 4 h (it should be noted that under these conditions, adventitious Al doping from the crucible could be possible [59, 147], but no Al signal was observed from the energy-dispersive X-ray spectroscopy analysis). Single crystals of LLZO with an average size of 0.05–0.10 mm could be obtained only when using Li_2CO_3 and $LiNO_3$ as self-fluxes, with the latter declared the most suitable solvent using a composition of Li:La:Zr = 20:3:2. In both of these examples of single salt fluxes, the LLZO crystals are subtly faceted, but not very uniform in size or shape.

While multicomponent salt mixtures have been exploited for quite some time for the synthesis of nanostructured oxides [144], these solvents have only been recently applied to the synthesis of LLZO. The advantage of multicomponent (often times, eutectic) mixtures is that the melting points and viscosity are lowered compared to the melts of the individual components [142]. Table 2.2 shows the melting points and compositions for multicomponent mixtures and eutectics that have been used in MSS.

In 2017, Reddy et al. [148] utilized a eutectic mixture of $LiNO_3$ and LiCl (88:12 mol%, melting point ~280 °C) as the molten medium, with La_2O_3, $ZrCl_4$, and Ta_2O_5 as precursors for the synthesis of Ta-doped LLZO (LLZTO) between 700–900 °C. Aggregates of micrometer-sized, irregularly faceted crystals of LLZTO were obtained using this method. The resultant materials exhibited a maximum ionic conductivity of 7.8×10^{-6} S/cm at room temperature, in contrast to the roughly 100× higher conductivity generally reported for LLZTO prepared using SSR [52, 56, 149]. The low ionic conductivity in the particles prepared using this MSS method was attributed to the post-synthesis, aqueous washing step used to dissolve and remove the eutectic salt medium [30]. This led to proton exchange with Li^+, as confirmed by the increase in lattice constant, which resulted in poor sinterability of the pellets and the uncharacteristically low Li^+ conductivity. Further, the morphology of the as-synthesized powders was somewhat irregular, whereas MSS commonly produces more uniform powders.

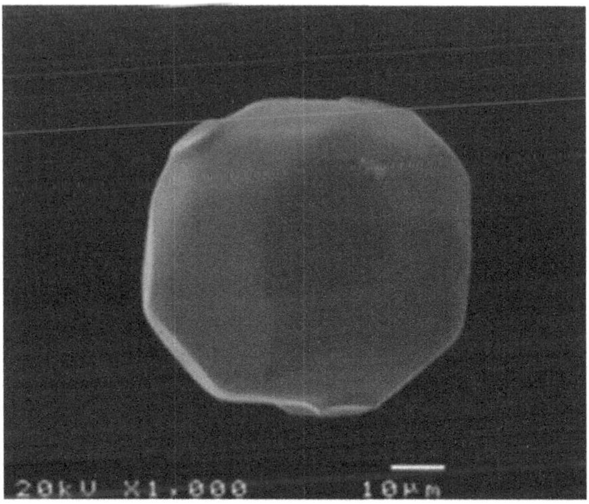

Fig. 2.18 SEM image of t-LLZO single crystal grown in Li_2CO_3 flux at 1040 °C; reproduced from Awaka et al. [145] with permission from Elsevier

Table 2.2 Melting points and compositions of some commonly used metal halides, hydroxides, and oxosalt systems. Adapted from Liu et al. [144] with permission from the Royal Society of Chemistry

Salt system		Composition (mol%)	Melting point (°C)
Halides	LiCl/KCl	59/41[a]	352
	NaCl/KCl	50/50	658
	$AlCl_3$/NaCl	50/50	154
	KCl/$ZnCl_2$	48/52	228
	LiF/NaF/KF	46.5/11.5/42[a]	459
	LiI/KI	63/37[a]	286
Oxosalts	NaOH/KOH	51/49[a]	170
	$LiNO_3$/KNO_3	43/57[a]	132
	Li_2SO_4/K_2SO_4	71.6/28.4[a]	535
	Li_2CO_3/K_2CO_3	50/50	503

[a]Eutectic mixtures

Weller et al. [47] reported the MSS of LLZO in a LiCl/KCl eutectic mixture (59:41 mol%, melting point 352 °C) using nitrate metal salt precursors to provide reactive oxygen species and promote the formation of a completely homogeneous salt melt (Fig. 2.19). This approach was very effective for obtaining phase-pure, cubic LLZO as faceted particles with side lengths between 0.5 and 2 μm using a temperature of 900 °C for 4 h, and could easily incorporate a wide variety of dopants (e.g., Al, Ga). Weller et al. also studied the formation mechanism of LLZO in the LiCl/KCl medium (Fig. 2.19a) and found that it proceeds via LaOCl at 500–600 °C (Fig. 2.19b), followed by the emergence of $La_2Zr_2O_7$ and the coexistence of these phases at 700 °C (Fig. 2.19c). At 800 °C, the product is mostly $La_2Zr_2O_7$ nanocrystals ~20 nm in size (Fig. 2.19d). The $La_2Zr_2O_7$ nanocrystals react further with the remaining Li and La sources to form LLZO at 900 °C (Fig. 2.19e) in the MSS medium.

The improved particle morphology in the LLZO prepared by MSS in the LiCl/KCl eutectic compared to the $LiNO_3$/LiCl medium was explained as follows: Due to the boiling and/or decomposition of $LiNO_3$ at around 600 °C, there is a high possibility that some, if not much, of the nitrate portion of the medium evaporates under conditions (i.e., temperatures between 700 and 800 °C) that form LLZO. The evaporation of a substantial portion of the reaction medium could have a deleterious effect on the uniformity and morphology of the resultant powders. Further, in contrast to the use of oxides by Reddy et al. for some of the precursors, nitrate salt precursors were used in the MSS method by Weller et al. so that potential concentration gradients due to dissolution and diffusion of the metal oxides were eliminated, which could promote the formation of a smaller particle size distribution due to improved initial homogeneity of all species in the reaction mixture.

The obtained LLZO fine particles displayed ionic conductivity ranging from 2.3–3.7 × 10^{-4} S/cm, depending on the dopant used, if excess LiOH (3–10 wt%) was

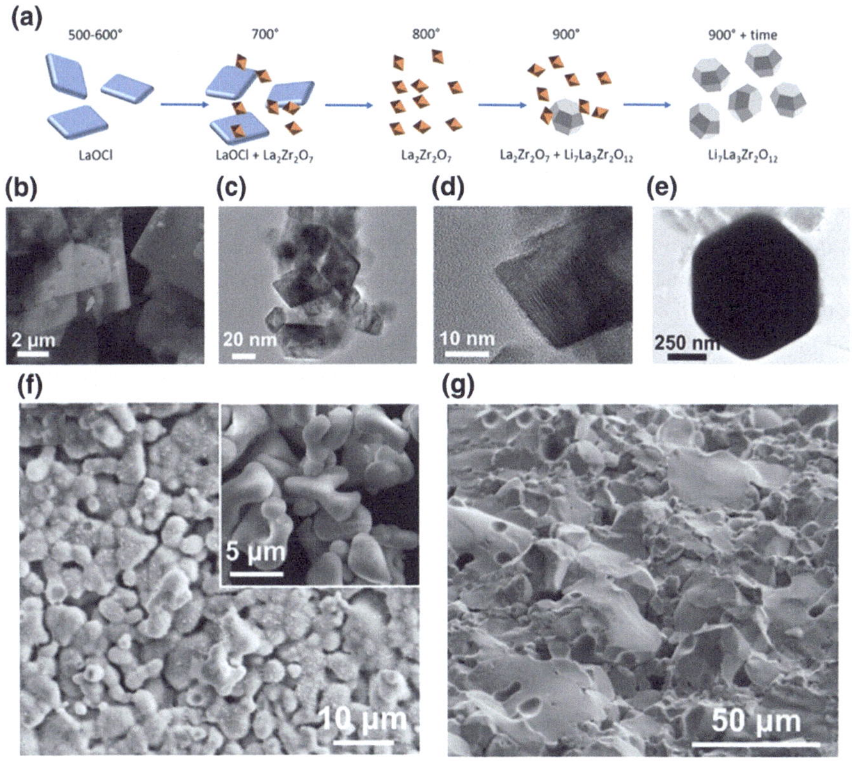

Fig. 2.19 **a** Schematic of overall MSS growth process of LLZO in eutectic LiCl/KCl; SEM or TEM images of reaction products formed in MSS at different temperatures: **b** LaOCl nanoplates formed at 500 °C, **c** mixture of LaOCl and $La_2Zr_2O_7$ formed at 700 °C, **d** $La_2Zr_2O_7$ nanocrystals formed at 800 °C, **e** c-LLZO particles formed at 900 °C; **f** SEM image of fracture surface of GLLZO pellet sintered for 18 h at 1100 °C without added LiOH; inset shows the as-synthesized GLLZO powder; **g** GLLZO pellet sintered with 10% excess LiOH for 12 h at 1100 °C

added to the powder prior to pellet sintering. Similar to the findings by Reddy et al., Weller et al. found that the process of washing the MSS products with a large volume of water to remove the salt media resulted in powders that could not be sintered (Fig. 2.19f); this likely results from the gradual dissolution and removal of any LiOH or Li_2CO_3 that forms as a consequence of proton exchange from the as-synthesized powders, which led to Li-deficiency and formation of the poorly sinterable $La_2Zr_2O_{12}$ phase. However, the simple addition of LiOH to reverse the hydration/proton exchange and obtain dense pellets (Fig. 2.19g) is an important result to motivate further exploration of MSS for the synthesis of LLZO, particularly since large batches (tens of grams) of powders can be prepared using this method. It should be noted that despite the lower melting points of the molten salt eutectics used so far, the temperatures needed to actually form LLZO are much higher. Hence, further studies focusing on the reduction of the LLZO formation temperature through modification of the salt melt properties would also be beneficial for improving the applicability of MSS to prepare LLZO.

2.9 Summary and Outlook

In the decade or so since LLZO was first reported as a Li^+ ion conductor, preparation methods for nanostructured LLZO have developed drastically, particularly as issues related to the mechanical robustness, critical current density, phase stability, and processibility of LLZO for application in all-solid-state batteries have come to light. Despite a number of numerous inconvenient traits (e.g., the high temperatures required to crystallize LLZO, the multiple poly-types and need for extrinsic dopants, the facile reaction with water and CO_2), it is quite impressive that so many successful synthesis methods for nanostructured LLZO have already been reported. There is still much to be learned about the effect of particle and grain size on the ionic conductivity, mechanical properties, electrochemical stability, and resistance to Li dendrite propagation in LLZO-based solid electrolytes. Future efforts to realize the scalable synthesis of uniform LLZO powders with excellent shape and morphology control will not only help to elucidate the fundamental properties of LLZO, but will also greatly facilitate the implementation of LLZO as a solid electrolyte in high-energy density batteries for practical applications of commercial importance.

References

1. Arico AS, Bruce P, Scrosati B, van Schalkwijk W (2005) Nanostructured materials for advanced energy conversion and storage devices. Nat Mater 4:366–377
2. Wang Y, Cao G (2008) Developments in nanostructured cathode materials for high-performance lithium-ion batteries. Adv Mater 20:2251–2269
3. Ji L, Lin Z, Alcoutlabi M, Zhang X (2011) Recent developments in nanostructured anode materials for rechargeable lithium-ion batteries. Energy Environ Sci 4:2682–2699
4. Sun Y, Liu N, Cui Y (2016) Promises and challenges of nanomaterials for lithium-based rechargeable batteries. Nat Energy 1:16071
5. Murugan R, Thangadurai V, Weppner W (2007) Fast lithium ion conduction in garnet-type $Li_7La_3Zr_2O_{12}$. Angew Chem Int Ed 46:7778–7781
6. Thangadurai V, Narayanan S, Pinzaru D (2014) Garnet-type solid-state fast li ion conductors for li batteries: critical review. Chem Soc Rev 4714:4714–4727
7. Teng S, Tan J, Tiwari A (2014) Recent developments in garnet based solid state electrolytes for thin film batteries. Curr Opin Solid State Mater Sci 18:29–38
8. Ramakumar S, Deviannapoorani C, Dhivya L et al (2017) Lithium garnets: synthesis, structure, Li^+ conductivity, Li^+ dynamics and applications. Prog Mater Sci 88:325–411
9. Huang M, Liu T, Deng Y et al (2011) Effect of sintering temperature on structure and ionic conductivity of $Li_{7-x}La_3Zr_2O_{12-0.5x}$ ($x = 0.5 \sim 0.7$) ceramics. Solid State Ionics 204–205:41–45
10. Rangasamy E, Wolfenstine J, Sakamoto J (2012) The role of Al and Li concentration on the formation of cubic garnet solid electrolyte of nominal composition $Li_7La_3Zr_2O_{12}$. Solid State Ion 206:28–32

11. Bernstein N, Johannes MD, Hoang K (2012) Origin of the structural phase transition in $Li_7La_3Zr_2O_{12}$. Phys Rev Lett 109:205702
12. Wang Y, Lai W (2012) High ionic conductivity lithium garnet oxides of $Li_{7-x}La_3Zr_{2-x}Ta_xO_{12}$ compositions. Electrochem Solid-State Lett 15:A68–A71
13. Bates JB, Dudney NJ, Gruzalski GR et al (1993) Fabrication and characterization of amorphous lithium electrolyte thin films and rechargeable thin-film batteries. J Power Sources 43:103–110
14. Kerman K, Luntz A, Viswanathan V et al (2017) Review-practical challenges hindering the development of solid state li ion batteries. J Electrochem Soc 164:A1731–A1744
15. Chiang Y-M, Lavik EB, Kosacki I et al (1996) Defect and transport properties of nanocrystalline CeO_{2-x}. Appl Phys Lett 69:185
16. Lavik EB, Kosacki I, Tuller HL et al (1997) Nonstoichiometry and electrical conductivity of nanocrystalline CeO_{2-x}. J Electroceramics 1:7–14
17. Xu X, Wen Z, Yang X et al (2006) High lithium ion conductivity glass-ceramics in Li_2O-Al_2O_3-TiO_2-P_2O_5 from nanoscaled glassy powders by mechanical milling. Solid State Ion 177:2611–2615
18. Wen Z, Xu X, Li J (2009) Preparation, microstructure and electrical properties of $Li_{1.4}Al_{0.4}Ti_{1.6}(PO_4)_3$ nanoceramics. J Electroceram 22:342–345
19. Jackman SD, Cutler RA (2012) Effect of microcracking on ionic conductivity in LATP. J Power Sources 218:65–72
20. Tenhaeff WE, Rangasamy E, Wang Y et al (2014) Resolving the grain boundary and lattice impedance of hot-pressed $Li_7La_3Zr_2O_{12}$ garnet electrolytes. ChemElectroChem 11:375–378
21. Cheng L, Park JS, Hou H et al (2014) Effect of microstructure and surface impurity segregation on the electrical and electrochemical properties of dense Al-substituted $Li_7La_3Zr_2O_{12}$. J Mater Chem A 2:172–181
22. Cheng L, Chen W, Kunz M et al (2015) Effect of surface microstructure on electrochemical performance of garnet solid electrolytes. ACS Appl Mater Interfaces 7:2073–2081
23. Sakamoto J, Rangasamy E, Kim H et al (2013) Synthesis of nano-scale fast ion conducting cubic $Li_7La_3Zr_2O_{12}$. Nanotechnology 24:424005
24. Cheng L, Wu CH, Jarry A et al (2015) Interrelationships among grain size, surface composition, air stability, and interfacial resistance of Al-substituted $Li_7La_3Zr_2O_{12}$ solid electrolytes. ACS Appl Mater Interfaces 7:17649–17655
25. Albertus P, Babinec S, Litzelman S, Newman A (2018) Status and challenges in enabling the lithium metal electrode for high-energy and low-cost rechargeable batteries. Nat Energy 3:16–21
26. Xia W, Xu B, Duan H et al (2016) Ionic conductivity and air stability of Al-doped $Li_7La_3Zr_2O_{12}$ sintered in alumina and Pt crucibles. ACS Appl Mater Interfaces 8:5335–5342
27. Sharafi A, Haslam CG, Kerns RD et al (2017) Controlling and correlating the effect of grain size with the mechanical and electrochemical properties of $Li_7La_3Zr_2O_{12}$ solid-state electrolyte. J Mater Chem A 5:21491–21504
28. Groza R (1999) Nanosintering. Nanostructured Mater 12:987–992
29. Fang ZZ, Wang H, Wang X, Kumar V (2010) Grain growth during sintering of nanosized particles. In: Bordia RK, Olevsky EA (eds) Advances in sintering science and technology: ceramic transactions. Wiley, Hoboken, pp 389–400
30. Bowen P, Carry C (2002) From powders to sintered pieces: forming, transformations and sintering of nanostructured ceramic oxides. Powder Technol 128:248–255
31. Mayo MJ (1993) Superplasticity of nanostructured ceramics. In: Nastasi M, Parkin D, Gleiter H (eds) Mechanical properties and deformation behavior of materials having ultra-fine microstructures. Springer, Dordrecht, pp 361–380
32. McCloskey BD (2015) Attainable gravimetric and volumetric energy density of Li-S and Li ion battery cells with solid separator-protected Li metal anodes. J Phys Chem Lett 6:4581–4588
33. Kumar B, Scanlon LG (1994) Polymer-ceramic composite electrolytes. J Power Sources 52:261–268

34. Wieczorek W, Florjanczyk Z, Stevens JR (1995) Composite polyether based solid electrolytes. Electrochim Acta 40:2251–2258
35. Croce F, Appetecchi GB, Persi L, Scrosati B (1998) Nanocomposite polymer electrolytes for lithium batteries. Nature 394:456–458
36. Croce F, Persi LL, Scrosati B, Serraino-Fiory F, Plichta E, Hendrickson MA (2001) Role of the ceramic fillers in enhancing the transport properties of composite polymer. Electrochim Acta 46:2457–2461
37. Kim J-W, Ji K-S, Lee J-P, Park J-W (2003) Electrochemical characteristics of two types of PEO-based composite electrolyte with functional SiO_2. J Power Sources 119–121:415–421
38. Fu K, Gong Y, Dai J et al (2016) Flexible, solid-state, ion-conducting membrane with 3D garnet nanofiber networks for lithium batteries. Proc Natl Acad Sci 113:7094–7099
39. Yang T, Zheng J, Cheng Q et al (2017) Composite polymer electrolytes with $Li_7La_3Zr_2O_{12}$ garnet-type nanowires as ceramic fillers: mechanism of conductivity enhancement and role of doping and morphology. ACS Appl Mater Interfaces 9:21773–21780
40. Choi JH, Lee CH, Yu JH et al (2015) Enhancement of ionic conductivity of composite membranes for all-solid-state lithium rechargeable batteries incorporating tetragonal $Li_7La_3Zr_2O_{12}$ into a polyethylene oxide matrix. J Power Sources 274:458–463
41. Zhang J, Zhao N, Zhang M et al (2016) Flexible and ion-conducting membrane electrolytes for solid-state lithium batteries: dispersion of garnet nanoparticles in insulating polyethylene oxide. Nano Energy 28:447–454
42. Chan CK, Yang T, Mark Weller J (2017) Nanostructured garnet-type $Li_7La_3Zr_2O_{12}$: synthesis, properties, and opportunities as electrolytes for Li-ion batteries. Electrochim Acta 253:268–280
43. Kokal I, Somer M, Notten PHL, Hintzen HT (2011) Sol–gel synthesis and lithium ion conductivity of $Li_7La_3Zr_2O_{12}$ with garnet-related type structure. Solid State Ionics 185:42–46
44. Xie H, Li Y, Goodenough JB (2012) Low-temperature synthesis of $Li_7La_3Zr_2O_{12}$ with cubic garnet-type structure. Mater Res Bull 47:1229–1232
45. Yang T, Gordon ZD, Li Y, Chan CK (2015) Nanostructured garnet-type solid electrolytes for lithium batteries: electrospinning synthesis of $Li_7La_3Zr_2O_{12}$ nanowires and particle size-dependent phase transformation. J Phys Chem C 119:14947–14953
46. Gordon ZD, Yang T, Gomes Morgado GB, Chan CK (2016) Preparation of nano- and microstructured garnet $Li_7La_3Zr_2O_{12}$ solid electrolytes for Li-ion batteries via cellulose templating. ACS Sustain Chem Eng 4:6391–6398
47. Weller JM, Whetten JA, Chan CK (2018) Synthesis of fine cubic $Li_7La_3Zr_2O_{12}$ powders in molten LiCl-KCl eutectic and facile densification by reversal of Li^+/H^+ exchange. ACS Appl Energy Mater 1:552–560
48. Garvie RC (1978) Stabilization of the tetragonal structure in zirconia microcrystals. J Phys Chem 82:218–224
49. Zhang H, Banfield JF (1998) Thermodynamic analysis of phase stability of nanocrystalline titania. J Mater Chem 8:2073–2076
50. Navrotsky A (2005) Thermochemical insights into refractory ceramic materials based on oxides with large tetravalent cations. J Mater Chem 15:1883–1890
51. Allen JL, Wolfenstine J, Rangasamy E, Sakamoto J (2012) Effect of substitution (Ta, Al, Ga) on the conductivity of $Li_7La_3Zr_2O_{12}$. J Power Sources 206:315–319
52. Thompson T, Wolfenstine J, Allen JL et al (2014) Tetragonal vs. cubic phase stability in Al-free Ta doped $Li_7La_3Zr_2O_{12}$ (LLZO). J Mater Chem A 2:13431–13436
53. Matsui M, Takahashi K, Sakamoto K et al (2014) Phase stability of a garnet-type lithium ion conductor $Li_7La_3Zr_2O_{12}$. Dalt Trans 43:1019–1024
54. Toda S, Ishiguro K, Shimonishi Y et al (2013) Low temperature cubic garnet-type CO_2-doped $Li_7La_3Zr_2O_{12}$. Solid State Ionics 233:102–106
55. Larraz G, Orera A, Sanjuán ML (2013) Cubic phases of garnet-type $Li_7La_3Zr_2O_{12}$: the role of hydration. J Mater Chem A 1:11419
56. Buschmann H, Berendts S, Mogwitz B, Janek J (2012) Lithium metal electrode kinetics and ionic conductivity of the solid lithium ion conductors "$Li_7La_3Zr_2O_{12}$" and $Li_{7-x}La_3Zr_{2-x}Ta_xO_{12}$ with garnet-type structur. J Power Sources 206:236–244

57. Imagawa H, Ohta S, Kihira Y, Asaoka T (2014) Garnet-type $Li_{6.75}La_3Zr_{1.75}Nb_{0.25}O_{12}$ synthesized by coprecipitation method and its lithium ion conductivity. Solid State Ionics 262:609–612
58. Ohta S, Kobayashi T, Asaoka T (2011) High lithium ionic conductivity in the garnet-type oxide $Li_{7-x}La_3(Zr_{2-x}, Nb_x)O_{12}$ (x=0–2). J Power Sources 196:3342–3345
59. Geiger CA, Alekseev E, Lazic B et al (2011) Crystal chemistry and stability of "$Li_7La_3Zr_2O_{12}$" garnet: a fast lithium-ion conductor. Inorg Chem 50:1089–1097
60. Kingery WD, Bowen HK (1976) Introduction to ceramics, 2nd edn. Wiley, New York
61. Tsai CL, Dashjav E, Hammer EM et al (2015) High conductivity of mixed phase Al-substituted $Li_7La_3Zr_2O_{12}$. J Electroceramics 35:25–32
62. Tsai CL, Roddatis V, Chandran CV et al (2016) $Li_7La_3Zr_2O_{12}$ interface modification for Li dendrite prevention. ACS Appl Mater Interfaces 8:10617–10626
63. Rao RP, Gu W, Sharma N et al (2015) In situ neutron diffraction monitoring of $Li_7La_3Zr_2O_{12}$ formation: toward a rational synthesis of garnet solid electrolytes. Chem Mater 27:2903–2910
64. Rangasamy E, Wolfenstine J, Allen J, Sakamoto J (2013) The effect of 24c-Site (A) cation substitution on the tetragonal-cubic phase transition in $Li_7La_3Zr_2O_{12}$ garnet-based ceramic electrolyte. J Power Sources 230:261–266
65. Hellstrom EE, van Gool W (1981) Li ion conduction in Li_2ZrO_3, Li_4ZrO_4, and $LiScO_2$. Solid State Ionics 2:59–64
66. Rao RP, Reddy MV, Adams S, Chowdari BVR (2012) Preparation and mobile ion transport studies of Ta and Nb doped $Li_6Zr_2O_7$ Li-fast ion conductors. Mater Sci Eng B Solid-State Mater Adv Technol 177:100–105
67. Vassen R, Cao X, Tietz F et al (2000) Zirconates as new materials for thermal barrier coatings. J Am Ceram Soc 83:2023–2028
68. Li JY, Dai H, Li Q et al (2006) Lanthanum zirconate nanofibers with high sintering-resistance. Mater Sci Eng B Solid-State Mater Adv Technol 133:209–212
69. Huang X, Lu Y, Jin J et al (2018) A method using water-based solvent to prepare $Li_7La_3Zr_2O_{12}$ solid electrolytes. ACS Appl Mater Interfaces 10:17147–17155
70. Ma J, Lim LC (2002) Effect of particle size distribution on sintering of agglomerate-free submicron alumina powder compacts. J Eur Ceram Soc 22:2197–2208
71. Yi E, Wang W, Kieffer J, Laine RM (2017) Key parameters governing the densification of cubic-$Li_7La_3Zr_2O_{12}$ Li^+ conductors. J Power Sources 352:156–164
72. Lee J-M, Kim T, Baek S-W et al (2014) High lithium ion conductivity of a method using water-based solvent to prepare $Li_7La_3Zr_2O_{12}$ solid electrolytes synthesized by solid state reaction. Solid State Ionics 258:13–17
73. Yoshima K, Harada Y, Takami N (2016) Thin hybrid electrolyte based on garnet-type lithium-ion conductor $Li_7La_3Zr_2O_{12}$ for 12 V-class bipolar batteries. J Power Sources 302:283–290
74. Jeevan Kumar P, Nishimura K, Senna M et al (2016) A novel low-temperature solid-state route for nanostructured cubic garnet $Li_7La_3Zr_2O_{12}$ and its application to Li-ion battery. RSC Adv 6:62656
75. Brinker C, Scherer G (1990) Sol–gel science: the physics and chemistry of sol–gel processing. Adv Mater 3:912
76. Danks AE, Hall SR, Schnepp Z (2016) The evolution of "sol–gel" chemistry as a technique for materials synthesis. Mater Horiz 3:91–112
77. Stöber W, Fink A (1968) Controlled growth of monodispersed silica spheres in the micron size range. J Colloid Interface Sci 26:62–69
78. Janani N, Ramakumar S, Dhivya L et al (2011) Synthesis of cubic $Li_7La_3Zr_2O_{12}$ by modified sol–gel process. Ionics (Kiel) 17:575–580
79. El Shinawi H, Janek J (2013) Stabilization of cubic lithium-stuffed garnets of the type "$Li_7La_3Zr_2O_{12}$" by addition of gallium. J Power Sources 225:13–19
80. Yow ZF, Oh YL, Gu W et al (2016) Effect of Li^+/H^+ exchange in water treated Ta-doped $Li_7La_3Zr_2O_{12}$. Solid State Ionics 292:122–129
81. Buannic L, Orayech B, López Del Amo J-M et al (2017) Dual substitution strategy to enhance Li^+ ionic conductivity in $Li_7La_3Zr_2O_{12}$ solid electrolyte. Chem Mater 29:1769–1778

82. Il'Ina EA, Andreev OL, Antonov BD, Batalov NN (2012) Morphology and transport properties of the solid electrolyte Sol–Gel synthesis and lithium ion conductivity of $Li_7La_3Zr_2O_{12}$ with garnet-related type structure prepared by the solid-state and citrate-nitrate meth. J Power Sources 201:169–173
83. Xie H, Alonso JA, Li Y et al (2011) Lithium distribution in aluminum-free cubic $Li_7La_3Zr_2O_{12}$. Chem Mater 23:3587–3589
84. Tan J, Tiwari A (2012) Garnet-type $Li_7La_3Zr_2O_{12}$ electrolyte prepared by a solution-based technique for lithium ion battery. Mater Res Soc Symp Proc 1440:61–66
85. Kotobuki M, Koishi M (2014) Preparation of $Li_7La_3Zr_2O_{12}$ solid electrolyte via a sol–gel method. Ceram Int 40:5043–5047
86. Shimonishi Y, Toda A, Zhang T et al (2011) Synthesis of garnet-type $Li_{7-x}La_3Zr_2O_{12-1/2x}$ and its stability in aqueous solutions. Solid State Ionics 183:48–53
87. Jin Y, McGinn PJ (2011) Al-doped $Li_7La_3Zr_2O_{12}$ synthesized by a polymerized complex method. J Power Sources 196:8683–8687
88. Hubaud AA, Schroeder DJ, Key B, et al (2013) Low temperature stabilization of cubic $(Li_{7-x}Al_{x/3})\,La_sZr_2O_{12}$: role of aluminum during formation. J Mater Chem A 1:8813–8818
89. Takano R, Tadanaga K, Hayashi A, Tatsumisago M (2014) Low temperature synthesis of Al-doped $Li_7La_3Zr_2O_{12}$ solid electrolyte by a sol–gel process. Solid State Ionics 255:104–107
90. Yoo AR, Yoon SA, Kim YS, Sakamoto J, Lee HC (2016) A comparative study on the synthesis of Al-doped $Li_{6.2}La_3Zr_2O_{12}$ powder as a solid electrolyte using sol–gel synthesis and solid-state processing. J Nanosci Nanotechnol 16:11662–11668
91. Raskovalov AA, Il'Ina EA, Antonov BD (2013) Structure and transport properties of $Li_7La_3Zr_{2-0.75x}Al_xO_{12}$ superionic solid electrolytes. J Power Sources 238:48–52
92. Janani N, Deviannapoorani C, Dhivya L, Murugan R (2014) Influence of sintering additives on densification and Li^+ conductivity of Al doped $Li_7La_3Zr_2O_{12}$ lithium garnet. RSC Adv 4:51228–51238
93. van den Broek J, Afyon S, Rupp JLM (2016) Interface-engineered all-solid-state Li-ion batteries based on garnet-type fast Li^+ conductors. Adv Energy Mater 6:1–11
94. van den Broek J, Rupp JLM, Afyon S (2017) Boosting the electrochemical performance of Li-garnet based all-solid-state batteries with $Li_4Ti_5O_{12}$ electrode: routes to cheap and large scale ceramic processing. J Electroceramics 38:182–188
95. El-Shinawi H, Paterson GW, MacLaren DA et al (2017) Low-temperature densification of Al-doped $Li_7La_3Zr_2O_{12}$: a reliable and controllable synthesis of fast-ion conducting garnets. J Mater Chem A 5:319–329
96. Bae J, Li Y, Zhao F et al (2018) Designing 3D nanostructured garnet frameworks for enhancing ionic conductivity and flexibility in composite polymer electrolytes for lithium batteries. Energy Storage Mater 15:46–52
97. Rosenkiewitz N, Schuhmacher J, Bockmeyer M, Deubener J (2015) Nitrogen-free sol–gel synthesis of Al-substituted cubic garnet $Li_7La_3Zr_2O_{12}$ (LLZO). J Power Sources 278:104–108
98. Chen R-J, Huang M, Huang W-Z et al (2014) Sol–gel derived Li–La–Zr–O thin films as solid electrolytes for lithium-ion batteries. J Mater Chem A 2:13277
99. Tadanaga K, Takano R, Ichinose T et al (2013) Low temperature synthesis of highly ion conductive $Li_7La_3Zr_2O_{12}$–Li_3BO_3 composites. Electrochem Commun 33:51–54
100. Afyon S, Krumeich F, Rupp JLM (2015) A shortcut to garnet-type fast Li-ion conductors for all-solid state batteries. J Mater Chem A 3:18636–18648
101. Liu X, Li Y, Yang T et al (2017) High lithium ionic conductivity in the garnet-type oxide $Li_{7-2x}La_3Zr_{2-x}Mo_xO_{12}$ (x = 0–0.3) ceramics by sol–gel method. J Am Ceram Soc 100:1527–1533
102. Rosero-Navarro NC, Yamashita T, Miura A et al (2016) Preparation of $Li_{7-x}La_3(Zr_{2-x},Nb_x)O_{12}$ (x = 0–1.5) and $Li_3BO_3/LiBO_2$ composites at low temperatures using a sol–gel process. Solid State Ionics 285:6–12
103. Yoon SA, Oh NR, Yoo AR et al (2017) Preparation and characterization of Ta-substituted $Li_7La_3Zr_{2-x}O_{12}$ garnet solid electrolyte by sol–gel processing. J Korean Ceram Soc 54:278–284

104. Han X, Gong Y, Fu K et al (2017) Negating interfacial impedance in garnet-based solid-state Li metal batteries. Nat Mater 16:572–579
105. Tao X, Liu Y, Liu W et al (2017) Solid-state lithium-sulfur batteries operated at 37 °C with composites of nanostructured $Li_7La_3Zr_2O_{12}$/carbon foam and polymer. Nano Lett 17:2967–2972
106. Deviannapoorani C, Dhivya L, Ramakumar S, Murugan R (2012) Synthesis of garnet structured $Li_{7+x}La_3Y_xZr_{2-x}O_{12}$ (x = 0–0.4) by modified sol–gel method. J Sol–Gel Sci Technol 64:510–514
107. Bitzer M, Van Gestel T, Uhlenbruck S, Hans-Peter-Buchkremer (2016) Sol–gel synthesis of thin solid $Li_7La_3Zr_2O_{12}$ electrolyte films for Li-ion batteries. Thin Solid Films 615:128–134
108. Chen X, Cao T, Xue M et al (2018) Improved room temperature ionic conductivity of Ta and Ca doped $Li_7La_3Zr_2O_{12}$ via a modified solution method. Solid State Ionics 314:92–97
109. Pechini MP (1967) Method of preparing lead and alkaline earth titanates and niobates and coating method using the same to form a capacitor
110. Li Y, Han JT, Wang CA et al (2012) Ionic distribution and conductivity in lithium garnet $Li_7La_3Zr_2O_{12}$. J Power Sources 209:278–281
111. Li D, McCann JT, Xia Y (2006) Electrospinning: a simple and versatile technique for producing ceramic nanofibers and nanotubes. J Am Ceram Soc 89:1861–1869
112. Ramaseshan R, Sundarrajan S, Jose R, Ramakrishna S (2007) Nanostructured ceramics by electrospinning. J Appl Phys 102:111101
113. Wang HG, Yuan S, Ma DL et al (2015) Electrospun materials for lithium and sodium rechargeable batteries: from structure evolution to electrochemical performance. Energy Environ Sci 8:1660–1681
114. Shao C, Yu N, Liu Y, Mu R (2006) Preparation of $LiCoO_2$ nanofibers by electrospinning technique. J Phys Chem Solids 67:1423–1426
115. Yu N, Shao C, Liu Y et al (2005) Nanofibers of $LiMn_2O_4$ by electrospinning. J Colloid Interf Sci 285:162–166
116. Zhu N, Liu W, Xue M et al (2010) Graphene as a conductive additive to enhance the high-rate capabilities of electrospun $Li_4Ti_5O_{12}$ for lithium-ion batteries. Electrochim Acta 55:5813–5818
117. Liu W, Liu N, Sun J et al (2015) Ionic conductivity enhancement of polymer electrolytes with ceramic nanowire fillers. Nano Lett 15:2740–2745
118. Yang T, Li Y, Chan CK (2015) Enhanced lithium ion conductivity in lithium lanthanum titanate solid electrolyte nanowires prepared by electrospinning. J Power Sources 287:164–169
119. Mayo MJ, Hague DC, Chen DJ (1993) Processing nanocrystalline ceramics for applications in superplasticity. Mater Sci Eng, A 166:145–159
120. Huang J, Kunitake T (2003) Nano-precision replication of natural cellulosic substances by metal oxides. J Am Chem Soc 125:11834–11835
121. Schattka JH, Wong EHM, Antonietti M, Caruso RA (2006) Sol–gel templating of membranes to form thick, Porous Titania, Titania/Zirconia and Titania/Silica films. J Mater Chem 16:1414–1420
122. Greil P (2010) Templating approaches using natural cellular plant tissue. Mater Res Bull 35:145–149
123. Shigapov AN, Graham GW, McCabe RW, Plummer HK (2001) Preparation of high-surface area, thermally-stable, metal-oxide catalysts and supports by a cellulose templating approach. Appl Catal A Gen 210:287–300
124. Xie H, Yang C, Fu KK et al (2018) Flexible, scalable, and highly conductive garnet-polymer solid electrolyte templated by bacterial cellulose. Adv Energy Mater 1703474:1–7
125. Gong Y, Fu K, Xu S et al (2018) Lithium-ion conductive ceramic textile: a new architecture for flexible solid-state lithium metal batteries. Mater Today 21:594–601
126. Messing GL, Zhang S-C, Jayanthi GV (1995) Ceramic powder synthesis by spray pyrolysis. J Am Ceram Soc 76:2707–2726
127. Djenadic R, Botros M, Benel C et al (2014) Nebulized spray pyrolysis of Al-doped $Li_7La_3Zr_2O_{12}$ solid electrolyte for battery applications. Solid State Ionics 263:49–56

128. Yi E, Wang W, Kieffer J, Laine RM (2016) Flame made nanoparticles permit processing of dense, flexible, Li^+ conducting ceramic electrolyte thin films of cubic-$Li_7La_3Zr_2O_{12}$ (c-LLZO). J Mater Chem A 12947–12954
129. Teoh WY, Amal R, Mädler L (2010) Flame spray pyrolysis: an enabling technology for nanoparticles design and fabrication. Nanoscale 2:1324–1347
130. Bickmore CR, Waldner KF, Treadwell DR, Laine RM (1996) Ultrafine spinel powders by flame spray pyrolysis of a magnesium aluminum double alkoxide. J Am Ceram Soc 79:1419–1423
131. Bickmore CR, Waldner KF, Baranwal R et al (1998) Ultrafine titania by flame spray pyrolysis of a titanatrane complex. J Eur Ceram Soc 18:287–297
132. Sutorik AC, Neo SS, Treadwell DR, Laine RM (1998) Synthesis of ultrafine β-alumina powders vis flame spray pyrolysis of polymeric precursors. J Am Ceram Soc 81:1477–1486
133. Marchal J, John T, Baranwal R et al (2004) Yttrium aluminum garnet nanopowders produced by liquid-feed flame spray pyrolysis (LF-FSP) of metalloorganic precursors. Chem Mater 16:822–831
134. Hinklin T, Toury B, Gervais C et al (2003) Liquid-feed flame spray pyrolysis of metalloorganic and inorganic alumina sources in the production of nanoalumina powders. Chem Mater 16:21–30
135. Yi E, Wang W, Mohanty S et al (2014) Materials that can replace liquid electrolytes in li batteries: superionic conductivities in $Li_{1.7}Al_{0.3}Ti_{1.7}Si_{0.4}P_{2.6}O_{12}$. processing combustion synthesized nanopowders to free standing thin. J Power Sources 269:577–588
136. Langer F, Glenneberg J, Bardenhagen I, Kun R (2015) Synthesis of single phase cubic Al-substituted $Li_7La_3Zr_2O_{12}$ by solid state lithiation of mixed hydroxides. J Alloys Compd 645:64–69
137. Kim K-W, Yang S-H, Kim MY et al (2016) Cubic phase behavior and lithium ion conductivity of $Li_7La_3Zr_2O_{12}$ prepared by co-precipitation synthesis for all-solid batteries. J Ind Eng Chem 36:279–283
138. Shao C, Liu H, Yu Z et al (2016) Structure and ionic conductivity of cubic $Li_7La_3Zr_2O_{12}$ solid electrolyte prepared by chemical co-precipitation method. Solid State Ionics 287:13–16
139. Wolfenstine J, Rangasamy E, Allen JL, Sakamoto J (2012) High conductivity of dense tetragonal $Li_7La_3Zr_2O_{12}$. J Power Sources 208:193–196
140. Zhang Y, Cai J, Chen F et al (2015) Preparation of cubic $Li_7La_3Zr_2O_{12}$ solid electrolyte using a nano-sized core-shell structured precursor. J Alloys Compd 644:793–798
141. Bugaris DE, Zur Loye HC (2012) Materials discovery by flux crystal growth: quaternary and higher order oxides. Angew Chem Int Ed 51:3780–3811
142. Boltersdorf J, King N, Maggard PA (2015) Flux-mediated crystal growth of metal oxides: synthetic tunability of particle morphologies, sizes, and surface features for photocatalysis research. CrystEngComm 17:2225–2241
143. Elwell D, Neate BW (1971) Mechanisms of crystal growth from fluxed melts. J Mater Sci 6:1499–1519
144. Liu X, Fechler N, Antonietti M (2013) Salt melt synthesis of ceramics, semiconductors and carbon nanostructures. Chem Soc Rev 42:8237–8265
145. Awaka J, Kijima N, Hayakawa H, Akimoto J (2009) Synthesis and structure analysis of tetragonal $Li_7La_3Zr_2O_{12}$ with the garnet-related type structure. J Solid State Chem 182:2046–2052
146. Awaka J, Takashima A, Hayakawa H et al (2011) Single crystal synthesis of cubic garnet related-type $Li_7La_3Zr_2O_{12}$ by a self-flux method. Key Eng Mater 485:99–102
147. Buschmann H, Doïle J, Berendts S et al (2011) Structure and dynamics of the fast lithium ion conductor "$Li_7La_3Zr_2O_{12}$". Phys Chem Chem Phys 13:19378–19392
148. Reddy MV, Adams S (2017) Molten salt synthesis and characterization of fast ion conductor $Li_{6.75}La_3Zr_{1.75}O_{12}$. J Solid State Electrochem 21:2921–2928
149. Ishiguro K, Nemori H, Sunahiro S et al (2014) Ta-Doped $Li_7La_3Zr_2O_{12}$ for water-stable lithium electrode of lithium-air batteries. J Electrochem Soc 161:A668–A674

Chapter 3
Air Stability of LLZO Electrolytes

Oluwatemitope Familoni, Ying Zhou and Huanan Duan

Abstract Garnet-type oxide, $Li_7La_3Zr_2O_{12}$ (LLZO), is a promising solid electrolyte material for all-solid-state lithium-ion batteries. This chapter reviews the latest research efforts on the air stability of lithium garnets, which includes the origin of the air stability and the reaction mechanisms of LLZO electrolytes in ambient air; the factors affecting the air stability such as synthesis conditions, materials properties, and storage conditions; and finally the improvement strategies such as mechanical polishing, using additives, synthesis optimization, and etching.

3.1 Introduction

Lithium-ion (Li^+) conduction has been reported in a wide range of materials including crystalline metal oxides and sulfides as electrolyte in all-solid-state lithium-ion batteries (ASSLIBs). Among them, the oxides hold some advantages such as the ease of preparation, better mechanical properties, chemical and thermal stability. Garnet-type oxide, $Li_7La_3Zr_2O_{12}$ (LLZO) in particular, has drawn increasing interest since it was first reported by Murugan and coworkers [1] in 2007 because of its chemical stability with respect to reduction versus lithium, and high ionic conductivity. After a decade of intensive research, great advances have been achieved in the aspects of material synthesis, cation doping, structure characterization, Li^+ transport mechanism, and conductivity improvement of LLZO. Several excellent review papers have been published discussing these aspects [2–6].

Recently, room-temperature ionic conductivity as high as 1.6 mS cm^{-1} has been reported for LLZO pellets, approaching that of conventional carbonate-based liquid electrolyte. However, compared to Li^+ conductivity itself, the improvement of LLZO stability, especially the stability against different environmental conditions and electrode materials is of equal, if not more, importance. The scope of this chapter is to discuss the air stability of LLZO electrolytes. Section 3.2 discusses the origin of air

O. Familoni · Y. Zhou · H. Duan (✉)
State Key Laboratory of Metal Matrix Composites, School of Materials Science and Engineering, Shanghai Jiao Tong University, Shanghai 200240, People's Republic of China
e-mail: hd1@sjtu.edu.cn

R. Murugan and W. Weppner (eds.), *Solid Electrolytes for Advanced Applications*,
https://doi.org/10.1007/978-3-030-31581-8_3

stability and the reaction mechanisms of LLZO electrolytes in ambient air; Sect. 3.3 summarizes the factors affecting air stability; Sect. 3.4 introduces the improvement strategies employed to address the issue.

3.2 Origin of Air Stability Problems

The application of LLZO in ASSLIBs has been hindered by high interfacial resistance at the electrolyte/electrode interface. According to Buschmann et al. poor electrode/electrolyte contact, slow charge transfer, and sluggish carrier transport in the interfacial region can cause an increase in interfacial resistance [7]. In the quest to understand the origin and provide solution to this resistive interface that limited the device to low current density cycling and also dominated the cell behavior, it was discovered that instability of LLZO in the atmosphere also contributed to the high interfacial resistance [8]. Though, it was initially believed that garnet-type solid-state electrolytes were chemically stable in ambient air, but both computational and experimental results demonstrate that its air stability is susceptible to atmospheric moisture and carbon (IV) oxide (CO_2), consequently, leading to lithium carbonate (Li_2CO_3) formation, ionic conductivity degradation, and high interfacial resistance. Since lithium garnet will inevitably be exposed to ambient air during the preparation process and the subsequent cell-assembly process, it becomes imperative to fully understand the origin of the stability of lithium garnet in air and the reaction mechanism of the stability-altering process, and to figure out ways to improve it.

Based on the density functional theory (DFT) calculations, Kang et al. analyzed the stability of $Li_7La_3M_2O_{12}$ (M = Zr, Sn, Hf) in environments containing gaseous H_2O and CO_2 [9]. According to his study, the lithium garnets exposed to humid air thermodynamically tend to form hydroxides and carbonates according to the following reactions:

$$2Li_7La_3M_2O_{12} + 7H_2O \leftrightarrow 14LiOH + 3La_2O_3 + 4MO_2 \tag{3.1}$$

$$2Li_7La_3M_2O_{12} + 7CO_2 \leftrightarrow 7Li_2CO_3 + 3La_2O_3 + 4MO_2 \tag{3.2}$$

The calculation results showed that among these three garnets, $Li_7La_3Sn_2O_{12}$ showed higher chemical stability than the other two; all three materials are more stable with respect to H_2O than to CO_2. However, it is worth noting that the DFT calculations are based solely on thermodynamics, and cannot provide information on the potential role of reaction kinetics in practical applications.

Similar first-principles calculations were carried out specifically for $Li_7La_3Zr_2O_{12}$ by Cheng et al. based on density functional theory to evaluate also the thermodynamic tendency of Al-substituted and Al-free LLZO to form Li_2CO_3 with possible reactants present in the air [10]. The calculations show that LLZO can react via two different pathways to form Li_2CO_3. The first is believed to be more rapid and involves a

Table 3.1 Gibbs free energy of possible chemical reactions[10]

Chemical reaction	Gibbs energy (eV)
$Li_{56}La_{24}Zr_{16}O_{96}(s) + H_2O\ (g) = Li_{54}La_{24}Zr_{16}O_{95}\ (s) + 2LiOH\ (s)$ (3.3)	−0.53
$Li_{56}La_{24}Zr_{16}O_{96}(s) + CO_2\ (g) = Li_{54}La_{24}Zr_{16}O_{95}\ (s) + Li_2CO_3\ (s)$ (3.4)	−1.99
$Li_{50}Al_2La_{24}Zr_{16}O_{96}(s) + H_2O\ (g) = Li_{48}Al_2La_{24}Zr_{16}O_{95}\ (s) + 2LiOH\ (s)$ (3.5)	0.42
$Li_{50}Al_2La_{24}Zr_{16}O_{96}(s) + CO_2\ (g) = Li_{48}Al_2La_{24}Zr_{16}O_{95}\ (s) + Li_2CO_3\ (s)$ (3.6)	−1.04

reaction with moisture in air to form LiOH, which then form Li_2CO_3 upon absorption of CO_2. The second pathway was a direct reaction with CO_2 believed to be slow and favored when surface lithium contents are lower, as with the small-grained samples. The chemical reactions of Al-substituted and Al-free LLZO considered by this group and their Gibb's free energies are given in Table 3.1.

Obviously in Table 3.1, the formation LiOH or Li_2CO_3 from these reactions is always accompanied by the loss of Li in LLZO and presumable compensation with concurrent loss of oxygen atom so as to balance the charge. Comparison of the reaction Gibbs free energies in Table 3.1 also indicates that for both Al-free and Al-substituted LLZO, the reactions (3.4) and (3.6) with gas phase CO_2 are more energetically favorable than the reactions (3.3) and (3.5). Though it was pointed out in this paper that the formation of Li_2CO_3 via direct reaction with CO_2, (3.4 and 3.6) are likely to be kinetically slower than the formation of LiOH by reactions (3.3) and (3.5), neither experimental observation nor computational prediction was made. Conclusively, these thermodynamic calculation results stress the role of CO_2 over moisture in degrading lithium garnet and forming Li_2CO_3.

In contrast, to clarify the effect of water/moisture on forming Li_2CO_3 on LLZO, Xia et al. recently designed comparison experiments on 0.5-mol Ta-doped LLZO (0.5 Ta-LLZO) pellets [11]. They compared different Ta-LLZO pellets exposed to dry (humidity ~5%) or humid (humidity ~80%) air for 6 weeks, and found that much more amount of Li_2CO_3 was formed on the surface of Ta-LLZO exposed to humid air (Fig. 3.1). Consequently, the ionic conductivity of pellets exposed to humid air suffered severe decrease from 6.45×10^{-4} to 3.61×10^{-4} S cm^{-1}, while that of pellets exposed to dry air had barely changed. These results confirm the role of water/moisture in degrading the structural stability of lithium garnet and verify the sluggish kinetics of the direct reaction between CO_2 and LLZO.

It is worth noting that the formation of LiOH on the LLZO surface can be regarded as the loss of Li in pristine LLZO and concurrent addition of hydrogen atom to balance the charge, which is the so-called Li^+/H^+ exchange process and has been reported previously. This replacement of Li^+ in garnet lattice by H^+ from water was first reported by Nyman et al. when powder samples of $Li_5La_3Nb_2O_{12}$ or $Li_5La_3Ta_2O_{12}$ were dispersed in water and a rise in the pH value of the aqueous media was observed [12]. Similar observation of the Li^+/H^+ exchange reaction was reported by Galven et al. for

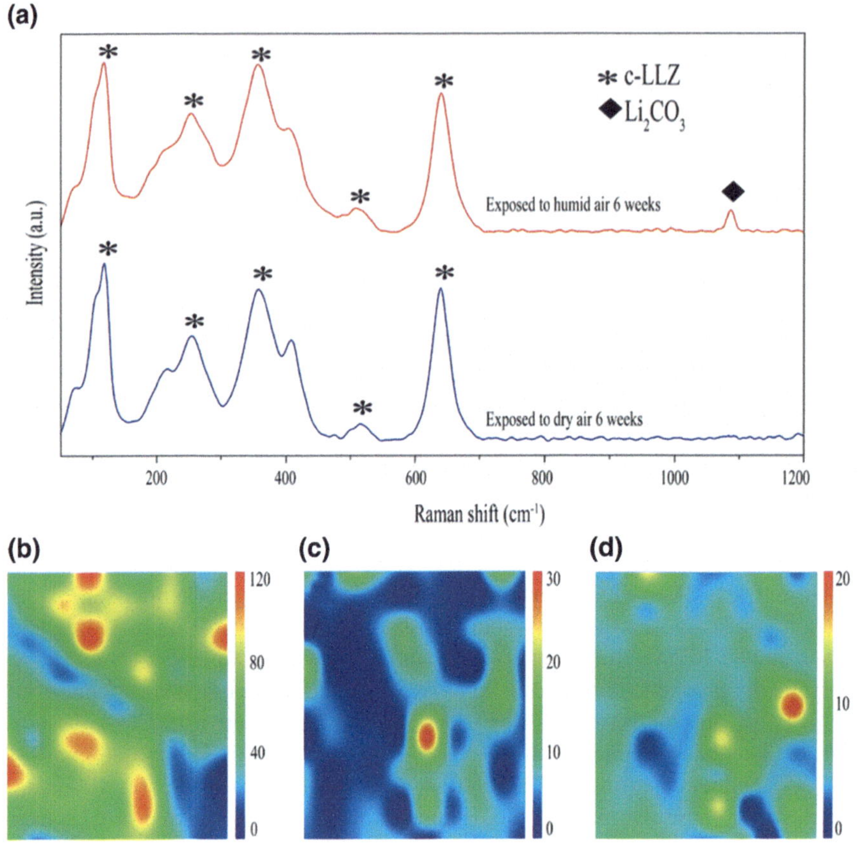

Fig. 3.1 **a** Typical Raman spectra of the 0.5Ta–LLZO pellets exposed to dry and humid air. Raman mapping images of top surface of the 0.5Ta–LLZO pellets exposed in **b** humid air, **c** dry air, and **d** in humid air and being polished [11]

single crystal $Li_7La_3Sn_2O_{12}$ in ambient air [13]. By using powder neutron diffraction, solid-state nuclear magnetic resonance, and chemical analysis, Galven et al. further showed that the Li^+/H^+ exchange feasibility is directly connected to the lithium stoichiometry—only when the number of Li^+ per formular is more than three does the garnet become sensitive to humidity [14]. Recently, Ma et al. used the scanning transmission electron microscopy (STEM) and electron energy loss spectroscopy (EELS) analysis to graphically illustrate that the Li^+/H^+ exchange reaction of LLZO in water was reversible and did not change the garnet structure even under a high exchange rate of 63.6% [15]. Based on DAADF-STEM and HRTEM observation in Fig. 3.2, it was claimed that the Li^+/H^+ exchange increases the sensitivity of LLZO to the electron beam, probably preventing the acquisition of atomic-resolution HAADF-STEM images of the LLZO exposed to water; meanwhile, the Li^+/H^+ exchange does not form any impurities or amorphous phase near the grain boundary [15].

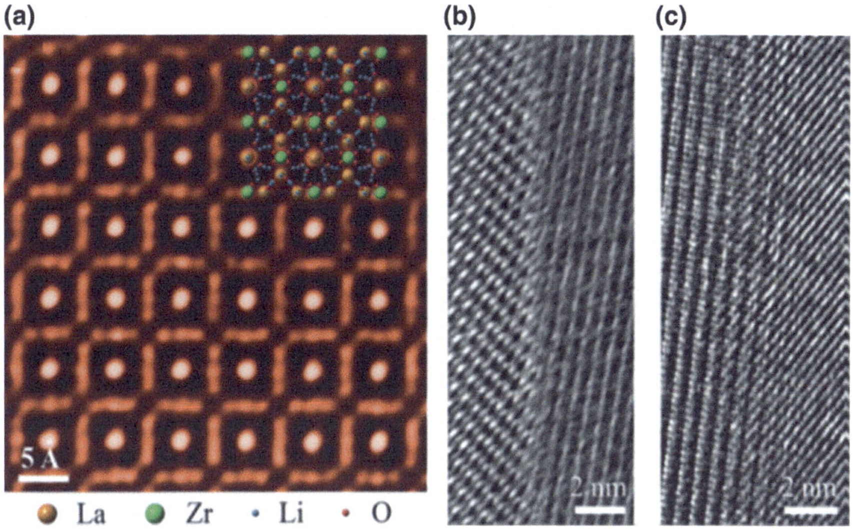

Fig. 3.2 **a** HAADF-STEM image of the pristine cubic LLZO along the [001] zone axis (image quality improved with the Butterworth filter). The atomic model is shown as an overlay. **b** HRTEM image of a grain boundary in the pristine specimen. **c** HRTEM image of a grain boundary after Li^+/H^+ exchange in water [15]

To further study the reaction mechanism of LLZO electrolytes in ambient air, Xia et al. [11] decoupled the reactions of LLZO with water and CO_2 by immersing the pellets in deionized water for 1 week and then exposing them to ambient air for another week. As shown in Fig. 3.3, after immersion in deionized water, similar Li^+/H^+ exchange occurred with $LiOH{\cdot}H_2O$ formed on the surface with negligible Li_2CO_3. While after exposure to ambient air, considerable amount of Li_2CO_3 was observed on the surface, verifying the reaction between LiOH and CO_2. Based on the experimental observations and literature, the overall reaction mechanism of Li_2CO_3 formation process is schematically illustrated in Fig. 3.4. The stepwise reaction is summarized with the following equations:

$$Li_{6.5}La_3Zr_{1.5}Ta_{0.5}O_{12} + xH_2O = Li_{6.5-x}H_XLa_3Zr_{1.5}Ta_{0.5}O_{12} + xLiOH \quad (3.7)$$

$$LiOH + H_2O = LiOH \cdot H_2O \quad (3.8)$$

$$2LiOH \cdot H_2O + CO_2 = Li_2CO_3 + 3H_2O \quad (3.9)$$

An overall reaction is given as:

$$2Li_{6.5}La_3Zr_{1.5}Ta_{0.5}O_{12} + xH_2O + xCO_2 = 2\ Li_{6.5-x}H_XLa_3Zr_{1.5}Ta_{0.5}O_{12} + xLi_2CO_3 \quad (3.10)$$

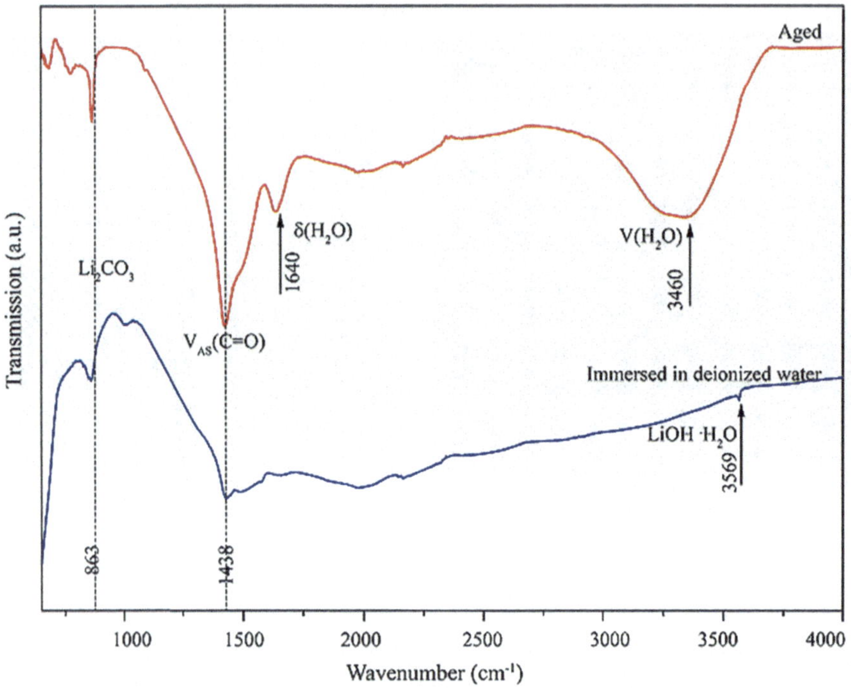

Fig. 3.3 FT-IR spectra of the 0.5Ta–LLZO pellets immersed in DI water for 1 week and subsequently left in ambient air for another week to demonstrate the Li^+/H^+ exchange evolution when exposed to moist environment [11]

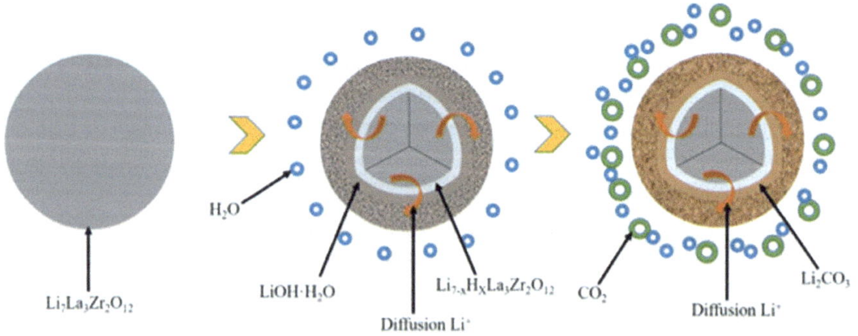

Fig. 3.4 A schematic model illustrating the Li_2CO_3 formation process on the surface of lithium garnets [11]

When the pristine garnet pellets are in contact with humid air, water from the ambient air will be adsorbed on the pellet surface and the Li^+/H^+ exchange occurs. As a result, a layer of $LiOH{\cdot}H_2O$ is formed on the pellet surface based on (3.7 and 3.8). $LiOH{\cdot}H_2O$ is a necessary intermediate and required precursor for the subsequent formation of Li_2CO_3 because negligible amount of Li_2CO_3 formed on the sample surface (Fig. 3.1c) for the pellets exposed in dry air. Next, the reaction between $LiOH{\cdot}H_2O$ and CO_2 leads to a top layer of Li_2CO_3 that protects the LLZO underneath (3.9). Further Li_2CO_3 formation requires Li^+ ions to extract from the inner part of LLZO and diffuse through the Li_2CO_3 shell to meet H_2O and CO_2 from the air. As a result, the thickness of Li_2CO_3 layer increases with exposure time and it is directly related to factors such as aging time, air humidity, and the porosity of the pellets. Overall, the reaction of lithium garnets and air can be expressed by (3.10).

3.3 Factors Affecting Air Stability

There are numerous factors that influence the stability of lithium garnet in air, among which are synthesis conditions, material properties (chemical composition, microstructure, etc.) and environmental conditions playing important roles. Figure 3.5 categorizes these factors into three levels, namely, synthesis, materials

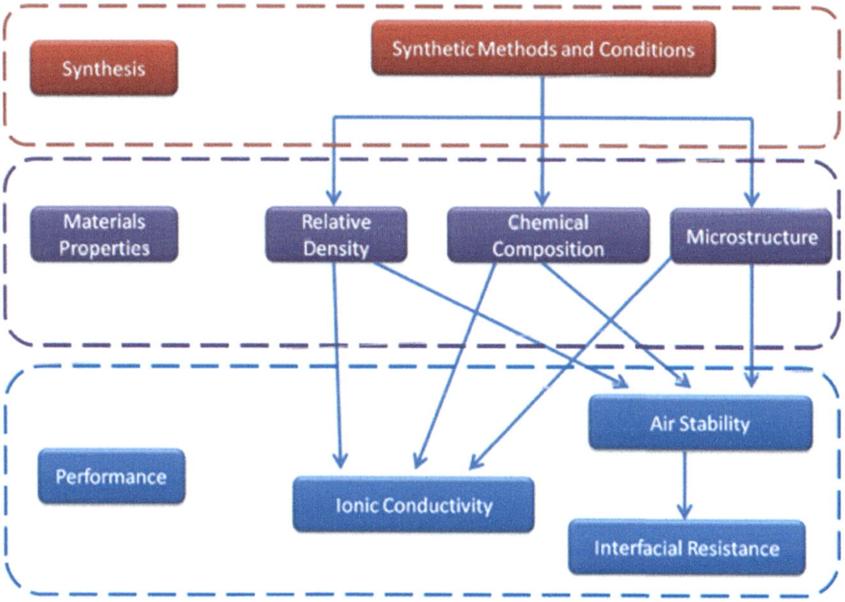

Fig. 3.5 Schematic of the interdependency garnet-type electrolyte characteristics on synthesis methods and conditions

properties, and performance. It clearly shows that the synthesis methods and conditions are at the heart of every parameter of a garnet-type system. Materials properties such as relative density, chemical composition, microstructure, etc. are directly determined by the synthesis; meanwhile, the materials properties have profound influence on the LLZO performance parameters including air stability, ionic conductivity, and the interfacial impedance.

3.3.1 Synthetic Methods and Conditions

Many approaches have been adopted to synthesize Li garnets-type solid electrolytes emphasizing on the importance of the synthetic routes and conditions toward having an air-stable LLZO solid electrolyte. This also affects the ionic transport properties because of its interrelation to the microstructures, relative density and chemical compositions of polycrystalline ceramics electrolytes [16, 17]. Al-substituted LLZO, for example was densified through a sintering process involving high temperatures (above 1000 °C) and lengthy heating periods, with both lithium and aluminum contents highly sensitive to the processing conditions [7, 18]. Attempts to better the performance of garnet-type solid electrolyte through modification of the microstructure—altering either synthetic routes or processing conditions— inevitably lead to variations in the composition of the resulting densified ceramic. This accounts for the wide distribution of chemical compositions and microstructures reported for LLZO samples [7, 18–27]. From the synthesis perspective, Xia et al. [17] studied the influence the crucibles used on the ionic conductivity and air stability of LLZO prepared through conventional solid-state reaction route. They studied the correlation between the chemical composition, the relative density, the amount of grain boundary, the ionic conductivity, and the surface reactions in the atmosphere for the LLZO samples sintered in alumina crucibles and Pt crucibles. The use of Pt crucibles holds the benefit of preventing unwanted Al contamination from alumina crucibles while also inhibiting excessive Li loss during high-temperature sintering. Additionally, the use of Pt crucibles gave rise to a well-densified LLZO with high relative density (96%), high conductivity (4.48×10^{-4} S cm^{-1}), and good air stability when compared to that synthesized in alumina crucible as shown in Fig. 3.6b [17]. This work stresses the importance of synthesis parameters, crucibles included, toward obtaining LLZO electrolyte with high ionic conductivity and good air stability.

Another method available is the use of hot pressing which is efficient in increasing the relative density of garnet up to 96–98%, with high conductivity (8.7×10^{-4} S cm^{-1}) recorded [24, 28]. Sintering lithium garnets in flowing oxygen carried out by Guo et al. is another approach with which a high density of 96% and excellent conductivity of 7.4×10^{-4} S cm^{-1} were achieved [26]. There are several other technologies like field-assisted sintering and spark plasma sintering adopted to prepare high conducting LLZO with great stability [29, 30]. As stated earlier that various characteristics of garnet and performance are interrelated and dependent on

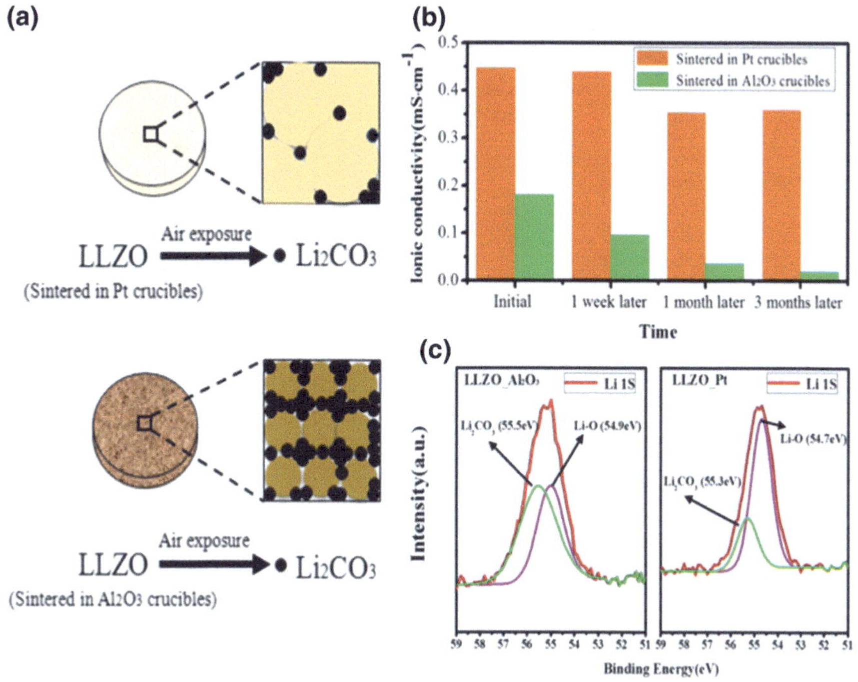

Fig. 3.6 **a** Schematic representation of the surface microstructure of the garnet both in Pt and Al_2O_3 crucible **b** The change of ionic conductivity of LLZO sintered in Pt and alumina crucibles over time. **c** Li 1 s XPS spectra collected from LLZO pellet surfaces sintered in Pt and alumina crucibles after 3-month air exposure [17]

the synthesis method, Table 3.2 summarizes the effect of synthesis method on the composition, relative density, activation energy and ionic conductivity of garnets.

3.3.2 Materials Properties

3.3.2.1 Relative Density

Densification process helps increase the relative density of LLZO in order to maximize its electrochemical properties, so high relative density is a crucial factor for LLZO to improve the ionic conductivity. According to Xia et al. [17] who examined the effect of crucibles on the ionic conductivity and air stability of Al-doped LLZO pellets in Pt crucibles and alumina crucibles, the outcome was in agreement with the findings of others that composition and microstructure of the pellets play important roles in influencing the ionic conductivity, relative density and air stability of LLZO [7, 18–27]. The reported LLZO pellets sintered in Pt crucibles exhibited high relative density (~96%), high ionic conductivity (4.48×10^{-4} S cm^{-1}) and had

Table 3.2 The synthesis methods, relative density, total Li conductivity, and activation energy of typical lithium garnets [31]

Samples	Synthesis method	Relative density (%)	RT ionic conductivity (mS cm^{-1})	Activation energy (eV)	Reference
LLZO pellets	Solid-state reaction, 1230 °C, 36 h, in air	92	0.774	0.30	[1]
0.25 Ga-LLZO pellets	Solid-state reaction, 1100 °C, 24 h, in air	94.1	1.46	0.25	[32]
0.6 Ta-LLZO pellets	Hot pressing, 1150 °C, 1 h, 20 MPa in Ar	99.6	1.6	0.26	[33]
LLZO pellets	Sol-gel, 1180 °C, 36 h	–	0.139	–	[34]
0.25 Ga-LLZO thin films	Flame spray pyrolysis, 1130°C, 0.3 h	95 ± 1	1.3	0.28 ± 0.01	[35]
Ta– and Al-LLZO thin films	Radio frequency magnetron sputtering (RF-MS)	–	0.12	0.47	[36]

good air stability. However, the ionic conductivity dropped a bit while maintaining 3.6×10^{-4} S cm^{-1} after 3 months of air exposure. In contrast, the ionic conductivity of the pellets from alumina crucibles was about 1.81×10^{-4} S cm^{-1} but dropped to 2.39×10^{-5} S cm^{-1} after 3 months of air exposure [17]. The use of Pt crucibles gave rise to a well-densified LLZO as seen in Fig. 3.6a. The effect of relative density on air stability is rooted in its effect on the microstructure, which will be discussed as follows.

3.3.2.2 Microstructure

The microstructure— the pores, grain size, and grain boundaries— has been proven to be vital toward preparing an air-stable LLZO. Jin et al. showed that the grain boundaries are more susceptible to moisture and CO_2 in the air in comparison to the grains due to the high interfacial energy near the grain boundaries, and the resultant reaction taking place at the grain boundaries causes drastic decline of total conductivity of LLZO after a week of exposure [20]. This is consistent with the observation by Xia et al. [17], where LLZO pellets were prepared in both Al_2O_3 and Pt crucibles. The pellets sintered in Pt crucibles have higher relative density, larger grains, and less grain boundaries than those in alumina crucibles (as shown in Fig. 3.6a; key points for better air stability. The large grains and the reduced grain boundaries in the pellets sintered in Pt crucibles are favorable in obtaining high ionic conductivity and good air stability. The results from the X-ray photoelectron spectroscopy (XPS, Fig. 3.6c) show a weaker peak for Li_2CO_3 for LLZO sintered in Pt-crucible and a more intense peak for LLZO sintered in alumina crucible. This suggests that the formation of Li_2CO_3 on the surface of garnet pellet is one of the reasons causing high interfacial resistance, interrelated with the relative density and the amount of grain boundary within the pellets [17].

Interestingly prior to the work of Xia et al. [17], Cheng et al. [10, 37] claimed that the air stability for the small-grained (grain size of ~20 μm) samples were better than that for the large-grained (grain size of 150–200 μm) ones. As shown in the normalized O K-edge X-ray absorption spectroscopy (XAS) in Fig. 3.7, a very intense LLZO peak is observed in the spectrum of the exposed E_LLZO_S24 h sample, which also has a relatively weak Li_2CO_3 feature, while in the spectrum of E_LLZO_L24 h, the two peaks have similar intensities. They attributed this difference to the variation in the Al and Li distribution at sample surfaces and explained that Al tends to segregate at grain boundaries resulting in an Al depleted surface, but Al is restrained within the grains of the small-grained samples resulting in an Al-rich surface. The small-grained samples have limited reaction during air exposure and thus produces less Li_2CO_3 impurity on the LLZO surface compared to those with

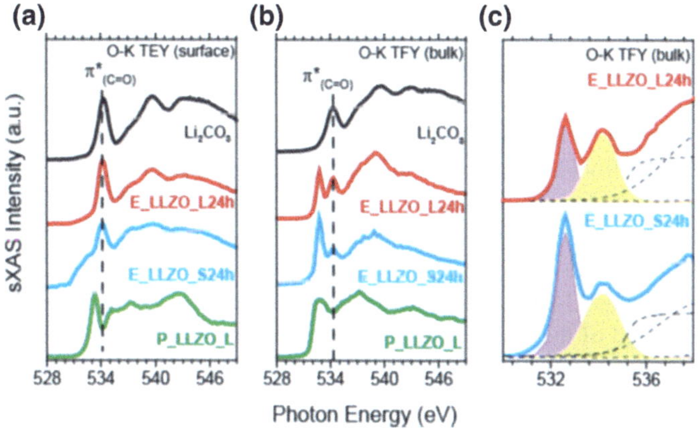

Fig. 3.7 **a** and **b** Normalized O K-edge XAS spectra of P_LLZO_L reference (bottom), air-exposed E_LLZO_S24 h (second from bottom), E_LLZO_L24 h (second from top), and Li_2CO_3 reference (top) collected in TEY and TFY modes. **c** Expanded region of the OK-edge spectra of E_LLZO_L24 h and E_LLZO_S24 h samples with fitted LLZO and Li_2CO_3 peaks [10]

large grains. Obviously, there exist some forms of contradictions between Cheng et al. and Xia et al.'s work, more work remains to be done to elucidate this gray area.

3.3.2.3 Chemical Composition

In an attempt to achieve a stabilized cubic LLZO, both intended and unintended cations (Al^{3+}, Fe^{3+}, Ga^{3+}, Nb^{4+}, Ta^{5+}, Te^{6+}, etc.) doping have been reported by researchers [23, 38–42]. The use of the right stoichiometric amount of lithium salts precursors during the synthesis of garnets is also demonstrated. Narayanan S. et al., for instance, studied the effect of excess Li on the structural and electrical properties of LLZO [43]. This study made use of excess $LiNO_3$ (2.5 to 15 wt%) during the synthesis and the powder x-ray diffraction (PXRD) results confirmed that cubic phase was formed with no significant variation in lattice parameter arising from excess $LiNO_3$ used. However, it was believed excess lithium decreased interparticle contact and increased grain growth during sintering, producing sharply varied microstructures. Furthermore, the results from PXRD showed no secondary phase and the scanning electron microscopy (SEM) analysis showed uniform morphology and the absence of "glassy" materials at the grain boundaries. Although, increased bulk Li-ion conductivity was established with excess lithium, the Raman microscopy study (Fig. 3.8) indicated the presence of Li_2CO_3 in all aged $Li_6La_3Ta_{1.5}Y_{0.5}O_{12}$ samples they prepared with excess $LiNO_3$. The lithium dependence of microstructure is evident from the SEM images in Fig. 3.9. The sample with less than or equal to 5 wt% excess $LiNO_3$ had significant porosity of which the microstructure indicated liquid phase sintering while increase from 7 to 15 wt% excess $LiNO_3$ had increase in grain size. There exists an outlier in the trend where 10 wt% excess $LiNO_3$ had smaller particle size. Overall, the inter-particle contact was decreasing with increasing $LiNO_3$ content while the grain size equally increased.

As mentioned in Sect. 3.2, Galven et al. showed that the Li^+/H^+ exchange feasibility and subsequently the air stability are directly connected to the lithium stoichiometry— only when more than three Li^+ per formular is present then the garnet becomes sensitive to humidity [14]. Liu et al. showed that excess lithium (0–50 mol%) added during synthesis of $Li_{6.5}La_3Ta_{0.5}Zr_{1.5}O_{12}$ increase the relative density of the garnet and Li-ion conductivity by the liquid phase formation during sintering but the air stability reduced [44]. Cheng et al. reported a lower area-specific resistance (ASR) of 540 Ω cm^2 for an Al-substituted LLZO synthesized using a stoichiometric amount of Li_2CO_3 instead of the usual excess [18]. The value for ASR obtained was lower to the initial values earlier reported and this was attributed to fact that there was less lithium ion available for reaction with atmospheric moisture which leads to the formation of Li_2CO_3. Therefore, he believed that the Li_2CO_3 content critically affects the ASR.

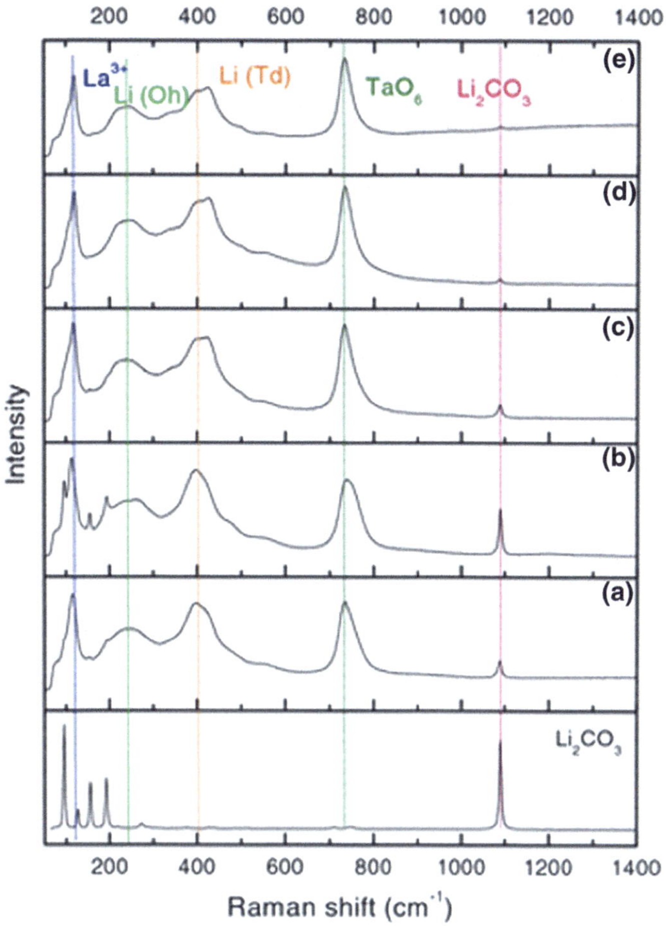

Fig. 3.8 Raman spectroscopy of $Li_6La_3Ta_{1.5}Y_{0.5}O_{12}$ with varying amounts of excess **a** 2.5, **b** 5, **c** 7, **d** 10 and **e** 15 wt% $LiNO_3$. For comparison, Raman spectrum of commercial Li_2CO_3 is also shown [43]

3.3.3 Storage Conditions

For environmental conditions, the detrimental effects of humidity on electronics, energy devices, consumable products, etc. are well-documented, but controlling these conditions can help prevent deterioration in the performance of the LLZO samples. Xia et al. [17] stored Ta-doped LLZO in dry air (humidity $\sim$ 5%) and humid air (humidity $\sim$ 80%) for 6 weeks and found that the room-temperature ionic conductivity decreased from 6.45×10^{-4} to 3.61×10^{-4} S cm^{-1} in humid air, whereas it barely dropped in dry air. Clearly, humidity is one critical factor affecting air stability. The likes of Cheng et al. [8, 10, 37] and other scientists working on garnet-type

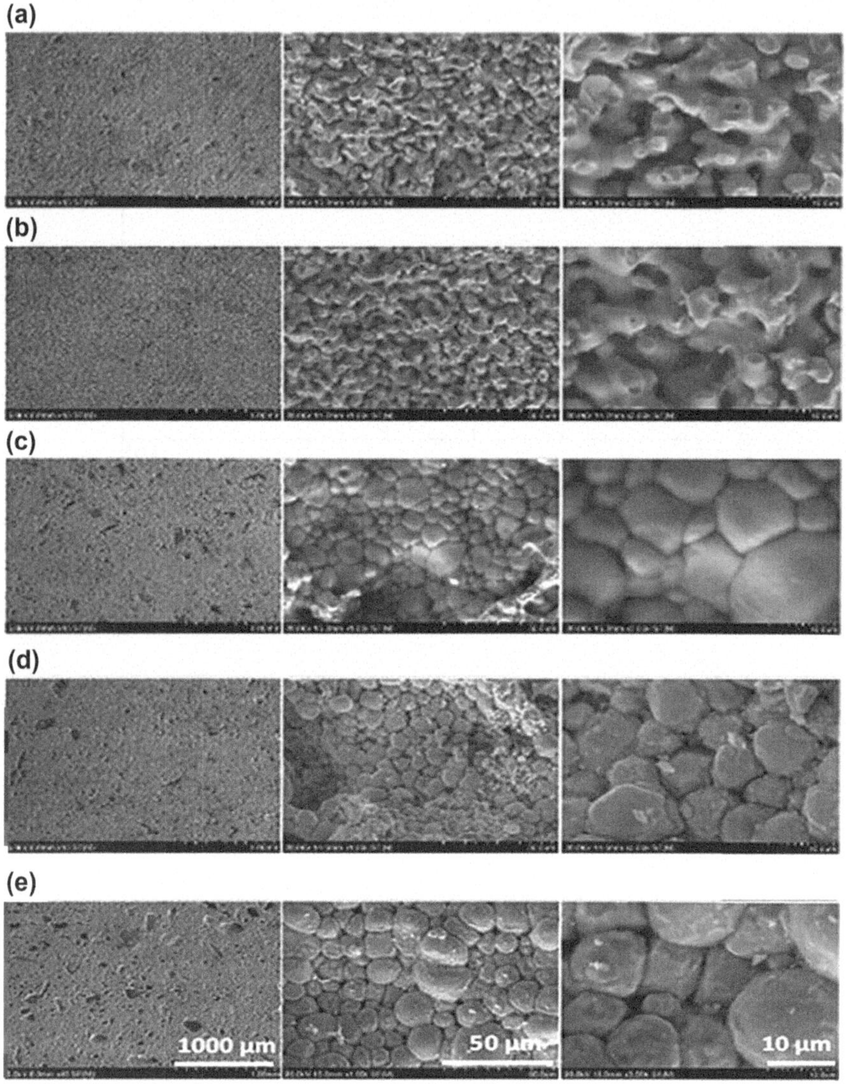

Fig. 3.9 Scanning electron micrographs of $Li_6La_3Ta_{1.5}Y_{0.5}O_{12}$ with varying amounts of excess **a** 2.5, **b** 5, **c** 7, **d** 10, and **e** 15 wt% $LiNO_3$ (corresponding higher magnification images are shown in the same row) [43]

solid electrolyte exposed their various samples to the atmosphere and determined the effect of this environment parameter on the stability of garnet in air; it turned out that Li garnets-type solid electrolytes are not stable in humid air.

3.4 Improvement Strategies

3.4.1 Mechanical Polishing

Post-processing surface polishing can effectively remove the top Li_2CO_3 layer on the LLZO surface and substantially lower the interfacial impedance since the Li-ion insulating Li_2CO_3 layer has a definite thickness. As suggested by Cheng et al., the ASR of LLZO dropped from 960 to 109 Ω cm^2 after the surfaces of the sintered pellets were dry-polished using several pieces of polishing paper in an Ar-filled glove box to avoid contact with air, water, and contamination from liquid polishing media [8]. The SEM morphological image of the pre- and post-processing surfaces are shown in Fig. 3.10. It was made known that the procedure produced very low interfacial impedance for Al-substituted LLZO cells, and more stable galvanostatic cycling than that of unprocessed material. In Xia et al's work, this technique was proven effective too, as shown in Fig. 3.1c [11]. However, special care is needed in polishing the surfaces of the hetero-structured garnet pellets to fully remove the surface Li_2CO_3 layers because different Li_2CO_3 thickness ranging from 3 nm to 25 μm has been reported due to factors such as aging time, air humidity, and the relative density of the LLZO samples [8, 11].

3.4.2 Use of Additives

This approach, also known as second-phase doping, is different from lattice doping; it is the use of compounds that could inhibit the reaction and formation of lithium carbonate on the surface of LLZO. Some of the additives (LiF, Li_3BO_3, Li_2O, Li_3PO_4, Li_4SiO_4, etc.) serve multipurpose functions like increasing the ionic conductivity, engineering the grain boundary properties or tailoring the surface microstructures.

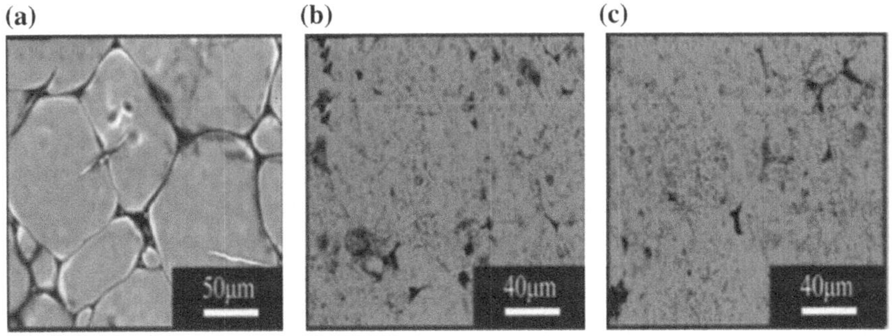

Fig. 3.10 SEM surface morphologies of **a** an as-sintered LLZO pellet **b** a LLZO pellet polished in an Ar glove box (LLZO_Ar) and **c** a LLZO pellet polished in air (LLZO_air) and exposed to the ambient atmosphere for several days [8]

Previously, the use of Li_3BO_3 with Nb-substituted LLZO-based electrolyte was reported to have high conductivity at room temperature after sintering at 790 °C. It was further proven that an improved lithium ionic conductivity was achievable with the Ca- and Al- substituted, and Li_3BO_3 additives at lower sintering temperature [45, 46]. However, most of the works on Li_3BO_3 did not optimize the amount of Li_3BO_3 additives used; [45–49] and for those with optimized quantities, no rationale was provided for the usage of such quantity.

To overcome the disadvantages of the Nb-substitution and clarify the inconsistent air stability of the LLZO-based electrolyte, Zhang et al. demonstrated the effectiveness of the use of Li_3BO_3 additives in inhibiting the formation of lithium carbonate while enhancing the air stability of garnets [50]. The introduction of lithium borate (Li_3BO_3) to the cubic Ca/Ta-substituted LLZO electrolyte at a temperature of 800 °C saw the need to control the synthesis parameters while using additives to engineer the grain boundaries of lithium garnets. A molar ratio of 4:5 (Li_3BO_3:LLZO) resulted in high Li^+ conductivity at room temperature and suppressed $LiCO_3$ formation when exposed to the atmosphere. In conclusion, they believe Li_3BO_3 additives can efficiently improve the stability of the electrolyte in air [50].

Recently, Li et al. [51] proposed an improved strategy by introducing 2 wt% LiF into the garnets to increase the stability of the garnet electrolyte against moist air. As shown in Fig. 3.11, the prepared sample had less Li_2CO_3 on the surface and showed small interfacial resistance with Li metal, a solid polymer electrolyte, and organic-liquid electrolytes. Indeed, 2 wt% LiF was found to be able to effectively suppress the Li_2CO_3 formation and decrease the garnet/Li interfacial resistance from 1260 to 345 Ω cm^{-2}. The rationale for the use of LiF was due to the fact that LiF has a strong ionic bond and is insoluble in water and if used within the grain boundary of the garnet, so it prevents diffusion of moisture and carbon dioxide into the pellet.

3.4.3 Synthesis Methods and Environmental Control

The synthesis of Li garnets has been performed generally using the conventional solid-state reactions at elevated temperature in air. New routes are currently being proposed toward decreasing the sintering temperature or improving the reaction conditions [2, 47, 48, 52]. The synthesis methods and its conditions, being at the heart of obtaining desired material properties, performance and other characteristics, have been demonstrated to be vital in obtaining an air-stable and highly conducting garnet-type electrolyte. Another possible approach to decrease the boundary resistance and reduce the susceptibility of garnets to humid air is to synthesize the electrolyte positive electrode double-layer structure in a single sintering process. This will not only ensure that both the electrode and the electrolyte are tightly coupled during the sintering process which will lead to lower boundary resistance, but also prevent the formation of Li_2CO_3 on the electrolyte surface because the contact interface of the electrolyte with the positive electrode is buried within the electrode/electrolyte composite [53]. Generally, during the processing of garnet materials and assembling of

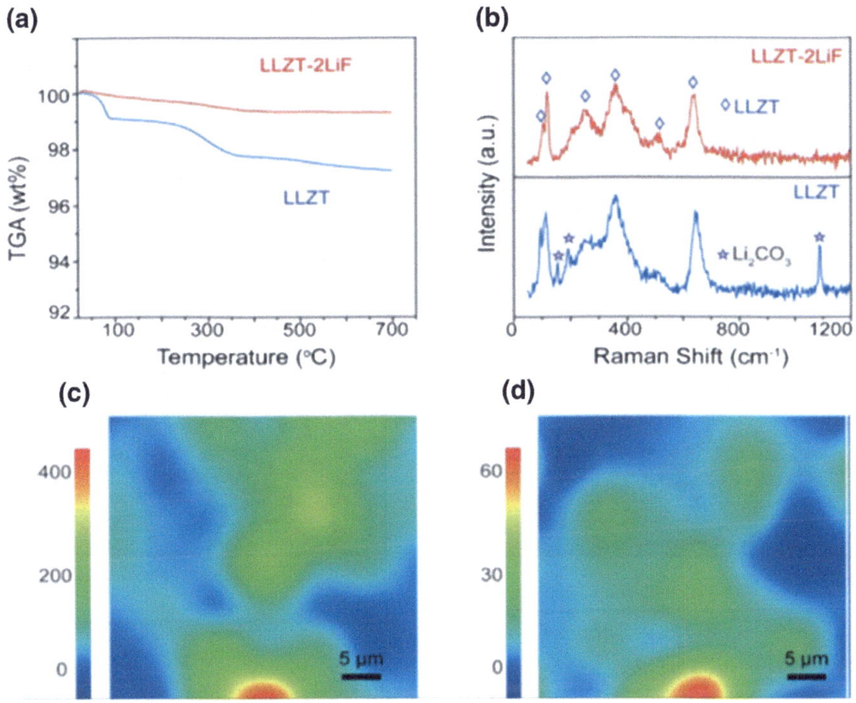

Fig. 3.11 **a** Thermogravimetric analysis (TGA) curves of LLZT and LLZT-2LiF, **b** Raman spectra of LLZT and LLZT-2LiF; Raman mapping of LLZT **c** and LLZT-2LiF **d** after two weeks of exposure in air [51]

battery cells, scientists take extra measures in ensuring that their working environment is inert by using argon-filled compartment or other means of excluding moisture from the system. This is not new since it has been widely applied to the manufacture of liquid-electrolyte-based lithium-ion batteries. The storage condition for garnet electrolytes also requires a moisture-free environment just as the synthesis.

3.4.4 Etching

Thermal and chemical etching has been proven to be another efficient means to address the instability of garnets in air. According to Abdel-Basset et al., the Li_2CO_3 on the air-exposed $Li_{5.3}La_3Ta_{1.85}Sm_{0.15}O_{12}$ pellets can either be removed through thermally etching at 900 °C for 6 h or chemically etching using 0.5 M HNO_3 for 15 min [54]. However, the Li_2CO_3 layer formation was removed totally by acid etching while thermal etching partially removed the Li_2CO_3 layer, as seen in the SEM images (Fig. 3.12) and there was no observable change in the structure after the etching treatments which implies that only Li_2CO_3 as secondary phase was removed.

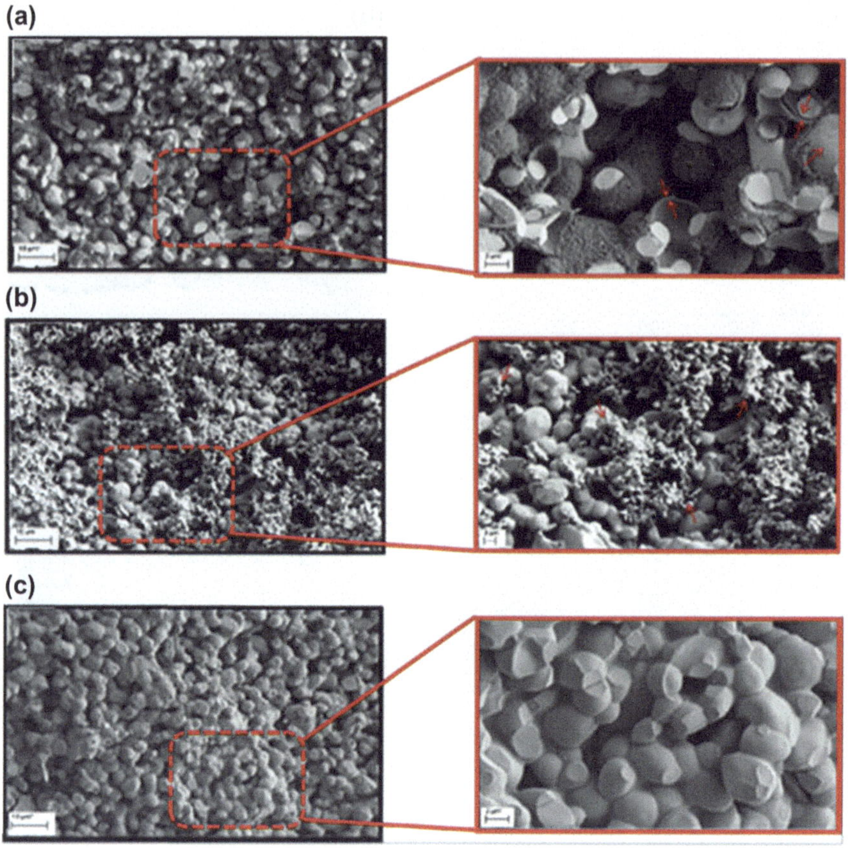

Fig. 3.12 SEM images show the cross-section morphology of **a** 2-day air-exposed, **b** thermal-etched, and **c** 0.5 M HNO_3 etched (15 min) $Li_{5+2x}La_3Ta_{2-x}Sm_xO_{12}$ electrolyte, respectively. Higher magnification images on the right-hand side are given for better illustration [54]

The reasons may be that carbonate reforms at a temperature less or equal to 600 °C (CO_2 at $\leq$600 °C) during the cooling process [54].

3.5 Summary

Garnet-type lithium-ion conductor, a candidate electrolyte for solid-state lithium-ion batteries has suffered setbacks due to lack of air stability which potentially contribute to the high interfacial resistance hampering its development. The air stability is closely related to the material properties, synthetic conditions and the humidity in ambient air. The proton/Li^+ exchange and the formation of $LiOH{\cdot}H_2O$ play an important role in the formation of surface impurity, Li_2CO_3. On the other hand, various approaches like the use of additives or new synthetic conditions have been developed to mitigate the formation of Li_2CO_3 and improve its air stability, which have been proven to be effective. Much effort has also been dedicated to optimizing the microstructure in order to enhance the ionic conductivity toward reaching 1.6 mS cm^{-1}, close to that of conventional liquid electrolyte.

Conclusively, the high interfacial impedances resulting from the chemical instability of LLZO in the atmosphere are not intrinsic of the garnet system but arise from the complications of the processing technique of its materials into thin films needed for electrochemical cells. These complications can be avoided by taking necessary precautions and carefully engineering the surface microstructure, especially the grain boundaries, which is critical to achieving low interfacial resistances.

Acknowledgements This work is financially supported by the Natural Science Foundation of China (no. 11304198), the SJTU Materials Genome Initiative Center grant (15X190030002), the SAST-SJTU fund (USCAST-2015-40), and SMC-Chen Xing Young Scholar Award of SJTU (15X100080047).

References

1. Murugan R, Thangadurai V, Weppner W (2007) Fast lithium ion conduction in garnet-type $Li_7La_3Zr_2O_{12}$. Angewand Chem Int Ed 46(41):7778–7781
2. Kerman K, Luntz A, Viswanathan V et al (2017) Practical challenges hindering the development of solid state Li ion batteries. J Electrochem Soc 164(7):A1731–A1744
3. Ramakumar S, Deviannapoorani C, Dhivya L et al (2017) Lithium garnets: synthesis, structure, Li^+ conductivity, Li^+ dynamics and applications. Prog Mater Sci 88:325–411
4. Bizhu Z, Hongchun W, Jialin M et al (2017) A review of inorganic solid electrolyte/electrode interface in all-solid-state lithium batteries. Sci Sin Chim 47(5):579–593
5. Ren Y, Chen K, Chen R et al (2015) Oxide electrolytes for lithium batteries. J Am Cera Soc 98(12):3603–3623
6. Bachman JC, Muy S, Grimaud A et al (2015) Inorganic solid-state electrolytes for lithium batteries: mechanisms and properties governing ion conduction. Chem Rev 116(1):140–162
7. Buschmann H, Dölle J, Berendts S et al (2011) Structure and dynamics of the fast lithium ion conductor "$Li_7La_3Zr_2O_{12}$". Phys Chem Chem Phys 13(43):19378–19392
8. Cheng L, Crumlin EJ, Chen W et al (2014) The origin of high electrolyte–electrode interfacial resistances in lithium cells containing garnet type solid electrolytes. Phys Chem Chem Phys 16(34):18294–18300
9. Kang SG, Sholl DS (2014) First-principles study of chemical stability of the lithium oxide garnets $Li_7La_3M_2O_{12}$ (M = Zr, Sn, or Hf). J Phys Chem C 118(31):17402–17406
10. Cheng L, Wu CH, Jarry A et al (2015) Interrelationships among grain size, surface composition, air stability, and interfacial resistance of Al-substituted $Li_7La_3Zr_2O_{12}$ solid electrolytes. ACS Appl Mater Interfaces 7(32):17649–17655
11. Xia W, Xu B, Duan H et al (2017) Reaction mechanisms of lithium garnet pellets in ambient air: The effect of humidity and CO_2. J Am Cera Soc 100(7):2832–2839
12. Nyman M, Alam TM, McIntyre SK et al (2010) Alternative approach to increasing Li mobility in Li-La-Nb/Ta garnet electrolytes. Chem Mater 22(19):5401–5410
13. Galven C, Fourquet J-L, Crosnier-Lopez M-P et al (2011) Instability of the lithium garnet $Li_7La_3Sn_2O_{12}$: Li^+/H^+ exchange and structural study. Chem Mater 23(7):1892–1900

14. Galven C, Dittmer J, Suard E et al (2012) Instability of lithium garnets against moisture. Structural characterization and dynamics of $Li_{7-x}H_xLa_3Sn_2O_{12}$ and $Li_{5-x}H_xLa_3Nb_2O_{12}$. Chem Mater 24(17):3335–3345
15. Ma C, Rangasamy E, Liang C et al (2015) Excellent stability of a lithium-ion-conducting solid electrolyte upon reversible Li $^+$/H $^+$ exchange in aqueous solutions. Angewand Chem Int Ed 54(1):129–133
16. Knauth P (2009) Inorganic solid Li ion conductors: an overview. Solid State Ion 180(14–16):911–916
17. Xia W, Xu B, Duan H et al (2016) Ionic conductivity and air stability of Al-doped $Li_7La_3Zr_2O_{12}$ sintered in alumina and Pt crucibles. ACS Appl Mater Interfaces 8(8):5335–5342
18. Cheng L, Park JS, Hou H et al (2014) Effect of microstructure and surface impurity segregation on the electrical and electrochemical properties of dense Al-substituted $Li_7La_3Zr_2O_{12}$. J Mate Chem A 2(1):172–181
19. Sakamoto J, Rangasamy E, Kim H et al (2013) Synthesis of nano-scale fast ion conducting cubic $Li_7La_3Zr_2O_{12}$. Nanotechnology 24(42):424005
20. Jin Y, McGinn PJ (2013) $Li_7La_3Zr_2O_{12}$ electrolyte stability in air and fabrication of a Li/$Li_7La_3Zr_2O_{12}$/$Cu_{0.1}V_2O_5$ solid-state battery. J Power Sources 239:326–331
21. Buschmann H, Berendts S, Mogwitz B et al (2012) Lithium metal electrode kinetics and ionic conductivity of the solid lithium ion conductors "$Li_7La_3Zr_2O_{12}$" and $Li_{7-x}La_3Zr_{2-x}Ta_xO_{12}$ with garnet-type structure. J Power Sources 206:236–244
22. Huang M, Liu T, Deng Y et al (2011) Effect of sintering temperature on structure and ionic conductivity of $Li_{7-x}La_3Zr_2O_{12-0.5x}$ ($x = 0.5 \sim 0.7$) ceramics. Solid State Ion 204:41–45
23. Rettenwander D, Geiger CA, Tribus M et al (2014) A synthesis and crystal chemical study of the fast ion conductor $Li_{7-3x}Ga_xLa_3Zr_2O_{12}$ with $x = 0.08$ to 0.84. Inorg chem 53(12):6264–6269
24. Rangasamy E, Wolfenstine J, Sakamoto J (2012) The role of Al and Li concentration on the formation of cubic garnet solid electrolyte of nominal composition $Li_7La_3Zr_2O_{12}$. Solid State Ion 206:28–32
25. Murugan R, Thangadurai V, Weppner W (2008) Effect of lithium ion content on the lithium ion conductivity of the garnet-like structure Li_{5+x} $BaLa_2Ta$ $_2O_{11.5+0.5x}$ ($x = 0$–2). Appl Phys A 91(4):615–620
26. Li Y, Wang Z, Li C et al (2014) Densification and ionic-conduction improvement of lithium garnet solid electrolytes by flowing oxygen sintering. J Power Sources 248:642–646
27. Rettenwander D, Blaha P, Laskowski R et al (2014) DFT study of the role of Al_{3+} in the fast ion-conductor $Li_{7-3x}Al_{3+x}La_3Zr_2O_{12}$ Garnet. Chem Mater 26(8):2617–2623
28. Allen JL, Wolfenstine J, Rangasamy E et al (2012) Effect of substitution (Ta, Al, Ga) on the conductivity of $Li_7La_3Zr_2O_{12}$. J Power Sources 206:315–319
29. Zhang Y, Chen F, Tu R et al (2014) Field assisted sintering of dense Al-substituted cubic phase $Li_7La_3Zr_2O_{12}$ solid electrolytes. J Power Sources 268:960–964
30. Baek S-W, Lee J-M, Kim TY et al (2014) Garnet related lithium ion conductor processed by spark plasma sintering for all solid state batteries. J Power Sources 249:197–206
31. Duan H, Zheng H, Zhou Y et al (2018) Stability of garnet-type Li ion conductors: an overview. Solid State Ion 318:45–53
32. Wu J-F, Chen E-Y, Yu Y et al (2017) Gallium-doped $Li_7La_3Zr_2O_{12}$ garnet-type electrolytes with high lithium-ion conductivity. ACS Appl Mater Interfaces 9(2):1542–1552
33. Du F, Zhao N, Li Y et al (2015) All solid state lithium batteries based on lamellar garnet-type ceramic electrolytes. J Power Sources 300:24–28
34. Shimonishi Y, Toda A, Zhang T et al (2011) Synthesis of garnet-type $Li_{7-x}La_3Zr_2O_{12-1/2x}$ and its stability in aqueous solutions. Solid State Ion 183(1):48–53
35. Yi E, Wang W, Kieffer J et al (2017) Key parameters governing the densification of cubic-$Li_7La_3Zr_2O_{12}$ Li^+ conductors. J Power Sources 352:156–164
36. Lobe S, Dellen C, Finsterbusch M et al (2016) Radio frequency magnetron sputtering of $Li_7La_3Zr_2O_{12}$ thin films for solid-state batteries. J Power Sources 307:684–689

37. Cheng L, Chen W, Kunz M et al (2015) Effect of surface microstructure on electrochemical performance of garnet solid electrolytes. ACS Appl Mater Interfaces 7(3):2073–2081
38. Geiger CA, Alekseev E, Lazic B et al (2010) Crystal chemistry and stability of "$Li_7La_3Zr_2O_{12}$" garnet: a fast lithium-ion conductor. Inorg Chem 50(3):1089–1097
39. Rettenwander D, Geiger CA, Amthauer G (2013) Synthesis and crystal chemistry of the fast Li-ion conductor $Li_7La_3Zr_2O_{12}$ doped with Fe. Inorg Chem 52(14):8005–8009
40. Jalem R, Rushton M, Manalastas W Jr et al (2015) Effects of gallium doping in garnet-type $Li_7La_3Zr_2O_{12}$ solid electrolytes. Chem Mater 27(8):2821–2831
41. Tong X, Thangadurai V, Wachsman ED (2015) Highly conductive Li garnets by a multielement doping strategy. Inorgan Chem 54(7):3600–3607
42. Deviannapoorani C, Dhivya L, Ramakumar S et al (2013) Lithium ion transport properties of high conductive tellurium substituted $Li_7La_3Zr_2O_{12}$ cubic lithium garnets. J Power Sources 240:18–25
43. Narayanan S, Hitz GT, Wachsman ED et al (2015) Effect of excess Li on the structural and electrical properties of garnet-type $Li_6La_3Ta_{1.5}Y_{0.5}O_{12}$. J Electrochem Soc 162(9):A1772–A1777
44. Liu K, Ma J-T, Wang C-A (2014) Excess lithium salt functions more than compensating for lithium loss when synthesizing $Li_{6.5}La_3Ta_{0.5}Zr_{1.5}O_{12}$ in alumina crucible. J Power Sources 260:109–114
45. Ohta S, Seki J, Yagi Y et al (2014) Co-sinterable lithium garnet-type oxide electrolyte with cathode for all-solid-state lithium ion battery. J Power Sources 265:40–44
46. Ohta S, Komagata S, Seki J et al (2013) All-solid-state lithium ion battery using garnet-type oxide and Li_3BO_3 solid electrolytes fabricated by screen-printing. J Power Sources 238:53–56
47. Tadanaga K, Takano R, Ichinose T et al (2013) Low temperature synthesis of highly ion conductive $Li_7La_3Zr_2O_{12}$–Li_3BO_3 composites. Electrochem Commun 33:51–54
48. Takano R, Tadanaga K, Hayashi A et al (2014) Low temperature synthesis of Al-doped $Li_7La_3Zr_2O_{12}$ solid electrolyte by a sol–gel process. Solid State Ion 255:104–107
49. Janani N, Deviannapoorani C, Dhivya L et al (2014) Influence of sintering additives on densification and Li^+ conductivity of Al doped $Li_7La_3Zr_2O_{12}$ lithium garnet. RSC Adv 4(93):51228–51238
50. Zhang L, Yang J, Gao Y et al (2017) Influence of Li_3BO_3 additives on the Li^+ conductivity and stability of Ca/Ta-substituted $Li_{6.55}(La_{2.95}Ca_{0.05})(Zr_{1.5}Ta_{0.5})O_{12}$ electrolytes. J Power Sources 355:69–73
51. Li Y, Xu B, Xu H et al (2017) Hybrid polymer/garnet electrolyte with a small interfacial resistance for lithium-ion batteries. Angewand Chem Int Ed 56(3):753–756
52. Chen R-J, Huang M, Huang W-Z et al (2014) Sol–gel derived Li–La–Zr–O thin films as solid electrolytes for lithium-ion batteries. J Mater Chem A 2(33):13277–13282
53. Thangadurai V, Weppner W (2005) Investigations on electrical conductivity and chemical compatibility between fast lithium ion conducting garnet-like $Li_6BaLa_2Ta_2O_{12}$ and lithium battery cathodes. J Power Sources 142(1–2):339–344
54. Abdel-Basset DM, Mulmi S, El-Bana MS et al (2017) Structure, ionic conductivity, and dielectric properties of Li-rich garnet-type $Li_{5+2x}La_3Ta_{2-x}Sm_xO_{12}$ ($0 \leq x \leq 0.55$) and Their Chemical Stability. Inorg Chem 56(15):8865–8877

Chapter 4
Influence of Strain on Garnet-Type Electrolytes

Hirotoshi Yamada

Abstract All-solid-state batteries (ASSBs) using inorganic solid electrolytes are one of the candidates of next-generation batteries. However, ASSBs suffer from various issues, most of which do not matter in conventional lithium-ion batteries with liquid electrolytes. In this chapter, mechanical stress in solid electrolytes are focused. First, we reveal that garnet-type solid electrolytes, $Li_{6.5}La_3Zr_{1.5}Ta_{0.5}O_{12}$ (LLZT), prepared by the spark plasma sintering (SPS) method, exhibit a residual tensile stress of more than 100 MPa in the direction of uni-axial pressure during the SPS process, which was revealed by XRD (side-inclination method). Then, the influence of the stress on ionic conduction is studied. Detailed analyses reveal that the stress mainly influences grain boundary resistance with little change in the bulk resistance of LLZT. The results suggest the importance of mechanically strong grain boundaries (including interfaces) for practical ASSBs.

4.1 Introduction

Recently, all-solid-state batteries (ASSBs) using inorganic solid electrolytes attract many interests, because of their potential performance for next-generation batteries [1]. For advanced energy storage systems in large-scale systems such as electric vehicles and the load-leveling of renewable energies in electrical grids, next-generation batteries are required to exhibit various severe performances: high energy density, high power density, low cost, high reliability, cycle life, shell life, etc. There have been various solid electrolytes developed so far, and in the last decades, they are evaluated as solid electrolyte in battery systems. Among them, sulfide-based solid electrolytes exhibit good performance not only because of their high ionic conductivity but also due to plasticity [2, 3]. However, it is supposed that a critical drawback of the sulfide-based solid electrolytes hinders their practical application, especially in large-scale systems. When the sulfide-based solid electrolytes are exposed to moisture, they hydrolyze to produce strong toxic H_2S. On the contrary, oxide-based solid

H. Yamada (✉)
Graduate School of Engineering, Nagasaki University, Nagasaki, Japan
e-mail: h-yama@nagasaki-u.ac.jp

R. Murugan and W. Weppner (eds.), *Solid Electrolytes for Advanced Applications*,
https://doi.org/10.1007/978-3-030-31581-8_4

electrolytes exhibit relatively lower ionic conductivity and little plasticity, which causes difficulty in assembling ASSBs. Among them, garnet-type solid electrolytes are one of the most fascinating materials, because of their rather high ionic conductivity and high chemical stability as well as high electrochemical stability. Since the garnet-type solid electrolytes are reduced by lithium metal [4–6], these materials may enable ASSBs using lithium metal anode. Lithium metal exhibits the lowest electrode potential and the largest theoretical capacity as an anode, and thus, it may be possible to increase the energy density of ASSBs. One of the drawbacks of the garnet-type solid electrolytes is that they require high temperatures (>1200 °C) and a long sintering time (>10 h) to densify them by conventional solid-state sintering [7–9]. In conventional solid-state sintering, powder with the same composition is necessary to cover pellets to suppress evaporation of lithium. Recently, several groups employed spark plasma sintering (SPS) technique to obtain sintered pellets of the garnet-type solid electrolytes, which can decrease sintering temperature below 1000 °C with shorter duration time (~10 min) [10–16]. We have reported that the garnet-type solid electrolytes prepared by SPS exhibit high relative density (>95%) and ionic conductivity (~7×10^{-4} S cm^{-1} @25 °C), and rather high short-circuit tolerance on lithium plating/stripping [13–15].

For SPS-prepared garnet-type solid electrolytes, slight changes in X-ray diffraction (XRD) profiles were confirmed: broadening and shifts toward lower angle [13, 16]. Botros et al. suggested the formation of a distorted tetragonal phase owing to the stress during the SPS [16]. The influence of the distortion in the crystalline lattice of ionic conductors is not only scientifically interesting, but also important from the viewpoint of the application. During charge/discharge of batteries, most of the active materials change their volume, which cause internal stress. There are a few reports on the influence of stress on ionic conductivity [17–19]. In the case of Y-doped ZrO_2, a compressive stress/strain decreased oxygen ionic conductivity [17], and vice versa [18]. Regarding lithium-ion conductors, Okumura et al. reported the influence of stress on the interfacial resistance between $LiCoO_2$ thin film deposited on a NASICON-type lithium-ion conducting sheet [19]. They attributed the strain to one of the reasons for high charge transfer resistance at the interface. While the influence of strain on the ionic conductivity has been pointed out, the mechanisms are not clear yet. It is not known whether the strain influences bulk resistance, grain boundary resistance, or both of them. In this study, a detailed analysis of the XRD profiles of SPS-prepared garnet-type solid electrolytes revealed that the profile changes are not caused by the tetragonal phase, but are due to residual stress (residual strain) [20]. It was also demonstrated that the influence of the stress on the ionic conduction is little on the bulk, while the grain boundary resistance depended on the stress.

4.2 Experimental*

4.2.1 Preparation of Materials

Garnet-type solid electrolyte of $Li_{6.5}La_3Zr_{1.5}Ta_{0.5}O_{12}$ (LLZT) was prepared by conventional solid-state reaction and pelletized by SPS, which are described in previous papers [13–15, 20]. First, LiOH, La_2O_3, ZrO_2, and Ta_2O_5 were mixed and calcined at 900 °C for 12 h in air. Then, the obtained LLZT powder was sintered using graphite die by the SPS method on a CSP-I-02121 (SS Alloy Co. Ltd.). SPS was conducted under vacuum (<10 Pa) at 950 °C for 10 min. During the sintering, a uni-axial pressure of 12.5–50 MPa was applied, which was reduced to 2.5 MPa on cooling. Finally, the LLZT pellets were ground using emery paper (#400) to remove graphite sheets and other impurities produced during the SPS (amorphous carbon on the cathode side and $La_2Zr_2O_7$ on the anode side) [13]. The LLZT pellets were typically 10 mm and 0.8–0.9 mm in diameter and thickness, respectively.

It is known that a grinding process using a fixed abrasive such as emery papers causes stress in the subsurface of specimens [21, 22]. Therefore, the stress in the subsurface was released by further polishing using loose abrasive (diamond slurry with an average diameter of 0.125 μm dispersed in paraffin oil).

4.2.2 Structural Analysis

XRD was conducted to study the crystalline structure of the LLZT powders and pellets on a SmartLab (Rigaku Corp., 45 kV, 200 mA, $\lambda = 1.5418$ Å) with a parallel beam arrangement. For a Rietveld analysis, diffraction profiles were recorded in the 2θ range from 15° to 140° with a 0.02° step and accumulation time of 0.5 s. For the higher resolution, several slits were used (incident slit: 1 mm, two receiver slits: 0.4 mm, soller slits: 5.0°, parallel slit analyzer: 0.5°). To suppress the fluorescent X-ray, a diffracted X-ray was monochromated with graphite. Synchrotron XRD was also carried out to obtain high-resolution data, which was performed at beamline BL15 of the SAGA Light Source with the approval of the Kyusyu Synchrotron Light Research Center (Proposal No. 1607057R/BL15). The obtained data was analyzed by the Rietveld method using a Rietan-FP [23] with several structural models.

Residual stress of the LLZT pellets was analyzed by a side-inclination method of XRD. As shown in Fig. 4.1, XRD profiles were recorded with tilting the pellets at various Ψ angles (the angle between a normal vector of the specimen plane and a scattering vector). With Ψ ranging from 0° to 60°, an X-ray source and a detector were scanned $2\theta/\theta$ angles from 15° to 40°. The diffraction angles (2θ) of peaks were plotted as a function of $\sin^2 \Psi$, and the residual stress (σ_{stress}) was obtained using the following equation:

*Most of the Experimental section is reproduced by permission from [20]. Copyright 2017 Elsevier B.V.

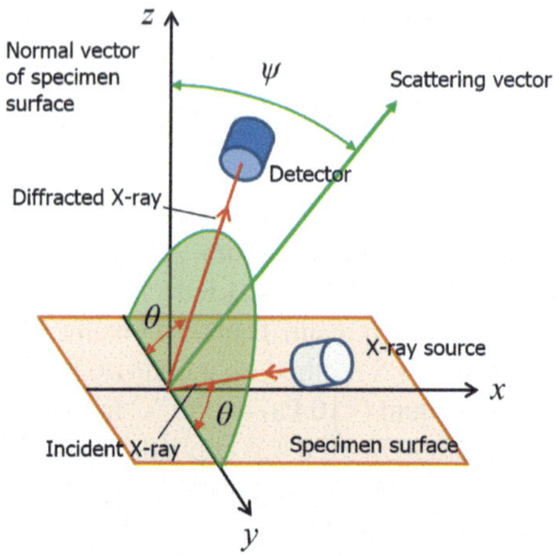

Fig. 4.1 Arrangement of side-inclination method for residual stress analysis. Reproduced by permission from [18]. Copyright 2017 Elsevier B.V.

$$\sigma_{\text{stress}} = \frac{E}{2(1+\nu)} \cot\theta_0 \frac{\partial(2\theta)}{\partial(\sin^2\Psi)} \tag{4.1}$$

where E, ν, and θ_0 are Young's modulus, Poisson's ratio, and the diffraction angle for a specimen without stress, respectively. The derivative is the slope of the $2\theta - \sin^2\Psi$ plot.

The local structure of LLZT was investigated by energy dispersive X-ray spectroscopy (EDS) and electron diffraction (ED) on a scanning transmission electron microscope (STEM; JEM-ARM200F, JEOL Ltd.). Prior to the STEM observation, thin specimen was prepared from an LLZT pellet by a focused ion beam (FIB; JIB-4501, JEOL Ltd.).

4.2.3 Electrochemical Analysis

Ionic conductivity was estimated from electrochemical impedance spectroscopy (EIS). For the EIS, Au electrodes (ca. 300 nm in thickness) were deposited on both sides of the LLZT pellets. The EIS was recorded on an impedance analyzer HP4192A with a four-terminal pair configuration and an oscillation voltage of 10 mV in a frequency range from 10 MHz to 10 Hz. The LLZT pellets were placed in a one-side closed glass tubular cell with four coaxial cables connected to the impedance analyzer. Two pairs of the cables (high voltage and high current, low potential and low current) were connected to each other at the electrodes in order to minimize noises and parasitic capacitance. The glass cell, in which Ar gas flowed, was placed in a temperature-controlled oven (SU-222, ESPEC Corp.). Cell temperature was raised to and kept at 110 °C for 3 h, then was very slowly reduced down to −20 °C at −0.25°/min. EIS of the specimen was recorded every 10 min. A thermocouple was placed near the specimen in the cell to monitor the temperature. Because of the very

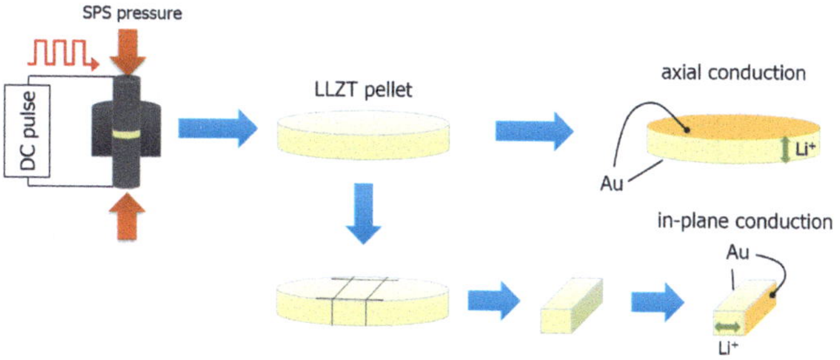

Fig. 4.2 Preparation of specimens for ionic conductivity measurements in the horizontal and axial direction. Reproduced by permission from [18]. Copyright 2017 Elsevier B.V.

slow ramp rate and small heat capacity of the cell, the cell temperature was supposed to reach the pseudo-equilibrium state. This is supported by the fact that the resistance values recorded on cooling and heating processes agreed well with each other within the difference of 1%. The EIS data was analyzed using a software Z-View (Scribner Inc.).

In the above measurements, the direction of the ionic conduction was parallel to the pressure axis of the SPS. To investigate the direction dependence of the ionic conductivity, an LLZT pellet was cut into a cuboid (ca. 9.0 × 0.9 × 1.5 mm^3) and Au electrodes were deposited on the cut sides (9.0 × 0.9 mm^2) of the cuboid by sputtering (see Fig. 4.2). Using this specimen, lithium ionic conductivity perpendicular to the SPS axis was also studied.

4.3 Results and Discussion

4.3.1 Structural Analysis

XRD profiles of an LLZT pellet and fitted results are shown in Fig. 4.3. Here, fitting results using two different models are demonstrated. Figure 4.3a employs a model of mixture of a cubic garnet phase and a tetragonal garnet phase, which was suggested by Botros et al. [16]. On the other hand, a model of mixture of two cubic garnet phases with different lattice parameters is shown in Fig. 4.3b. As listed in Table 4.1, *S* value ($=R_{wp}/R_e$), which indicate the plausibility of the models, were 2.5 and 2.8, suggesting both models are improper models (*S* decreases toward 1 with more plausible model, and in general, *S* is less than 1.5 for reliable models.). When both fitting profiles are carefully studied (see insets of Fig. 4.3), the calculated curves (green lines) seem to well fit the experimental data (brown dots) for some diffraction peaks (e.g., (3 2 1) at 25.7°). However, some peaks couldn't be fitted well (e.g., (4 0 0) at 27.5°). We tried several other models including composition change, and distributed lattice parameters, etc., but no models could fit the experimental result.

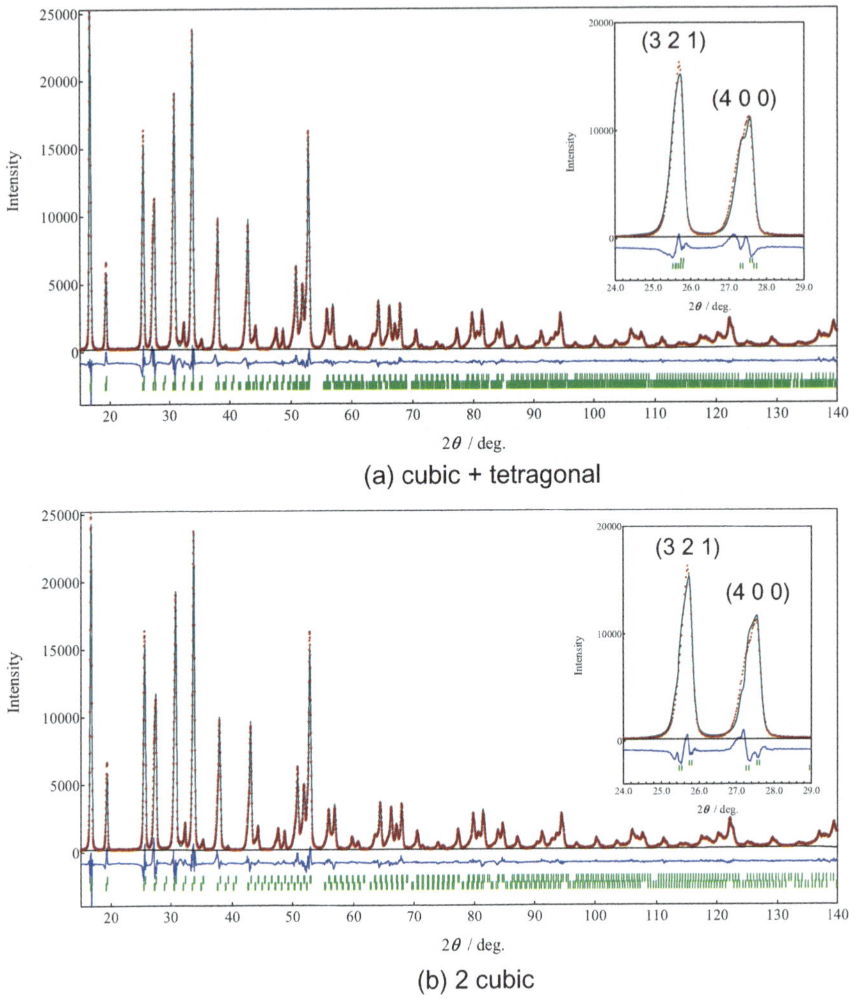

Fig. 4.3 Rietveld analysis results of a SPS-prepared LLZT pellet sintered at 950 °C, 25 MPa for 10 min. **a** Fitted with cubic and tetragonal phases. **b** Fitted with two cubic phases with different lattice parameters. Brown dots and green lines are experimental data and calculated curves, respectively. Reproduced by permission from [18]. Copyright 2017 Elsevier B.V.

STEM-EDS and ED (Fig. 4.4) was conducted to investigate local structure of the LLZT pellets. As indicated by the EDS mapping images (Fig. 4.4b–f), Zr and Ta show inhomogeneous distribution. Note that not all regions in the pellets exhibit inhomogeneous Zr/Ta distribution. The inhomogeneous distribution is supposed to be already included in the LLZT powder [16], and the sintering time of the SPS (10 min) is too short for the heavy elements to diffuse. ED patterns were taken at a point where Ta is slightly concentrated (indicated by a red circle in Fig. 4.4b). Figure 4.4g, h show simulated patterns for cubic phase ($Ia\bar{3}d$) and tetragonal phase

Table 4.1 Refinement results of Rietveld analysis of a SPS-prepared LLZT pellet

			(a) Cubic + Tetragonal	(b) Two cubic
Fitting results	R_{wp}		7.471	8.512
	R_e		2.991	2.992
	S		2.498	2.845
Fitting parameters	Phase 1	Phase	Cubic $Li_{6.5}La_3Zr_{1.5}Ta_{0.5}O_{12}$	Cubic $Li_{6.5}La_3Zr_{1.5}Ta_{0.5}O_{12}$
		Space group	$Ia\bar{3}d$	$Ia\bar{3}d$
		Lattice parameter	a = 12.9354(3) Å	a = 12.9357(2) Å
		Weight ratio	65.0%	77.0%
	Phase 2		Tetragonal $Li_{6.5}La_3Zr_{1.5}Ta_{0.5}O_{12}$	Cubic $Li_{6.5}La_3Zr_{1.5}Ta_{0.5}O_{12}$
		Space group	$I4_1/acd$	$Ia\bar{3}d$
		Lattice parameter	a = 13.055(1) Å c = 12.885(2) Å	a = 13.0702(6) Å
		Weight ratio	35.0%	23.0%

($I4_1/acd$), respectively. Apparently, the patterns are similar to each other and both patterns can fit the observed one (Fig. 4.4f) well.

However, when we compare the intensity of the spots, there are some differences between tetragonal and cubic phases. Spots yellow circles in the observed pattern are all very weak, which agree with simulated spots for the cubic symmetry. This indicates that the local structure at this point of the LLZT pellet is cubic. STEM-EDS and ED were carried out at the Zr-rich point as well (Fig. 4.5). For this point, difference between simulated patterns with cubic and tetragonal phases was so little that it is difficult to decide the phase. In Fig. 4.5g, the observed ED pattern is compared with the simulated one for cubic phase. Some spots shown with yellow circles correspond to forbidden diffraction according to the extinction rule. However, these are forbidden not by the point group ($m3m$) but by the space group ($Ia\bar{3}d$) that includes glide plane symmetry. In the case of the extinction rule based on the space group symmetry, forbidden reflections appear clearly due to the multiple scattering as shown by yellow allows in Fig. 4.5g. Therefore, it is impossible to decide the symmetry for this point. However, as far as the investigation goes, clear evidence of the tetragonal phase is not obtained.

The difficulty in fitting with the conventional Rietveld analysis may suggest anisotropic strain [24]. One of the possible reasons for the anisotropic strain is residual stress. When stress remains and the stress is distributed inside the LLZT pellet, the d-spacing is not unique for one lattice plane, and diffraction peaks are distorted. Further, because of the anisotropic Young's modulus, the strain depends on the orientation of the crystalline lattice of each grain, resulting in the different degree of the strain with the lattice orientation. As a result, it is impossible to obtain unique

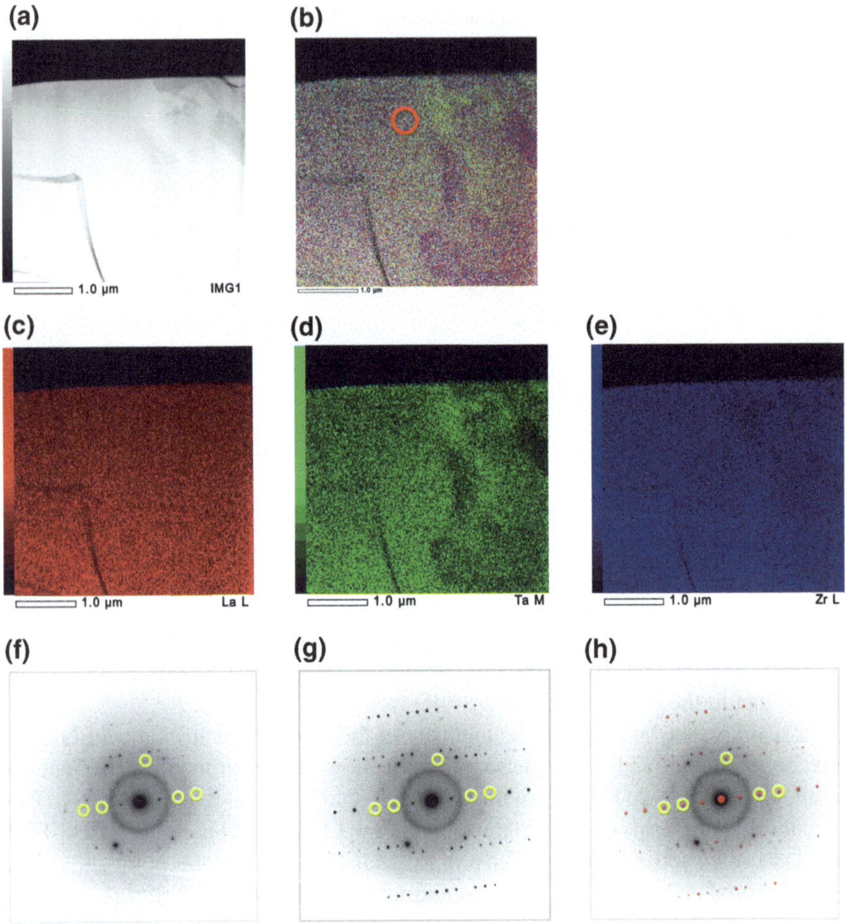

Fig. 4.4 **a** STEM dark field image and **b–e** EDS mapping images of an LLZT pellet. **b** Overlapped EDS images for **c** La, **d** Ta, and **e** Zr. **f** ED pattern at point indicated by red circle in (**b**). Simulated ED patterns using **g** cubic ($Ia\bar{3}d$) and **h** tetragonal ($I4_1/acd$) LLZT structure models with projection vector tilted slightly from [3 2 2] and thickness of 100 nm. Yellow circles indicate diffraction at which intense spots should be observed for tetragonal phase. Reproduced by permission from [18]. Copyright 2017 Elsevier B.V.

lattice parameters, and an error of the lattice parameters is increased. Figure 4.6a shows the (4 0 0) diffraction peak recorded with different Ψ angles, which clearly demonstrates the diffraction peak shift toward higher angle with increasing Ψ. The peak position, obtained by fitting the peak with a pseudo-Voigt function, is plotted as a function of $\sin^2\Psi$. Obviously, these two parameters show a linear correlation as indicated by (4.1). And the positive correlation indicates the tensile stress in the direction of the axis perpendicular to the pellet surface. Using (4.1) with Young's

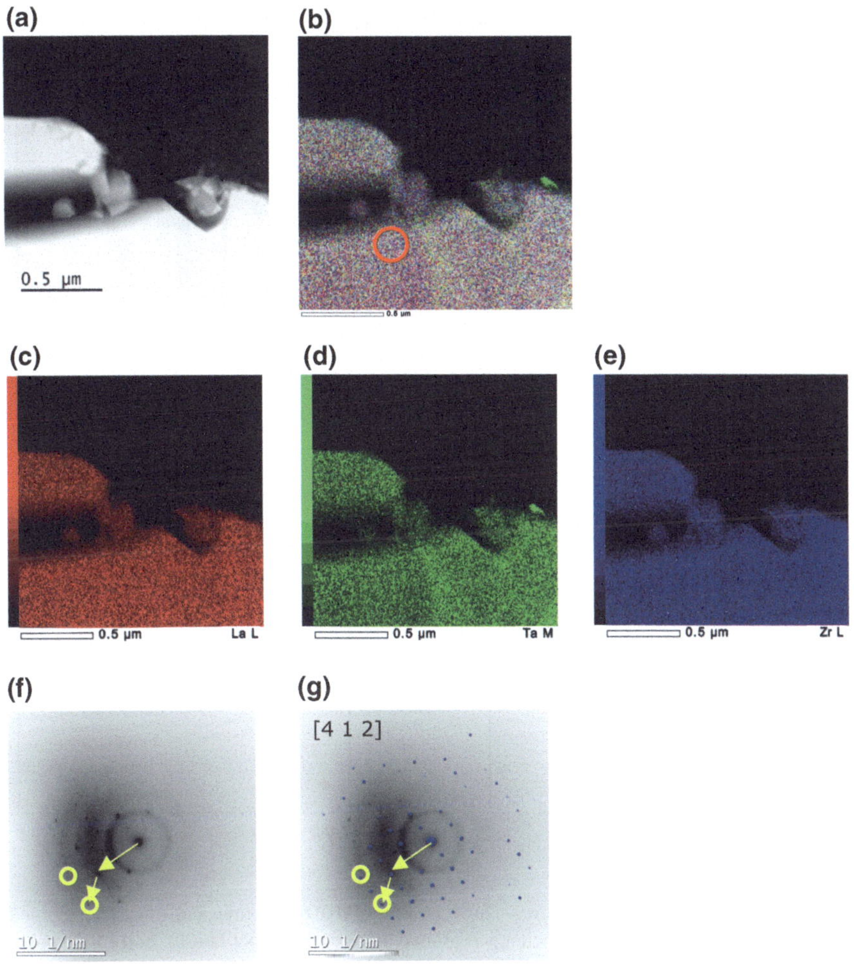

Fig. 4.5 **a** STEM dark field image and **b**–**e** EDS mapping images of an LLZT pellet. **b** overlapped EDS images for **c** La, **d** Ta, and **e** Zr. **f** ED pattern at the point in Ta-poor region indicated by a red circle in (**b**). **g** Simulated ED pattern using the cubic LLZT structure with the projection vector of [4 1 2]. Reproduced by permission from [18]. Copyright 2017 Elsevier B.V.

modulus ($E[1\,0\,0] = 135.6$ GPa) and Poisson's ratio ($\nu = 0.24$) [25, 26], the residual stress is estimated to be −603 MPa (a negative value indicates tensile stress).

Before the detailed investigation, we must note that the influence of grinding on the residual stress. It is known that grinding processes using fixed abrasive like emery papers leave residual stress in subsurface region of pellets [21, 22]. In addition, SPS also causes residual stress inside pellets [27, 28]. It is expected that ionic conductivity would be influenced by internal stress, and thus, residual stress in the subsurface region should be excluded. For that purpose, first, the LLZT pellets was ground and polished with emery papers #400, #1000, #1500, and #4000. Then, the pellets

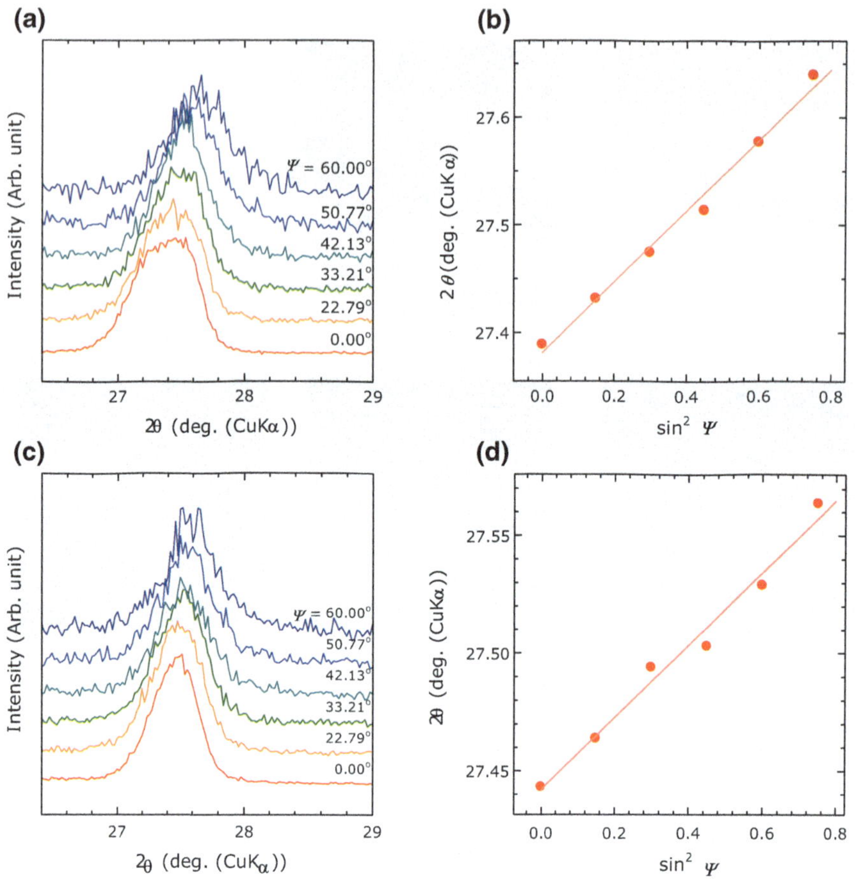

Fig. 4.6 **a**, **c** XRD profiles and **b**, **d** $2\theta - \sin^2\Psi$ plot for (4 0 0) diffraction peak of LLZT recorded with different Ψ angles. **a**, **b** LLZT pellet was ground using emery paper (#400). **b**, **d** LLZT pellet was polished using diamond slurry (0.125 μm). Reproduced by permission from [18]. Copyright 2017 Elsevier B.V.

were further polished using a cloth on which diamond suspension was dropped. The diamond particles in the suspension were 0.125 μm in average diameter. As the diamond particles are not fixed, less stress is left on the subsurface region.

Figure 4.7a shows the change in the shape of diffraction peak (4 0 0) during polishing with diamond suspension. With increasing polishing time, the intensity and the width of the peak increases and decreases, respectively. In addition, the diffraction angle shifts toward higher angle, and the profile gets symmetric. After polishing for 30 min., the pellet was thinned by ca. 5 μm. As indicated by Fig. 4.7b, the change is saturated after polishing for 10 min, which indicates that grinding causes stress for the subsurface region with a thickness of ca. 2 μm.

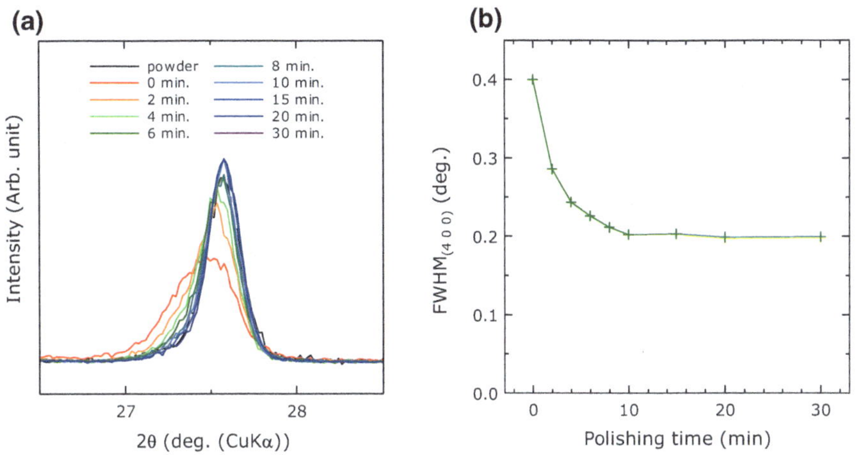

Fig. 4.7 Polishing-time dependence of XRD profiles and **b** full width at half maximum of (4 0 0) of LLZT polished using diamond slurry (0.125 μm). Reproduced by permission from [18]. Copyright 2017 Elsevier B.V.

After the surface was well-polished with the diamond slurry, the residual stress was again investigated. As shown in Fig. 4.6c, d, residual stress is still confirmed. The residual stress is estimated to be −281 MPa after polishing with diamond slurry, which is induced by the SPS.

On the SPS-induced stress, we can note the following interesting facts: (1) the residual stress (−281 MPa) is much larger than the SPS pressure (25 MPa), and (2) the tensile stress remains in the pellet in spite of the compressive uni-axial load. As indicated by grain shape before (Fig. 4.8a) and after (Fig. 4.8b) the SPS, LLZT particles are plastically deformed accordingly in the direction of the SPS pressure: the compressive stress in the axial direction and the tensile stress in the radial direction. However, observed residual stress is in the opposite direction, which has been reported for SPS and hot press [29, 30].

Two mechanisms for the residual stress may be proposed. The first mechanism is thermal shrinkage after the sintering. However, the linear thermal expansion coefficient of LLZT (15.5×10^{-6} K^{-1}) is larger than that of graphite ($2.5–3.0 \times 10^{-6}$ K^{-1} for *c* axis). This means that LLZT reduces its volume than graphite. Thus, the graphite die cannot apply radially compressive force to LLZT pellets. The second mechanism is the difference of the plastic deformation and the elastic deformation. During the sintering process, the LLZT exhibits plastic deformation as shown in Fig. 4.8b. The plastic deformation releases applied stress. On the other hand, the graphite die is radially pressed by the internal LLZT powder and is elastically deformed. During the cooling process, the graphite die shrinks due to the elastic recovery, while the stress in the LLZT pellet is already relaxed by the plastic deformation. Therefore, the shrinking graphite die causes radially compressive stress to the LLZT pellet on the cooling, which results in the tensile stress in the axial direction. This mechanism is supported by the diameter of the LLZT pellets. After the SPS, the diameters of the

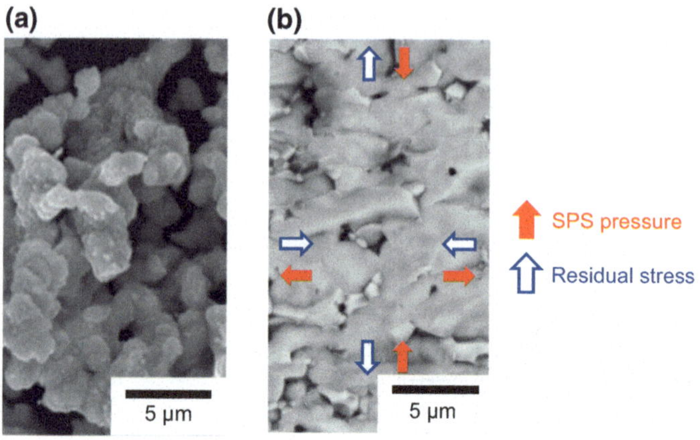

Fig. 4.8 FE-SEM images of **a** LLZT powder and **b** fractured cross-section of LLZT pellet. On SPS, pressure was applied in vertical direction of image. Reproduced by permission from [18]. Copyright 2017 Elsevier B.V.

LLZT pellets are typically 10.1–10.2 mm, which is larger than the inner diameter of the graphite die (10.0 mm). Therefore, the LLZT pellets are radially pressed by the graphite die. The stress can be roughly estimated using Young's modulus of the graphite ($E = 4.8$ GPa) and the strain ($\varepsilon \sim 0.2$ mm/10 mm = 0.02). The stress σ_{stress} is calculated as follows:

$$\sigma_{\text{stress}} = E\varepsilon = 96\,\text{MPa}.$$

The estimated value is higher than the SPS pressure and on the same order of magnitude of the observed residual stress (281 MPa). As the recovery force of the graphite die is not one-dimensional but two-dimensional (or radial), the above value would be underestimated. Numerical simulation of SPS by the finite element method suggested residual tensile stress of ca. 300 MPa for Ti–TiB with graphite die [29].

4.3.2 *Electrochemical Analysis*

In this section, the influence of the strain on the ionic conductivity is discussed. Figure 4.9a shows the total ionic conductivity of the LLZT pellet as a function of reciprocal temperature. The ionic conductivity was measured for a cuboid specimen prepared from a pellet (Fig. 4.2). The ionic conductivity in the axial direction (tensile strain) was slightly higher than in-plain conduction (compressive strain).

A typical Nyquist plot, shown in Fig. 4.9b, apparently consists of one semi-circle and one straight line. However, it could not be fitted with one electrochemical process, i.e., parallelly connected one resistance (R) and one constant phase element (CPE). The fitting was improved by adding another R//CPE component, suggesting EIS of the

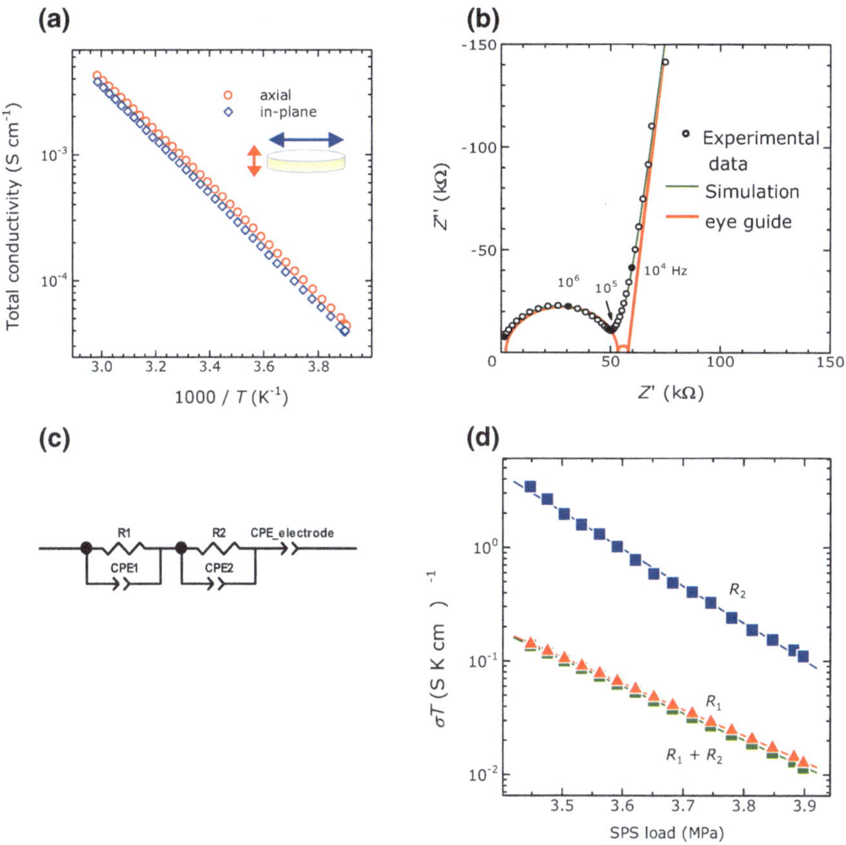

Fig. 4.9 **a** Temperature dependence of total ionic conductivity of SPS-prepared LLZT pellet in axial and in-plane directions. **b** Nyquist plot of SPS-prepared LLZT pellet recorded at −15 °C (axial direction). Red line is an eye guide showing two electrochemical processes with changed capacitance for CPE_2. **c** Equivalent circuit used for fitting. **d** Temperature dependence of σT obtained for R_1, R_2 and $R_1 + R_2$ of (**b**). Reproduced by permission from [18]. Copyright 2017 Elsevier B.V.

LLZT pellet consists of the two *R*//CPE with different time constants (Fig. 4.9c). Chi-square (χ^2) for the fitting was lower than 10^{-4} for EIS data below 15 °C, suggesting reliable analyses for this temperature range. The temperature dependence of the conductivity ($\sigma_{\mathrm{cond.}}$) is plotted in Fig. 4.9d in the form of $\sigma_{\mathrm{cond.}}T$, where T is absolute temperature. R_1 and R_2 are obtained using the two *R*//CPE model (Fig. 4.9c). Each plot was fitted with an Arrhenius-type equation:

$$\sigma_{\mathrm{cond.}}T = A_0 \exp\left(-\frac{E_a}{R_g T}\right) \quad (4.2)$$

Table 4.2 Parameters estimated by fitting complex impedance with an equivalent circuit (inset of Fig. 4.10b). Capacitance C was obtained from constant phase element and resistance

	R_1//CPE$_1$			R_2//CPE$_2$		
	R_1		C_1 (F cm^{-2})	R_2		C_2 (F cm^{-2})
	A_0 (S K cm^{-1})	E_a (eV)		A_0 (S K cm^{-1})	E_a (eV)	
Axial	1.2×10^7	0.46	4×10^{-11}	6.1×10^{11}	0.65	7×10^{-8}
In-plane	1.2×10^7	0.46	3×10^{-11}	2.6×10^9	0.54	1×10^{-8}

where A_0, E_a, and R_g are the pre-exponential factor, activation energy, and gas constant, respectively. The fitting parameters A_0 and E_a are summarized in Table 4.2. The characteristic capacitances (C_1 and C_2) are obtained using parameters R and CPE [31], which are also shown in Table 4.2. The characteristic capacitances for R_1//CPE$_1$ and R_2//CPE$_2$ are in the order of 10^{-11} F cm^{-2} and 10^{-8} F cm^{-2}, respectively, indicating that R_1 and R_2 correspond to bulk and grain boundary resistance, respectively [32, 33].

The parameters A_0 and E_a for the different conduction directions (axial and in-plain) are compared in Table 4.2. Bulk resistance (R_1) shows similar values for both directions, showing that the local distortion does not influence the ionic conduction. This result is interesting because it is expected that the strain of the crystalline framework would change the ionic conductivity by modulating local structure. For example, the expansion of the crystal lattice would have positive and negative effects: for larger crystal lattice, bottleneck of the ion migration is wider and jump distance is longer, which might facilitate higher ionic conductivity. On the other hand, with expanding the lattice, the lithium-ion sites become larger. If the space of the sites is too large for lithium ions, lithium ions would be localized on a certain oxygen ion. The localization of Li would stabilize Li at a site, i.e., decrease the site energy, which would increase the activation energy and reduce the diffusivity of lithium ions. As for the influence of the stress/strain on the ionic conductivity, strain is anisotropic, and further, the crystalline particles are randomly orientated. Thus, both positive and negative effects are averaged in the bulk ionic conductivity.

While the bulk resistance is independent of stress, grain boundary resistance (R_2) is affected by anisotropic stress not only in the pre-exponential factor (A_0), but also in the activation energy (E_a). Grain boundary resistance exhibits higher values both for A_0 and E_a in the axial direction than the in-plain direction. The influence of the strain on the grain boundary resistance will be discussed in the next section.

4.3.3 *Influence of SPS Pressure*

Figure 4.10a compares residual stress and relative density of LLZT pellets sintered with different SPS pressure. The residual stress is obtained for pellets polished

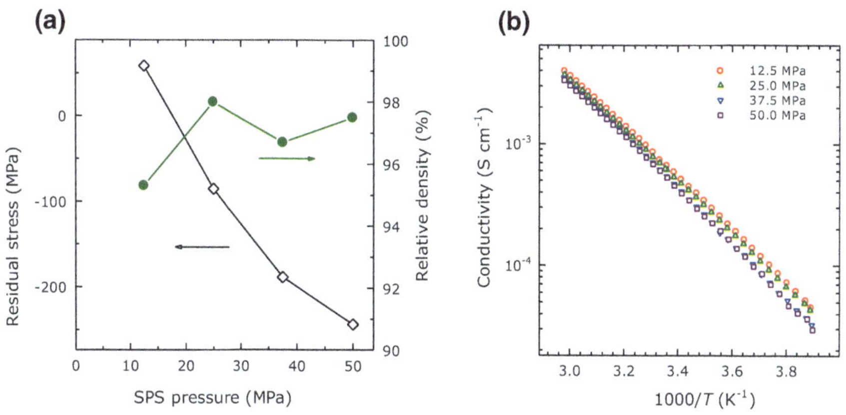

Fig. 4.10 **a** Residual stress and relative density, and **b** temperature dependence of ionic conductivity of LLZT pellets (parallel direction) prepared by SPS at different pressures. Reproduced by permission from [18]. Copyright 2017 Elsevier B.V.

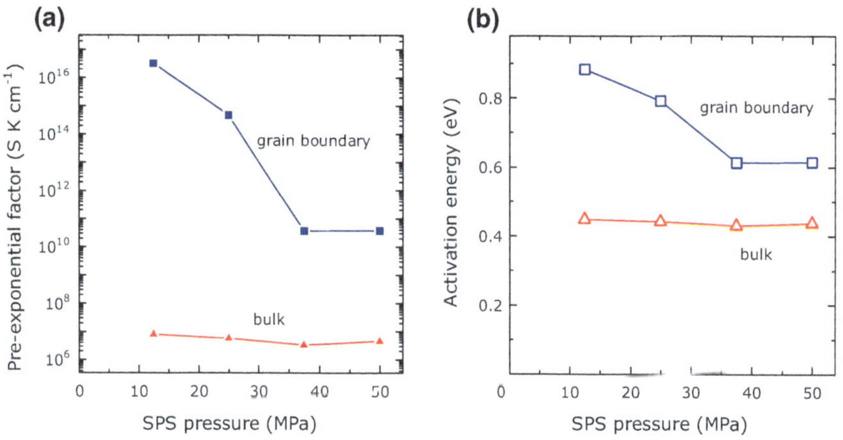

Fig. 4.11 **a** Pre-exponential factors and **b** activation energies of ionic conductivity of LLZT pellets prepared by SPS at different pressures. Reproduced by permission from [18]. Copyright 2017 Elsevier B.V.

using the diamond slurry, and thus, the grinding-induced stress on the subsurface is released. Except for the pellet prepared at 12.5 MPa, the LLZT pellets exhibit a high relative density of 97–98%, and the residual tensile stress (negative value) is increased with the increasing SPS pressure. Figure 4.10b shows the total ionic conductivity in the axial direction decreased with increasing the SPS pressure. As analyzed in the previous section, the Nyquist plots are separated into bulk and grain boundary resistances and analyzed with (4.2) to obtain the parameters

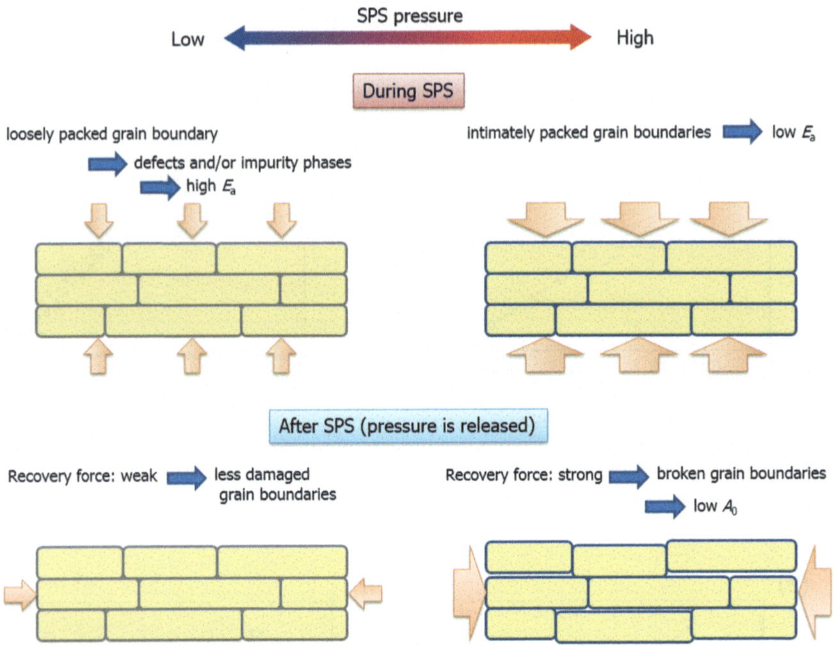

Fig. 4.12 Schematic images of estimated influence of SPS process on grain boundary of LLZT pellets. Reproduced by permission from [18]. Copyright 2017 Elsevier B.V.

A_0 and E_a (Fig. 4.11a, b). While A_0 and E_a for bulk resistance are independent of the SPS pressure, both of these parameters for grain boundary resistance decrease with increasing the SPS pressure.

By using the mechanism of the residual stress that is proposed in Sect. 4.3.1, the influence of the SPS pressure on the grain boundary resistance is explained as follows (Fig. 4.12). When the SPS pressure is low, grain boundaries are loosely sintered, and defects and/or impurity phases remain on the grain boundaries, which results in high activation energy. For high SPS pressure, grain boundary is intimately packed with fewer defects and impurities, showing low activation energy. When the SPS pressure is released, the graphite die causes radial recovery force to the pellets. The recovery force of the graphite die results in the compressive radial stress in the pellets, which causes fracture and microcracks on grain boundaries along the radial direction. The radial microcracks increase grain boundary resistance in the axial direction. Lithium ions migrate through grain boundaries without cracks (low A_0) with intimately packed interface (low E_a).

Figure 4.13 compares XRD profiles of the pellets before and after annealing at 700 °C for 2 h in Ar flow. After the annealing, the diffraction peak shifted to a higher angle and the shape is more symmetric. Then, the profile is similar to the LLZT powder (no residual stress), which suggests the residual stress is released by the annealing process. The influence of the annealing on the parameters A_0 and E_a

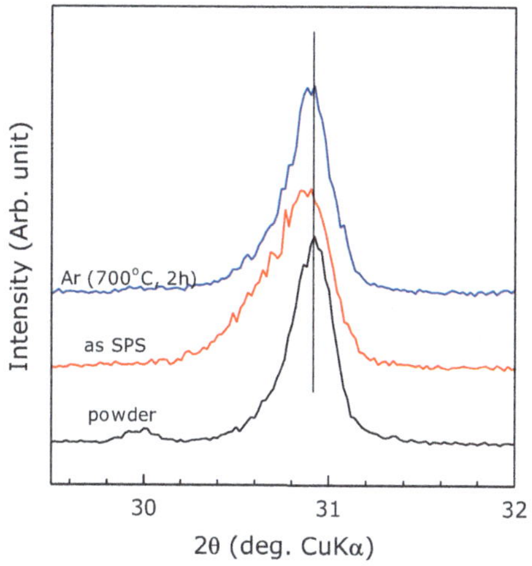

Fig. 4.13 Influence of annealing on the XRD profile of an LLZT pellet sintered by SPS by annealing. Reproduced by permission from [18]. Copyright 2017 Elsevier B.V.

Table 4.3 Effect of annealing on residual stress and ionic conductivity of SPS-prepared LLZT pellet (25 MPa). Numbers in parentheses for residual stress are standard deviations

	Residual stress (MPa)	Total conductivity at 25 °C ($S\ cm^{-1}$)	Bulk		Grain boundary	
			A_0 ($S\ K\ cm^{-1}$)	E_a (eV)	A_0 ($S\ K\ cm^{-1}$)	E_a (eV)
Before anneal	−75(34)	7.26×10^{-4}	1.4×10^7	0.46	3.8×10^{10}	0.58
After anneal	7(18)	7.85×10^{-4}	1.4×10^7	0.46	4.7×10^{11}	0.62

of bulk and grain boundary resistance are compared in Table 4.3. The annealing caused no change in the bulk, while grain boundary resistance is reduced owing to the improved contact, by releasing the residual stress.

4.3.4 *Influence of Stress on Performance of Solid-State Electrochemical Devices*

In this section, the discussions are extended to the application of the solid electrolytes in the ASSBs. It is revealed that the mechanical stress results in little change in the bulk resistance, while the grain boundary resistance is deteriorated. In the ASSBs, active materials change their volume during charging/discharging processes, which causes stresses inside electrode composites and at the interface between electrode layers and solid electrolyte layers. Although the bulk resistance would not change during the charging/discharging, microcracks would be formed on brittle grain boundaries and interfaces between active materials and solid electrolytes, which would increase grain boundary resistance and charge transfer resistance, respectively.

In bulk-type ASSBs consisting of composite electrodes of active materials and solid electrolytes, grain boundary resistance is dominant in the composite electrodes. This is because of the difficulty in co-sintering of active materials and solid electrolytes without inter-diffusion of elements in each phase [34]. At low process temperature, grain boundaries and interfaces are less sintered, and thus, the resistances across the grain boundaries and the interfaces are larger in comparison with homogeneous systems. In addition to originally large grain boundary/interface resistances, the mechanical stress further increases these resistances, which would cause poor cycle-ability of ASSBs.

4.4 Conclusion

It is clearly demonstrated that the distortion of XRD profiles of SPS-prepared LLZT pellets was caused by the residual stress, which was induced by the grinding process and SPS process. Further analysis revealed that the strain increases grain boundary resistance, while bulk resistance remains unchanged. Therefore, at interfaces of active materials and solid electrolytes in ASSBs, strain at the interfaces owing to volume changes of active materials would not change ionic conduction in the crystalline. However, much attention should be paid to deterioration of contact of grain boundaries and interfaces in composite electrodes.

Acknowledgments Part of this work was financially supported by the research project "Advanced Low Carbon Technology Research and Development Program for Specially Promoted Research for Innovative Next Generation Batteries" of the Japan Science and Technology Agency (JST-ALCA SPRING), which is gratefully acknowledged. The author is also thankful to Ms. T. Ito (Nagasaki University) for her synthesis and preparation of specimens, and to Dr. R. Bekarevich (National Institute for Materials Science) and Dr. K. Mitsuishi (National Institute for Materials Science) for advanced skills and helpful discussions on the STEM observation and EDS-ED analyses.

References

1. Takada K (2013) Progress and prospective of solid-state lithium batteries. Acta Mater 61:759–770
2. Kato Y, Hori S, Saito T, Suzuki K, Hirayama M, Mitsui A, Yonemura M, Iba H, Kanno R (2016) High-power all-solid-state batteries using sulfide superionic conductors. Nat Energy 4:16030-1–16030-7
3. Sakuda A, Hayashi A, Tatsumisago M (2010) Interfacial observation between $LiCoO_2$ electrode and $Li_2S–P_2S_5$ solid electrolytes of all-solid-state lithium secondary batteries using transmission electron microscopy. Chem Mater 22:949–956
4. Thangadurai V, Kaack H, Weppner W (2003) Novel fast lithium ion conduction in garnet-type $Li_5La_3M_2O_{12}$ (M = Nb, Ta). J Am Ceram Soc 86:437–440
5. Murugan R, Thangadurai V, Weppner W (2007) Fast lithium ion conduction in garnet-type $Li_7La_3Zr_2O_{12}$. Angew Chem Int Ed 46:7778–7781

6. Murugan R, Thangadurai V, Weppner W (2008) Lattice parameter and sintering temperature dependence of bulk and grain-boundary conduction of garnet-like solid Li-electrolytes. J Electrochem Soc 155:A90–A101
7. Shimonishi Y, Toda A, Zhang T, Hirano A, Imanishi N, Yamamoto O, Takeda Y (2011) Synthesis of garnet-type $Li_{7-x}La_3Zr_2O_{12-1/2x}$ and its stability in aqueous solutions. Solid State Ion 183:48–53
8. Ishiguro K, Nemori H, Sunahiro S, Nakata Y, Sudo R, Matsui M, Takeda Y, Yamamoto O, Imanishi N (2014) Ta-doped $Li_7La_3Zr_2O_{12}$ for water-stable lithium electrode of lithium-air batteries. J Electrochem Soc 161:A668–A674
9. Ren Y, Deng H, Chen R, Shen Y, Lin Y, Nan CW (2015) Effects of Li source on microstructure and ionic conductivity of Al-contained $Li_{6.75}La_3Zr_{1.75}Ta_{0.25}O_{12}$ ceramics. J Eur Ceram Soc 35:561–572
10. Kali R, Mukhopadhyay A (2014) Spark plasma sintered/synthesized dense and nanostructured materials for solid-state Li-ion batteries: Overview and perspective. J Power Sources 247:920–931
11. Baek SW, Lee JM, Kim TY, Song MS, Park Y (2014) Garnet related lithium ion conductor processed by spark plasma sintering for all solid state batteries. J Power Sources 249:197–206
12. Zhang Y, Chena F, Tu R, Shen Zhang L (2014) Field assisted sintering of dense Al-substituted cubic phase $Li_7La_3Zr_2O_{12}$ solid electrolytes. J Power Sources 268:960–964
13. Yamada H, Ito T, Hongahally Basappa R (2016) Sintering mechanisms of high-performance garnet-type solid electrolyte densified by spark plasma sintering. Electrochim Acta 222:648–656
14. Hongahally Basappa R, Ito T, Yamada H (2017) Contact between garnet-type solid electrolyte and lithium metal anode: influence on charge transfer resistance and short circuit prevention. J Electrochem Soc 164:A666–A671
15. Hongahally Basappa R, Ito T, Yamada H (2017) Grain boundary modification to suppress lithium penetration through garnet-type solid electrolyte. J Power Sources 363:145–152
16. Botros M, Djenadic R, Clemens O, Möller M, Hahn H (2016) Field assisted sintering of fine-grained $Li_{7-3x}La_3Zr_2Al_xO_{12}$ solid electrolyte and the influence of the microstructure on the electrochemical performance. J Power Sources 309:108–115
17. Sato K, Suzuki K, Narumi R, Yashiro K, Hashida T, Mizusaki J (2011) Ionic conductivity in uniaxial micro strain/stress fields of yttria-stabilized zirconia. Jpn J Appl Phys 50:055803
18. Araki W, Imai Y, Adachi T (2009) Mechanical stress effect on oxygen ion mobility in 8 mol% yttria-stabilized zirconia electrolyte. J Eur Ceram Soc 29:2275–2279
19. Okumura T, Nakatsutsumi T, Ina T, Orikasa Y, Arai H, Fukutsuka T, Iriyama Y, Uruga T, Tanida H, Uchimoto Y, Ogumi Z (2011) Depth-resolved X-ray absorption spectroscopic study on nanoscale observation of the electrode–solid electrolyte interface for all solid state lithium ion batteries. J Mater Chem 21:10051–10060
20. Yamada H, Ito T, Hongahally Basappa R, Bekarevich R, Mitsuishi K (2017) Influence of strain on local structure and lithium ionic conduction in garnet-type solid electrolyte. J Power Sources 368:97–106
21. Tanaka K (1993) X-ray stress measurement of alumina/zirconia composites. In: Tanaka K, Kodama S, Goto T (ed) X-ray diffraction studies on the deformation and fracture of solids, North Holland, p 1–27
22. Murata K, Mizutani K, Tanaka Y (1992) An approach to residual stress in the ground layer of ceramics from material removal by ground. J Soc Mat Sci Jpn 41:624–630
23. Izumi F, Momma K (2007) Three-dimensional visualization in powder diffraction. Solid State Phenom 130:15–20
24. Stephens PW (1999) Phenomenological model of anisotropic peak broadening in powder diffraction. J Appl Cryst 32:281–289
25. Ni JE, Case ED, Sakamoto JS, Rangasamy E, Wolfenstine JB (2012) Room temperature elastic moduli and Vickers hardness of hot-pressed LLZO cubic garnet. J Mater Sci 47:7978–7985
26. Yu S, Schmidt RD, Garcia-Mendez R, Herbert E, Dudney NJ, Wolfenstine JB, Sakamoto J, Siegel DJ (2016) Elastic properties of the solid electrolyte $Li_7La_3Zr_2O_{12}$ (LLZO). Chem Mater 28:197–206

27. Angerer P, Artner W, Neubauer E, Yu LG, Khor KA (2009) Residual stress of ruthenium powder samples compacted by spark-plasma-sintering (SPS) determined by X-ray diffraction. Int J Refract Met Hard Mater 27:105–110
28. Benabdallah F, Elissalde C, Chung Seu UC, Michau D, Poulon-Quintin A, Gayot M, Garreta P, Khemakhem H, Maglione M (2015) Structure–microstructure–property relationships in lead-free BCTZ piezoceramics processed by conventional sintering and spark plasma sintering. J Eur Ceram Soc 35:4153–4161
29. Wei S, Zhang ZH, Shen XB, Wang FC, Lee SK (2012) Numerical simulation for residual stresses of the spark plasma sintered Ti–TiB composites. J Comp Theor Nanoscience 9:1180–1184
30. Logéat A, Köhler T, Eisele U, Stiaszny B, Harzer A, Tovar M, Senyshyn A, Ehrenberg H, Kozinsky B (2012) From order to disorder: the structure of lithium-conducting garnets $Li_{7-x}La_3Ta_xZr_{2-x}O_{12}$ ($x = 0$–2). Solid State Ion 206:33–38
31. Fleig J (2002) The grain boundary impedance of random microstructures: numerical simulations and implications for the analysis of experimental data. Solid State Ion 150:181–193
32. Shirpour M, Merkle R, Lin CT, Maier J (2012) Nonlinear electrical grain boundary properties in proton conducting Y–$BaZrO_3$ supporting the space charge depletion model. Phys Chem Chem Phys 14:730–740
33. Adams TB, Sinclair DC, West AR (2002) Giant barrier layer capacitance effects in $CaCu_3Ti_4O_{12}$ ceramics. Adv Mater 14:1321–1323
34. Yamada H, Tsunoe D, Shiraishi S, Isomichi G (2015) Reduced grain boundary resistance by surface modification. J Phys Chem C 119:5412–5419

Chapter 5
Sintering Additives for Garnet-Type Electrolytes

Nataly C. Rosero-Navarro and Kiyoharu Tadanaga

Abstract Producing densified garnet-type solid electrolytes by lowering sintering temperature is an important target, which can prevent not only the lithium loss (controlling chemical stoichiometry) but also make it more compatible with cathode electrode materials. In this chapter, the use of sintering additives for enhancing the densification and microstructure of high conductive garnet-type solid electrolytes at low temperatures of $\leq$900 °C is reviewed. Sintering additives can modify the grain and grain boundary, both contributing to the optimization of the chemical and electrochemical properties of garnet-type solid electrolytes.

Keywords Garnet-type solid electrolyte · Control chemical stoichiometry · Low sintering temperature · Sintering additives · Glass sintering additives

5.1 Introduction

Clean energy technology is poised to play an important role in overcoming fossil fuel exhaustion and global pollution. Lithium secondary batteries have an exceptional combination of high energy and power density, making them the technology of choice for portable electronics, power tools, and hybrid/full electric vehicles.

Currently, lithium secondary batteries contain a graphite anode, a cathode formed by a lithium-transition metal oxide (e.g., $LiCoO_2$) and electrolytes consisting of a solution of a lithium salt and organic solvents, imbedded in a separator felt. In this configuration, the chemical and electrochemical stability, between liquid electrolyte and electrodes, has been the limiting fact of the battery performance in terms of maximus power and energy density.

Solid electrolytes (SEs) are considered a potential alternative to substitute the liquid electrolytes because there is no risk of leakage. The garnet-type (e.g.,

N. C. Rosero-Navarro (✉) · K. Tadanaga
Division of Applied Chemistry, Faculty of Engineering, Hokkaido University, Sapporo 060-8628, Japan
e-mail: rosero@eng.hokudai.ac.jp

K. Tadanaga
e-mail: tadanaga@eng.hokudai.ac.jp

R. Murugan and W. Weppner (eds.), *Solid Electrolytes for Advanced Applications*,
https://doi.org/10.1007/978-3-030-31581-8_5

$Li_7La_3Zr_2O_{12}$, LLZ) solid electrolytes have received attention because of high ionic conductivity of ~10^{-3}–10^{-4} S/cm, high chemical stability against Li metal and wide potential windows with electrochemical decomposition voltage of ~6 V versus elemental Li anode [1, 2]. High chemical stability at low and high temperatures is also expected. In addition, the garnet-type solid electrolytes present an advantage in that they are easy to handle and can be kept nonreactive under an ambient atmosphere. LLZ solid electrolyte can be present in two phases, tetragonal or cubic phase. The cubic phase obtained at 1230 °C led to higher ionic conductivity [1]. The control of the chemical stoichiometry of garnet material at high temperature, used to stabilize the cubic phase and promote adequate ceramic densification, is challenging. In these conditions, the volatile nature of lithium has been mitigated by the addition of an excess of lithium source to the stoichiometric ratio of around 10–15 wt%. To prevent lithium loss and to control the garnet structure, reduction of the sintering temperature as well as processing periods are desirable.

The formulation of new chemical composition by element doping has resulted in the reduction of processing periods. However, relatively high sintering temperatures above 1100–1200 °C are still required to densify the ceramic material. Some examples include the doping of Ta, Nb, and Te, where the substitution is expected on the Zr site. Co-doping with elements such as Ca or Ba, where the substitution is expected on the La sites has been also prepared. The substitution of these elements leads to the formation of Li vacancies to balance the extra positive charge of the doping elements, contributing to higher lithium mobility.

Recent reviews [2–4] collect extended information of different chemical formulations of garnet-type solid electrolytes developed in the last years, including studies of ionic conductivity mechanisms, electrochemical properties, and chemical stability. As an example, Table 5.1 shows some of these garnet-type solid electrolytes modified by elements doping. The sintering temperature and processing period are included to display the effect of element doping on sintering behavior.

Sintering temperatures of 1100 °C and shorter processing periods of 6 h have been achieved to prepare garnet solid electrolytes with high ionic conductivity up to 10^{-3} S cm^{-1}. The high ionic conductivity has been ascribed to optimized lithium content in octahedral and tetrahedral sites of the garnet structure created by Li vacancies and modification of the lattice parameter of LLZ structure.

Table 5.1 Conductivity and sintering temperatures and processing periods of element doped LLZ

Compound	σ (S cm^{-1})	T Sintering (°C)
$Li_7La_3Zr_2O_{12}$ (LLZ)	2.4×10^{-4}	1230 (36 h) [1]
$Li_{6.75}La_3Zr_{1.75}Nb_{0.25}O_{12}$	8.0×10^{-4}	1200 (30 h) [5]
$Li_{6.4}La_3Zr_{1.4}Ta_{0.6}O_{12}$	10.0×10^{-4}	1140 (16 h) [6]
$Li_{6.4}La_3Zr_{1.75}Te_{0.25}O_{12}$	10.2×10^{-4}	1100 (15 h) [7]
$Li_{6.5}La_{2.5}Ba_{0.5}ZrTaO_{12}$	2.0×10^{-4}	1100 (6 h) [8]

Further reduction of sintering temperatures around 800–900 °C with suitable densification and high ionic conductivity have been developed by means of microstructural modification of garnet-type solid electrolyte through the incorporation of sintering additives.

5.2 Sintering Additives Background

Figure 5.1 illustrates the role of the sintering additives of ceramic electrolyte. The distribution of sintering additives in the ceramic garnet electrolyte can be divided into two categories: sintering additive (a) exists at grain boundaries or (b) are incorporated into the ceramic grains. In the first case, the sintering additive remains as an intergranular crystalline or glassy phase in the sintered material. In the second case, the sintering additives promote densification at the initial and middle stage of sintering and subsequently are incorporated into ceramic grains.

A typical example of sintering additive used for LLZ is the incorporation of γ-Al_2O_3. It can be considered as a sintering additive, modifying both grain and grain boundary. It was found that Al substitution per formula unit of LLZ can stabilize the cubic phase and creates more Li-ion vacancies, improving the Li-ion conductivity. Quantities of aluminum of 0.2–0.24 mol of Al have been proved to be sufficient in stabilizing the cubic phase [10].

The role of Al as a sintering aid at high sintering temperatures (1200 °C) involves the precipitation of conductor compounds at the grain boundary. During sintering, the formation of a liquid phase may initially occur by a reaction between Li_2O and Al_2O_3 at 1055 °C (corresponding to the eutectic liquid of the Li_2O–Al_2O_3 system for 23 mol% Al_2O_3) [11–13]. Dissolution and reprecipitation process occur in the presence of the liquid phase which results in the incorporation of La and Zr in the liquid phase [13]. Simultaneously, the wetting of particle surfaces by the liquid phase results in capillary forces that facilitate particle rearrangement and enhanced

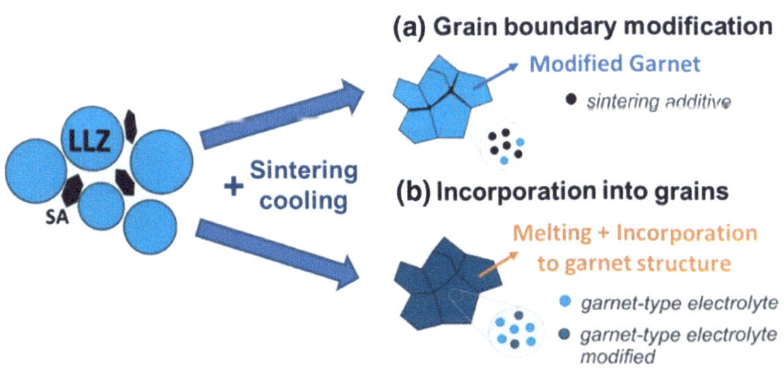

Fig. 5.1 Distribution of sintering additives (SA) after sintering [9]

diffusion that accelerates densification. After sintering, some Al is expected to remain along grain boundaries in the frozen liquid [13].

Kumazaki et al. [14] have confirmed the presence of Li–Al–Si–O glass phase with nano-crystalline $LiAlSiO_4$ at grain boundaries in the LLZ garnet material obtained at 1230 °C (36 h). Al and Si were incorporated into the LLZ garnet structure as contamination from ball milling and crucibles used to prepare the solid electrolyte. Figure 5.2 shows a TEM image at a triple point grain boundary of Al–Si-doped LLZ, STEM image and energy dispersive X-ray analysis (EDX) of Al, Si, Zr, La, and O are also included. The presence of Al in the grains confirms a partial substitution of Al with the constituent elements of LLZ, it is the probable substitution of Al^{3+}

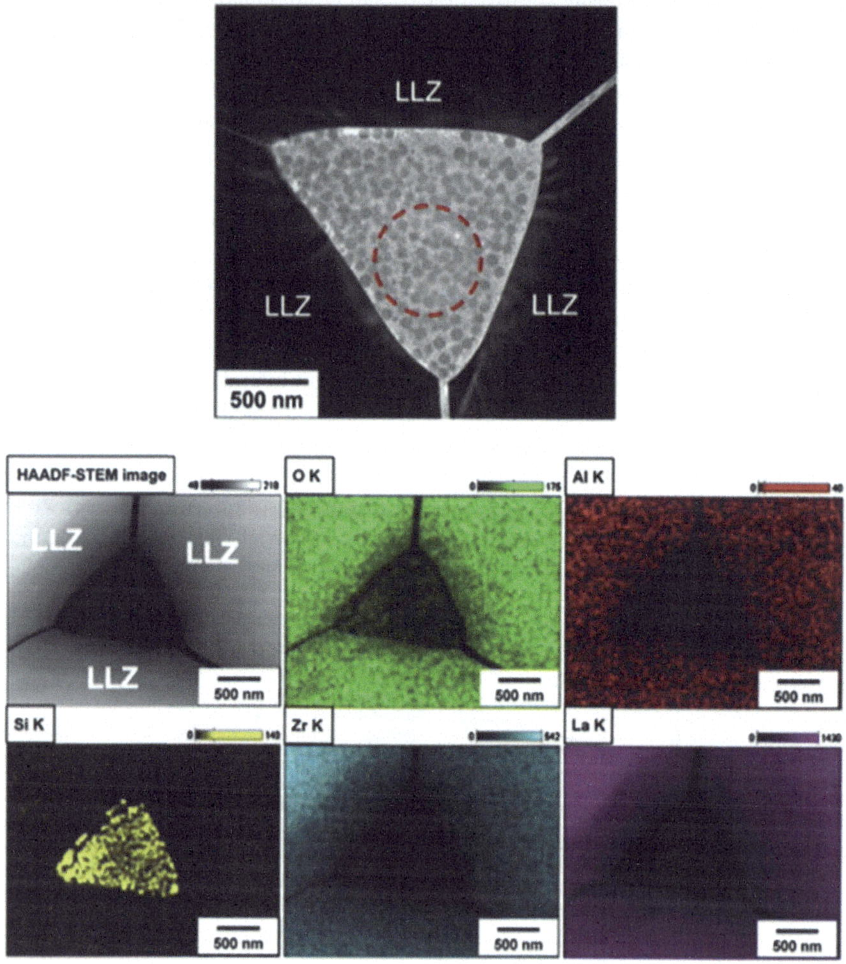

Fig. 5.2 TEM image at a triple point grain boundary of Al–Si doped LLZ. STEM image and EDX analysis of Al, Si, Zr, La, and O [14]

with Li^+. Then, the Li^+ extracted from the bulk LLZ is able to react and form the glassy-ceramics compound at the grain boundary, confirmed by the presence of Si and Al in Fig. 5.2 [14].

The ionic conductivity at 25 °C achieves 6.8×10^{-4} S/cm, which is 3 times higher than that LLZ without this structural modification (2.1×10^{-4} S/cm) [1]. The improved conductivity has been attributed to the low resistance at the grain boundary because of its lithium enrichment compounds and the high lithium mobility promoted by Al substitution in LLZ structure.

Based on this statement, the question arises: What kind of compounds are suitable to produce similar structural modification of LLZ at lower sintering temperatures? The following sections will address this question and the effect of sintering additives will be described, emphasizing the microstructure and ionic conductivity of garnet electrolyte composite.

5.3 Sintering Additives Modifying Triple Point Grain Boundary and Grain Boundary

Pioneer results obtained by Tatsumisago group [15, 16] have allowed to sinter garnet-type solid electrolyte at significantly lower temperatures. They have used Li_3BO_3 as a sintering additive of Al-doped LLZ (cubic phase). The Al-doped LLZ pellet sintered with Li_3BO_3 at low temperature of 900 °C (30 h) achieved a high relative density of 92%, which is higher than those sintered without Li_3BO_3 of 58%. The Al-doped LLZ and Li_3BO_3 composite present a high ionic conductivity of 1×10^{-4} S/cm and activation energy of 0.36 eV, which is comparable to that of Al-doped LLZ pellet sintered at 1200 °C.

Li_3BO_3 forms a liquid phase at ~750 °C, a temperature below the sintering temperature of Al-doped LLZ solid electrolyte. The sintering of precursor powders of Al-doped LLZ and lithium borate was carried out in two steps [15, 16]. First, at ~700 °C: to promote the formation of the liquid phase of lithium borate which then acts as an accelerator for inter-reaction of grains at grain boundaries and the grain growth. Second, at 900 °C: to achieve an adequate densification where the lithium loss during heat-treatment is assumed to be suppressed by the presence of lithium borate.

Figure 5.3 shows TEM and electron diffraction images of the Al-doped LLZ pellet sintered with Li_3BO_3 at 900 °C. The grains correspond to the cubic phase of Al-doped LLZ, while the amorphous phase observed in the triple point grain boundary is assumed to be lithium borate glass (from the *d*-value of the halo in the electron diffraction image). Li_3BO_3 amorphous phase is also expected to form a very thin layer at the grain boundary between LLZ particles. The high total lithium-ion conductivity of the composite is assumed to be produced by the reduction of the grain boundary resistance of the glassy phases of lithium borate [16].

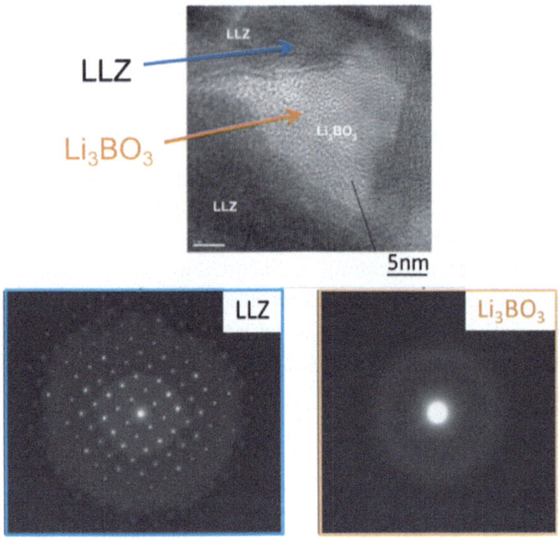

Fig. 5.3 TEM image and electron diffraction images of the sintered Al-doped LLZ pellet with Li_3BO_3 [16]

5.4 Sintering Additives of LLZ Garnet Solid Electrolytes Modified by Element Doping

The use of Li_3BO_3 and other sintering additives, such as $LiBO_2$, Li_3PO_4, and Li_4SiO_4, have been extended to sinter other LLZ garnet solid electrolytes modified by element doping. The doping of super valent cation (e.g., Al^{+3}, Nb^{+5}, Ta^{+5}, etc.) have been proven to improve the Li mobility (Table 5.1), thus it is expected that the incorporation of a sintering additive may produce composites with higher ionic conductivities.

In addition, garnet solid electrolyte powder with cubic phase and small particle size are preferred because these may facilitate the role of the sintering additives. Liquid phase processes such as sol–gel process or Pechini process (polymerized complex method) are alternative synthesis routes to prepare garnet-type solid electrolytes. In these processes, liquid precursors can react completely at lower temperatures and shorter times, without the need for intermittent grinding. Fine powders with a more uniform size distribution can be obtained after drying for the removal of solvents or organic phases [13, 17].

Figure 5.4 displays XRD patterns of $Li_7La_3(Zr_{2-x}Nb_x)O_{12}$ (named as Nb_x–LLZ, $x = 0.25$, 0.5 and 1.0) garnet-type solid electrolyte obtained by sol–gel process calcinated at 700 °C [18]. Cross-section SEM (scanning electron microscope) images of pellets sintered at 900 °C (10 h) are also included.

After heating at 700 °C, Nb0.25–LLZ showed the peaks which correspond to the crystallographic phase of $La_2Zr_2O_7$ pyrochlore, while Nb0.5–LLZ and Nb1–LLZ showed the formation of the cubic garnet phase. The presence of $La_2Zr_2O_7$ phase was identified, but a smaller peak was observed in Nb1–LLZ. High Nb doping LLZ obtained by sol–gel lead to a stabilized cubic phase at low temperature, of 600–700 °C

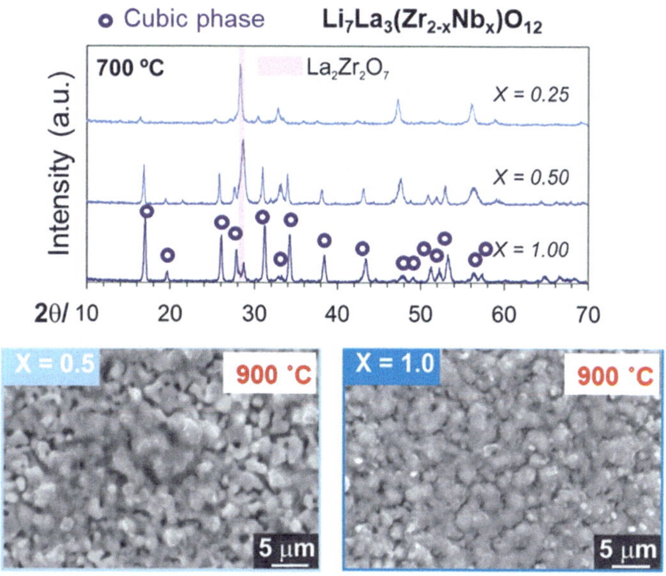

Fig. 5.4 XRD patterns of $Li_7La_3(Zr_{2-x}Nb_x)O_{12}$ with $x = 0.25$ and $x = 1.0$ calcined at 700 °C. Cross-section SEM images of pellets sintered at 900 °C [18]

for Nb1–LLZ [18]. Similar results have been obtained with Al-doped LLZ obtained by sol–gel process, where cubic phase has been stabilized at 600 °C [15, 19].

A porous structure was observed after sintering at 900 °C (10 h) of a cross-section pellet (Fig. 5.4), indicating that the materials are insufficiently sintered and a relative density of only ~55 ± 3% was obtained [18].

As was mentioned above, the densification of ceramic at low temperature is challenging, making it necessary to use sintering additives to reach sufficient densification of these low temperature cubic phase materials for practical applications.

Figure 5.5 shows XRD patterns of $Li_7La_3(Zr_{2-x}Nb_x)O_{12}$ sintered with 5 wt% $LiBO_2$ at 900 °C (10 h) and cross-section SEM images of Nb1–LLZ pellet sintered with 5 wt% $LiBO_2$ at 900 °C (10 h). $LiBO_2$ is expected to form a liquid phase of around 850 °C, below sintering temperature. XRD pattern showed the formation of the cubic phase for all compositions, even in the composition doped with low Nb content (Nb0.25–LLZ). An impurity phase corresponding to $La_2Zr_2O_7$ was also observed.

It is believed, that the reaction between garnet material and $LiBO_2$ is favored for the excess of Li in the garnet material and the bridging oxygen atoms present in the $LiBO_2$ structure. B–O–B bonds in a covalent network of $LiBO_2$ compound can be attacked by Li_2O during the thermal treatment, extracting Li ions from Nb–LLZ resulting in the formation of $La_2Zr_2O_7$ [15, 18].

Although, $La_2Zr_2O_7$ phase was present in Nb1–LLZ–$LiBO_2$ composite, a high relative density of ~70–80% was obtained, increasing up to 93% when the content of $LiBO_2$ was enlarged to 7.5 wt% [18]. A cross-section SEM image of Nb1–LLZ

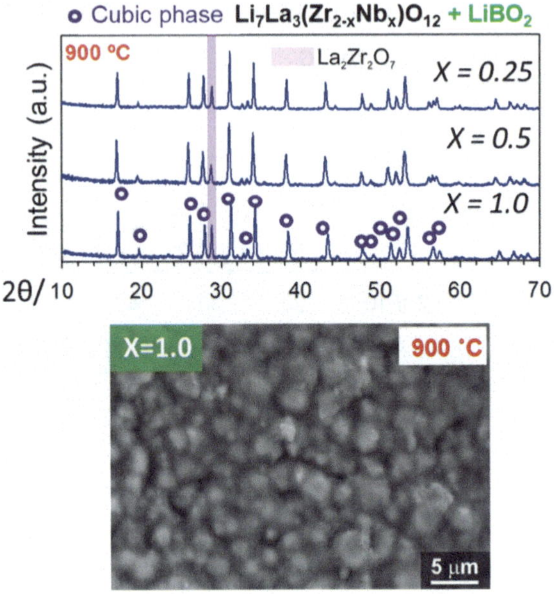

Fig. 5.5 XRD patterns of $Li_7La_3(Zr_{2-x}Nb_x)O_{12}$ sintered with 5 wt% $LiBO_2$ at 900 °C. Cross-section SEM image of Nb1–LLZ pellet sintered with 5 wt% $LiBO_2$ at 900 °C [18]

pellet sintered with 5 wt% $LiBO_2$ shows interconnected grains with less presence of porous due to the presence of $LiBO_2$ [18].

Figure 5.6 displays XRD patterns of Nb1–LLZ sintered with different content of Li_3BO_3 at 900 °C (10 h). Polished cross-section SEM images of Nb1–LLZ pellet sintered with 6.5 wt% Li_3BO_3 at 900 °C (10 h) were included. The XRD patterns showed the formation of the cubic phase. In contrast to $LiBO_2$ (Fig. 5.5), secondary phases were not detected confirming that Li_3BO_3 did not react with the Nb–LLZ solid electrolyte and it is present as an amorphous phase.

Polished cross-section SEM images (Fig. 5.6) of Nb1–LLZ–Li_3BO_3 composite shows a good linkage of Nb1–LLZ particles (bright areas), darker areas are assumed to be a lithium borate region. As the XRD pattern does not show peaks corresponding to the sintering additive, Li_3BO_3 was presumed to remain at the grain boundary as an amorphous phase [15, 18]. The relative density of 86% was obtained from Nb1–LLZ sintered with 6.5 wt% Li_3BO_3 at 900 °C, confirming good densification produced by the addition of sintering additive.

Figure 5.7 shows the Arrhenius plot of composites prepared with Nb1–LLZ and sintering additives, 7.5 wt% $LiBO_2$ (Nb1–LLZ–$LiBO_2$) and 6.5 wt% Li_3BO_3 (Nb1–LLZ–Li_3BO_3), sintered at 900 °C (10 h). Ionic conductivity at room temperature of Nb1–LLZ–$LiBO_2$ and Nb1–LLZ–Li_3BO_3 composites achieved 3×10^{-5} S cm^{-1} and 7×10^{-5} S cm^{-1}, respectively [18]. The activation energy of 0.3 eV and 0.35 eV for Nb1–LLZ–$LiBO_2$ and Nb1–LLZ–Li_3BO_3 composites were determined, respectively.

Although the relative density of Nb1–LLZ–$LiBO_2$ composite (93%) was higher than that of Nb1–LLZ–Li_3BO_3 composite (86%), presence of the non-conductor

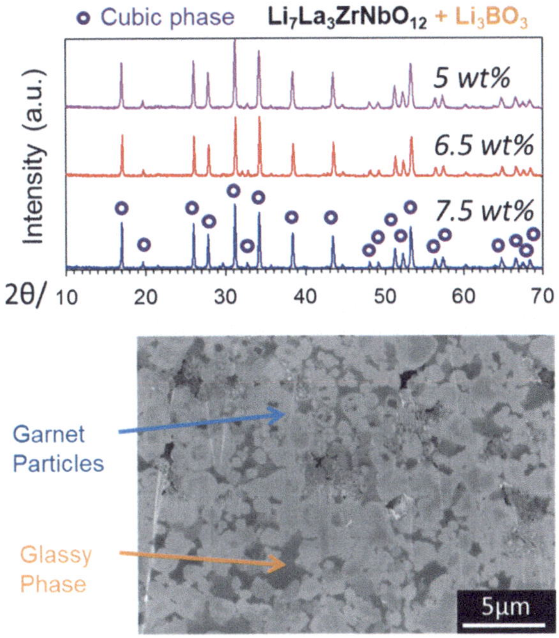

Fig. 5.6 XRD patterns of Nb1–LLZ sintered with Li_3BO_3 at 900 °C (10 h). Polished cross-section SEM images of Nb1–LLZ pellet sintered with 6.5 wt% Li_3BO_3 at 900 °C [18]

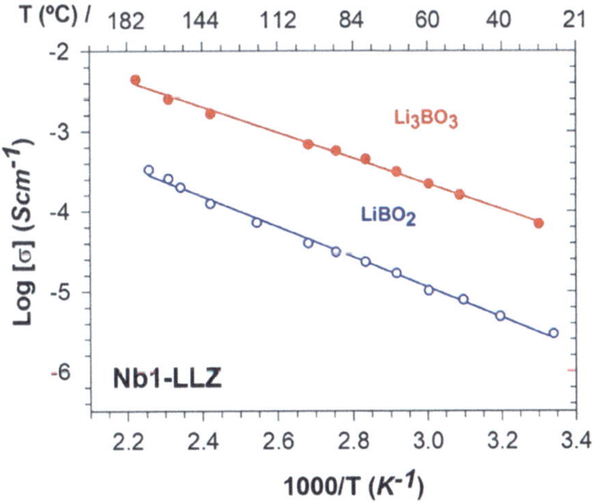

Fig. 5.7 Arrhenius plot of Nb1–LLZ–$LiBO_2$ and Nb1–LLZ–Li_3BO_3 composites [18]

$La_2Zr_2O_7$ phase was observed in the Nb1–LLZ sintered with $LiBO_2$ (Fig. 5.5) is assumed to produce a lower ionic conductivity. In addition, conductivity of Li_3BO_3 glass is higher than that of $LiBO_2$, since the conductivities of the lithium-ion conducting glasses are known to increase with an increase in the concentration of Li^+ ion [20].

The content of 6.5 wt% Li_3BO_3 was found to be sufficient to increase the relative density of ceramic composite and achieve high ionic conductivity. A larger content of sintering additives leads to a drop in the ionic conductivity [18]. This must be attributed to the relatively larger contribution of Li_3BO_3 to the total conductivity of the composite and ionic conductivity of Li_3BO_3 (2×10^{-6} S cm^{-1}) [21] is lower than that garnet material.

The activation energy of both composites, Nb1–LLZ–Li_3BO_3 and Nb1–LLZ–$LiBO_2$, are in good agreement with the values reported for garnet-type solid electrolytes sintered at high temperatures (0.3 eV) [1], confirming an adequate charge transfer of Li ions between the grains and the grain boundary.

The ionic conductivity of Nb1–LLZ–Li_3BO_3 composite was higher, around three times, than the composite of Al-doped LLZ sintered with Li_3BO_3 in the same conditions of 900 °C for 10 h [15]. Thus, it was confirmed that the element doping in garnet-type solid electrolytes played an important role in increasing the conductivity at lower temperatures using sintering additives.

Other sintering additives such as Li_3PO_4 and Li_4SiO_4 with higher melting points of 873 °C and 1255 °C, respectively, have been used to sinter Ta-doped LLZ [22] and Al-doped LLZ [23] garnet electrolytes. The composites prepared with Al-doped LLZ sintered with 1 wt% Li_4SiO_4 and 1 wt% Li_3PO_4 at 1200 °C (12 h) achieved an ionic conductivity of 6×10^{-4} S cm^{-1} (0.34 eV) and 3.5×10^{-4} S cm^{-1} (0.36) at 33 °C, respectively. The slight reduction of ionic conductivity using Li_3PO_4 was related to the precipitation of $AlPO_4$ phase during sintering [23]. In the case of Li_4SiO_4 since it does not melt at the sintering temperature, it is expected to be segregated in the grain boundary, facilitating the ionic conductivity. Although, high sintering temperatures of 1200 °C was used to prepare dense ceramic composites, a considerable short processing period of 12 h confirmed the role of the sintering additives.

Co-doped LLZ and incorporation of a second sintering additive, such as Al_2O_3, has allowed higher ionic conductivity of garnet solid electrolyte to be obtained when Li_3BO_3 is used as sintering additive, even at lower temperatures.

Ohta et al. [24] have used Al_2O_3 and Li_3BO_3 as sintering additives of Ca–Nb-doped LLZ ($Li_{6.8}La_{2.95}Ca_{0.05}Zr_{1.75}Nb_{0.25}O_{12}$). The relative density of 90% and high ionic conductivity of 3.6×10^{-4} S cm^{-1} were obtained after sintering at 790 °C (40 h). Figure 5.8 shows a cross-sectional electron probe microanalysis (EPMA) mapping images of Ca–Nb-doped LLZ with co-addition of Al_2O_3 and Li_3BO_3 sintered at 790 °C (40 h).

The results indicate that the Al element, initially located out of the CaNb–LLZ crystals, diffuses into the LLZ-based grains while the Ca element, initially located in the CaNb–LLZ crystals diffuses into the Li_3BO_3-based triple points during sintering [24]. Thus, the grain boundary is formed by Li–Ca–B–O phase while the grain was enriched with Al.

The relatively high density of the CaNb–LLZ sintered with Al_2O_3 and Li_3BO_3 has been attributed to the simultaneous interdiffusion between the garnet solid electrolyte and the sintering additives, in addition to the formation of a liquid phase by Li_3BO_3 [24]. The high ionic conductivity of the composite suggests that the conduction path

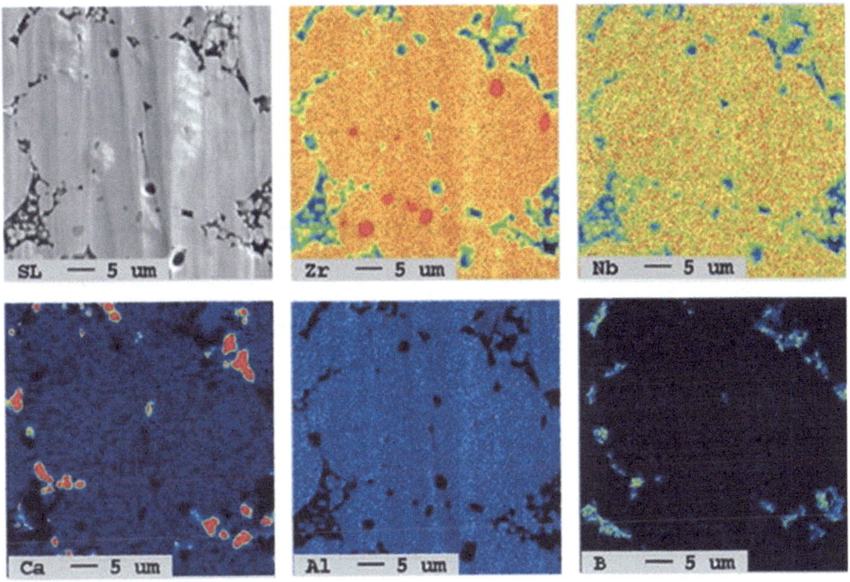

Fig. 5.8 Cross-sectional EPMA mapping images of Ca–Nb-doped LLZ with co-addition of Al_2O_3 and Li_3BO_3 sintered at 790 °C [24]

of Li ions between grains is sufficiently formed, despite the low sintering temperature of 790 °C.

The effective reduction of sintering temperature to 800 °C allowed the preparation of a bulk-type all-solid-state battery using co-sintering process. Chemical stability of solid electrolyte composite with $LiCoO_2$ during co-sintering was confirmed by SEM analysis, diffusion was not observed between cathode/electrolyte layers. Cathode composite using $LiCoO_2$, composite solid electrolyte and Li_3BO_3 was prepared in the ratio of 42:35:23 (wt%). A discharge capacity of 78 mA h g^{-1} was obtained, which corresponded to about 58% of the theoretical capacity [24]. The relatively low discharge capacity was due to the high resistance at the solid cathode/electrolyte interface, effecting an insufficient lithium-ion conduction path through the solid interface.

High densification and high ionic conductivity have been verified in other garnet-type solid electrolytes composites by the interdiffusion process combined with the formation of a liquid phase of sintering additives at low temperatures. Al_2O_3 and Li_3BO_3 were used as sintering additives of Ta–Ca doped LLZ ($Li_{7-x}La_{2.95}Ca_{0.05}ZrTaO_{12}$) at 900 °C (10 h) [25]. The effect of Al_2O_3 on composite properties has been studied.

Figure 5.9 shows XRD patterns of TaCa–LLZ sintered with 6.5 wt% Li_3BO_3 and different content of Al_2O_3 at 900 °C. XRD patterns show that the addition of Li_3BO_3 and lower concentration of Al_2O_3 (<0.22 mol) leads to the formation of the cubic phase. The absence of the secondary phase confirms that both sintering additives,

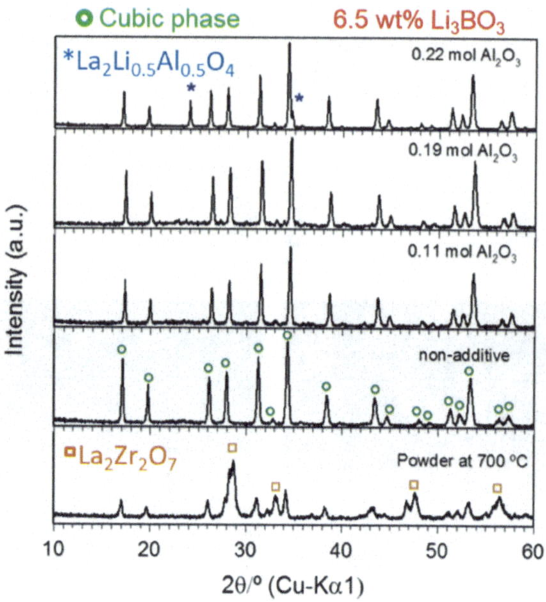

Fig. 5.9 XRD patterns of TaCa–LLZ sintered with 6.5 wt% Li_3BO_3 and different content of Al_2O_3 at 900 °C (10 h) [25]

Li_3BO_3 and Al_2O_3, form a glassy phase in the grain boundary or have been partially introduced to the garnet crystal. A content of 0.22 mol of Al_2O_3 leads to the formation of $La_2Li_{0.5}Al_{0.5}O_4$ phase. A content of 0.22 mol of Al_2O_3 corresponds to 0.44 mol of Al, which is an excess of more than double of the amount needed to stabilize the cubic phase of LLZ (0.204 mol of Al) [10]. However, a slight reduction of Al_2O_3 content to 0.23 mol of Al_2O_3 (0.42 mol of Al) lead the formation of the cubic phase without secondary phases [25]. Since no additional phases were observed, the high content of aluminum is assumed to be incorporated into the grains through an interdiffusion process and remains also at the grain boundary or triple point grain boundaries. Calcium is also expected to be diffused to the grain boundary or triple point grain boundary.

A high relative density of 95 and 94% were obtained from TaCa–LLZ pellets sintered with 6.5 wt% of Li_3BO_3 and 0.21 and 0.19 mol of Al_2O_3, respectively. Good interconnection between TaCa–LLZ grains distribution of the glassy phase formed by Li_3BO_3 and Al_2O_3 were confirmed by SEM-EDS analysis [25].

Figure 5.10 displays the Arrhenius plot of TaCa–LLZ pellet sintered with 6.5 wt% of Li_3BO_3 and different content of Al_2O_3. The total (bulk and grain boundary) ionic conductivity at 32 °C is estimated to be 1.0×10^{-4} S cm^{-1} for garnet material sintered with 0.21 mol of Al_2O_3.

Total ionic conductivity is three orders of magnitude higher than the material sintered without additives (1×10^{-7} S cm^{-1}), and it is higher than the Al-doped LLZ or Nb-doped LLZ pellets sintered with Li_3BO_3 at 900 °C (10 h) [15, 18]. The activation energies were estimated to be 0.35 and 0.34 eV for the garnet material sintered with 0.19 and 0.21 mol of Al_2O_3, respectively. These values are in good agreement with

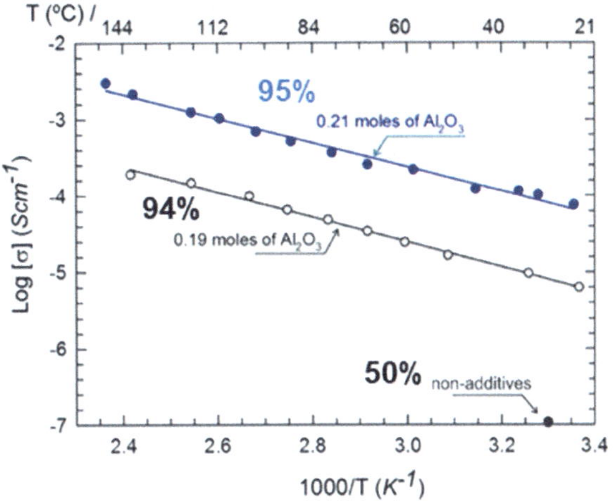

Fig. 5.10 Arrhenius plots for the total (bulk + grain boundary) ionic conductivity of TaCa–LLZ sintered with 6.5 wt% Li_3BO_3 and different concentrations of Al_2O_3 [25]

other Ta-doped LLZ sintered at higher temperatures such as $Li_{6.4}La_3Zr_{1.6}Ta_{0.6}O_{12}$ (1140 °C, 0.35 eV) [6], indicating an appropriate lithium pathway through grains and grain boundaries.

The effect of Al_2O_3 on ionic conductivity of solid electrolyte composites was clearly different, despite their relatively similar density of around 95%. The difference of ionic conductivity can be related to the effect of Al on the chemical composition of garnet crystal. This means that 0.23 mol of Al_2O_3 can be high enough to promote an adequate substitution inside of the grains garnet crystal, enhancing lithium mobility, and forming a stable Li–B–Al–O conductive glassy phase, ensuring also an ionic conductivity between grains.

The results indicate that higher conductivities can be achieved by optimized inter-diffusion of elements between liquid phase formed by sintering additives and garnet solid electrolyte material modified by element doping.

5.5 Suitable Sintering Additives Alternatives

To select suitable sintering additives, the following aspects must be addressed to obtain high ionic conductivity of garnet-type solid electrolytes sintered at low temperatures:

- The sintering additive should have good enough wetting/solubility properties to form an adequate liquid phase at low temperatures
- LLZ doped with strategic elements to promote simultaneous interdiffusion to optimize high ionic conductivity

The glassy phase containing lithium at triple point grain boundaries and grain boundaries is important to ensure lithium conduction in the garnet ceramic when low temperatures are used to sinter the ceramic garnet material.

In this section the effect of two glass systems, LiO_2–B_2O_3–SiO_2–CaO–Al_2O_3 (LBSCA) and BaO–B_2O_3–SiO_2–CaO–Al_2O_3 (BBSCA), as sintering additives of $Li_7La_3ZrNbO_{12}$ (Nb1–LLZ) garnet electrolyte are described [9]. LBSCA glass was previously used to promote sintering at low temperatures (920 °C, 2 h) of LiZnTi ferrites [26]. In lithium borosilicate glass, Li_2O–B_2O_3–SiO_2, the presence of SiO_2 produces an open random structure, decreasing the glass transition temperature and increasing the ionic conductivity, while B_2O_3 decreased the viscosity (increase fluidity) and crystallization tendency of the Li_2O–SiO_2 glass [27, 28]. On the other hand, the Al and Ca act as network formers and charge balancers in the glass. BBSCA glass had also been used as a sintering additive of ceramics at low temperature, and its efficiency had been proved at lower temperatures of 700–850 °C [29, 30].

The glass systems with softening temperature below 700 °C [26, 30] should form a liquid phase, which is able to penetrate the solid–solid interfaces and promotes good rearrangement of the particles leading to improving the densification of ceramics. During the process, grain growth is helped by grain coalescence and grain-reprecipitation from the small grains to the adjacent large grains. Simultaneous interdiffusion of elements between garnet material and additives is expected. The presence of Ba, Ca and Al from the liquid phase of the glasses could be beneficial since they can promote the interdiffusion with the garnet-type oxide resulting in an improvement of ionic conductivity [9].

Figure 5.11 displays XRD patterns of Nb1–LLZ sintered with different content of glass sintering additives at 900 °C (10 h). The formation of the cubic phase of the garnet material is clearly seen in the XRD patterns for LBSCA glass and low content of 4 wt% BBSCA glass, in a good agreement with the above results for LLZ sintered with sintering additives (e.g., Li_3BO_3 and Al_2O_3). Thus, glass sintering additives, as expected, are present as an amorphous phase in the composite.

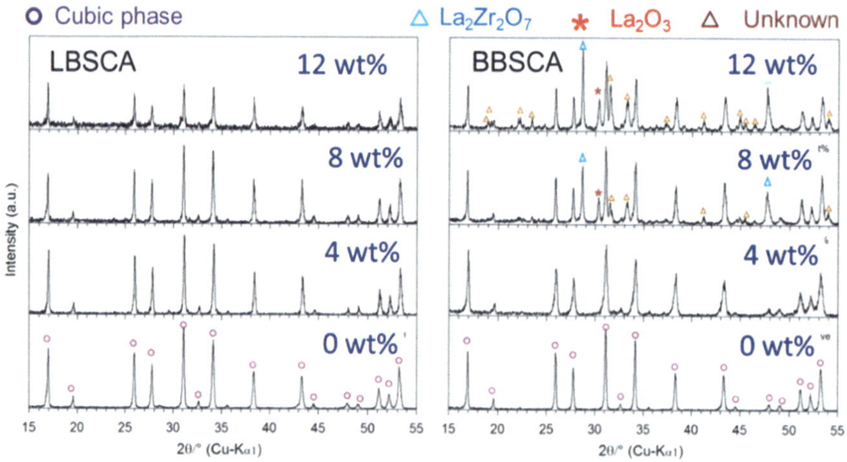

Fig. 5.11 XRD patterns of Nb1–LLZ sintered with different content of glass systems at 900 °C (10 h) [9]

Composites sintered with 8 wt% and 12% BBSCA glass shows the formation of secondary phases, $La_2Zr_2O_7$ and La_2O_3 phases, suggesting unfavorable reactions with the garnet material during sintering.

High relative densities were obtained for Nb1–LLZ sintered with glasses additives up to 92%, producing an enhancement of 35%, respectively, to Nb1–LLZ sintered without any additive. The microstructure of Nb1–LLZ sintered with glasses additives, studied by SEM analysis, showed an intimate contact between Nb1–LLZ grains and glassy phase [9].

The impedance spectra for Nb1–LLZ sintered with different content of LBSCA glass at 900 °C (10 h) is shown in Fig. 5.12. The total resistance was determined from the real component of impedance, Z, in the intersection point between the semicircles and the tail produced by the electrodes (dotted arrows). Total conductivity (bulk and grain boundary) of Nb1–LLZ sintered with 4 wt%, 8 wt% and 12 wt% of LBSCA were 0.78×10^{-4} S cm^{-1}, 0.46×10^{-4} S cm^{-1} and 0.15×10^{-4} S cm^{-1} at room temperature (30 °C), respectively. The contribution of high concentrations of sintering additives resulted in a drop of the total ionic conductivity. The glass phase is known to have a lower conductivity than garnet electrolyte, thus the excess of sintering additive produced a deterioration of the lithium ion transport.

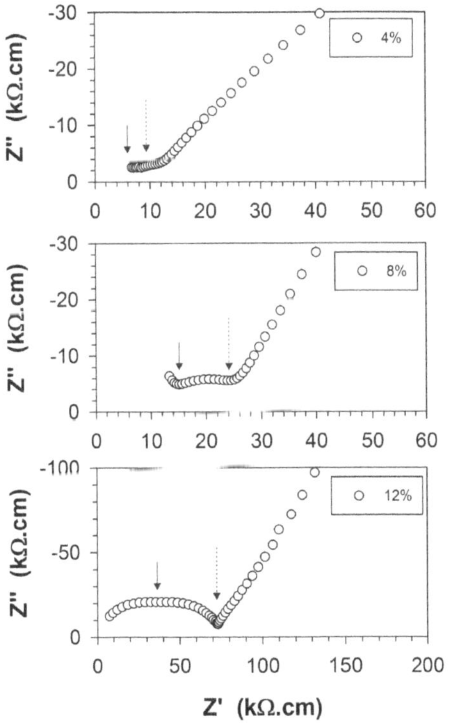

Fig. 5.12 Impedance spectra for Nb1–LLZ sintered with different content of LBSCA glass at 900 °C (10 h) [9]

Usually, the impedance spectra of ceramics are interpreted as the result of a series associations of the bulk resistance of the grains (at high frequency), inter-grains contacts or grain boundary (at intermediate frequency) and the tail corresponding to the polarization of electrodes. However, that separation cannot be applied to Nb1–LLZ sintered with LBSCA glass, where other contributions should be taken in account. The grain resistance involves Nb1–LLZ bulk and glass bulk, while grain boundary involves Nb1–LLZ/Nb1–LLZ grain boundary and Nb1–LLZ/LBSCA glass interface. The impedance spectra (Fig. 5.12) show wider semicircles at high frequencies, overlapping two (or more) contributions, when the content of the sintering additive was increased. Arrows clearly show the separation of these contributions. It can be inferred that the increasing sintering additive content produces higher resistance, which is verified by the larger semicircle at medium frequencies.

A similar tendency was observed for Al–LLZ or Nb1–LLZ sintered with Li_3BO_3 [15, 18], where content of around 6.5 wt% allowed a higher ionic conductivity to be obtained. Lower content of LBSCA glass was used to obtain a higher ionic conductivity, around 4 wt%. The addition of LBSCA glass allows not only lead to the rearrangement of the particles, but also to optimized ionic conductivity of the composite by the interdiffusion process of the constituent elements from garnet and glass additive. The designed glass additive allowed enhanced lithium in the grain structure and facilitated the lithium path through the grains.

5.6 Conclusions

The use of sintering additives plays an important role in reducing sintering temperatures of garnet-type solid electrolytes with high ionic conductivity. The adequate selection and content of sintering additives lead to sufficient densification of the ceramic electrolyte, keeping a high ionic conductivity with a controlled stoichiometry.

Li_3BO_3 is the sintering additive which extends to the sinter garnet-type solid electrolytes as it forms a liquid phase at low temperatures around 750 °C allowing the use of sintering temperature of 800–900 °C. Li_3BO_3 is nonreactive with garnet material during sintering and its ionic conductivity is high enough to form composites with high ionic conductivity. Sintering additives with high melting points such as Li_4SiO_4 and Li_3PO_4 resulted in being effective for producing composites with high density and high ionic conductivity at 1200 °C with a reduced period of sintering of 12 h.

The incorporation of two sintering additives (e.g., $Al_2O_3 + Li_3BO_3$) to sinter co-doped (e.g., Al^{+3}, Nb^{+5}, Ta^{+5}, and Ca^{2+}, etc.) LLZ garnet structure leads to composites with higher ionic conductivities. The liquid phase of sintering additives promotes the densification of the garnet material. During this process, an interdiffusion of the constituent elements from garnet structure and sintering additive occurred. This interdiffusion leads to composites with enhanced lithium mobility into the grains and through the grains.

New and suitable sintering additives have been studied and the properties such as wetting/solubility of liquid phase formed by sintering additive and strategic elements for co-doping LLZ to promote simultaneous interdiffusion have been taken into account.

Acknowledgments Appreciation is shown to the Japan Science & Technology Agency (JST) for their financial support under Japan Society for the Promotion of Science Program (JSPS) (2013–2015) with grant number P1337 and national project "Grants-in-Aid for Scientific Research (KAKENHI)" (2017–2020) with reference number 17K17559. Appreciation also goes to the Hokkaido University Global Networking Award 2017 for travel support to attend *1st World Conference on Solid Electrolytes for Advanced Applications: Garnets and Competitors* in India.

References

1. Murugan R, Thangadurai V, Weppner W (2007) Fast lithium ion conduction in garnet-type $Li_7La_3Zr_2O_{12}$. Angew Chem Int Ed 46:7778–7781
2. Thangadurai V, Narayanan S, Pinzaru D (2014) Garnet-type solid-state fast Li ion conductors for Li batteries: critical review. Chem Soc Rev 43:4714–4727
3. Cao C, Li Z, Wang X-L, Zhao X, Han W-Q (2014) Recent advances in inorganic solid electrolytes for lithium batteries. Front Energy Res 2
4. Ramakumar S, Deviannapoorani C, Dhivya L, Shankar LS, Murugan R (2017) Lithium garnets: synthesis, structure, Li^+ conductivity, Li^+ dynamics and applications. Prog Mater Sci 88:325–411
5. Ohta S, Kobayashi T, Asaoka T (2011) High lithium ionic conductivity in the garnet-type oxide Li_{7-x} $La_3(Zr_{2-x}, Nb_x)O_{12}$ ($x = 0$–2). J Power Sources 196:3342–3345
6. Li Y, Han JT, Wang CA, Xie H, Goodenough JB (2012) Optimizing Li^+ conductivity in a garnet framework. J Mater Chem 22:15357–15361
7. Deviannapoorani C, Dhivya L, Ramakumar S, Murugan R (2013) Lithium ion transport properties of high conductive tellurium substituted $Li_7La_3Zr_2O_{12}$ cubic lithium garnets. J Power Sources 240:18–25
8. Narayanan S, Epp V, Wilkening M, Thangadurai V (2012) Macroscopic and microscopic Li^+ transport parameters in cubic garnet-type "$Li_{6.5}La_{2.5}Ba_{0.5}ZrTaO_{12}$" as probed by impedance spectroscopy and NMR. RSC Adv 2:2553–2561
9. Rosero-Navarro NC, Yamashita T, Miura A, Higuchi M, Tadanaga K (2017) Effect of sintering additives on relative density and li-ion conductivity of Nb-doped $Li_7La_3ZrO_{12}$ solid electrolyte. J Am Ceram Soc 100:276–285
10. Rangasamy E, Wolfenstine J, Sakamoto J (2012) The role of Al and Li concentration on the formation of cubic garnet solid electrolyte of nominal composition $Li_7La_3Zr_2O_{12}$. Solid State Ion 206:28–32
11. Cook LP, Plante ER (1992) Phase diagram of the system lithia-alumina. Ceram Trans 27:193–222
12. Kulkarni NS, Besmann TM, Spear KE (2008) Thermodynamic optimization of lithia-alumina. J Am Ceram Soc 91:4074–4083
13. Jin Y, McGinn PJ (2011) Al-doped $Li_7La_3Zr_2O_{12}$ synthesized by a polymerized complex method. J Power Sources 196:8683–8687
14. Kumazaki S, Iriyama Y, Kim KH, Murugan R, Tanabe K, Yamamoto K, Hirayama T, Ogumi Z (2011) High lithium ion conductive $Li_7La_3Zr_2O_{12}$ by inclusion of both Al and Si. Electrochem Commun 13:509–512
15. Takano R, Tadanaga K, Hayashi A, Tatsumisago M (2014) Low temperature synthesis of Al-doped $Li_7La_3Zr_2O_{12}$ solid electrolyte by a sol–gel process. Solid State Ion 255:104–107

16. Tadanaga K, Takano R, Ichinose T, Mori S, Hayashi A, Tatsumisago M (2013) Low temperature synthesis of highly ion conductive $Li_7La_3Zr_2O_{12}$–Li_3BO_3 composites. Electrochem Commun 33:51–54
17. Rosero-Navarro NC, Tadanaga K (2016) Sol–gel processing of solid electrolytes for Li-ion Batteries. In: Klein L, Aparicio M, Jitianu A (eds) Handbook of sol–gel science and technology. Springer International Publishing, Cham, pp 1–18
18. Rosero-Navarro NC, Yamashita T, Miura A, Higuchi M, Tadanaga K (2016) Preparation of $Li_7La_3(Zr_{2-x}, Nb_x)O_{12}$ ($x = 0$–1.5) and $Li_3BO_3/LiBO_2$ composites at low temperatures using a sol–gel process. Solid State Ion 285:6–12
19. Sakamoto J, Rangasamy E, Kim H, Kim Y, Wolfenstine J (2013) Synthesis of nano-scale fast ion conducting cubic $Li_7La_3Zr_2O_{12}$. Nanotechnology 24:424005
20. Tatsumisago M, Hamada A, Minami T, Tanaka M (1983) Structure and properties of Li_2O–RO–Nb_2O_5 glasses (R = Ba, Ca, Mg) prepared by twin-roller quenching. J Non-Cryst Solids 56:423–428
21. Ohta S, Komagata S, Seki J, Saeki T, Morishita S, Asaoka T (2013) All-solid-state lithium ion battery using garnet-type oxide and Li_3BO_3 solid electrolytes fabricated by screen-printing. J Power Sources 238:53–56
22. Cao Y, Li Y-Q, Guo X-X (2013) Densification and lithium ion conductivity of garnet-type $Li_{7-x}La_3Zr_{2-x}Ta_x O_{12}$ ($x = 0.25$) solid electrolytes. Chinese Physics B 22:078201
23. Janani N, Deviannapoorani C, Dhivya L, Murugan R (2014) Influence of sintering additives on densification and Li^+ conductivity of Al doped $Li_7La_3Zr_2O_{12}$ lithium garnet. RSC Adv 4:51228–51238
24. Ohta S, Seki J, Yagi Y, Kihira Y, Tani T, Asaoka T (2014) Co-sinterable lithium garnet-type oxide electrolyte with cathode for all-solid-state lithium ion battery. J Power Sources 265:40–44
25. Rosero-Navarro NC, Miura A, Higuchi M, Tadanaga K (2017) Optimization of Al_2O_3 and Li_3BO_3 content as sintering additives of Li_{7-x} $La_{2.95}Ca_{0.05}ZrTaO_{12}$ at low temperature. J Electron Mater 46:497–501
26. Zhou T, Zhang H, Jia L, Liao Y, Zhong Z, Bai F, Su H, Li J, Jin L, Liu C (2015) Enhanced ferromagnetic properties of low temperature sintering LiZnTi ferrites with Li_2O–B_2O_3–SiO_2–CaO–Al_2O_3 glass addition. J Alloy Compd 620:421–426
27. Kim YH, Yoon MY, Lee EJ, Hwang HJ (2012) Effect of SiO_2/B_2O_3 ratio on Li ion conductivity of a Li_2O-B_2O_3-SiO_2 glass electrolyte. J Ceram Process Res 13:S37–S41
28. Maia LF, Rodrigues ACM (2004) Electrical conductivity and relaxation frequency of lithium borosilicate glasses. Solid State Ionics 168(1–2):87–92
29. Yuan L, Liu B, Shen N, Zhai T, Yang Da (2014) Synthesis and properties of borosilicate/AlN composite for low temperature co-fired ceramics application. J Alloys Compd 593:34–40
30. Heydari F, Maghsoudipour A, Hamnabard Z, Farhangdoust S (2013) Evaluation on properties of CaO–BaO–B_2O_3–Al_2O_3–SiO_2 glass-ceramic sealants for intermediate temperature solid oxide fuel cells. J Mater Sci Technol 29:49–54

Chapter 6
Deposition and Compositional Analysis of Garnet Solid Electrolyte Thin Films

Sandra Lobe and Christian Dellen

Abstract While the general working principle of all-solid-state batteries on a laboratory scale has nowadays already been frequently proven, one major improvement opportunity in optimizing the energy density still lies in the drastic reduction in the electrolyte thickness down to the range of a few micrometers. In this way, the overall inactive mass of the cell will be reduced and a potentially lower Li-ion conductivity compared to conventional liquid electrolytes can be counterbalanced. The focus of this chapter is on the thin film deposition of garnet structured Li-ion conductors. In the first part, different deposition approaches by wet chemical routes and gas-phase techniques are discussed. In the second part, an overview of the compositional analysis of Li-containing thin films, which is an elaborate and important part of the target-oriented development of garnet-structured thin films, is given.

6.1 Introduction

The interest in all-solid-state batteries is steadily increasing because of their potential to outperform today's commercial Li-ion batteries with respect to energy density and safety. Aiming for a maximum increase in the energy density two requirements have to be fulfilled:

(1) Lithium as anode material and
(2) A thin electrolyte

In order to achieve the first requirement, electrolyte materials are necessary, which show no reaction in contact with lithium metal. Furthermore, the electrolyte should be stable in the voltage range of the cell as well as showing a high Li-ion conductivity.

S. Lobe (✉) · C. Dellen
Forschungszentrum Jülich GmbH, Institute of Energy and Climate Research, Materials Synthesis and Processing (IEK-1), 52425 Jülich, Germany
e-mail: s.lobe@fz-juelich.de

C. Dellen
e-mail: c.dellen@fz-juelich.de

R. Murugan and W. Weppner (eds.), *Solid Electrolytes for Advanced Applications*,
https://doi.org/10.1007/978-3-030-31581-8_6

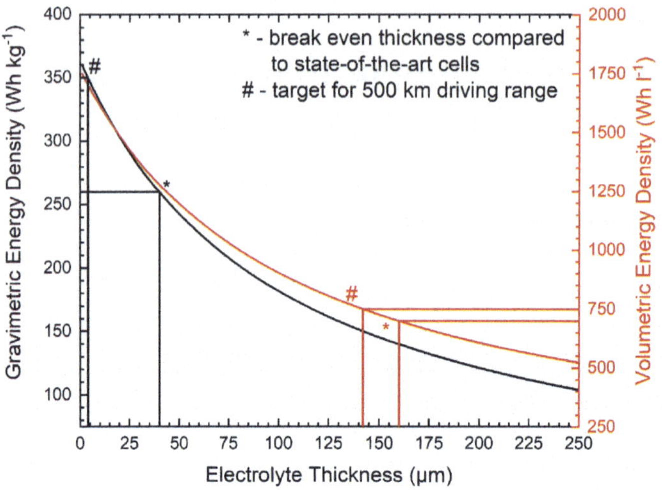

Fig. 6.1 Energy density in dependence of the electrolyte thickness calculated for a 100 μm cathode layer consisting of 67% $LiCoO_2$ and 33% $Li_7La_3Zr_2O_{12}$, an electrolyte layer with varying thickness and a lithium metal anode with a thickness of 5 μm. The breakeven thickness for achieving current Li-ion cell energy densities of 260 Wh kg^{-1} and 700 Wh l^{-1} and the targets for 350 Wh kg^{-1} and 750 Wh l^{-1} for a driving range of 500 km according to [1]

Garnets are a high-performance solid electrolyte material because they fulfill all the requirements mentioned here.

In Fig. 6.1 the energy density of an all-solid-state battery in dependence of the electrolyte thickness is shown. A hypothetical cell with a 100 μm thick mixed electrode (100% relative density), consisting of 67 wt% $LiCoO_2$ and 33 wt% $Li_7La_3Zr_2O_{12}$ (LLZ), a thin LLZ electrolyte layer and a 5 μm Li anode was used for the calculation. According to Schmuch et al. [1] current Li-ion cells have energy densities of around 260 Wh kg^{-1} and 700 Wh l^{-1}, whereas for future automobile applications a driving range of 500 km and therefore 350 Wh kg^{-1} and 750 Wh l^{-1} are targeted at the cell level. With a very thin electrolyte, i.e., a thickness of less than 4 μm the required energy densities are matched. By using high voltage cathodes, like 5 V-spinels, the energy density could be further increased enabling higher driving ranges in electric cars.

In order to achieve an electrolyte thickness of less than 4 μm a suitable processing of the thin film electrolyte is required. In this chapter we will emphasize on one hand the deposition of garnet thin films by wet-chemical methods, like sol–gel processes and on the other hand the thin film formation by processing from the gas phase, including physical as well as chemical vapor deposition processes.

So far, only lithium phosphorous oxynitride (LiPON) is commercialized as solid electrolyte material. It is used in thin film batteries for high current applications. The Li-ion conductivity of LiPON is about 2.3×10^{-6} S cm^{-1} [2], which can act as value for benchmarking the conductivity of garnet thin films.

Beside the application as solid electrolyte, garnet-structured thin films can also be used as a coating for electrode materials in conventional Li-ion batteries. These coatings can act as a protection layer against reactions between the liquid electrolyte and e.g., cathode materials like Mn-rich spinels [3]. This special application is beyond the scope of this chapter and not included in the discussion.

A main challenge for the development of garnet-structured thin films is the reduced availability of spatially and depth resolved analytics, especially for the determination of the Li content. The ionic conductivity and the activation energy (E_a) of the ion conduction process is highly dependent on the available Li in the garnet structure so that the interplay of thin film processing – thin film composition, thin film structure – ionic conductivity is from high importance for the development of a suitable deposition process.

For this reason, the chapter is divided into two parts; one part describes methods and developments of thin film deposition, mainly techniques based on wet-chemical processes and processes from the vapor phase. The second part describes methods and challenges of garnet thin film analytics.

6.2 Wet-Chemical Deposition Processes

Wet-chemical processes include solution based synthesis processes as well as sol–gel techniques. For garnet materials, only sol–gel processes were reported, so far. The sol–gel route is a common route for the synthesis of powders, fibers, aerogels, and thin films. In ceramic processing the fabrication of thin films by sol–gel route is used for deposition of dense layers with a defined particle size. Therefore, a sol, i.e., the stable dispersion of a solid material in a liquid, is synthesized by using soluble metal salts, e.g., nitrates, alkoxides, or acetates. Typically the dispersed phase has a particle size in the range of 1–100 nm and does not settle. To avoid particle agglomeration and settling of the particles normally the sol is stabilized by addition of stabilizing agents. Thin films are deposited by coating of the sol on a substrate, e.g., by a spin coating or dip-coating process and additional heat treatment, like drying, calcination, and sintering.

Ceramic thin film deposition using sol–gel process is already established for the deposition of electrolyte thin films for solid oxide fuel cell (SOFC) applications [4], gas separation membranes [5] or protective layers in batteries [6].

An advantage of the sol–gel based coating process is that the chemical composition and the particle size of the coating liquid can be controlled very precisely. So a variety of different substituents and dopants as well as the crystallite size in the resulting thin films can be scanned easily. A further advantage is that wet-chemical coating processes developed in the lab scale can be easily transferred into industrial scale.

In recent years several articles about the powder synthesis of garnet materials by sol–gel process were published. However, there are only a few publications about the thin film deposition by sol–gel process, so far.

The first publication of Li–La–Zr–O thin films is from Chen et al. published in 2014 [7]. They used a stable solution of lithium *tert*-butoxide (with 10% excess), lanthanum nitrate hexahydrate and zirconium *n*-propoxide in methoxyethanol as sol. The thin films were deposited via a spin-coating process on Pt/Ti/SiO_2/Si substrates. After drying the sol at 400 °C and calcination at 500 °C they annealed the films in the range of 600–800 °C. However, X-ray diffraction (XRD) showed only $La_2Zr_2O_7$ and no garnet phase. X-ray photoelectron spectroscopy (XPS) showed furthermore a decrease of Li content with increasing annealing temperature, which was attributed to the evaporation of lithium at elevated temperatures. Therefore, the ionic conductivity had the highest value for the thin films annealed at 600 °C, namely 1.67×10^{-6} S cm^{-1}. The influence of thin film thickness was also investigated for an annealing temperature of 600 °C, leading to the best conductivity of 1.87×10^{-6} S cm^{-1} after deposition of 8 layers which is equal to a thickness of 390 nm.

Tadanaga et al. followed a different heat treatment strategy [8]. Their sol was prepared from $LiNO_3$ (with 10% excess), $La(NO_3)_3 \cdot 6H_2O$, $Zr(O–n–C_3H_7)_4$ and Al(o-*sec*-$C_4H_9)_3$ as metal sources, ethyl acetoacetate as stabilizing agent and 1-propanol as solvent. The thin films were deposited via a dip-coating process on MgO substrates and calcined at 450 °C. High temperature annealing was done at 900 °C in the presence of Li_2CO_3 in order to compensate Li loss. Without Li_2CO_3 they determined $La_2Zr_2O_7$ to be the main phase in the thin films, whereas with Li_2CO_3 the garnet structure was proved by XRD. The microstructure of the thin films was improved by adding lithium dodecyl sulfate (LDS) as ionic surfactant to the coating liquid leading to dense microstructure. Ionic conductivity measurements showed an increase from 2.8×10^{-7} S cm^{-1} ($E_a = 0.61$ eV) for thin films annealed without LDS to 2.4×10^{-6} S cm^{-1} ($E_a = 0.52$ eV) for thin films annealed with LDS. The authors suppose a higher conductivity when thin films are annealed at temperatures higher than 900 °C but they report about reactions of the thin films with the substrate, e.g., for Al_2O_3, Pt and $Gd_3Ga_5O_{12}$.

Bitzer et al. [9] used an ethanol-based sol with $LiNO_3 \cdot 3H_2O$, $La(NO_3)_3 \cdot 6H_2O$ $(CH_3CO_2)_xZr(OH)_y$ $(x + y \approx 4)$ and $Y(NO_3)_3 \cdot 6H_2O$ as metal sources and methoxyethanol as stabilizing agent. The sol concentration was between 0.02 and 0.08 mol/l, for higher concentration they observed cracking and peeling-off of the coated thin films. Thin film coating was followed by drying in air and additional rapid thermal processing at 800 °C in different atmospheres. First coating experiments by dip-coating on silicon and steel substrates with 20% Li-excess in the sol did not show the cubic LLZ phase. They increased the Li-excess to 1000% so that on steel substrates the cubic garnet phase and $LaFeO_3$ as secondary phase was detected by XRD. Calcination of the thin films in argon atmosphere showed a cubic garnet structure without $LaFeO_3$ secondary phase. Impedance spectroscopy showed an ionic conductivity of 3.4×10^{-8} S cm^{-1} at room temperature. As a further substrate $LiCoO_2$ coated steel was tested. First experiments revealed an unwanted reaction between the garnet phase and $LiCoO_2$ leading to the formation of $La_2Li_{0.5}Co_{0.5}O_4$. By using an improved calcination process, i.e., rapid thermal processing without holding time, no reaction phase was observed.

Another study about the compatibility of $LiCoO_2$ and sol–gel processed Li–La–Zr–O thin films was published by Zarabian et al. [10]. They prepared a gel from a water-based sol which contained $LiNO_3 \cdot H_2O$ (with 10% excess), $La(NO_3)_3 \cdot 6H_2O$, and $ZrOCl_2 \cdot xH_2O$ as metal precursors and ethylene glycol as complexing agent. This sol was dried until a gel was formed. The gel was diluted in 2-butoxyethanol to get the coating liquid. As substrates single-crystalline MgO (100) and a $LiCoO_2$ pellet (pressed with PVA binder) were used. The precursor was dried on the substrates at 400 °C. The ability of the sol to form the garnet phase was proved after evaporation of water and thermal annealing at 900 °C by XRD. Thin films were processed at 400 °C so that a nanocrystalline Li–La–Zr–O phase was formed. From high temperature impedance measurements (72–250 °C) the room temperature Li-ion conductivity was estimated to be 1×10^{-6} S cm^{-1} with an activation energy in the range of 0.6 eV. The investigation of the deposition on $LiCoO_2$ pellet by XPS shows already after the drying process a broad range of Co-species indicating a reaction between the coated Li–La–Zr–O layer and the cathode material.

Uhlenbruck et al. used a different synthesis method to produce Ta-substituted $Li_7La_3Zr_2O_{12}$ thin films [11]. They used a microemulsion based synthesis method to form an ink containing Li–La–Zr–Ta–O nanoparticles. Calcination of the ink at different temperatures in air showed a single-phase garnet structure at 800 °C, at 700 and 900 °C impurities of $La_2Zr_2O_7$ and La_2O_3 were observed. The ink was used to print garnet thin films on steel and $LiCoO_2$-coated steel substrates with a commercial ink-jet printer. Calcination of Li–La–Zr–Ta–O in different atmospheres showed that no garnet phase was obtained in air, but under conditions with low oxygen partial pressure, like in Ar/H_2 the cubic garnet phase was obtained. These results are similar to the results of Bitzer et al. [9] who concluded that an oxide layer at the surface of the steel substrate can react with the components of the sol leading to unwanted side reactions. If the calcination atmosphere does not contain oxygen, fewer side reactions are observed. Similar thin films printed on $LiCoO_2$/steel substrates also showed a strong dependence on the used atmosphere during calcination. If air is used for annealing, a strong formation of $LaCoO_3$, indicating a reaction between the cathode layer and the printed film is observed. In argon atmosphere a cubic garnet phase and the cathode material could be identified by XRD. However, there are still some unidentified secondary phases.

In summary, the main challenge in deposition of thin films by sol–gel route is the sintering process of garnets, which usually needs high temperatures in the range of 1100–1200 °C [12]. In bulk material processing, the material is covered with powder of the same composition to compensate the Li loss during sintering. At this point, thin films produced by the sol–gel route have a fundamental disadvantage, due to their high surface to volume ratio, which can lead to a high loss of Li during sintering. Coverage of the sample with garnet powder would lead to a bad surface quality, because it will be difficult to remove the powder after the sintering process. The addition of excess lithium to the coating liquid can help to solve the problem. However, large amounts of additional lithium are necessary and the influence of this additional mobile species during the calcination and sintering of the thin film on the substrate is not clear yet.

A further challenge regarding the sol–gel processing is choice of a suitable substrate. Tadanaga et al. [8], Bitzer et al. [9], and Zarabian et al. [10] reported about reactions with the used substrates already at temperatures which are not exceeding 1000 °C. A suitable heat treatment process, like it was shown by Bitzer et al. [9] for the deposition of Y-doped $Li_7La_3Zr_2O_{12}$ thin film on $LiCoO_2$ substrate is necessary for further development of wet-chemical coating of garnets on cathode materials.

6.3 Deposition Processes from Vapor Phase

Deposition processes from the vapor phase can be divided into two sub-sections – physical and chemical vapor deposition processes. Physical vapor deposition (PVD) processes are processes where the condensation of the material takes place without an additional chemical reaction, whereas in chemical vapor deposition processes (CVD) the gaseous precursors react while forming the thin film.

In general, thin film processing techniques from the vapor phase have an advantage compared to conventional ceramic processing techniques: the deposition temperature of the dense thin films is far below the normally used sintering temperature of the material. As an example from Solid oxide fuel cell (SOFC) development, the deposition of the electrolyte YSZ (yttria stabilized zirconia) can be taken. Ceramic processing techniques e.g., screen printing, need a sintering temperature of around 1400 °C in order to get a dense electrolyte layer [13]. In comparison to this high value, CVD [14] and PVD [15] processes can lower the temperature for the deposition of dense layers down to 800 °C.

A disadvantage of vapor-based processes is the requirement of expensive vacuum and processing equipment.

6.3.1 Physical Vapor Deposition (PVD)

6.3.1.1 Pulsed Laser Deposition (PLD)

Pulsed laser deposition is a technique, in which the target material is ablated by a pulsed laser. During ablation, a plasma containing the target atoms is generated above the surface of the target. The target atoms are condensing from the vapor phase on the substrate, which is normally at a distance of 3–8 cm in front of the target. Typical parameters for PLD processes are e.g., the wavelength of the laser, which should match the light absorption properties of the target material, laser fluence, and number of laser pulses as well as the chamber pressure and chamber atmosphere.

A big advantage of PLD is the ability to deposit stoichiometric multicomponent thin films from one single target so that the desired material composition is easily transferable into a thin film. At the same time, epitaxial growth as well as large grain-oriented films and complex stoichiometries are possible. Therefore, PLD has already been used in a lot of different research fields like high T_c-superconductors, piezoelectric and ferroelectric materials and also in battery research several electrode, e.g., $LiCoO_2$ and $LiMn_2O_4$ [16] and electrolyte, e.g., LiPON [17], materials already have been deposited by PLD. The disadvantage of PLD is the small substrate area that can be coated. By using a movable substrate holder an area up to several cm^2 can be covered with a thin film, which is sufficient for applications in fundamental research, but a limiting factor for industrial application.

First garnet thin films grown by the PLD process were published by Tan and Tiwari [18]. The films were deposited from a ceramic $Li_7La_3Zr_2O_{12}$ target and a Li_2O target to compensate Li loss in oxygen atmosphere with a pressure of 0.4 mbar on $SrTiO_3$ (100) and sapphire (0001) substrates. Their conducted XRD measurements showed cubic garnet structure for thin films annealed at 800 °C and 1000 °C. The ionic conductivities were determined in dependence of different process parameters. The lowest conductivity of 7.19×10^{-8} S cm^{-1} was measured for an as-deposited sample deposited without additional Li source. This value increased to 3.35×10^{-7} S cm^{-1} if Li_2O was used as additional target during the deposition. Samples annealed in a conventional annealing process showed a decrease in conductivity to 1.78×10^{-7} S cm^{-1}. As reason for this behavior they proposed a lower ionic conductivity due to an increased amount of cracks and defects in the thin film. The best conductivity of 7.36×10^{-7} S cm^{-1} with an activation energy of 0.32 eV was achieved for a laser annealed sample. Furthermore, they tested the stability of the thin films against metallic lithium indicating a good stability of thin films.

Kim et al. [19] showed the epitaxial growth of $Li_7La_3Zr_2O_{12}$ thin films on $Gd_3Ga_5O_{12}$ (001) and (111) single crystal substrates with garnet structure. Therefore, they used a ceramic target and also oxygen atmosphere with a pressure of 3.3×10^{-2} mbar. XRD revealed the epitaxial growth of the thin films. ICP-MS analysis showed a depletion of Li in the deposited thin films. Impedance spectroscopy measurements showed an ionic conductivity of 2.5×10^{-6} S cm^{-1} and E_a to be 0.55 eV for the (001) oriented thin film and 1.0×10^{-5} S cm^{-1} and 0.52 eV for the (111) oriented thin film.

Li loss was observed for garnet thin films processed by PLD by other researchers, too. In order to compensate for the loss during deposition, Rawlence et al. [20] investigated the post-annealing of Al-doped $Li_7La_3Zr_2O_{12}$ thin films on MgO (100) substrates in a crucible with LiOH as additional Li source at 600 °C. They showed that with increasing film thickness the amount of secondary phases, like $La_2Zr_2O_7$ is increasing. The decreasing lithium content per thin film volume is supposed to reflect this behavior. For thin films with thicknesses of 90 nm and 110 nm the crystal structure is reported to be cubic and a mixture of cubic and tetragonal phases, respectively. Due to the insufficient thin film morphology of these films they only reported the ionic conductivity of thin films with a thickness of 380 nm. Their conductivity is given to be 1.2×10^{-3} S cm^{-1} at 325 °C with an activation energy of 0.6 eV between 300 and 400 °C and activation energy of 1.39 eV in the range lower than 300 °C. The low conductivity and high activation energy are most likely a consequence of the $La_2Zr_2O_7$-rich thin film, which only contained low amounts of cubic garnet phase.

Park et al. [21] studied the substrate-thin film interaction of Li–La–Zr–O thin films deposited by PLD. Therefore, they deposited $Li_7La_3Zr_2O_{12}$ and Al-doped $Li_7La_3Zr_2O_{12}$ on different substrates, like silicon, glass, MgO (100), sapphire and anodized aluminum. The layers deposited on the Al_2O_3 substrates showed a diffusion of Al in the thin film, resulting in an Al-rich layer on top of the sample. In contrast to that, the deposition on the other substrates led to a Li depleted thin film ($Li_{4.79}La_3Zr_{1.92}O_{10.75}$ determined by Laser-induced breakdown spectroscopy) with an amorphous structure up to a deposition temperature of 600 °C and a clear formation of $La_2Zr_2O_7$ at 750 °C. By grazing incidence X-ray diffraction with a small incidence angle, they were able to show that on top of the sample there is a thin layer of tetragonal LLZ formed at 600 °C as well as at 700 °C. Ionic conductivities were determined by impedance spectroscopy in the range from 175 to 375 °C, showing activation energies of 0.41 eV for the film deposited at room temperature, 0.35 eV for the film deposited at 600 °C and 0.67 eV for the film deposited at 700 °C. The room temperature conductivity was extrapolated to be 1.61×10^{-6} S cm^{-1} for the film deposited at 600 °C.

A recent publication by Saccoccio et al. [22] showed that the garnet phase is already detectable by XRD at 50 and 300 °C deposition temperatures. They used sintered $Li_{6.4}La_3Zr_{1.4}Ta_{0.6}O_{12}$ as target material and MgO (100) as substrates. The as-deposited thin films exhibit a poor crystallinity, leading to low ionic conductivities, which are even lower than the values reported by Rawlence et al. [20] and Park et al. [21] and exhibit similar high activation energies.

Reinacher et al. [24] had a different concept to solve the problems which are related to the growth of $Li_7La_3Zr_2O_{12}$ thin films, like the formation of $La_2Zr_2O_7$ as secondary phase or the high crystallization temperature of $Li_7La_3Zr_2O_{12}$. They investigated the deposition of the Zr-free high ion conducting $Li_6BaLa_2Ta_2O_{12}$ by PLD. In this process, they used a ceramic target with 5% Li-excess, MgO as substrate and as atmosphere oxygen with a pressure of 5×10^{-2} mbar. Instead of a post-annealing process, an elevated substrate temperature of 550 °C during deposition was used. The cubic garnet structure of the deposited thin films was revealed by XRD. A detailed investigation of the electrical properties by impedance spectroscopy gives a total conductivity of 1.7×10^{-6} S cm^{-1} and an activation energy of 0.47 eV. The stability of the deposited thin films in contact with metallic lithium was also investigated by coating Ohara glass with the garnet thin films. Regions of the glass, which were not coated, showed a dark color after contact with lithium whereas the coated areas remained unchanged.

6.3.1.2 Sputtering

Sputter deposition is a technique to deposit thin films from a metallic or ceramic target. The sputter deposition process relies on the interaction of the target material with plasma, which contains electrons and positively charged ions. Mostly argon or another noble gas is used for plasma generation because of the inert nature of these elements. The Ar-ions are accelerated by an electric field (direct current for conducting or radio frequency (r.f.) for insulating targets) to the target and due to momentum transfer atoms are knocked out from the target, transferred to the gas phase and deposited on an opposite substrate.

The advantage of sputter deposition is that nearly all compounds can be deposited. Furthermore, by careful design of the deposition parameters, sputter deposition can also be used to deposit crystal structures far from thermodynamic equilibrium, e.g., the deposition of the high temperature phase of $LiCoO_2$ [47]. The adhesion of sputter-deposited thin films is in general better compared to other PVD techniques, because the atoms have a comparable high energy when they hit the substrate. A disadvantage of sputter deposition is the comparable low deposition rate, especially for r.f. sputter processes.

In general, the sputtering process is known to deposit the stoichiometry of the target. However, the deposition of complex and difficult materials like Li-stuffed garnets has not been studied very often yet.

The first sputter deposition experiments from a ceramic Li–La–Zr–O target were reported by Kalita et al. [25]. They deposited amorphous thin films on stainless steel electrodes by an r.f. sputter deposition process. XPS results showed that the amount of Li in the deposited films was decreasing with increasing r.f. power. The thin film with the highest Li content showed also the highest ionic conductivity, numerical 4×10^{-7} S cm^{-1} and $E_a = 0.70$ eV.

Up to now, the garnet-structured thin film with the highest ionic conductivity was deposited by a sputtering process [26]. As target Ta- and Al-substituted $Li_7La_3Zr_2O_{12}$ was used. The deposition was carried out via an r.f. magnetron sputter deposition process at different temperatures in the range from 150 to 800 °C on stainless steel foils as substrates. Structural (XRD) and chemical (ToF-SIMS) analysis showed an increased amount of Li and Li-containing phases with increasing substrate temperature leading to a single-phase garnet thin film at a deposition temperature of 700 °C. At 700 and 800 °C a reaction between the deposited thin film and the stainless steel substrate can be observed by SEM and Time-of-flight secondary-ion-mass-spectrometry (ToF-SIMS). The formed interlayer consists mainly of Al- and Li-containing species, most likely $LiAlO_2$, which was also confirmed by XRD at 800 °C. The interlayer is insulating for electrons as well as for Li-ions leading to a low conductivity of only 2×10^{-9} S cm^{-1} perpendicular to the thin film surface. Lateral conductivity measurements showed an ionic conductivity of 1.2×10^{-4} S cm^{-1} and an activation energy of 0.47 eV.

6.3.2 Chemical Vapor Deposition

CVD is the general abbreviation for all chemical vapor deposition techniques including miscellaneous processes which have in common that the deposition of the substrate comes along with a chemical reaction or a pyrolysis process. In the simplest

CVD process, a precursor, e.g., a metal-organic compound or a metal salt is vaporized and transported to a heated substrate, where cracking or pyrolysis takes place and the material is deposited.

The biggest advantage of CVD techniques compared to wet-chemical or PVD processes is the possibility to coat conformal thin films on 3-dimensional structures. Further advantages are that the deposition rates can be quite high and that there are also techniques available which are working with atmospheric or low vacuum, so that the requirements for the vacuum parts are not so strict.

A disadvantage of most CVD techniques is that the precursors have to be volatile near room temperature so that often highly toxic and/or expensive metal-organic precursors are used.

The first publication of garnet thin films by chemical vapor deposition method was published by Katsui and Goto [27]. They used a metal-organic CVD process, which means that their precursors are metal-organic compounds. As metal ion sources for the deposition of $Li_7La_3Zr_2O_{12}$ Li(DPM), $La(DPM)_3$ and $Zr(DPM)_4$ (DPM = dipivaloylmethanato) were used. Their work concentrated on the temperature dependence of the thin film deposition process on Al_2O_3 substrates. They showed, that only in a certain deposition temperature range the garnet structure can be observed. XRD confirmed that at 700 and 1000 °C $La_2Zr_2O_7$ was the main phase of the thin film, whereas at 800 °C tetragonal LLZ and at 950 °C cubic LLZ were received. The chemical compositions determined by XPS showed a stoichiometry close to the theoretical values and at the same time no Al was detected in the layers, which means that there was no interdiffusion of Al from the substrate into the thin film. The deposition rate of 20 $\mu m\ h^{-1}$ was higher than for comparable PLD processes.

In 2017 Kazyak et al. [28] showed the deposition of ultrathin Al-doped $Li_7La_3Zr_2O_{12}$ coatings by using an atomic layer deposition (ALD) process. In ALD processes the coating is deposited monolayer by monolayer. In order to deposit garnet-structured thin films lithium *tert*-butoxide (LiO^tBu), tris(*N*,*N*′-diisopropylformamidinato) lanthanum (La(FAMD)), TDMAZ tetrakis(dimethylamido)zirconium (TDMAZ) and trimethylaluminium (TMA) were used as metal ion sources and O_3 as oxidizer. They deposited alternating layers of Li_2O, La_2O_3, ZrO_2, and Al_2O_3 after optimizing the single processes. Intermixing of the layers was facilitated by post-annealing. The deposition of the single oxide layers was strongly influenced by the metal oxide layer deposited before. As an example, they reported that the deposition of La was decreased when Li was deposited before and increased when Zr was deposited in the previous subcycle. The as-deposited films were amorphous. The composition was revealed to be close to the theoretical value by XPS. Furthermore they also showed that the amount of Li in the as-deposited films can be increased by increasing the amount of Li subcycles. The amorphous thin film showed ionic conductivities of 7.8×10^{-5} S cm^{-1} at 200 °C and 1.2×10^{-6} S cm^{-1} at 100 °C. The activation energy given by these values is 0.63 eV. Post-annealing at 600 °C showed $La_2Zr_2O_7$ as main phase, either a rapid thermal annealing process or a conventional muffle furnace was used. Garnet structure was only observed by synchrotron XRD when the thin film showed a Li-excess between 400 and 900% and $LiMnO_2$ as a Li source during annealing was used. They also reported interactions with the substrate during heat treatment. Reactions with the substrate, Li loss and delamination of the thin film was observed with many substrates, e.g., Si, Si_3N_4, Cu, stainless steel, Ni, Mo, Pt, and Ni thin films on Si, soda–lime glass, fused quartz, Al_2O_3, and ZrO_2. Only single crystal MgO did not show unwanted behavior.

A special kind of CVD is the evaporation of material by a laser instead of the usually used methods for material evaporation, which enables cheaper metal precursors. Loho et al used a CO_2-Laser assisted method, where the laser can evaporate even solid materials. In order to deposit $Li_7La_3Zr_2O_{12}$ thin films [29], they used $LiC_{11}H_{19}O_2$, $La(C_5H_7O_2)_3{\cdot}xH_2O$, and $Zr(C_5H_7O_2)$ as metal precursors and oxygen as a reactive gas. The thin films were deposited on Pt-foil at different temperatures and oxygen partial pressures. They report a high sensitivity of the phase formation to the deposition parameters, so that only for a deposition temperature of 700 °C and 2 mbar oxygen single-phase tetragonal LLZ films could be observed. At the same time, a pronounced reaction with the Pt-foil could be observed leading to a Li_2PtO_3 interlayer. The formation of the interlayer was further revealed by ToF-SIMS and XPS measurements. Conductivity measurements showed a lithium-ion conductivity of 4.2×10^{-6} S cm^{-1} at 25 °C and an activation energy of 0.50 eV. The same deposition method was also used for the deposition of $Li_5La_3Ta_2O_{12}$ which was investigated with regard to processing-structure–property relations [30]. In a first step the deposition in dependence of substrate temperature and total gas pressure was studied. Similar to previous results they found a minimum substrate temperature of 700 °C is required for the garnet structure. The microstructure of the thin films was more coarse-grained for increased deposition temperature and increased gas pressure. Furthermore, the samples deposited at 700 °C and 5 mbar total pressure (similar conditions to the LLZ growth in the paper discussed before) were post-annealed at 550 and 650 °C leading to a more dense and uniform microstructure. The annealing process also improved the ionic conductivity from 7.8×10^{-6} S cm^{-1} (as deposited, $E_a = 0.66$ eV) to 1.6×10^{-5} S cm^{-1} (annealed at 550 °C, $E_a = 0.61$ eV) and 3.8 $\times$ 10^{-5} S cm^{-1} (annealed at 650 °C, $E_a = 0.52$ eV). The latter value is close to the values observed for bulk $Li_5La_3Ta_2O_{12}$ with high relative density.

In summary, the composition and crystal structure of complex garnets like $Li_7La_3Zr_2O_{12}$ or $Li_5La_3Ta_2O_{12}$ can be well controlled by CVD techniques. However, the choice of the substrate is difficult and also a proper deposition and post-annealing procedure is necessary in order to achieve high-conducting garnet thin films.

6.4 Compositional Analysis of Garnet Electrolytes

6.4.1 General Considerations

Besides the choice of a suitable deposition method, as already discussed in the previous sections, the proper analysis of the thin film composition is also essential for a fast and target-oriented development of an eligible garnet thin film electrolyte.

Therefore this section will give a general overview of selected analytical techniques, which are commonly used for the compositional characterization of garnet thin films. However, this discussion does not claim to include all possible methods, as well as the detailed explanation of the measurement principle for every technique, which would be beyond the scope of this chapter.

Depending on the thin film properties under investigation the sample geometry and the scale on which the information should be resolved already sets a first limitation in the selection of suitable analytical techniques. For example, if the lateral and especially in case of thin films the depth resolution is of interest, techniques like inductively coupled plasma – optical emission spectroscopy or – mass spectrometry, which involve the dissolution of the sample in the preparation process, are non-applicable. On the other hand, when it comes to the analysis of a very confined, small sample volume with a finite amount of atoms, the physical detection limit of the different analytical techniques is another point that has to be taken into account. In addition to these geometrical restrictions, the physical properties of lithium itself also limit the selection of convenient analytical techniques. One well-known issue is, for example, the detection of lithium in thin films by electron dispersive X-ray spectroscopy, which is discussed in more detail in the next section of this chapter. In the successive sections the analysis by X-ray based techniques and the use of ions as well as neutrons as probing particles for the compositional analysis of garnet thin films is described. Besides the short descriptions of the different analytical techniques also the application of these specific techniques for garnet electrolytes is pointed out using various examples from literature.

6.4.2 Electron Beam Analytical Techniques

In principle, there are a variety of electron beam-based techniques for the compositional analysis of thin films. Commonly, a scanning electron microscope (SEM) equipped with an energy dispersive X-ray spectroscopy (EDS) detector can be used to analyze the element composition of different kind of thin film samples. However, this analytical method is generally not applicable for the quantification of lithium. While the technical issue given by the absorption of the Li X-rays through the detector window can nowadays be tackled by the use of windowless detectors, the accurate detection of lithium in compounds for battery application is still challenging. The low energy Li X-rays can for instance be easily absorbed by heavier elements in the investigated sample or by an oxide layer on top of the thin film [31]. Furthermore, when the elemental composition of thin films with only a few hundred nanometers is of interest, the depth resolution of the EDS, which is correlated to the energy-dependent interaction volume of the electrons, can also be a limiting factor.

Experimental evidence for the fact that the detection of lithium within a bulk sample of LLZ is nowadays in principle possible was recently published by Burgess et al. [31]. The published EDS spectrum (Fig. 6.2) was collected using a windowless detector with a detector area of 100 mm^2 and an acceleration voltage of 3.5 kV.

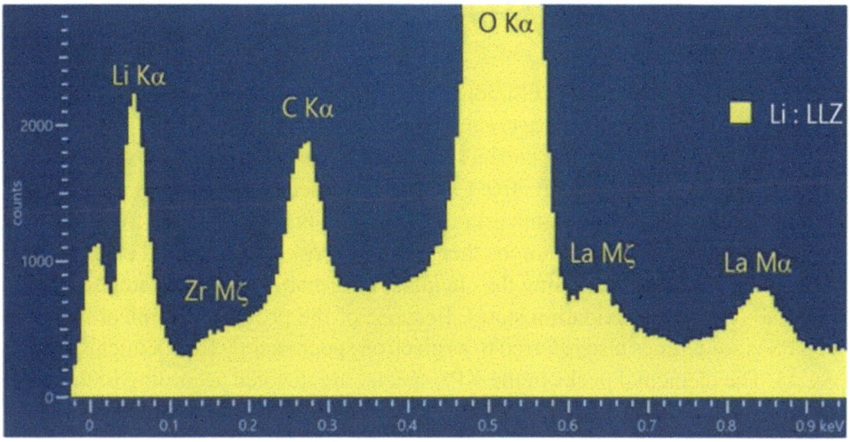

Fig. 6.2 EDS Spectrum of a bulk lithium lanthanum zirconate (LLZ) sample measured with a windowless detector. Besides the low energy lines of the heavy elements (La, Zr) the spectrum also shows the K_{α}-lines of oxygen, carbon, and especially lithium. Reprinted from [31] with permission from Cambridge University Press

Besides the low energy lines of the heavy elements (La, Zr) the spectrum clearly shows the K-lines of oxygen, carbon and also especially lithium. While this proves the detection of lithium in bulk LLZ, so far there was no report about EDS analysis of garnet thin films published.

Another electron beam-based analytical technique, which was already used to measure depth profiles with the elemental information of lithium containing thin film systems, is the so-called Auger electron spectroscopy (AES) [23, 32]. For the AES analysis the sample is usually bombarded with a focused electron beam, which has enough energy to kick out core electrons and ionize the elements within the sample. In the subsequent relaxation process this hole in the inner shell is filled up with an electron from an outer shell leading to the release of excess energy as an X-ray photon or by emitting a so-called Auger electron. While the X-rays are used in the EDS analysis, the AES is based on the detection of emitted Auger electrons. The information about the elemental composition is then given by measuring the characteristic kinetic energy of the emitted Auger electrons. The typical nomenclature for an Auger transition is based on the X-ray notation giving first the core level of the generated hole, second the state of the relaxed electron and third the state of the emitted Auger electron, e.g., $KL_1L_{2,3}$. While AES by itself is a very surface-sensitive technique with a shallow information depth, for the measurement of a depth profile the sample has to be etched e.g., by an ion beam. The information depth is given by the mean free path of the Auger electron in the solid because after inelastic scattering the electron loses its characteristic energy and hence the elemental information.

While there is so far no report on the use of AES on LLZ thin films, but Uhart et al. already showed the general feasibility of this technique for lithium containing thin films by analyzing a thin film battery [32].

6.4.3 Analytical Techniques Based on X-Rays

Whereas X-ray diffraction is commonly used to determine the crystal structure of garnet thin films, the quantitative compositional analysis as well as the characterization of interfaces is frequently carried out using X-ray photoelectron spectroscopy (XPS). This generally nondestructive technique is based on the ionization of the elements within the sample by bombardment with high energetic X-rays. With respect to the energy of the incident photons the binding energy of the emitted electrons can afterward be calculated revealing the chemical information of the different elements within the solid (e.g., oxidation state). Because of the provided chemical information XPS is sometimes also referred to as electron spectroscopy for chemical analysis (ESCA). The elemental peaks in the XPS spectra are denoted according to the spectroscopic notation, e.g., Li 1s. Due to the short mean free path of the electrons in the solid, XPS is generally a very surface-sensitive technique with an information depth of a few nanometers. Therefore, to gain information about elemental composition in deeper layers, too, the analysis has to be combined for example with the erosion of the sample via sputtering. The increase of the photon energy or a change in the angle of the detection system are also additional options to raise the information depth.

With respect to lithium containing compounds XPS is widely used today not only in the case electrolyte thin films [7, 10, 21, 22, 25, 27, 29], but also to analyze for example decomposition products on electrode materials [33]. One critical point for lithium analysis by XPS is the fact that the Li 1s signal can overlap with other signal like Zr 4s hampering a reliable quantification of the elements [22, 29].

Zarabian et al. for example, analyzed the interface between LLZ thin films and $LiCoO_2$ by XPS. In their study they deposited the LLZ thin film via spin coating on top of a $LiCoO_2$ pellet and then analyzed the materials interaction by collecting XPS spectra at different sample depths [10]. Figure 6.3 exemplary shows a series of Co 3p and Li 1s XPS spectra taken at different sample depths. The depth of 32.2 nm corresponds to a region in the LLZ thin film close to the surface and the depth of 3606.4 nm corresponds to a region close to the interface of the LLZ thin film and the $LiCoO_2$ pellet. In addition there is also a reference spectrum for $LiCoO_2$. Starting at a depth of 32.2 nm inside the LLZ thin film there is the spectra only a contribution of the Li 1s peak visible in the spectrum. With increasing sample depth the intensity of the Li 1s peak is decreasing and the contribution of the Co 3p peak is increasing, according to the transition from the LLZ thin film into the $LiCoO_2$ pellet. The decrease of the Li 1s intensity up to a depth of 2769.2 and 3606.4 nm, where in comparison to the $LiCoO_2$ reference spectra nearly no Li 1s peak is detected, indicates a depletion of Li at the interfaces between the LLZ and the $LiCoO_2$ [10].

The changing intensity of different elemental peaks in the depth series of successive XPS spectra can be also directly plotted against the sputter time or sputter depth. An example for such a XPS depth profile is given in Fig. 6.4, showing the elemental distribution of a LLZ thin film with an Au layer on top and a Pt current collector below [29]. The Au and Pt layers are represented by the Au 4f, respectively the Pt 4f peak intensities plotted over the sputter time. Within the region of the LLZ thin

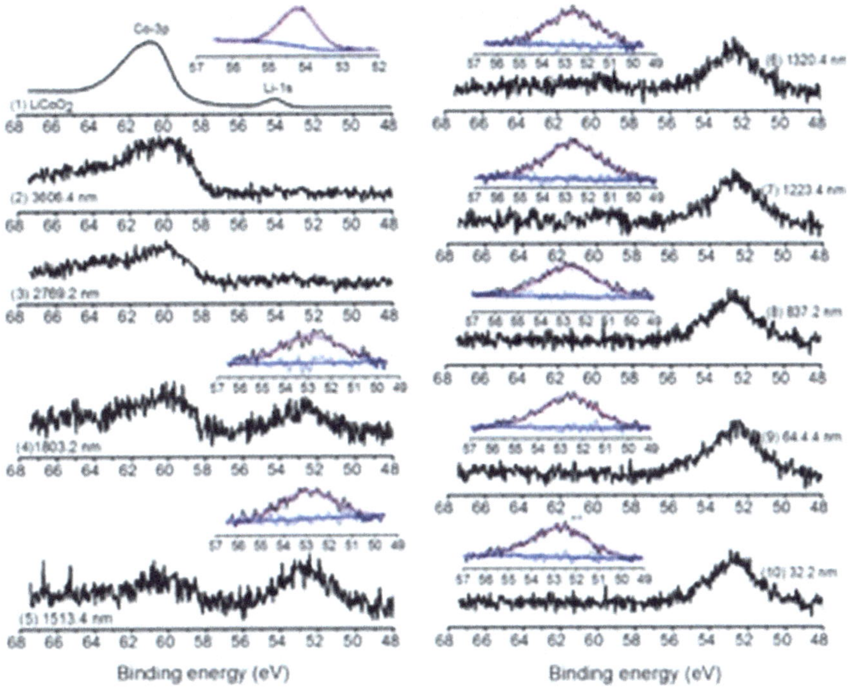

Fig. 6.3 XPS spectra of Co 3p and Li 1s for an LLZ thin film deposited on a $LiCoO_2$ pellet at different sample depths. The depth of 32.2 nm corresponds to a region in the LLZ thin film close to the surface and the depth of 3606.4 nm corresponds to a region within the $LiCoO_2$ pellet. The inset shows a magnification of the Li 1s contribution to the spectra and the first spectra is a reference for $LiCoO_2$. Graph reproduced from [10] (https://doi.org/10.1149/2.0621706jes) © Mina Zarabian et al. (Licensed under CC BY 4.0) https://creativecommons.org/licenses/by/4.0/

film only the peak intensities of O 1s, Zr 3d and La 3d are plotted for the respective elements. The Li elemental distribution was not plotted, due to the already discussed overlap with the Zr 4s peak [29].

6.4.4 *Ion Beam Analysis*

In principle, ion beam based analytical techniques are very common in R&D for the general compositional characterization of thin films independent of their material composition. As exemplary ion beam techniques for the analysis of Li-containing garnet thin films time of flight secondary ion mass spectrometry (ToF-SIMS) as well as Rutherford backscattering (RBS) and nuclear reaction analysis (NRA) are discussed in detail within this section. While these three techniques are all based on

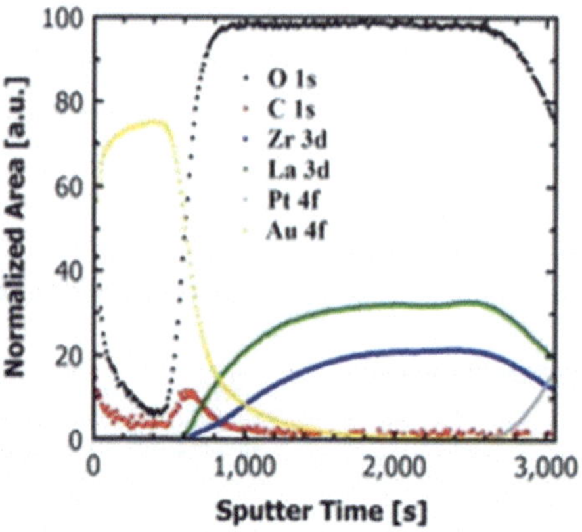

Fig. 6.4 XPS profile of an LLZ (O 1s, Zr 3d, La 3d) thin film grown by CO_2-laser assisted CVD with an Au layer on top (Au 4f) and a Pt (Pt 4f) current collector below. The lithium distribution is not plotted because of the overlap with the Zr 4s peak. Republished with permission of Journal of Electrochemical Society, from [29], Copyright 2016; permission conveyed through Copyright Clearance Center, Inc

the probing of the investigated sample with an ion beam, the respective analytical principle for determining the elemental composition is different.

In case of ToF-SIMS a focused ion beam (e.g., Bi^+, Ga^+) is scanned over the sample surface and the sputtered secondary ions are analyzed according to their mass to charge ratio. In addition to the analysis ion beam a further ion beam (e.g., Cs^+, O_2^+) can be used to erode the sample between two consecutive analysis ion pulses and hence providing a depth-resolved elemental distribution of the investigated thin film. In principle ToF-SIMS is capable to detect all elements in parallel with a good lateral resolution and an excellent depth resolution in the nanometer range. Another advantage of ToF-SIMS is its low detection limit, which results in high sensitivity for lithium. On the other hand, disadvantages of ToF-SIMS are the destructive analysis principle and the fact that the measurements in general only provide semi-quantitative information on the elemental composition. A direct quantification of SIMS depth profiles is not possible, due to so-called matrix effects, like for example the sputter yield or the ionization probability, which are element specific and also dependent on the surrounding sample matrix.

The use of ToF-SIMS is already very common for the elemental analysis of garnet thin films and there are several publications using this technique for the compositional analysis [8, 11, 22, 26, 29, 34, 35]. The impact of the processing temperature on the elemental distribution in sputter deposited [26] or sol–gel garnet thin films [8] is for example one aspect, that was investigated by ToF-SIMS. In case of the sputter deposited garnet thin films it was possible to show by a qualitative comparison of ToF-SIMS depth profiles that the lithium content in the thin films is increasing when the applied temperature during the deposition is increased from 150 °C up to 800 °C [26]. Figure 6.5 shows the respective depth profiles of the thin films deposited at

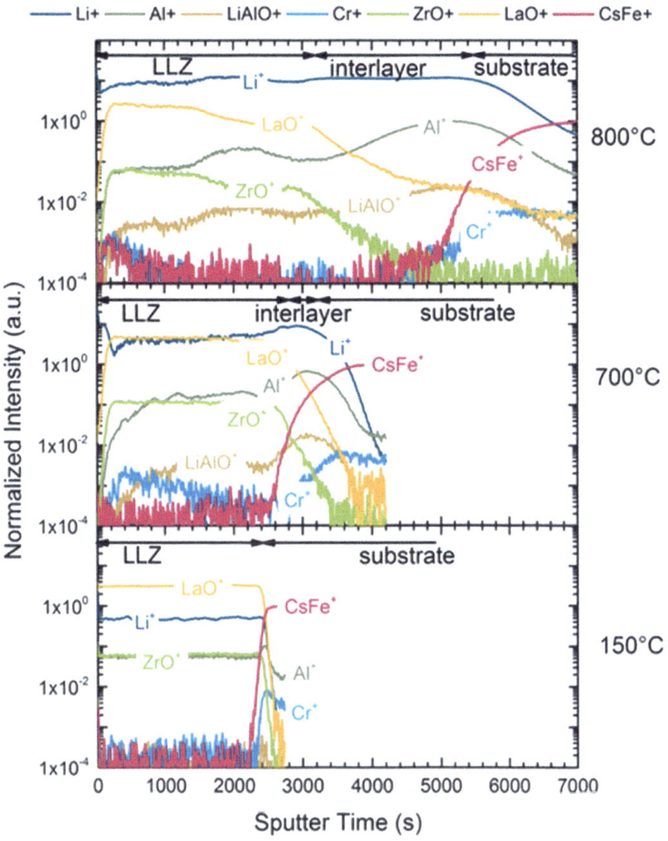

Fig. 6.5 Elemental composition of garnet thin film deposited at 150, 700 and 800 °C analyzed by ToF-SIMS. The changing ration of the Li^+/LaO^+ secondary ions indicates an increase of lithium concentration with increasing temperature. Reprinted from [26], Copyright 2016, with permission from Elsevier

different temperatures. Though it is not possible to directly quantify the lithium concentration out of this data, the change in the ratio of the Li^+/LaO^+ secondary ions is a qualitative indication for the increase in the lithium content. In addition, the depth profiles also nicely illustrate the growth of an interlayer between the thin film (Li^+, LaO^+, ZrO^+ secondary ions) and the metal foil substrate ($CsFe^+$ secondary ions) with increasing deposition temperature. This growing interlayer is also in agreement with the published scanning electron microscopy images and the increased intensity of $LiAlO^+$ secondary ions is in line with the detection of a γ-$LiAlO_2$ impurity phase in the XRD of the 800 °C sample.

Another aspect which can be visualized by ToF-SIMS is the formation of surface layers on top of the garnet thin films due to e.g., reaction with moisture or air [22, 29, 35]. Saccoccio et al. [22] and Loho et al. [29], for example, observed in a combined analysis by ToF-SIMS and XPS the formation of a Li_2CO_3 surface layer on garnet thin films due to a reaction with atmospheric CO_2 during sample handling. Similar observations were also published by Uhlenbruck et al. where they used ToF-SIMS, RBS, and NRA to visualize the formation of a Li_2CO_3 surface layer [35].

The analytical methods of RBS and NRA are based on the interaction of a high energetic ion beam (e.g., H^+, He^+) with the elemental nuclei inside the investigated sample. RBS is based on the energetical analysis of the backscattered ions. While this method can nicely be used to derive the concentration of relatively heavy elements within the sample, it is not very sensitive for the quantification of light elements like lithium. Therefore the lithium concentration is derived by measuring in parallel to the backscattered ions also the particles which are generated by an ion beam induced nuclear reaction. For the quantification of lithium by means of NRA commonly the reaction $^7Li(p,\alpha_0)^4He$ is used. The impinging protons are reacting with the ^7Li isotopes in the sample generating a charged alpha particle with characteristic energy and a ^4He atom. By detecting the alpha particles the lithium concentration can be calculated taking into account the energy-dependent cross-section of the particular nuclear reaction. For the depth dependency of the concentration profile the energy loss of the incident proton and of the outcoming alpha particle due to interaction with the solid has to be taken into account. In this way a probing depth in the μm range can be achieved, which is very feasible especially for the analysis of thin film samples.

An example for the analysis of Ta-doped LLZ thin films (LLZT) by means of NRA was recently published by Uhlenbruck et al. [35]. In combining the analytical techniques of NRA, RBS and also the previously discussed ToF-SIMS they show in their depth-resolved study how the lithium distribution in the thin films is changing when the material reacts with water. By ToF-SIMS they were able to qualitatively investigate the change in the lithium depth profile and visualize the formation of a reaction layer at the surface. Accordingly to the ToF-SIMS results the RBS and NRA spectrum, which are shown in Fig. 6.6, confirmed the formation of this reaction layer and also provided quantitative insight into its composition. In the NRA spectrum, where the counts of the detected alpha particles of the $^7Li(p,\alpha_0)^4He$ reaction are plotted against the detector channel number, a Li-rich surface layer was detected.

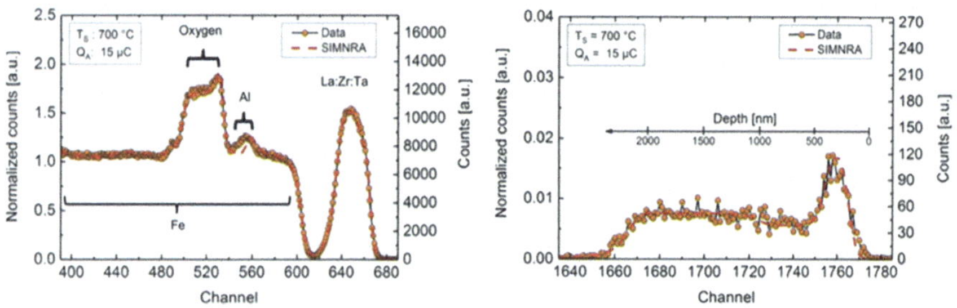

Fig. 6.6 RBS (left) and NRA (right) spectrum of an LLZT thin film deposited at 700 °C. The NRA shows the formation of a lithium-rich surface layer. Reprinted from [35], Copyright 2018, with permission from Elsevier

Here the depth information is given by the channel number, which is proportional to the particle energy, starting at the thin film surface with around a channel number of 1770. In combination with the in parallel detected RBS spectrum the surface layer composition was derived to be $Li_{1.9}C_{1.1}O_3$ indicating the formation of lithium carbonate as already described in literature [36]. With the depth resolution of the NRA and the complimentary ToF-SIMS they could also visualize, a minimum in the lithium concentration below the surface layer indicating that the lithium is taken out from the LLZT during the reaction process.

6.4.5 Neutrons for Lithium Quantification

While the application of neutron based techniques for the structural investigation of different materials is well established, in recent years also the usage of neutrons for quantitative, nondestructive depth profiling of lithium compounds gained a lot of interest. This so-called neutron depth profiling (NDP) of lithium containing materials is based on the interaction of neutrons with the ^{6}Li atoms in the sample. As products of this reaction a monoenergetically charged alpha (2055 keV) and triton (2727 keV) particles are released [37]. The number of detected alpha or triton particles can be used to calculate the lithium concentration within the sample. Through the calculation of the energy loss of the detected particles, which originates from the interaction with the solid matrix, the depth dependency of the concentration profile can be derived. Usually, the energy spectrum of the triton instead of the alpha particles is used for quantification, because of two advantages. First, the achievable probing depth for the tritons is larger due to the higher initial energy. Second, the energy spectrum of the tritons is not overlapped, while the energy spectrum of the alpha particles is overlapped by tritons with lower energies, which originate from deeper sample depths [38]. A general drawback of the NDP method is the fact that the neutrons only react with the less abundant ^{6}Li isotope (7.5% abundance), which has a great impact on the needed measurement time. Furthermore, while having high sensitivity for light elements like lithium, it is not possible to quantify in parallel other heavy elements of the sample matrix like, e.g., Zr, La.

Until now there is no report in literature about the usage of NDP for garnet thin films, but Wang et al. already showed the feasibility of the technique to analyze in situ the lithium metal–garnet interface of a bulk $Li_{6.75}La_{2.75}Ca_{0.25}Zr_{1.75}Nb_{0.25}O_{12}$ (LLZCN) sample in combination with Li metal and a carbon nanotube electrode [39]. Figure 6.7 shows exemplary ^{3}H energy spectra and a 2D projection of the collected NDP spectra. In order to cancel out the overlap of the alpha and triton particles in the energy spectra, the cell setup is covered with a 7.5 μm Kapton film, which is sufficient to shield the alpha particles from the detector. The energy spectra show the count rate of detected ^{3}H particles plotted against the particle energy at different cycling times. The plotted particle energy reflects in this graphic the sample depth from which the particle originated and hence the depth information of the lithium profile. The pairs of plotted spectra (0 and 0.5 h, 7 and 8 h, 19 and 19.5 h) correspond to the beginning

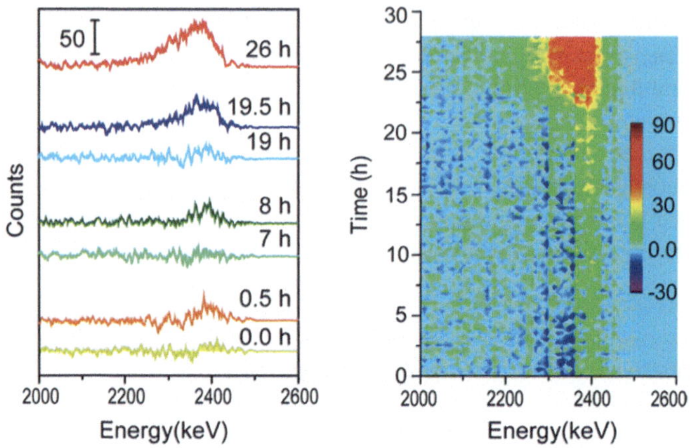

Fig. 6.7 ^{3}H energy spectra (left) and 2D projection (right) of the collected NDP spectra showing Li plating in a symmetric Li/LLZCN/Li cell. The paired spectra, which correspond to the beginning and the end of a plating cycle, show a reversible Li plating and striping. Shortcutting the cell at increased current density is visible by higher counts in energy spectra (26 h). Reprinted with permission from [39]. Copyright 2017 American Chemical Society

and the end of a plating cycle. Accordingly to the good reversibility of Li plating and stripping the peak around 2400 keV is close to zero at the beginning of each cycle. Only when the current density is further increased with ongoing cycling time to a point where the cell shows a short circuit the NDP counts at the outer layers start to increase. This is shown in Fig. 6.5 by the ^{3}H energy spectra after 26 h and also nicely visible in the 2D projection of the NDP spectra. The 2D projection shows a permanent increase in the count rate (reflected by the color map) after the short circuit occurs after 21 h of cycling [39].

While the paper by Wang et al. [39] demonstrates the feasibility of NDP for the garnet material class there is also another publication by Oudenhoven et al. [40] which shows that the techniques can be used to monitor the lithium movement in a thin film all-solid-state battery [40].

6.5 Summary

In Fig. 6.8 the ionic conductivities and activation energies of the selected different methods mentioned in the sections before are summarized in comparison to the ionic conductivity of bulk-$Li_7La_3Zr_2O_{12}$ and LiPON as thin film benchmark material.

The ionic conductivities of the garnet thin films are higher than for LiPON, which is the only commercialized solid-state electrolyte, so far.

For Zr-free garnets, the conductivities and activation energy of the CVD-grown $Li_5La_3Ta_2O_{12}$ are close to the data for bulk $Li_5La_3Ta_2O_{12}$, whereas the conductivity

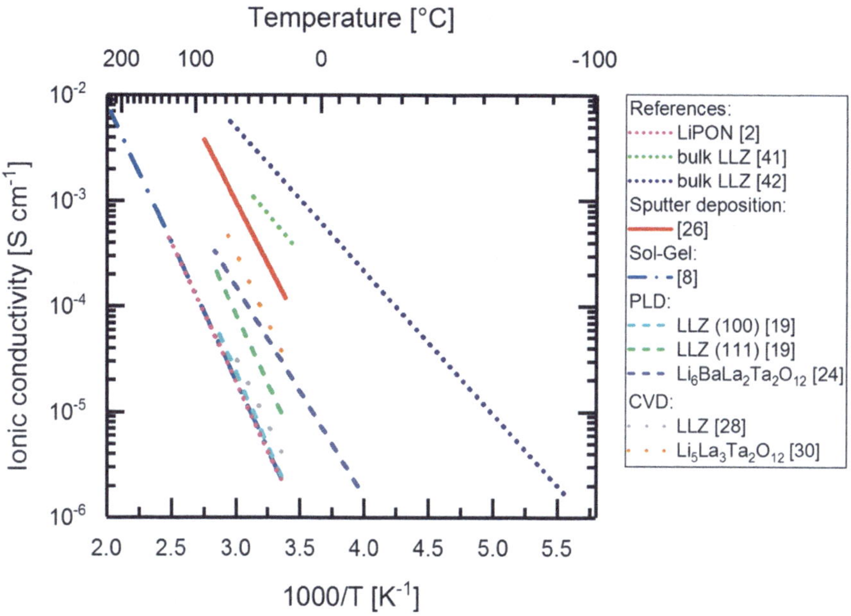

Fig. 6.8 Temperature dependence of ionic conductivities of garnet-structured thin films deposited by different deposition techniques. As a comparison, the values of LiPON and unsubstituted and optimized $Li_7La_3Zr_2O_{12}$ are added for reference

of the $Li_6BaLa_2Ta_2O_{12}$ thin film is one order of magnitude lower than for comparable bulk material.

In case of Zr-containing thin films, the conductivities of bulk-$Li_7La_3Zr_2O_{12}$, which has conductivities in the range from 3.97×10^{-4} to 1.3×10^{-3} S cm^{-1} at room temperature [41, 42], are not reached yet. It can be seen, that the activation energies of the thin films are higher than the activation energies of the bulk material. In general, it is assumed that the thin films are consisting of smaller particles than the bulk material. For this reason, there are more grain boundaries in the thin films. The activation energy of the conduction mechanism across the grain boundaries was determined by Tenhaeff et al. [43] in Al-doped $Li_7La_3Zr_2O_{12}$ to be higher than the activation energy for the conduction across a grain. However, this publication can explain the trend, but the determined values for the thin films are in the range from 0.47 to 0.55 eV, which is still higher than the 0.44 eV determined for the conductivity across the grain boundaries by Tenhaeff et al. [43].

For all kinds of garnets and deposition methods there are several challenges regarding the development of an all-solid-state battery with a garnet thin film electrolyte.

Challenge 1: Deposition temperature

Depositions from the gas phase at room temperature, i.e., without substrate heating, do not show satisfying ionic conductivities. These films show an amorphous or nanocrystalline structure, and even though less loss of lithium is expected at this temperature, the deposition of stoichiometric thin films is already difficult, e.g., PLD processing showed a reduced amount of Li [21] and in sputter-deposited films, Li was not really detected [26]. Post-lithiation of thin films during an annealing process worked only for very thin films [20].

As a consequence, garnet thin films should be deposited at elevated temperatures. For example, from bulk processing it is known that temperatures of 650–700 °C and higher are required for the formation of a well crystallized garnet phase. At lower temperatures the pyrochlore phase $La_2Zr_2O_7$ is formed. This temperature range is confirmed by thin film deposition experiments from gas phase as well as with wet-chemical routes, if 700 °C or higher temperatures were used the garnet structure could be verified by XRD [8–11, 18, 19, 26–30]. The work of Reinacher et al. [24] showed that Zr-free garnets could be deposited at even lower temperatures with a sufficient degree of crystallinity.

These high deposition temperatures lead to further problems during processing. As already described in the literature, garnets have a low thermal stability with cathode materials [44–46], which causes secondary phase formation before sintering processes could form a suitable interface. Therefore, the processing has to be adjusted that these reactions could be avoided. Bitzer et al. already showed that a very fast wet-chemical processing could avoid unwanted side reactions, like the formation of $La_2Li_{0.5}Co_{0.5}O_4$ at the interface between $LiCoO_2$ and $Li_7La_3Zr_2O_{12}$ [9]. Another option would be a process design which is not exceeding the critical temperatures, e.g., sputter processes with an applied bias voltage, which enables the deposition of material far from thermodynamic equilibrium. It was already shown that such processes are suitable for the deposition of $LiCoO_2$ thin films with the high temperature crystal structure [47]. Another possibility is the deposition of the cathode on top of the electrolyte and the in situ formation of the anode during cycling. Therefore, the garnet is deposited on a suitable current collector. In a second step, the cathode material is deposited on top of the electrolyte. In the first charging cycle, the anode is formed between the current collector on the anode side and the garnet electrolyte. This concept was already shown to be working for LiPON-based thin film cells [48]. The disadvantage would be that the whole Li of the cell is provided by the cathode, so that losses during cathode processing could not be balanced by an excess of Li at the anode side. It should be also considered that the deposition of garnet thin films at elevated temperatures should always take into account that the coefficient of thermal expansion (CTE) of the substrate should match to the CTE of the garnets, which was determined for Al-substituted LLZ to be 15.498×10^{-6} K^{-1} [49]. Otherwise the formation of cracks in the garnet or faulting of the thin film would be the consequence.

Another important point which comes along with the deposition temperature is the crystallinity of the deposited thin films. Li-ion conducting garnets have high bulk conductivity and a low conductivity at the grain boundaries. Also the activation energy of the conduction process is influenced by the amount and thickness of the grain boundaries. Therefore, the deposited thin films should contain large grains, in ideal case with single-crystalline behavior from the surface of the thin film to the cathode interface.

Challenge 2: Surface to volume ratio of thin films

The second big challenge affects the geometry of the thin films. Thin films are known to have a high surface to volume ratio compared to bulk material. Therefore, lithium can evaporate easily from the surface of the thin film, leading once more to secondary phases, like the pyrochlore $La_2Zr_2O_7$. For bulk material the evaporation of Li is avoided by using a compensating powder bed. This method is not easily transferable to thin film processing. Powder of a Li source and the thin film has to be well separated in order to avoid contamination of the thin film with powder.

At last it can be concluded that the first steps toward the successful deposition of garnet-structured thin films are done. Different deposition methods already showed that dense thin films with a garnet structure can be formed. Currently, the main focus of the research is spent to increase the conductivity of the thin films as well as the realization of suitable laminates of electrode and electrolyte materials.

References

1. Schmuch R, Wagner R, Hörpel G, Placke T, Winter M (2018) Performance and cost of materials for lithium-based rechargeable automotive batteries. Nat Energy 3(4):267–278. https://doi.org/10.1038/s41560-018-0107-2
2. Yu X, Bates JB, Jellison GE, Hart FX (1997) A stable thin-film lithium electrolyte: lithium phosphorus oxynitride. J Electrochem Soc 144(2):524–532. https://doi.org/10.1149/1.1837443
3. Deng Y-F, Zhao S-X, Xu Y-H, Nan C-W (2014) Effect of the morphology of Li–La–Zr–O solid electrolyte coating on the electrochemical performance of spinel $LiMn_{1.95}Ni_{0.05}O_{3.98}F_{0.02}$ cathode materials. J Mater Chem A 2(44):18889–18897. https://doi.org/10.1039/c4ta03772c
4. Van Gestel T, Sebold D, Buchkremer HP (2015) Processing of 8YSZ and CGO thin film electrolyte layers for intermediate- and low-temperature SOFCs. J Eur Ceram Soc 35(5):1505–1515. https://doi.org/10.1016/j.jeurceramsoc.2014.11.017
5. Bram M, Dornseiffer J, Hoffmann J, Gestel T, Meulenberg Wilhelm A, Stöver D, Bhandarkar S (2015) Inkjet printing of microporous silica gas separation membranes. J Am Ceram Soc 98(8):2388–2394. https://doi.org/10.1111/jace.13657
6. Teucher G, Van Gestel T, Krott M, Gehrke H-G, Eichel R-A, Uhlenbruck S (2016) Processing of Al-doped ZnO protective thin films on aluminum current collectors for lithium ion batteries. Thin Solid Films 619:302–307. https://doi.org/10.1016/j.tsf.2016.10.047
7. Chen R-J, Huang M, Huang W-Z, Shen Y, Lin Y-H, Nan C-W (2014) Sol–gel derived Li–La–Zr–O thin films as solid electrolytes for lithium-ion batteries. J Mater Chem A 2(33):13277–13282. https://doi.org/10.1039/C4TA02289K
8. Tadanaga K, Egawa H, Hayashi A, Tatsumisago M, Mosa J, Aparicio M, Duran A (2015) Preparation of lithium ion conductive Al-doped $Li_7La_3Zr_2O_{12}$ thin films by a sol–gel process. J Power Sources 273:844–847. https://doi.org/10.1016/j.jpowsour.2014.09.164
9. Bitzer M, Van Gestel T, Uhlenbruck S, Buchkremer HP (2016) Sol-gel synthesis of thin solid $Li_7La_3Zr_2O_{12}$ electrolyte films for Li-ion batteries. Thin Solid Films 615:128–134. https://doi.org/10.1016/j.tsf.2016.07.010

10. Zarabian M, Bartolini M, Pereira-Almao P, Thangadurai V (2017) X-ray photoelectron spectroscopy and AC impedance spectroscopy studies of Li–La–Zr–O solid electrolyte thin film/$LiCoO_2$ cathode interface for all-solid-state Li batteries. J Electrochem Soc 164(6):A1133–A1139. https://doi.org/10.1149/2.0621706jes
11. Uhlenbruck S, Dornseiffer J, Lobe S, Dellen C, Tsai C-L, Gotzen B, Sebold D, Finsterbusch M, Guillon O (2017) Cathode-electrolyte material interactions during manufacturing of inorganic solid-state lithium batteries. J Electroceram 38(2):197–206. https://doi.org/10.1007/s10832-016-0062-x
12. Ramakumar S, Deviannapoorani C, Dhivya L, Shankar LS, Murugan R (2017) Lithium garnets: synthesis, structure, Li^+ conductivity, Li^+ dynamics and applications. Prog Mater Sci 88:325–411. https://doi.org/10.1016/j.pmatsci.2017.04.007
13. Mücke R, Menzler Norbert H, Buchkremer Hans P, Stöver D (2009) Cofiring of thin zirconia films during SOFC manufacturing. J Am Ceram Soc 92(s1):S95–S102. https://doi.org/10.1111/j.1551-2916.2008.02707.x
14. Keuter T, Mauer G, Vondahlen F, Iskandar R, Menzler NH, Vaßen R (2016) Atomic-layer-controlled deposition of TEMAZ/O_2–ZrO_2 oxidation resistance inner surface coatings for solid oxide fuel cells. Surf Coat Technol 288:211–220. https://doi.org/10.1016/j.surfcoat.2016.01.026
15. Nédélec R, Uhlenbruck S, Sebold D, Haanappel VAC, Buchkremer HP, Stöver D (2012) Dense yttria-stabilised zirconia electrolyte layers for SOFC by reactive magnetron sputtering. J Power Sources 205:157–163. https://doi.org/10.1016/j.jpowsour.2012.01.054
16. Striebel KA, Deng CZ, Wen SJ, Cairns EJ (1996) Electrochemical behavior of $LiMn_2O_4$ and $LiCoO_2$ thin films produced with pulsed laser deposition. J Electrochem Soc 143(6):1821–1827
17. Zhao S, Fu Z, Qin Q (2002) A solid-state electrolyte lithium phosphorus oxynitride film prepared by pulsed laser deposition. Thin Solid Films 415(1):108–113. https://doi.org/10.1016/S0040-6090(02)00543-6
18. Tan JJ, Tiwari A (2012) Fabrication and characterization of $Li_7La_3Zr_2O_{12}$ thin films for lithium ion battery. ECS Solid State Lett 1(6):Q57–Q60. https://doi.org/10.1149/2.013206ssl
19. Kim S, Hirayama M, Taminato S, Kanno R (2013) Epitaxial growth and lithium ion conductivity of lithium-oxide garnet for an all solid-state battery electrolyte. Dalton Trans 42(36):13112–13117. https://doi.org/10.1039/c3dt51795k
20. Rawlence M, Garbayo I, Buecheler S, Rupp JLM (2016) On the chemical stability of post-lithiated garnet Al-stabilized $Li_7La_3Zr_2O_{12}$ solid state electrolyte thin films. Nanoscale 8(31):14746–14753. https://doi.org/10.1039/C6NR04162K
21. Park JS, Cheng L, Zorba V, Mehta A, Cabana J, Chen G, Doeff MM, Richardson TJ, Park JH, Son J-W, Hong W-S (2015) Effects of crystallinity and impurities on the electrical conductivity of Li–La–Zr–O thin films. Thin Solid Films 576:55–60. https://doi.org/10.1016/j.tsf.2014.11.019
22. Saccoccio M, Yu J, Lu Z, Kwok SCT, Wang J, Yeung KK, Yuen MMF, Ciucci F (2017) Low temperature pulsed laser deposition of garnet $Li_{6.4}La_3Zr_{1.4}Ta_{0.6}O_{12}$ films as all solid-state lithium battery electrolytes. J Power Sources 365:43–52. https://doi.org/10.1016/j.jpowsour.2017.08.020
23. Park HY, Lee SR, Lee YJ, Cho BW, Cho WI (2005) Bias sputtering and characterization of $LiCoO_2$ thin film cathodes for thin film microbattery. Mater Chem Phys 93(1):70–78. https://doi.org/10.1016/j.matchemphys.2005.02.024
24. Reinacher J, Berendts S, Janek J (2014) Preparation and electrical properties of garnet-type $Li_6BaLa_2Ta_2O_{12}$ lithium solid electrolyte thin films prepared by pulsed laser deposition. Solid State Ion 258:1–7. https://doi.org/10.1016/j.ssi.2014.01.046
25. Kalita DJ, Lee SH, Lee KS, Ko DH, Yoon YS (2012) Ionic conductivity properties of amorphous Li–La–Zr–O solid electrolyte for thin film batteries. Solid State Ion 229:14–19. https://doi.org/10.1016/j.ssi.2012.09.011
26. Lobe S, Dellen C, Finsterbusch M, Gehrke HG, Sebold D, Tsai CL, Uhlenbruck S, Guillon O (2016) Radio frequency magnetron sputtering of $Li_7La_3Zr_2O_{12}$ thin films for solid-state batteries. J Power Sources 307:684–689. https://doi.org/10.1016/j.jpowsour.2015.12.054

27. Katsui H, Goto T (2015) Preparation of cubic and tetragonal $Li_7La_3Zr_2O_{12}$ film by metal organic chemical vapor deposition. Thin Solid Films 584:130–134. https://doi.org/10.1016/j.tsf.2014.11.094
28. Kazyak E, Chen K-H, Wood KN, Davis AL, Thompson T, Bielinski AR, Sanchez AJ, Wang X, Wang C, Sakamoto J, Dasgupta NP (2017) Atomic layer deposition of the solid electrolyte garnet $Li_7La_3Zr_2O_{12}$. Chem Mater 29(8):3785–3792. https://doi.org/10.1021/acs.chemmater.7b00944
29. Loho C, Djenadic R, Bruns M, Clemens O, Hahn H (2017) Garnet-type $Li_7La_3Zr_2O_{12}$ solid electrolyte thin films grown by CO_2-laser assisted CVD for all-solid-state batteries. J Electrochem Soc 164(1):A6131–A6139. https://doi.org/10.1149/2.0201701jes
30. Loho C, Djenadic R, Mundt P, Clemens O, Hahn H (2017) On processing-structure-property relations and high ionic conductivity in garnet-type $Li_5La_3Ta_2O_{12}$ solid electrolyte thin films grown by CO_2-laser assisted CVD. Solid State Ion 313:32–44. https://doi.org/10.1016/j.ssi.2017.11.005
31. Burgess S, Sagar J, Holland J, Li X, Bauer F (2017) Ultra-low kV EDS—a new approach to improved spatial resolution, surface sensitivity, and light element compositional imaging and analysis in the SEM. Microscopy Today 25(2):20–29. https://doi.org/10.1017/S1551929517000013
32. Uhart A, Ledeuil JB, Pecquenard B, Le Cras F, Proust M, Martinez H (2017) Nanoscale chemical characterization of solid-state microbattery stacks by means of auger spectroscopy and ion-milling cross section preparation. ACS Appl Mater Interfaces 9(38):33238–33249. https://doi.org/10.1021/acsami.7b07270
33. Azmi R, Trouillet V, Strafela M, Ulrich S, Ehrenberg H, Bruns M (2017) Surface analytical approaches to reliably characterize lithium ion battery electrodes. Surf Interface Anal 50(1):43–51. https://doi.org/10.1002/sia.6330
34. Garbayo I, Struzik M, Bowman William J, Pfenninger R, Stilp E, Rupp Jennifer LM (2018) Glass-type polyamorphism in Li-garnet thin film solid state battery conductors. Adv Energy Mater:1702265. https://doi.org/10.1002/aenm.201702265
35. Uhlenbruck S, Dellen C, Möller S, Lobe S, Tsai C-L, Finsterbusch M, Bram M, Guillon O (2018) Reactions of garnet-based solid-state lithium electrolytes with water a depth-resolved study. Solid State Ion 320:259–265. https://doi.org/10.1016/j.ssi.2018.03.004
36. Xia W, Xu B, Duan H, Tang X, Guo Y, Kang H, Li H, Liu H (2017) Reaction mechanisms of lithium garnet pellets in ambient air: the effect of humidity and CO_2. J Am Ceram Soc 100(7):2832–2839. https://doi.org/10.1111/jace.14865
37. Downing RG, Lamaze GP, Langland JK, Hwang ST (1993) Neutron depth profiling: overview and description of NIST facilities. J Res Nat Inst Stand Technol 98(1):109–126. https://doi.org/10.6028/jres.098.008
38. Nagpure SC, Downing RG, Bhushan B, Babu SS, Cao L (2011) Neutron depth profiling technique for studying aging in Li-ion batteries. Electrochim Acta 56(13):4735–4743. https://doi.org/10.1016/j.electacta.2011.02.037
39. Wang C, Gong Y, Dai J, Zhang L, Xie H, Pastel G, Liu B, Wachsman E, Wang H, Hu L (2017) In situ neutron depth profiling of lithium metal–garnet interfaces for solid state batteries. J Am Chem Soc 139(40):14257–14264. https://doi.org/10.1021/jacs.7b07904
40. Oudenhoven JFM, Labohm F, Mulder M, Niessen RAH, Mulder FM, Notten PHL (2011) in situ neutron depth profiling: a powerful method to probe lithium transport in micro-batteries. Adv Mater 23(35):4103–4106. https://doi.org/10.1002/adma.201101819
41. Murugan R, Thangadurai V, Weppner W (2007) Fast lithium ion conduction in garnet-type $Li_7La_3Zr_2O_{12}$. Angew Chem Int Ed 46(41):7778–7781. https://doi.org/10.1002/anie.200701144
42. Buannic L, Orayech B, López Del Amo J-M, Carrasco J, Katcho NA, Aguesse F, Manalastas W, Zhang W, Kilner J, Llordés A (2017) Dual substitution strategy to enhance Li^+ ionic conductivity in $Li_7La_3Zr_2O_{12}$ solid electrolyte. Chem Mater 29(4):1769–1778. https://doi.org/10.1021/acs.chemmater.6b05369

43. Tenhaeff WE, Rangasamy E, Wang Y, Sokolov AP, Wolfenstine J, Sakamoto J, Dudney NJ (2014) Resolving the grain boundary and lattice impedance of hot-pressed $Li_7La_3Zr_2O_{12}$ garnet electrolytes. ChemElectroChem 1(2):375–378. https://doi.org/10.1002/celc.201300022
44. Miara L, Windmüller A, Tsai C-L, Richards WD, Ma Q, Uhlenbruck S, Guillon O, Ceder G (2016) About the compatibility between high voltage spinel cathode materials and solid oxide electrolytes as a function of temperature. ACS Appl Mater Interfaces 8(40):26842–26850. https://doi.org/10.1021/acsami.6b09059
45. Thangadurai V, Weppner W (2005) Investigations on electrical conductivity and chemical compatibility between fast lithium ion conducting garnet-like $Li_6BaLa_2Ta_2O_{12}$ and lithium battery cathodes. J Power Sources 142(1–2):339–344. https://doi.org/10.1016/j.jpowsour.2004.11.001
46. Ren Y, Liu T, Shen Y, Lin Y, Nan C-W (2016) Chemical compatibility between garnet-like solid state electrolyte $Li_{6.75}La_3Zr_{1.75}Ta_{0.25}O_{12}$ and major commercial lithium battery cathode materials. J Mater 2(3):256–264. https://doi.org/10.1016/j.jmat.2016.04.003
47. Tintignac S, Baddour-Hadjean R, Pereira-Ramos J-P, Salot R (2012) High performance sputtered $LiCoO_2$ thin films obtained at a moderate annealing treatment combined to a bias effect. Electrochim Acta 60:121–129. https://doi.org/10.1016/j.electacta.2011.11.033
48. Neudecker BJ, Dudney NJ, Bates JB (2000) "Lithium-free" thin-film battery with in situ plated li anode. J Electrochem Soc 147(2):517–523
49. Hubaud AA, Schroeder DJ, Ingram BJ, Okasinski JS, Vaughey JT (2015) Thermal expansion in the garnet-type solid electrolyte $(Li_{7-x}Al_{x/3})La_3Zr_2O_{12}$ as a function of Al content. J Alloy Compd 644:804–807. https://doi.org/10.1016/j.jallcom.2015.05.067

Chapter 7
Ultrathin Garnet-Type Electrolytes

Xufeng Yan and Weiqiang Han

Abstract Recent years, all solid-state battery (ASSB) has attracted much attention mainly due to the safety concerns. As the key factor of ASSB, the solid electrolyte determines the performance of the battery. Among the numerous electrolytes, the garnet type $Li_7La_3Zr_2O_{12}$ (LLZO) possesses the advantages of ionic conductivity and excellent chemical stability. Compared to the LLZO bulk ceramic electrolyte, LLZO thin films have advantages in better contact with electrodes, less lithium diffusion time in electrolytes and smaller resistance in ASSB. In this chapter, we will introduce LLZO films prepared by different thin-film preparation methods, including pulsed laser deposition (PLD), radio frequency (RF), sol–gel method, chemical vapor deposition (CVD), focused ion beam (FIB) milling, atomic layer deposition (ALD), and wet coating. Their methods, performance, and applications will be detailed expounded in each section.

7.1 Introduction

During the past few years, a lot of efforts have been made to improve the ionic conductivity by elements doping in ceramic bulk electrolyte after the first report of $Li_7La_3Zr_2O_{12}$ (LLZO) in 2007 [1–3]. Till now, the ionic conductivity of bulk LLZO has a value as high as 1.2×10^{-3} S·cm^{-1}, and much work has been put in the applications of bulk LLZO in the all solid state batteries [4–6].

Compared to the LLZO bulk ceramic electrolyte, only a few works have been made to the LLZO thin film. There are two advantages of ultrathin film. First, the lithium diffusion time will be decreased a lot. The lithium diffusion time t and the thickness of electrolyte layer L are proportional to the square, $t = L^2/D$ (D, diffusion constant). When L is reduced, t is more largely reduced. Second, the resistance of electrolytes will be much lower than ceramic bulk electrolyte. By using the ultrathin electrolyte, the total impedance of all-solid-state batteries will be reduced [7–9].

X. Yan · W. Han (✉)
Department of Materials Science and Engineering, Zhejiang University, Hangzhou 310027, China
e-mail: hanwq@zju.edu.cn

R. Murugan and W. Weppner (eds.), *Solid Electrolytes for Advanced Applications*,
https://doi.org/10.1007/978-3-030-31581-8_7

According to the reports, LLZO films have been prepared by different thin-film preparation methods, including pulsed laser deposition (PLD), radio frequency (RF), sol–gel method, and chemical vapor deposition (CVD). They are summarized in detail as follows.

7.2 Pulsed Laser Deposition (PLD)

Pulsed laser deposition (PLD) was first carried out as a method to prepare LLZO films in the year 2012 [10, 11]. The prepared amorphous LLZO had a large optical bandgap of 5.13 eV, and the value was reduced to 3.64 eV after the treatment of annealing [10]. With a laser-annealing treatment, the ionic conductivity was increased from 3.35×10^{-7} to 7.36×10^{-7} S cm^{-1} [11]. Rawlence et al. [12] prepared LLZO film on the MgO substrate by the PLD technique (Fig. 7.1c), with a thickness of 380 nm. The film showed an ionic conductivity of 1.2×10^{-3} S cm^{-1} at 325 °C. Reinacher et al. [13] reported that the garnet-type $Li_6BaLa_2Ta_2O_{12}$ thin films were prepared on MgO (100) single crystals by the way of pulsed laser deposition. The ionic conductivity of the film was 1.7×10^{-6} S cm^{-1} at room temperature, which was a little lower than the bulk value of 5.0×10^{-6} S cm^{-1}. The PLD technique was also successfully used in the deposition of epitaxial LLZO film [14]. On the $Gd_3Ga_5O_{12}$ (GGG) substrate, LLZO (001) and LLZO (111) were obtained. At room temperature, the ionic conductivities in the grains of the (001) and (111) films were 2.5×10^{-6} and 1.0×10^{-5} S cm^{-1}. On the basis of work [15] of preparing the LLZO film on AAO (Fig. 7.1d, e), Park et al. [16] studied the effects of substrate temperatures and impurities on electrochemical properties of the film, which was prepared by the PLD method. In the work, the film was deposited both on Si and AAO substrate and the SEM of the section in Fig. 7.1a, b. The result showed that the film was in a mixture of cubic and tetragonal phases of LLZO in the condition of deposition temperature above 600 °C.

The film, deposited at 600 °C, exhibited the highest room temperature conductivity of 1.61×10^{-6} S cm^{-1}. Saccoccio et al. [17] studied the low-temperature PLD method in fabricating pure cubic LLZO film on MgO (100) substrates. According to that work, having a range of the deposition temperature from 50 to 700 °C, the ionic conductivities of the films were not significantly influenced (Table 7.1).

7.3 Radio Frequency (RF)

Radio-frequency (RF) magnetron sputtering is also a common technique to prepare thin films. Kalita et al. [19] first deposited amorphous lithium–lanthanum–zirconium–oxide (Li–La–Zr–O) film by the method of RF magnetron sputtering. The thickness of the film could be changed by varying the RF power. The thickness, which could be seen in Fig. 7.2a–c, were 561 nm at 40 W, 861 nm at 60 W, and

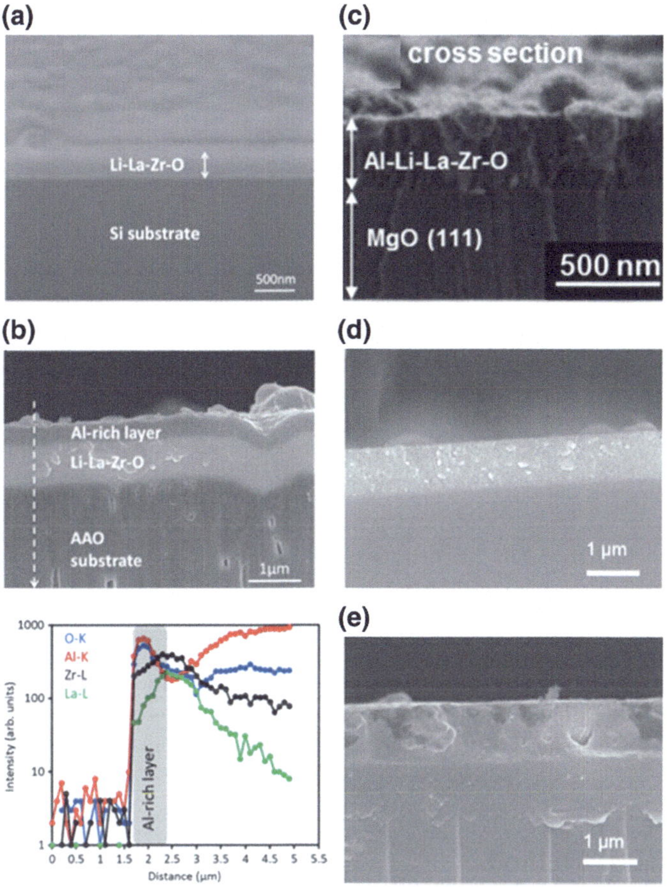

Fig. 7.1 **a** SEM cross-sectional image of a 200 nm thick Li–La–Zr–O film [16]; **b** SEM image of a polished cross-section of the Li–La–Zr–O film deposited on an AAO substrate at 700 °C and the corresponding EDS profile along the indicated line [16]; **c** An as-deposited $Li_7La_3Zr_2O_{12}$-based thin film deposited on a single-crystal MgO substrate at 50 °C by PLD, cross-sectional SEM micrograph [12]; **d**, **e** Cross-sectional SEM of LLZO films [15]; **d** Sapphire substrate at room temperature; **e** Sapphire substrate at 700 °C

941 nm at 80 W. The film which deposited at 40 W showed the highest ionic conductivity of 4×10^{-7} S cm^{-1}, compared to the 60 W and 80 W samples which showed the ionic conductivity values of 2×10^{-7} S cm^{-1} and 0.8×10^{-7} S cm^{-1}, respectively.

Compared with the amorphous lithium–lanthanum–zirconium–oxide (Li–La–Zr–O) film work [19], Lobe et al. [20] deposited Ta- and Al-substituted LLZO thin films on stainless steel substrates by radio-frequency (RF) magnetron sputtering at high temperature. According to the result of depositions at different temperatures, LLZO formation started at around 650 °C and a single-phase cubic thin film was obtained

Table 7.1 List of the garnet type films by the PLD technique with an outline of the material, year, substrate, thickness, room temperature ionic conductivity and activation energy

Material	Year	Substrate	Thickness	RT σ_{Li} [S cm^{-1}]	E_a [eV]	Ref.
$Li_7La_3Zr_2O_{12}$	2012	STO (100)	-	7.36×10^{-7}	0.32	[10]
		Sapphire (0001)		3.35×10^{-7}	0.36	[11]
Al:$Li_7La_3Zr_2O_{12}$	2016	MgO (100)	90-380 nm	1.2×10^{-3}	1.39	[12]
$Li_6BaLa_2Ta_2O_{12}$	2014	MgO (100)	~200 nm	2×10^{-6}	0.42	[13]
Al:$Li_7La_3Zr_2O_{12}$	2013	GGG (001)	26.2 nm	2.5×10^{-6}	0.55	[14]
		GGG (111)	30.4 nm	1.0×10^{-5}	0.52	
$Li_7La_3Zr_2O_{12}$	2015	MgO (100)	100–200 nm	1.61×10^{-6}	0.35	[16]
$Li_{6.4}La_3Zr_{1.4}Ta_{0.6}O_{12}$	2017	MgO (100)	~40 nm	-	-	[17]
Al:$Li_7La_3Zr_2O_{12}$	2013	AAO	1000 nm	-	-	[18]

at 700 °C. Figure 7.2d, e showed the films at the deposition temperature of 700 and 800 °C. The film showed the ionic conductivity values of 1.2×10^{-4} S cm^{-1} (in-plane, Fig. 7.2f) and 2.0×10^{-9} S cm^{-1} (perpendicular to the plane) (Table 7.2).

7.4 Sol–Gel Method

With the harsh vacuum condition, compared with pulsed laser deposition and radio frequency, the sol–gel method exhibits lots of advantages on low cost, easy synthesis, and flexible chemical components.

In the year 2014, Chen et al. [21] prepared Li–La–Zr–O film on the Si substrate by a sol–gel spin coating method. In the work, they studied the effects of the annealing temperature and the number of layers on the ionic conductivity of the films. With the annealing temperature increasing from 600 to 800 °C, the ionic conductivity of the films decreased from 1.67×10^{-6} to 8.53×10^{-7} S cm^{-1}. The thickness of the films could be controlled by the number of layers. Figure 7.3a shows the SEM images of cross-sectional Li–La–Zr–O thin films with different layers. According to the result, by adding the layer from 2 to 12, the thickness increased from 130 to 538 nm.

By a sol–gel dip-coating process, on the substrate of MgO, Tadanaga et al. [22] fabricated Al-doped $Li_7La_3Zr_2O_{12}$ using a precursor sol from Zr-alkoxide and Li, La, and Al nitrates. After the heat-treatment at 900 °C, the LLZO film showed an ionic conductivity of 2.4×10^{-6} S cm^{-1}.

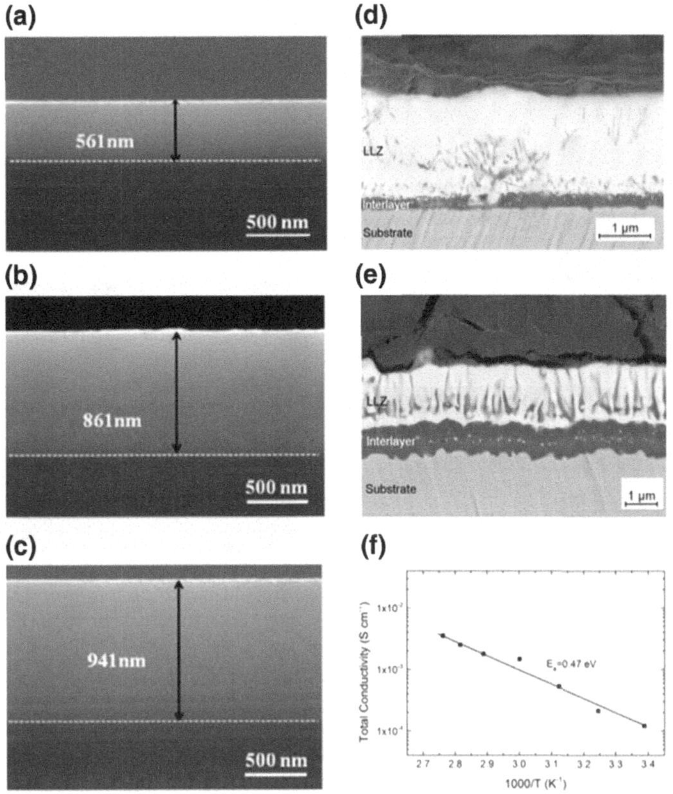

Fig. 7.2 SEM images of the cross section (**a**, **b**, **c**) [19] of Li–La–Zr–O thin films deposited at 40 W (**a**), 60 W (**b**), and 80 W (**c**); **d**, **e** Microstructure of the thin films deposited at 700 and 800 °C [20] **d** cross-sectional, BSE-image at 700 °C, **e** cross-sectional, BSE-image at 800 °C; **f** Impedance data of thin film deposited at 700 °C, Li-ion conductivity σ independence of temperature T

Table 7.2 Review of the literature on garnet-type film by the RF technique with an outline of the material, year, substrate, thickness, room temperature ionic conductivity, and activation energy

Material	Year	Substrate	Thickness	RT σ_{Li} [S cm^{-1}]	E_a [eV]	References
Li–La–Zr–O	2012	Si/SiO$_2$ wafer	561 nm	4×10^{-7}	0.7	[19]
			861 nm	2×10^{-7}	0.81	
			941 nm	0.8×10^{-7}	0.87	
$Li_7La_3Zr_2O_{12}$	2016	Stainless steel	~1 μm	1.2×10^{-4} (in-plane)	0.47	[20]
				2.0×10^{-9} (perpendicular to the plane)		

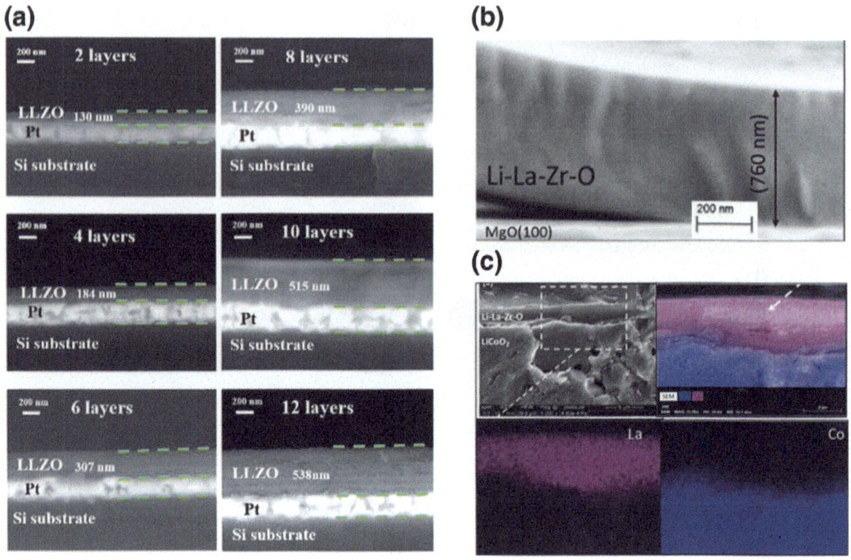

Fig. 7.3 **a** SEM images of cross-sectional Li–La–Zr–O thin films with different layers [21]; **b** cross-sectional SEM image of a broken layer shows dense and crack-free layer on MgO [23]; **c** SEM cross-sectional image of Li–La–Zr–O on $LiCoO_2$ substrate, along with mapping, showing dense electrolyte [23]

Zarabian et al. [23] prepared Li–La–Zr–O film by the sol–gel spin-coating method at a lower annealing temperature of 400 °C. The film was fabricated on the top surface of MgO (100) and $LiCoO_2$ pellets. The thickness of the films was 760 nm on the substrate of MgO (100) and 3606.4 nm on the $LiCoO_2$ pellets, which was in Fig. 7.3b, c The X-ray photoelectron spectroscopy analysis of the interface between $LiCoO_2$ and Li–La–Zr–O film showed that Co diffuses from the substrate toward the solid electrolyte (Table 7.3).

Table 7.3 Review of the literature on garnet-type film by the sol–gel technique with an outline of the material, year, substrate, thickness, room temperature ionic conductivity, and activation energy

Material	Year	Substrate	Thickness	RT σ_{Li} [S cm^{-1}]	E_a [eV]	References
Li–La–Zr–O	2014	Si	130–538 nm	1.87×10^{-6}	0.18–0.21	[21]
$Al:Li_7La_3Zr_2O_{12}$	2015	MgO	–	2.4×10^{-6}	–	[22]
Li–La–Zr–O	2017	MgO (100)	760 nm	–	0.6	[24]

7.5 Chemical Vapor Deposition (CVD)

This section is mainly composed of two parts: metal-organic chemical vapor deposition (MOCVD) and laser-assisted chemical vapor deposition (LACVD).

As a technique for the preparation of thin films, the MOCVD technique shows advantages of high deposition rates and moderate vacuum conditions. In addition, the morphology and phase formation can be controlled during the process of preparation of films by the MOCVD method. Equipped with the experience of the LCO film, Katsui and Goto [25] prepared cubic and tetragonal LLZO films on polycrystalline Al_2O_3 by metal-organic chemical vapor deposition with a high rate of 20 $\mu m\ h^{-1}$. Additionally, they investigated the influence of deposition temperature (T_{sub}), a key factor in the process, on the formation of phase, morphology, and deposition rates. When T_{sub} was 973 K, a pyrochlore type $La_2Zr_2O_7$ film having the (111) orientation was deposited. In a range from 1023 to 1173 K of T_{sub}, the phase of LLZO films was tetragonal. Cubic LLZO films with fine granular surface were obtained at T_{sub} = 1173–1223 K. Figure 7.4c, d showed the tetragonal LLZO film at 1073 K and the cubic LLZO film at 1223 K.

Loho et al. [7] prepared the tetragonal LLZO film on the Pt layer by a novel CO_2 laser-assisted chemical vapor deposition (LACVD) technique. The total ionic conductivity of the tetragonal LLZO film was 4.2×10^{-6} S cm^{-1} at 298 K with an activation energy of 0.50 eV, which was in good agreement with bulk ceramics tetragonal LLZO. In addition, the study presented that the deposition temperature and oxygen partial pressure had a great effect on the quality of the LLZO film. In the condition of 973 K and 40% O_2, the film, in Fig. 7.4a, is homogeneous, of high density, and free of cracks (Table 7.4).

7.6 Focused Ion Beam (FIB) Milling

Gong et al. [27] reported in situ atomic-scale observation of electrochemical delithiation in all-solid-state batteries. They constructed an all-solid-state LIB with a gold anode, a $LiCoO_2$ cathode, and Y- and Ta-doped LLZO ($Li_{6.75}La_{2.84}Y_{0.16}Zr_{1.75}Ta_{0.25}O_{12}$) as the solid-state electrolyte (thickness less than 2 μm) on a micro-electro-mechanical system (MEMS) device nanochip using FIB milling (in Fig. 7.5). Compared with the performance in liquid electrolyte batteries, they found that the pristine single-crystal $LiCoO_2$ became a nanosized polycrystal connected by coherent twin boundaries and antiphase domain boundaries after high-voltage delithiation [27].

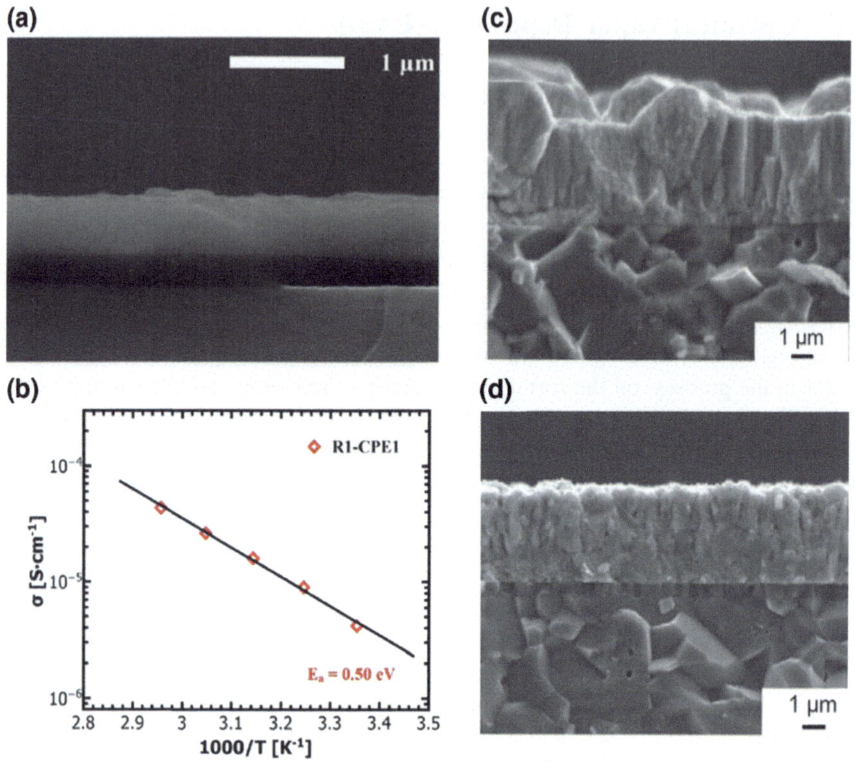

Fig. 7.4 **a** Secondary electron micrographs of LLZO thin films deposited at 973 K and 40% O_2 on Si, shows the LLZO cross section [7]; **b** Arrhenius plot in the temperature range 338–298 K [7]; **c**, **d** SEM images for Li–La–Zr–O films [25]; Cross-sectional morphologies of the tetragonal $Li_7La_3Zr_2O_{12}$ at 1073 K (**c**) and cubic $Li_7La_3Zr_2O_{12}$ at 1223 K (**d**) are also included

Table 7.4 Review of the literature on garnet-type film by the CVD technique with an outline of the material, year, substrate, thickness, room temperature ionic conductivity, and activation energy

Material	Year	Substrate	Thickness	RT σ_{Li} [S cm^{-1}]	E_a [eV]	Ref.
$Li_7La_3Zr_2O_{12}$	2015	Poly-Al_2O_3	~1 µm	-	-	[25]
$Li_7La_3Zr_2O_{12}$	2017	Pt	850 nm	4.2×10^{-6}	0.5	[26]

7.7 Atomic Layer Deposition

Atomic layer deposition (ALD) was applied as a powerful technique for interfacial modification of Li-ion and Li-metal electrodes and for the deposition of solid electrolytes in complex 3D architectures [28]. Kazyak et al. [28] demonstrated a viable ALD process for depositing Al-doped LLZO films with a low amount of impurities, tunable composition, and self-limiting behavior. The film, Fig. 7.6a, was prepared on the Si (1 0 0) substrate at a scale of 86 nm. In addition, ALD showed a good ability

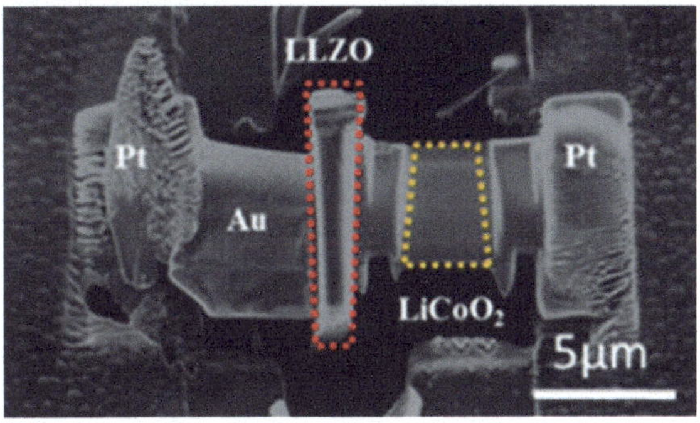

Fig. 7.5 SEM image of the FIB fabricated battery on the nanochip to apply the electric field [27]

to deposit LLZO films on 3D-structured substrates, such as Si wafer in Fig. 7.6b–d and single-crystal ZnO nanowires in Fig. 7.6e, f. The Al-doped stoichiometric film had an ionic conductivity of 7.8×10^{-5} S cm^{-1} at 200 °C and 1.2×10^{-6} S cm^{-1} at 100 °C, giving activation energy of 0.63 eV/atom and a calculated conductivity of approximately 1×10^{-8} S cm^{-1} at 25 °C [28].

7.8 Our Work-Wet Coating

In our work, we introduced an easy method of wet coating the nanoparticle slurry to prepare an LLZO ultrathin film with the thickness of several micrometers. The schematic illustration, in Fig. 7.7a, showed all the steps of the process. The thickness of the film was about 3 μm according to the SEM images in Fig. 7.7c, d. The ionic conductivity of film, shown in Fig. 7.7b, was 5.55×10^{-6} S cm^{-1} at 20 °C and 6.96 $\times$ 10^{-5} S cm^{-1} at 100 °C. In the report, LLZO was used as the solid electrolyte in the Li/LLZO/LFPO battery. As shown in Fig. 7.7e, the cell showed a discharge capacity of 136.8 mAh g^{-1} after 100 cycles at a rate of 0.1 C at room temperature. In Fig. 7.7f, at 60 °C, the cell showed a discharge capacity of 146.2 mAh g^{-1} after 100 cycles at 0.1 C, which was 99.4% of the second cycle. In addition, the LLZO film was also applied in the cell of Li/LLZO/LCO. Figure 7.7g, h showed the surface and the section of the solid electrolyte and cathode layer. The cycling performance of Li/LLZO/LCO at different rates was shown in Fig. 7.7i.

In conclusion, different methods have been developed to prepared LLZO films. According to the updated reports, LLZO film electrolytes have inferior ionic conductivity compared with bulk ceramic electrolytes. However, LLZO film electrolytes, especially for the ultrathin film ones, have the inherent advantage due to the thickness. With the increasing demand for the micro-sized power source and the development of the fabricating technique, the film-type electrolytes will be better and further improved.

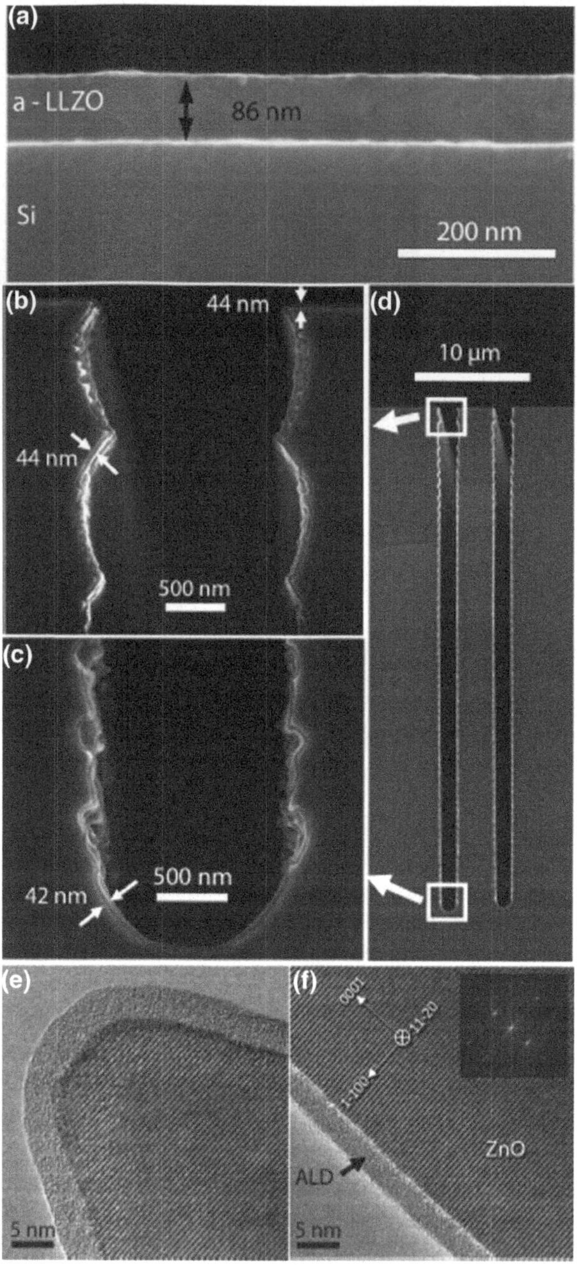

Fig. 7.6 **a** SEM image showing as-deposited ALD LLZO film on a Si substrate, **b** SEM image showing the ALD film from the top of a Si trench shown in (**c**), **d** SEM image showing the ALD coating at the bottom of a Si trench, **e** TEM image showing the ultrathin conformal coating at the tip of a ZnO nanowire, and **f** TEM image showing a sharp film/wire interface, with electron diffraction (inset) showing the amorphous nature of the film [28]

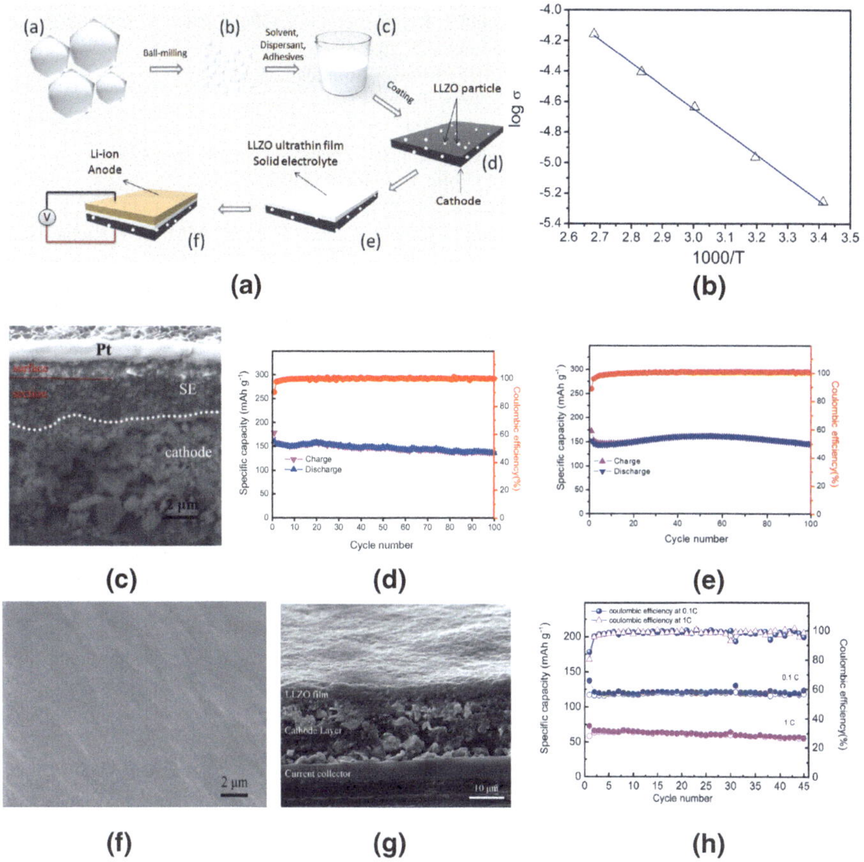

Fig. 7.7 **a** Schematic illustration of the synthesis procedure; **b** Arrhenius conductivity plots of LLZO film; **c** SEM of the section of LFPO cathode layer and LLZO film; **d** cycling performance and coulombic efficiency of Li/LLZO/LFPO at room temperature; **e** cycling performance and coulombic efficiency of Li/LLZO/LFPO at 60 °C; **a**–**e** in [29]; **f**–**h** in [30]; **f** SEM of the surface of LLZO film; **g** The SEM image of the section of cathode layer and the LLZO film; **h** cycling performance and coulombic efficiency of Li/LLZO/LCO

References

1. Murugan R, Thangadurai V, Weppner W (2007) Angew Chem Int Ed Engl 46:7778–7781
2. Ramakumar S, Deviannapoorani C, Dhivya L, Shankar LS, Murugan R (2017) Prog Mater Sci 88:325–411
3. Thangadurai V, Narayanan S, Pinzaru D (2014) Chem Soc Rev 43:4714–4727
4. Du F, Zhao N, Li Y, Chen C, Liu Z, Guo X (2015) J Power Sources 300:24–28
5. Ohta S, Kobayashi T, Seki J, Asaoka T (2012) J Power Sources 202:332–335
6. Wu JF, Pang WK, Peterson VK, Wei L, Guo X (2017) ACS Appl Mater Interfaces 9:12461–12468
7. Loho C, Djenadic R, Bruns M, Clemens O, Hahn H (2017) J Electrochem Soc 164:9

8. Su Y, Falgenhauer J, Polity A, Leichtweiß T, Kronenberger A, Obel J, Zhou S, Schlettwein D, Janek J, Meyer BK (2015) Solid State Ionics 282:63–69
9. Nowak S, Berkemeier F, Schmitz G (2015) J Power Sources 275:144–150
10. Tan J, Tiwari A (2012) Mater Res Soc Symp Proc 1471
11. Tiwari A, Tan J (2012) ECS Solid State Lett 1:4
12. Rawlence M, Garbayo I, Buecheler S, Rupp JL (2016) Nanoscale 8:14746–14753
13. Reinacher J, Berendts S, Janek J (2014) Solid State Ion 258:1–7
14. Kim S, Hirayama M, Taminato S, Kanno R (2013) Dalton Trans 42:13112–13117
15. Hoon Park J, Hong W-S, Park JS, Cheng L, Cabana J, Chen G, Doeff MM, Li Y, Richardson TJ, Son J-W (2013) ECS Trans 53:4
16. Park JS, Cheng L, Zorba V, Mehta A, Cabana J, Chen G, Doeff MM, Richardson TJ, Park JH, Son J-W, Hong W-S (2015) Thin Solid Films 576:55–60
17. Saccoccio M, Yu J, Lu Z, Kwok SCT, Wang J, Yeung KK, Yuen MMF, Ciucci F (2017) J Power Sources 365:43–52
18. Park JH, Hong W-S, Park JS, Cheng L, Cabana J, Chen G, Doeff MM, Li Y, Richardson TJ, Son J-W (2013) Aluminum migration during deposition of $Li_7La_3Zr_2O_{12}$ thin films on aluminum oxide substrates. ECS Trans 53:1–4
19. Kalita DJ, Lee SH, Lee KS, Ko DH, Yoon YS (2012) Solid State Ion 229:14–19
20. Lobe S, Dellen C, Finsterbusch M, Gehrke HG, Sebold D, Tsai CL, Uhlenbruck S, Guillon O (2016) J Power Sources 307:684–689
21. Chen R-J, Huang M, Huang W-Z, Shen Y, Lin Y-H, Nan C-W (2014) J Mater Chem A 2:13277
22. Tadanaga K, Egawa H, Hayashi A, Tatsumisago M, Mosa J, Aparicio M, Duran A (2015) J Power Sources 273:844–847
23. Zarabian M, Bartolini M, Pereira-Almao P, Thangadurai V (2017) J Electrochem Soc 164:7
24. Zarabian M, Bartolini M, Pereira-Almao P, Thangadurai V (2017) J Electrochem Soc 164:A1133–A1139
25. Katsui H, Goto T (2015) Thin Solid Films 584:130–134
26. Loho C, Djenadic R, Bruns M, Clemens O, Hahn H (2017) J Electrochem Soc 164:A6131–A6139
27. Gong Y, Zhang J, Jiang L, Shi J-A, Zhang Q, Yang Z, Zou D, Wang J, Yu X, Xiao R, Hu Y-S, Gu L, Li H, Chen L (2017) J Am Chem Soc 139:4274–4277
28. Kazyak E, Chen K-H, Wood KN, Davis AL, Thompson T, Bielinski AR, Sanchez AJ, Wang X, Wang C, Sakamoto J, Dasgupta NP (2017) Chem Mater 29:3785–3792
29. Yan X, Li Z, Wen Z, Han W (2017) J Phys Chem C 121:1431–1435
30. Yan X, Li Z, Ying H, Nie F, Xue L, Wen Z, Han W (2018) Ionics 24:1445–1551

Chapter 8
Composite Electrolytes Based on Tetragonal $Li_7La_3Zr_2O_{12}$ for Lithium Batteries

E. A. Il'ina and S. V. Pershina

Abstract All-solid-state batteries are known to be in great demand both on an industrial scale (electric vehicles, stand-alone devices) and for special population needs (pacemakers), and therefore attract scientific interest. The difficulty of high-capacity lithium-ion batteries creation consists in the development of suitable solid electrolytes that would have high lithium conductivity and would be chemically and thermodynamically resistant to metallic lithium (the most energy-intensive anode). Among the well-known oxide conductors, only the $Li_7La_3Zr_2O_{12}$ compound satisfies all the requirements. However, there is a problem of high porosity and low density of this electrolyte with tetragonal structure. The introduction of inorganic binders, in particular glasses, into the ceramic electrolyte can solve the indicated problem. The total conductivities of two $Li_7La_3Zr_2O_{12}$ modifications (tetragonal and cubic) are comparable in the region of medium temperatures (above 200 °C). However, the synthesis of tetragonal $Li_7La_3Zr_2O_{12}$ is realized without the dopants' introduction. The $50Li_2O{\cdot}50P_2O_5$, $65Li_2O{\cdot}27B_2O_3{\cdot}8SiO_2$, $40.2Li_2O{\cdot}5.7Y_2O_3{\cdot}54.1SiO_2$ glasses were chosen and the effect of various additives on the functional properties of the tetragonal $Li_7La_3Zr_2O_{12}$ was studied. By a complex of modern techniques, it was established that the additives have a good effect on the electrical conductivity and the density of solid electrolytes. Thus, the proposed approach allows an increase electrical conductivity of the electrolytic membrane for all-solid-state batteries by several orders of magnitude.

8.1 Introduction

8.1.1 All-Solid-State Batteries

Energy plays an important role in many aspects of human life. The increase in population and the desire for a higher standard of living gives rise to rapid growth in energy

---

E. A. Il'ina (✉) · S. V. Pershina
Institute of High Temperature Electrochemistry of the Ural Branch of the RAS, Akademicheskaya st., 20, Yekaterinburg 620137, Russian Federation
e-mail: koksharova.zh@mail.ru

R. Murugan and W. Weppner (eds.), *Solid Electrolytes for Advanced Applications*,
https://doi.org/10.1007/978-3-030-31581-8_8

resources. At the present time, lithium-ion batteries have a wide range of applications due to their high specific energy intensity. They are used in laptops, mobile phones, electro tools, and other compact devices. It seems that the next application step of these sources is electrotransport (electrocars, electrobuses, etc.) employment. However, it is hard to scale such power sources. The lithium-ion batteries lose safety with increase in their size. Because the liquid electrolytes have a tendency to gas formation and volume increase, power sources with big size can warm up, make the electrolyte to bubble, and even explode. Moreover, the liquid electrolyte usage restricts the operating temperature range: at high temperature, the electrolyte can bubble, at low—it can freeze and become a dielectric. This fact prevents the application of the lithium-ion batteries in conditions of extreme temperatures: outer space, deep bowels of the earth, and arctic. So the next generation storage systems require high capacity and energy density, which are typical for all-solid-state batteries (ASSB) [1–5]. A conception of all-solid-state battery has been proposed for solving the mentioned problems. Along with high safety and absence of electrolyte leakage in comparison to liquid electrolytes, the ASSB can work at elevated temperatures and over a wide temperature range [1–12]. In such a battery, anode, cathode, and electrolyte are solids.

One of the main problems of the successful application of ASSB is the resistance at the solid–solid interface of electrode with solid electrolyte because as a rule, it is higher than that at the solid–liquid interface [13, 14]. A low specific surface of the contact area leads to high polarization resistances in such power sources and to more rapid degradation processes on the electrodes (because of high overvoltage values). Hence, it is important to form a tight contact between electrodes and the solid electrolyte [15]. The problem of interface contact can be solved by applying solid polymer electrolytes. The polymer acts like a highly viscous liquid, so it is possible to fit any shape and provide good contact with the solid interface. Unfortunately, polymer electrolytes are metastable systems, which split into a lithium salt and a nonconductive polymer under external condition changes. Despite their good characteristics—145 mA h kg^{-1} and 3.45 V—cycling of these cells is unstable [16]. Further research in this field was shifted to gel polymer electrolytes, which has previously pointed disadvantages: tight working temperatures range, large thickness of electrolyte membrane, and still, actual possibility of liquid electrolyte leakage in extreme conditions. The amount of articles considering all-solid-state lithium batteries with inorganic electrolytes is rapidly rising. Early papers were devoted to development of the battery separated units while nowadays, there are many articles concerning the investigation of model cell properties [9, 10, 15]. For example, the comparison of all-solid-state secondary batteries with different charge carriers (lithium, sodium, and silver ions) was presented in Kim's overview [8]. Among them, lithium-ion battery systems are in high demand, because they possess much larger theoretical capacity. But still, the choice of suitable electrolyte and electrode materials with respect to their compatibility remains the main problem in creating such devices.

Despite the fact that thin-film batteries with Li anodes and LiPON separators demonstrated thousands of cycles, the mechanical stability of bulk-type batteries based on this electrolyte material design seemed to be insufficient. In addition, LiPON has a low electrical conductivity at room temperature—3.3×10^{-6} S cm^{-1} [17]. All-solid-state lithium–air batteries were created with glass–ceramic $Li_{1.5}Al_{0.5}Ge_{1.5}(PO_4)_3$, which showed good characteristics. The discharge and charge capacities were about 1700 mA h g^{-1} and 900 mA h g^{-1}, respectively, at a current density of 500 mA g^{-1} in the voltage range of 2.0–4.2 V (vs. Li/Li^+) [18]. All-solid-state phosphate symmetric batteries with $Li_3V_2(PO_4)_3$ as both the positive and negative electrodes and the phosphate $Li_{1.5}Al_{0.5}Ge_{1.5}(PO_4)_3$ as the solid electrolyte were investigated in the work [19]. Glassy $Li_{1.5}Al_{0.5}Ge_{1.5}(PO_4)_3$ was added into the electrode to increase the interface area between the active materials and the electrolyte. The discharge capacity could reach 92 mA h g^{-1} at a current density of 22 μA cm^{-2} at 80 °C, and 38 mA h g^{-1} at 25 °C, respectively. For comparison, the Li–LiPON–$Li_{0.05-3x}La_{0.5+x}TiO_3$–$LiCoO_2$ all-solid-state cell showed the discharge capacitance of 50 μA h cm^{-2} with the capacitance loss of 0.5% per cycles after 100 cycles at a current density of 7 μA cm^{-2}. The main difficulty in creating the all-solid-state lithium batteries is the polarization resistance due to the point contact of the two solid phases. Effective ways to reduce the resistance at the electrodes–electrolyte boundary, such as sputtering of the buffer layer (Ni), introduction of low melting additives (Li_3BO_3) or glasses coating of the cathode material grains have already been developed. The interface at the anode side, however, has rarely been explored, especially when metal Li is used. It should be noted that molten Li metal leaves microscopic gaps at the interface with solid electrolytes due to poor wetting properties of the material, which leads to a decrease in the characteristics of the battery [13]. The interface modification at the anode (lithium metal)–solid electrolyte interface was shown to be a useful technique for improving the cycle stability and the rate of a charge–discharge reaction. The problem of high interface resistance is solved by Al_2O_3 coating on $Li_7La_{2.75}Ca_{0.25}Zr_{1.75}Nb_{0.25}O_{12}$ electrolyte via atomic layer deposition. In the work [20], the Li–In/glass–ceramic $70Li_2S$–$29P_2S_5$–$1P_2S_3$/$Li_4Ti_5O_{12}$ cell was obtained; indium was sprayed for better contact of the Li metal with the electrolyte. Indium after heating with the lithium foil forms the intercalated indium–lithium alloy. However, there are only a few publications on this subject since a number of solid electrolytes stable against Li metal are not numerous. Some authors successfully used aluminum oxide (Al_2O_3) [13], Au [21], or Ge [22] deposition on $Li_7La_3Zr_2O_{12}$ (LLZO) substrate to develop a good contact between the metal anode and garnet electrolyte. Moreover, it has been shown that Li_3PO_4 introducing as an additive in garnet-type $Li_{6.5}La_3Zr_{1.5}Ta_{0.5}O_{12}$ improves the interfacial compatibility and suppresses Li-dendrite formation [23]. However, reduction of the interfacial resistance at the cathode–lithium garnet solid electrolyte interface remains a challenge [24, 25]. For example, it was found in [26] that the high-temperature process of the $LiCoO_2$ and LLZ fusion induced the cross-diffusion of atoms and the formation of a less-conductive phase at the interface. Some researchers use traditional cathode materials mixed with compounds having a low melting point to reduce the resistance at the cathode–lithium garnet interface (for example, Li_3BO_3 powder [27] or electrolytic salt [28]). In several works, some compounds were artificially delivered as a layer between the cathode and electrolyte (Nb [29], liquid organic electrolyte [13] or gel membrane [15]). Presumably, these problems can be solved by using glassy cathodes or by adding conducting glass into the interface between the conventional cathode and solid electrolyte [30–33].

The solid electrolytes have much less conductivity (10^{-4} S cm^{-1}) as compared to the liquid electrolytes (10^{-2} S cm^{-1}). So another of the key aspects for the all-solid-state batteries operation is a high ionic conductivity of the solid electrolyte (SE) ~10^{-4}–10^{-3} S cm^{-1} at room temperature. Moreover, such electrolytes must have good compatibility and chemical stability to cathode and anode materials, high density of the ceramic membrane, and energy-efficient method of the synthesis.

8.1.2 Lithium-Ion Solid Electrolytes

In the past decades, lithium-ion-conducting solid electrolytes have been extensively studied. This interest is confirmed by numerous reviews and experimental articles. For use in a lithium-ion battery, they must possess not only high lithium-ion conductivity, but also low electronic conductivity, air resistance, chemical stability with electrode materials, a wide voltage window, good mechanical properties, low porosity, be easy to manufacture for large-scale applications and have a low cost. The proper electrolyte can be found among sulfide solid electrolytes and oxide solid electrolytes [34], and also can be selected among crystalline conductors, amorphous materials (glasses and polymers) or composites. As for the lithium-ion conductors with a crystal structure, the $Li_{10}GeP_2S_{12}$ [11, 35–37], $Li_7La_3Zr_2O_{12}$ (LLZO) [38–40], and $Li_{1.3}Al_{0.3}Ti_{1.7}(PO_4)_3$ [41–44] solid electrolytes have the largest Li-ion conductivity (~10^{-2}–10^{-5} S cm^{-1} at room temperature). Commercially used LiPON is an example of the amorphous conductor with high conductivity at 25 °C (2.3×10^{-6} S cm^{-1}) [45–47]. The conductivity of chalcogenide (often sulfide) glasses is much higher, for example, the $0.03Li_3PO_4$–$0.59Li_2S$–$0.38SiS_2$ glass has a total conductivity of 6.9×10^{-4} S cm^{-1} at room temperature [7, 48]. Unfortunately, the sulfide glasses are susceptible to hydrolysis in the air atmosphere with the release of toxic products [49], that is why their usage in ASSB is limited. Thus, the oxide electrolytes remain more attractive in the background of these glasses. Perovskite $Li_{0.05-3x}La_{0.5+x}TiO_3$ showed a bulk conductivity ~1×10^{-3} S cm^{-1} at 25 °C, but the total conductivity in the cell was lower (2×10^{-5} S cm^{-1}) due to a higher grain boundary resistance [50]. Garnet-like structural compounds are frequently proposed to be solid electrolytes for lithium and lithium-ion power sources because of their high conductivity (~10^{-4} S cm^{-1} at 25 °C) and chemical stability versus metallic lithium [51].

Promising lithium-ion electrolytes are NASICON-type conductors such as $LiTi_2(PO_4)_3$, $Li_{1+x}Al_xTi_{2-x}(PO_4)_3$, and $Li_{1.5}Al_{0.5}Ge_{1.5}(PO_4)_3$. However, the $LiTi_2(PO_4)_3$ compound has no stability in contact with metallic lithium. Partial substitution of titanium for aluminum and lithium increases its stability versus Li metal and substantially increases the electrical conductivity to 10^{-3} S cm^{-1} at 25 °C. However, a low relative density is a disadvantage of ceramics based on $LiTi_2(PO_4)_3$. One way to increase the density of a solid electrolyte can be to obtain it in the form of glass–ceramics. Glass–ceramic materials are obtained by directional crystallization of glasses of the corresponding compositions by heat treatment. The high density of glass–ceramics is due to the fact that glass can "heal" the pores at the volume changes during crystallization. Also, a significant advantage of crystallized glass is its manufacturability. It is possible to obtain glass–ceramics with specified properties when the composition of the glass changes. Glass–ceramics of the composition $Li_{1.5}Al_{0.5}Ge_{1.5}(PO_4)_3$ with a conductivity of 5.2×10^{-4} S cm^{-1} at 25 °C were obtained in the work [52]. This electrolyte has a wide electrochemical window, chemical stability against Li metal, and high density.

The oxide conductor $Li_7La_3Zr_2O_{12}$ is a promising electrolyte for ASSB [53–57], as it has a relatively high ionic conductivity, electrochemical stability, and better chemical compatibility with the electrode materials than other highly conductive electrolytes—$Li_{10}GeP_2S_{12}$ and $Li_{1,3}Al_{0,3}Ti_{1,7}(PO_4)_3$ [58]. It is known that $Li_7La_3Zr_2O_{12}$ has two modifications—tetragonal (t-LLZO) and cubic (c-LLZO). The total conductivity of t-LLZO depends on the synthesis conditions [59, 60] and varies in the range of 10^{-6}–10^{-7} S cm^{-1} at room temperature while the total conductivity of c-LLZO is ~10^{-4} S cm^{-1} at 25 °C [38, 40]. However, the values of their conductivities are close at 300 °C [60, 61]. According to the works [38, 39, 62], the highly conducting cubic modification can be stabilized by introducing some dopants, in particular, ions of aluminum [38, 39], gallium [63–66], tellurium [67], tantalum [68], yttrium [69], chrome [70], niobium [71, 72], silicon [73, 74], or germanium [74–77].

8.2 Composite Electrolytes

In a number of works, it has been shown [78–81] that it is possible to increase the electrical conductivity and reduce the porosity of ceramic materials using different techniques: (1) by preparing composite polymer electrolytes; (2) by obtaining glass–ceramics (the crystallized glasses) [27, 80, 81]; and (3) by the introduction of sintering additives—the lithium salt, Li_3PO_4, $LiBO_2$, Li_3BO_3 [79, 82–84], $LiNO_3$ [85], LiF [86] or glassy powder (the glass–ceramic composites) [78, 87]. The method of increasing the conductivity by preparing glass–ceramic composites is less studied although it is well-known that the inorganic glasses are able to fill pores in ceramic materials. Moreover, they have good thermal and mechanical characteristics, and also high values of the lithium-ion conductivity [79, 88]. The conductivity and density of composites can be increased due to the formation of the liquid phase during the synthesis.

8.2.1 Composite Polymer Electrolytes

The creation of electrolytes based on a polymer and a lithium-conducting compound can be considered as one of the promising areas for the successful realization of all-solid-state batteries. Polymer-based electrolytes are attractive because they can more easily form interfaces to electrodes compared to inorganic electrolytes. Al-substituted $Li_7La_3Zr_2O_{12}$ ceramic particles were dispersed in a poly(ethylene oxide) ($P(EO)_{20}$–$LiClO_4$) matrix [89]. The total conductivity of the composite with 40 vol% LLZO was 5×10^{-5} S cm^{-1} at 80 °C. Composite electrolytes (PEO)—$Li_{6.4}La_3Zr_{1.4}Ta_{0.6}O_{12}$ with 10, 50, and 80 wt% LLZTO were investigated in the work [90]. Electrolyte with 5 wt% PEO had the highest conductivity, 1.17×10^{-4} S cm^{-1} at 30 °C. The ceramic particles of $Li_7La_3Zr_2O_{12}$ were also added in the PEO by F. Chen et al. [91]; the total conductivity of the composite with 7.5 wt% LLZO was 5.5×10^{-4} S cm^{-1} at 22 °C. In the work [92], it was shown that the incorporation of 5 wt% of the ceramic filler (c-LLZO nanowires) to polyacrylonitrile–$LiClO_4$ increased three orders of magnitude; the total conductivity at room temperature (1.3×10^{-4} S cm^{-1}). It was established using solid-state NMR that increasing of LLZO content in PEO leads to ion mobility decreases because of LLZO-blocked pathways in the polymer phase [93].

Lithium garnet composite polymer electrolyte membrane based on $Li_{6.28}Al_{0.24}La_3Zr_2O_{12}$ with PEO and $LiClO_4$ was prepared in the work [94] by standard solution casting method. The fabricated cell with $LiCoO_2$ and metallic lithium delivers an initial charge–discharge capacity of 146 mA h g^{-1}/142 mA h g^{-1} at 25 °C with 0.06 C rate. A cubic garnet-type $Li_7La_3Zr_2O_{12}$ with 4 wt% polyacrylonitrile (PAN) addition was prepared as the thin electrolyte layer by K. Yoshima et al. [95]. The cells based on this electrolyte show the reduction of the internal resistance with $LiMn_{0.8}Fe_{0.2}PO_4$ and $Li_4Ti_5O_{12}$ electrode materials and enable discharge at low temperatures. Hybrid ceramic polymer electrolyte based on $Li_7La_3Zr_2O_{12}$ (70 wt% LLZO and 30 wt% P(EO)·15LiTFSI) was prepared in the work [96]. The improved compatibility with lithium anode was observed for LLZO–P(EO) 15LiTFSI hybrid electrolytes with respect to P(EO)·15LiTFSI.

Thus, the composite polymer electrolytes based on LLZO with high lithium-ion conductivity was obtained; it was shown that the application of such electrolytes reduce the resistance at the interface between the electrolyte and electrodes and improve the electrochemical characteristics of the cells.

8.2.2 Glass–Ceramics

According to the literature data, there are attempts to increase the glass conductivity by precipitating fast ionic crystallites in a glass matrix. Glass–ceramics are usually obtained by glass crystallization. The parent glass corresponding to the glass–ceramics composition is made by fast cooling, very often called melting quench [97]. The electrical properties of some glass–ceramics were found to be quite better than the precursor glassy system. For example, the $70Li_2S–30P_2S_5$ glass and glass–ceramic melt-quenched at 750 °C showed ionic conductivities of 9.2×10^{-5} and 2.1×10^{-3} S cm^{-1} at room temperature, respectively [25]. The total conductivity of glass and glass–ceramics of $Li_{1.5}Al_{0.5}Ge_{1.5}(PO_4)_3$ compound was 9.8×10^{-7} and 7.6×10^{-5} S cm^{-1} at 25 °C, respectively [98]. It should be noted that the electrical properties of the glass–ceramic electrolyte significantly depend on the crystallization parameters of the annealing process [99, 100]. Melting-quench process, thermal behavior, crystalline kinetics, and the influence of crystalline content on ionic conductivity have been extensively investigated in the works [101–103]. For example, lithium-ion conductivity at room temperature of $Li_{1.5}Al_{0.5}Ge_{1.5}(PO_4)_3$ glass–ceramics obtained at 571 and 618 °C for 30 min was 2×10^{-5} and 7.6×10^{-5} S cm^{-1}, respectively [98]. Other researchers obtained the electrolyte of similar composition with total conductivity of ~10^{-3}–10^{-4} S cm^{-1} at 25 °C [99, 100, 104]. Thokchom et al. [99] reported the highest total conductivity of 5.08×10^{-3} S cm^{-1} at 27 °C for this lithium-ion solid electrolyte.

Besides, glass–ceramics has a number of advantages over sintered polycrystalline ceramics in terms of flexibility, homogeneity, dense microstructure, very low grain-boundary resistance, and easy preparation in the required size and shape [105]. During the glass crystallization, the amorphous phase is transformed into the crystalline ones; therefore, the size of the ceramic grain can be controlled by crystallization parameters, which determine the microstructure of the final glass–ceramics. So the important advantage of glass–ceramic electrolytes is homogeneous microstructure and low porosity [98].

Thus, the glass–ceramic route of obtaining solid electrolytes is a promising way; however, the glass-forming ability of oxide systems is limited.

8.2.3 Sintering Additives and Glass–Ceramic Composites

In the work [84], Li_3PO_4 and Li_3BO_3 additives were used as a binder to obtain high-density $LiTi_2(PO_4)_3$-based solid electrolytes. It was shown that the porosity of the obtained samples decreases from 35 to 1% after Li_3BO_3 introduction; at the same time, the total conductivity growth by two orders of magnitudes was observed because of the contact area increasing between $LiTi_2(PO_4)_3$ grains. It was established that the additive nature (Li_3PO_4 or Li_3BO_3) does not have a significant influence on the bulk activation energy unlike grain-boundary one, therefore, Li^+ migration on the grain boundaries has a predominant character. In the work [85], it was shown that the ionic conductivity of composite electrolytes depends on the sintering temperature. Thus, the room temperature conductivity of $LiTi_2(PO_4)_3$–$LiNO_3$ composites sintered at 900 °C was higher by two orders of magnitude than for pure $LiTi_2(PO_4)_3$ obtained under the same conditions. In contrast, annealing at 700 °C did not lead to an increase in conductivity. The increase in conductivity above 750 °C was observed because of partial melting of the lithium salt and solidification of the samples.

LiF additive with the melt temperature of 848 °C was used to increase the density and conductivity of NASICON-type $LiTi_2(PO_4)_3$ [86]. The relative densities of composites obtained at 900 °C for 12 h reached ~95% compared to 71.2% for pure $LiTi_2(PO_4)_3$. The $LiTi_2(PO_4)_3$—0.5 mol LiF composite has the highest lithium-ion conductivity of 2.3×10^{-4} S cm^{-1} at room temperature. It is noted that significantly enhanced density means much better contact between the grains and improved lithium-diffusion ability. C. Leo and coworkers [106] enhanced the Li^+ conductivity of $LiGe_2(PO_4)_3$ and $Li_{1.4}Al_{0.4}Ge_{1.6}(PO_4)_3$ by the introduction of MgO and Li_2O additives. They demonstrated that these additives act as binders rather than being incorporated into the parent system.

Moreover, the sintering additives were successfully used for solid electrolytes with Na-ion conductivity. For example, a dense $Na_3Zr_2Si_2PO_{12}$-based ceramic with total conductivity of ~1×10^{-3} S cm^{-1} at room temperature was successfully obtained by sintering at 900 °C with 4.8 wt% Na_3BO_3 [107]. The room temperature conductivity is identical to that of conventional high-temperature-sintered NASICON. $Na_{1+x}Zr_2Si_xP_{3-x}O_{12}$-based solid electrolytes with 9.1 wt% Na_3BO_3 sintering at 700 °C exhibited the room temperature conductivity by three orders of magnitude higher than that of an Na_3BO_3-free sample [108]. Thus, lithium (sodium) salts concentrate on the grain boundaries, contributing to an increase in density and conductivity if the intergranular phase does not act as a barrier to the Li^+ migration.

The application of glasses as sintering additives are described to reduce the temperature and time of sintering and to increase the conductivity of LLZO with a cubic structure. In the work [88], $Li_{6.16}Al_{0.28}La_3Zr_2O_{12}$ compound was doped by 1 wt% of glasses with different compositions (Li_3BO_3, Li_3PO_4, and Li_4SiO_4). An addition of Li_4SiO_4 glass results in maximal total conductivity (6.1×10^{-4} S cm^{-1} at 33 °C) and relative density (96%). Tadanaga and coworkers [109] reduced the final annealing temperature to 900 °C by the addition of amorphous Li_3BO_3 to c-LLZO in the ratio of Li_3BO_3/LLZO equal to 0.68. The total conductivity of the obtained LLZO–Li_3BO_3 composite was 1×10^{-4} S cm^{-1} at 30 °C. Takano et al. [78] sintered c-LLZO (with the addition of Al) at 900 °C using $LiBO_2$ and Li_3BO_3 glasses as the sintering additives. Using $LiBO_2$ led to the formation of the $La_2Zr_2O_7$ impurity phase while a pure c-LLZO was obtained by addition of Li_3BO_3. Monophase sample was annealed at 900 °C and its total electrical conductivity was 1.9×10^{-5} S cm^{-1} at room temperature while conductivity of c-LLZO obtained under the same conditions reached only 1.6×10^{-6} S cm^{-1}. In the work [110], the sintering mechanism of the LLZO–8 wt% Li_3BO_3 composite was investigated. It was established that the sintering of the composite is due to the viscous flow of the glass at low temperatures while the overall interactions including the particle rearrangement of LLZO by molten Li_3BO_3 at high temperatures and the solid-state sintering of LLZO contribute to the densification process.

The $LiBO_2$ and Li_3BO_3 additives were used to increase the density of $Li_7La_3(Zr_{2-x}Nb_x)O_{12}$ ($x = 0$–1.5) ceramic electrolytes sintered at low temperature [111]. The density of composites sintered at 900 °C after the addition of the glasses reached 90% of the theoretical value. The addition of 6.5% Li_3BO_3 resulted in an increase in electrical conductivity up to 7×10^{-5} S cm^{-1} at 30 °C which was two orders of magnitudes higher compared with pure material. Through addition of 1–6 wt% of Li_3PO_4 glass to $Li_{6.75}La_3Zr_{1.75}Ta_{0.25}O_{12}$ (LLZTO) electrolyte, sintering time reduces dramatically from 36 to 6 h and grain size noticeably decreases. These changes lead to an increase in total conductivity LLZTO up to 7.2 × 10^{-4} S cm^{-1} at the Li_3PO_4 concentration of 1 wt% [112]. The LiO_2–B_2O_3–SiO_2–CaO–Al_2O_3 (LBSCA) and BaO–B_2O_3–SiO_2–CaO–Al_2O_3 (BBSCA) multicomponent glasses with a low softening point were used in the work [113] in order to reduce the grain boundary resistance of LLZO. The sample with 4 wt% of LBSCA sintered at 900 °C for 10 h had the highest total Li-ion conductivity—8 × 10^{-5} S cm^{-1} at 30 °C.

Moreover, insertion of such sintering additives into electrolytes is considered as one of the methods to improve the interface between solid electrolyte and electrode, which shows unacceptable resistance in case of contact between two solids [23, 114]. In the work [115], a new garnet-based composite electrolyte $Li_{6.25}La_3Zr_2Al_{0.25}O_{12}$–LiBr was developed. It was shown that the interfacial resistance of electrode–composite electrolyte is smaller (nearly 30%), than for the initial LLZO compound. The composite electrolytes based on $Li_{6.75}La_3Zr_{1.75}Ta_{0.25}O_{12}$ with the amorphous Li_3OCl addition were obtained in [116] and present high ionic conductivity at room temperature (2.27 × 10^{-4} S cm^{-1}, for composition with 2 wt% Li_3OCl) and low interfacial resistance between the composite electrolyte and lithium metal (from 1850 to 90 Ω cm^2). The addition of Li_3PO_4 to $Li_{6.5}La_3Zr_{1.5}Ta_{0.5}O_{12}$ solid electrolyte also improves the interfacial compatibility and suppresses Li-dendrite formation during Li-metal plating/stripping [23].

8.3 Composite Electrolytes Based on Tetragonal $Li_7La_3Zr_2O_{12}$

The tetragonal $Li_7La_3Zr_2O_{12}$ as well as its analogues with a garnet structure is characterized by poor compressibility of ceramic powders. This leads to the low density of the LLZO ceramic membrane, and as a result, to the high total resistance of a sample. The relative density of the tetragonal LLZO obtained by the static compression method was ~60% [117, 118]. A high density can be achieved by using a hot-pressing technique [75, 117]. However, this method leads to the evaporation of lithium oxide from LLZO and, consequently, to a stoichiometric deviation. In the work [119], the effect of hydrostatic pressing on the density and conductivity of tetragonal LLZO was also investigated. The relative density of the samples was ranged from 76 to 85% at a pressure of 300–500 MPa, total conductivity of these samples was about ~10^{-6} S cm^{-1} at room temperature. In the work [120] the dense tetragonal LLZO with high relative density about 93% and high total conductivity ~5.7 × 10^{-5} S cm^{-1} at 30 °C has been prepared successfully by auto-consolidation method. In spite of extensive research into the influence of glasses as sintering additives on the properties of $Li_7La_3Zr_2O_{12}$ with the cubic structure, the ones for LLZO with a tetragonal structure have not been studied. The glass addition to the tetragonal

LLZO can reduce the sintering time and/or temperature and finally increase the density and conductivity of the ceramic. Moreover, the important advantages of the glass additive introduction into the solid electrolyte are the following: an increase in the mechanical strength of the ceramic product, a reduction of both the grain-boundary resistance and the resistance at the interface [31], a decrease in the thickness of the electrolyte membrane, and an increase in the fraction of active components, which enhances the specific energy of the cell. However, the effect of different glass addition on the physicochemical and transport properties of the tetragonal LLZO has not been systematically studied yet. Besides, binary glasses were used as sintering additives in most works although in most cases, ternary glasses have higher electrical conductivity than binary [121, 122]. Different glass additives to LLZO have different effects on the morphology and phase composition of their composites. These parameters primarily impact the transport properties of electrolytes. Thus, an effect of the $LiPO_3$, $65Li_2O{\cdot}27B_2O_3{\cdot}8SiO_2$, and $40.2Li_2O{\cdot}5.7Y_2O_3{\cdot}54.1SiO_2$ glass additives on microstructure, conductivity, and porosity of the tetragonal modification of $Li_7La_3Zr_2O_{12}$ has been studied by our group. In the work [123], a systematic study of a series of borosilicate glasses with different content of lithium oxide was carried out; the composition $65Li_2O{\cdot}27B_2O_3{\cdot}8SiO_2$ (LBS) had the highest total conductivity (3.6×10^{-6} S cm^{-1} at room temperature). Lithium yttrium-silicate glass with the composition $40.2Li_2O{\cdot}5.7Y_2O_3{\cdot}54.1SiO_2$ (LYS) has been selected for further investigations [124]. The total conductivity of this glass at room temperature is comparable to that of tetragonal LLZO (~10^{-7} S cm^{-1}). The advantage of this glass consists of its chemical stability when it is in contact with molten lithium and air. The choice of metaphosphate glass was determined by its low glass transition temperature and melting temperature compared to the silicon-containing glasses and also a relatively high lithium-ion conductivity at room temperature [125, 126].

8.3.1 Synthesis of the Composite Electrolytes

Li_2CO_3, La_2O_3, and an aqueous solution of $ZrO(NO_3)_2$ were used as starting components for the citrate–nitrate synthesis of tetragonal $Li_7La_3Zr_2O_{12}$. The reagents were mixed in the stoichiometric ratio except Li_2CO_3 which was taken with 10 wt% excess as shown in [40, 61]. These components were dissolved in the mixture of dilute nitric and citric acids. The resulting solution was evaporated at 80 °C to a transparent gel. Then the gel was dried and pyrolyzed at ~200 °C. The synthesis was performed by increasing the temperature stepwise (from 700 to 900 °C for 3 h). The final temperature step was carried out under argon atmosphere. After each stage of the synthesis, the mixture of reagents was thoroughly ground in an agate mortar for 0.5 h.

The $LiPO_3$ glass was prepared in the air by a conventional melting-quench method, which is described in more detail in the works [125–128]. The lithium carbonate (Li_2CO_3) and ammonium hydrogen phosphate $(NH_4)_2HPO_4$ were the original materials. These materials were proportionally weighed and mixed. Then, the mixture was slowly heated and melted in air in a Pt crucible for 1 h at 900 °C. The melt was quenched by pressing two metallic plates. As a result, the colorless, transparent plane-parallel plates without any crystalline inclusions were obtained. The $65Li_2O{\cdot}27B_2O_3{\cdot}8SiO_2$ and $40.2Li_2O{\cdot}5.7Y_2O_3{\cdot}54.1SiO_2$ glasses were obtained by conventional quenching method which was described in [123, 124]. The amorphous structure of the obtained samples was confirmed by XRD for the lack of Bragg peaks.

In order to obtain $Li_7La_3Zr_2O_{12}$–x wt% glass ($x = 1$–10) composites, the synthesized products were ground and homogenized in a planetary mill (FRITSCH, Germany) with agate balls of 10 mm diameter (mass ratio between the balls and the product was of 3:1). The grinding was performed in 5 ml of ethanol per 10 g of the product for 30 min at a speed of 750 rpm. After drying at a temperature of 200 °C, the mixture was pressed into pellets at 240 MPa. Pellets with 1 wt% of glass were covered with powder of the same composition and then sintered at temperatures from 900 to 1200 °C for 1 h in argon atmosphere. It was found that an increase in the sintering temperature was accompanied by an increase in the relative density of the composite electrolytes. The temperatures of 1150 °C (for LLZO–x wt% LBS) and 1200 °C (for LLZO–x wt% LYS and LLZO–x wt% $LiPO_3$) were chosen for subsequent annealing of samples as at this temperature, the solid electrolytes had the highest density and total conductivity.

The XRD analysis of the samples was completed using a Rigaku D/MAX-2200 V diffractometer with a vertical goniometer and a graphite monochromator in Cu Kα radiation, $2\theta = 10$–$60°$ at room temperature. The identification of compounds was performed by means of PDF-2 database (release 2008 RDB 2.0804). The obtained data were processed using the FullProf program. The XRD patterns of the LLZO–x $LiPO_3$ ($x = 1$–7) composites with $1 \leq x \leq 4$ lithium phosphate glass correspond to a single phase of the tetragonal $Li_7La_3Zr_2O_{12}$ with the space group $I4_1/acd$. However, the introduction of 7 wt% $LiPO_3$ leads to the appearance of an impurity phase—Li_3PO_4. The addition of LBS glass ($x = 1$–10) to tetragonal LLZO did not lead to impurity formation, according to XRD. Probably the added glass is in the amorphous state or the glass is partly or completely crystallized and reflexes of impurity compounds overlap with the reflections of the main phase. In the case of composite electrolytes with LYS addition, the introduction of more than 3 wt% of glass leads to the appearance of additional peaks of $La_2Zr_2O_7$ and ZrO_2. This trend can be associated with the interaction between LLZO and added glass.

The structure changes were investigated by Raman spectroscopy in order to further elucidate the effect of the added glass on the phase composition and properties of the composite electrolytes. The molecular structure of solid electrolytes was studied using a Raman spectrometer U 1000 (Renishaw, England). Raman spectra were collected by an Ar-ion laser with a wavelength of 532 nm and a power of 40 mW at a scanning angle of 90°. Raman spectra were recorded in the static and extended modes in the range of 1200–50 cm^{-1} with a spectral resolution of 1 cm^{-1}. The intensities were normalized to the maximum value. Raman spectra of the tetragonal LLZO garnet as a reference is shown in very similar patterns to the spectra given in the literature [129–132]. The vibrational frequencies of the collected spectra and the literature data virtually are the same; this confirms that the sintered samples have a single phase of the tetragonal garnet structures. Raman spectra of the LLZO–x $LiPO_3$ composites were fully described in [133]. The obtained data confirms the impurity Li_3PO_4 formation for composite electrolytes with $x > 3$ wt%, which was established by XRD. So it can be concluded that for the composites with $x > 3$ wt%, the glass is at crystallized state and the glassy addition is not completely crystallized at low glass.

Raman spectra of the LLZO–x LBS composites and $65Li_2O{\cdot}27B_2O_3{\cdot}8SiO_2$ glass are presented in Fig. 8.1. Raman spectra of LBS glass contain the bands at 95, 768, 839, and 928 cm^{-1}. The band near 768 cm^{-1} appears due to the groups of triborates or tetraborates [134]. The band at 928 cm^{-1} is linked to the vibrations in BO_3^{3-} orthoborates, and the band near 839 cm^{-1} is associated with the vibrations of the $B_2O_5^{4-}$ pyroborate units. Boron in the glass is simultaneously in the coordination of III and IV, which is confirmed by the presence of the band at 768 cm^{-1}. The obtained data have a good consistency with the results of the works [123, 134]. Thus, the lithium borosilicate glass predominantly consists of the borates that do not form bridging bonds with the oxygen; this might be due to the high content of Li_2O modifier, which destroys the glass network.

According to the Raman studies of LLZO–x wt% LBS composites, changes in the structure of the tetragonal modification of LLZO does not occur by increasing the glass content because the bands responsible for the vibrations in LLZO are not shifted and there is no redistribution of their intensities [131]. Raman spectra of composites with $1 \leq x \leq 3$ differ from the pure LLZO spectrum only by the presence of an additional band at 1090 cm^{-1}, which is also present in LLZO–$LiPO_3$ composites and is associated with vibrations of carbonate groups. In addition, at $x > 3$, the bands that appeared at 840 and 930 cm^{-1} are associated with the vibrations in the pure

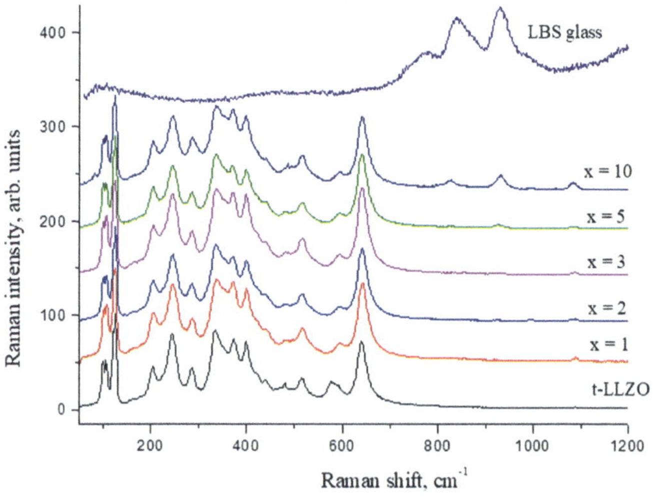

Fig. 8.1 Raman spectra for the LLZO–x wt% LBS composites and individual glass $65Li_2O{\cdot}27B_2O_3{\cdot}8SiO_2$

LBS glass. Thus, the intensity of the bands near 840 and 930 cm^{-1} of the LLZO–*x* LBS composites grow by increasing the glass additive content (Fig. 8.1). So it can be concluded that the glassy additive is not completely crystallized according to Raman spectroscopy.

Raman spectra of composites with LYS glass are shown in Fig. 8.2. It can be noted that the bands at 100 and 106 cm^{-1} are merged, as well as the band in the 125 cm^{-1} region is broadened. According to the literature [130, 135], Raman spectra of two different modifications of $Li_7La_3Zr_2O_{12}$ have the same frequency bands, but the cubic LLZO spectrum is characterized by a smaller number of bands. The structure of composites changes even with the introduction of 1 wt% of glass, and the LLZO ceramic is a mixture of tetragonal and cubic phases in the obtained composites. Strong bands near 299, 394, 492, and 516 cm^{-1} at $x \geq 5$ are linked to the vibrations in $La_2Zr_2O_7$ [133]. It should be noted there is an insignificant shift of the band at 640 cm^{-1} to lower frequencies (643 cm^{-1}) with the introduction of x up to 3 wt%, but the reverse shift of this band to 640 cm^{-1} is observed with a further increase in x (up to 10 wt%). According to [130, 131], the band at 640 cm^{-1} is linked to the vibrations in ZrO_6. It might be due to the distortion of the zirconium sublattice in the composites or the appearing of ZrO_2 impurity phase. The obtained data from Raman spectroscopy are in good agreement with the XRD results. In order to evaluate the state of the additive in the composites, Raman spectra of glassy and crystallized LYS were obtained (Fig. 8.3). The most intensive band of the crystallized glass is observed in the 600 cm^{-1} region, which can be observed on the spectra of composites with $x \leq 3$. For $x \geq 6$, the additional bands appear in the region of 820 and 930 cm^{-1} which characterize the vibrations in lithium silicates [136].

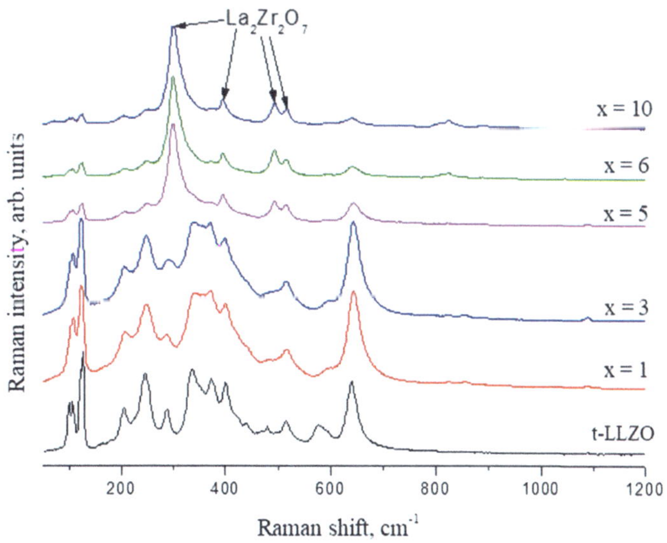

Fig. 8.2 Raman spectra for the LLZO–*x* wt% LYS composites

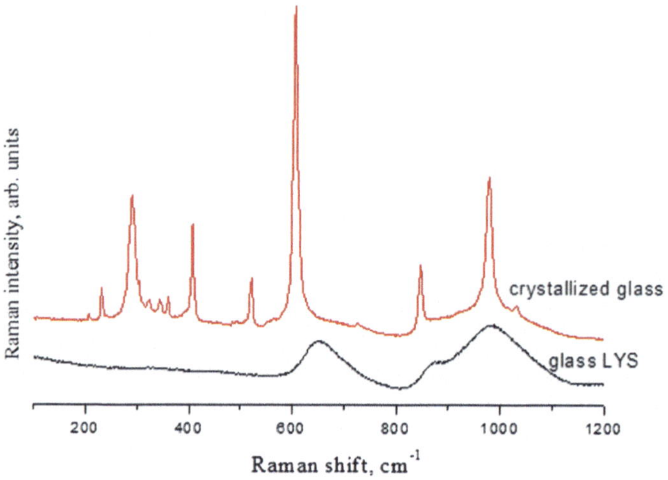

Fig. 8.3 Raman spectra for the glass LYS and crystallized glass LYS

8.3.2 Composite Electrolytes Microstructure, Thermal Properties, and Conductivity

The size and morphology of the ceramic grains, the $LiPO_3$, LBS, and LYS glass distribution in the matrix of the LLZO ceramic electrolyte were determined by the SEM method using an auto-emission electron microscope TESCAN MIRA 3 LMU (TESCAN, Czech Republic). The morphology of the sample cross-section was investigated using a scanning electron microscope (Merlin, Carl Zeiss, Germany). The solid electrolyte without the glass additive ($x = 0$) has a high porosity and is formed by particles with size from 1 to 5 μm [133]. But already 1 wt% glass addition leads to the formation of a thin film on the grains of LLZO. The added glass fills the interparticle space by covering all particles of the solid electrolyte with an increase of additional concentration, thereby reducing the material porosity and increasing its density. The glass additive is located on the surface of the ceramic particles for composite electrolytes with LBS and LYS, however, it also forms areas with a high content of lithium phosphate glass in $Li_7La_3Zr_2O_{12}$–$LiPO_3$ composite electrolytes [133].

The cross-section of the $Li_7La_3Zr_2O_{12}$–3 wt% $65Li_2O{\cdot}27B_2O_3{\cdot}8SiO_2$ composite electrolyte is shown in Fig. 8.4. The structure of the composite electrolyte with $x =$ 3 is not uniform and several segments can be identified. The dark areas correspond to the regions between the ceramic grains where the additive in the glassy state is concentrated while the nanosized crystals grow at the interface between glass and pore. Such selective behavior of the glassy component is probably due to the predominance of its surface crystallization over the bulk one.

The thermal properties of samples were obtained using the simultaneous thermal analysis (STA), combining methods of differential scanning calorimetry (DSC), and

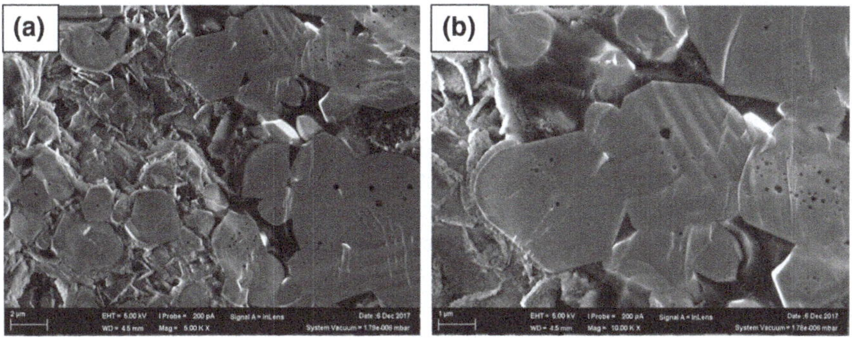

Fig. 8.4 SEM micrographs of the cross-section of LLZO–3 wt% LBS composite at various magnifications: ×5000 (**a**) and ×10000 (**b**)

thermogravimetry (TG). The STA-curves were obtained by means of the thermal analyzer STA 449 C Jupiter (NETZSCH, Germany) in the temperature range of 20–850 °C with a heating and cooling rate of 10 °C min^{-1}. A quadrupole mass spectrometer «QMS 403 C Aeolos» (NETZSCH, Germany) coupled with a thermal analyzer was used to detect and identify the evolved gaseous products. The DSC measurements were carried out in the Pt pans with a heating rate of 10 °C min^{-1} in a dry Ar atmosphere at an expulsion rate of 30 ml min^{-1} in the temperature range of 35–600 °C using a DSC 204 F1 Phoenix calorimeter (NETZSCH, Germany). The obtained results were processed by the computer software NETZSCH Proteus.

The differential scanning calorimetry measurements were carried out in order to determine the presence of the glass in composite electrolytes based on tetragonal LLZO with $LiPO_3$, LBS, and LYS glass addition. The DSC-curves of the obtained electrolytes are present in Fig. 8.5. It is shown that there are no peaks associated with the glass transition temperature and crystallization of the $LiPO_3$ and LYS glass on the DSC-curves of LLZO–x $LiPO_3$ and LLZO–x LYS composites. So,

Fig. 8.5 DSC-curves of composites electrolytes LLZO–x wt% glass ($LiPO_3$, LBS, and LYS)

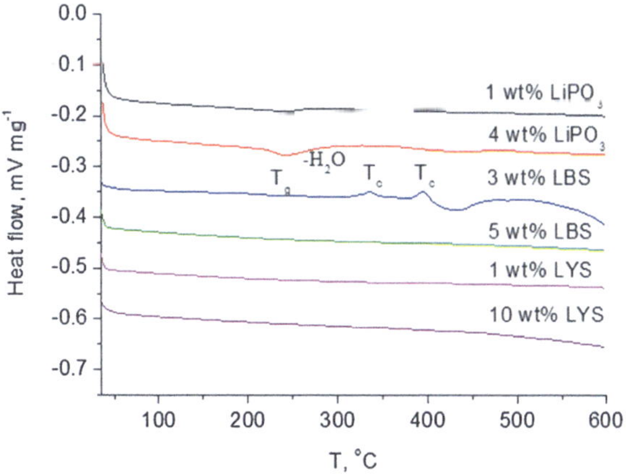

one can assume that the $LiPO_3$ and $40.2Li_2O{\cdot}5.7Y_2O_3{\cdot}54.1SiO_2$ in composites are not in a glassy state; it is partly or fully crystallized. However, the curve of sample with $x = 3$ wt% LBS shows a bend corresponding to the glass transition and two endothermic peaks corresponding to the crystallization temperatures; these characteristic temperatures are closed to glass transition and crystallization temperatures of $65Li_2O{\cdot}27B_2O_3{\cdot}8SiO_2$ [123]. So it can be concluded that the additive was not crystallized with the small concentration, but the LBS has partly or fully crystallized for the composite with $x > 3$ wt%. It is an interesting fact that composites with $LiPO_3$ are prone to water absorption from the air. The intensity of peak at ~250 °C increases with the glass content growing; this can be connected with the hygroscopicity of phosphates. For other composite electrolytes, such behavior has not been observed.

The electrical resistance of the samples was measured by the complex impedance method using the potentiostat/galvanostat Autolab PGSTAT 302 N (Metrohm, Netherlands). The obtained results are processed by the Nova program. The conductivity measurements were carried out in a two-probe cell with silver metal electrodes in the temperature range between 20 and 200 °C in air. The impedance spectra were recorded in the 0.025–1000 kHz frequency range. A gallium–silver paste was used as the electrodes. The electrical conductivity of the electrolytes was calculated taking into account the resistance and geometric dimensions of the samples.

The typical impedance plots of the LLZO composites with LYS addition are shown in Fig. 8.6. One can see two semicircles at the impedance spectrums at room temperature. However, it could not be separated into bulk and grain-boundary resistance because each semicircle can contain other contributions, as it was explained in the work [110]. The total resistance of the composites was determined from the real part of impedance Z' value at the minimum between the high-frequency arc and the low-frequency tail, as in the work [78]. At temperatures above 100 °C (Fig. 8.6b), impedance spectrum represents a ray that intersects the *X*-axis (the total resistance of the composite). Similar impedance plots were obtained for composite electrolytes with $LiPO_3$ and LBS addition [81, 133].

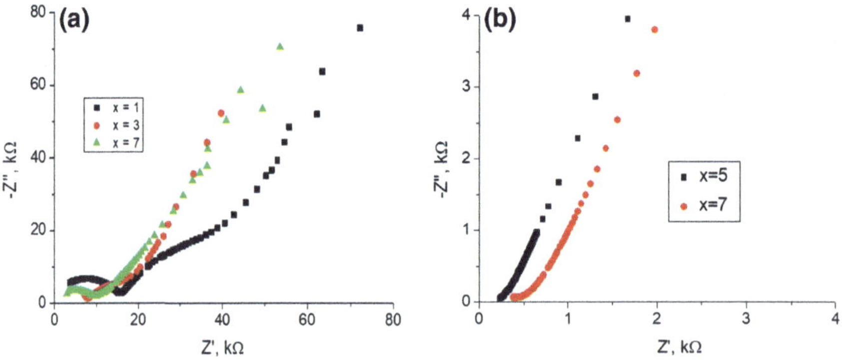

Fig. 8.6 Impedance plots of LLZO–*x* wt% LYS at 25 °C (**a**) and 110 °C (**b**)

The temperature dependences of the total conductivity of composite electrolytes based on the tetragonal modification of $Li_7La_3Zr_2O_{12}$ with $LiPO_3$, $65Li_2O{\cdot}27B_2O_3{\cdot}8SiO_2$, and $40.2Li_2O{\cdot}5.7Y_2O_3{\cdot}54.1SiO_2$ glass additives in Arrhenius coordinates are given in [81, 133, 137]. The activation energy of the total conductivity (E_a) of the solid electrolytes was calculated from the slope of the Arrhenius temperature dependences. E_a gradually decreases from 60.5 ± 0.7 ($x = 0$) to 48.9 ± 0.8 ($x = 5$) kJ mol^{-1} for the composites with $LiPO_3$ glass additive, to 42.5 ± 0.5 kJ mol^{-1} ($x = 3$) for LLZO–LBS system and to 37.5 ± 0.5 kJ mol^{-1} for LLZO–5 wt% LYS; see Fig. 8.7. The growth of the total conductivity correlates with the decrease in the activation energy for all the studied systems. Further growth of glass content leads to a slight increase in the activation energy.

The concentration dependences of the total lithium-ion conductivity of solid electrolytes based on tetragonal LLZO with different glassy additives are shown in Fig. 8.8. The 1 wt% of glass addition leads to an increase in the total conductivity of the solid electrolyte. With an increase in the additive content, a gradual increase of lithium-ion conductivity is observed until the maximum value is reached (for $LiPO_3$ and LYS, it is 5 wt%, and for LYS glass—3 wt%). The introduction of glasses of ternary systems has better effect on the conductivity of the LLZO solid electrolyte. It is most likely due to the higher values of lithium-ionic conductivity of the initial glasses, compared to $LiPO_3$. It should be noted that all synthesized composites had higher values of conductivity in the studied temperature range than initial LLZO. It can be suggested that the improvement of the transport properties of the obtained composites is caused by an increase in the samples' density, which monotonically grows with glass additive increasing. The glass fills the pores, and thereby a contact area between the ceramic particles increases. A drop in the total conductivity is caused by a decrease in the Li^+ mobility due to the appearance of the low-conductivity impurities or because the glass begins to crystallize at its high concentration.

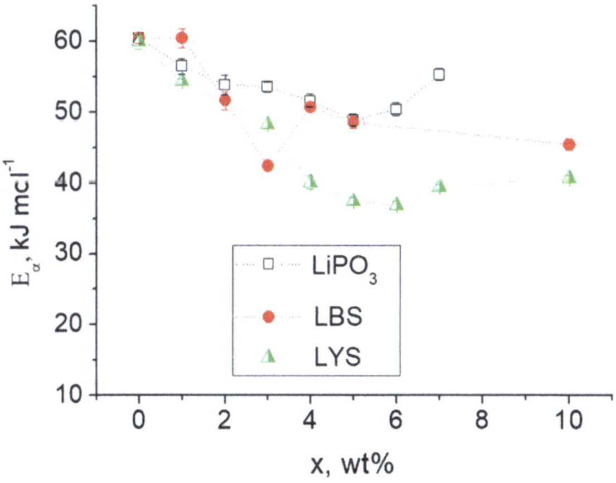

Fig. 8.7 Concentration dependences of activation energy of the obtained composites

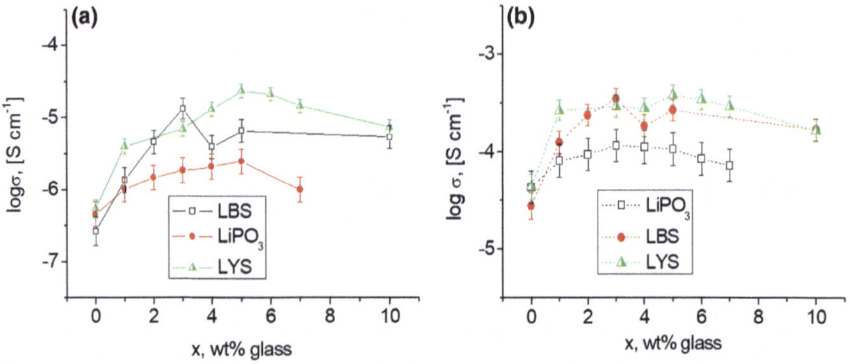

Fig. 8.8 The concentration dependences of the total conductivity of composites based on t-LLZO at 25 and 100 °C

8.4 Summarizing

The composite electrolytes based on the tetragonal modification of $Li_7La_3Zr_2O_{12}$ with different glass additives ($LiPO_3$, $65Li_2O{\cdot}27B_2O_3{\cdot}8SiO_2$, $40.2Li_2O{\cdot}5.7Y_2O_3{\cdot}54.1SiO_2$) were synthesized. First, the sintering temperatures of the glass–ceramic composites were chosen; the samples were single-phased under this condition and had the highest density and total conductivity.

The total conductivity of the tetragonal LLZO-based composites increased from 4.6×10^{-7} S cm^{-1} (ceramic electrolyte LLZO) to 2.5×10^{-6} S cm^{-1} (composite with 5 wt% $LiPO_3$) at 25 °C. It was found that the introduction of 7 wt% $LiPO_3$ glass in the LLZO led to the appearance of the Li_3PO_4 impurity phase in accordance with the XRD and Raman spectroscopy data. Despite the maximum density of the composition with $x = 7$, its conductivity is low in comparison to other compositions of the LLZO–x $LiPO_3$ system. The drop in the total conductivity is caused by a decrease in the Li^+ mobility due to the appearance of the low-conductive crystalline Li_3PO_4 at the grain boundary. The conductivity of composites LLZO–LBS increases with adding glass, reaching a maximum for composition with $x = 3$ wt% (1.5×10^{-5} S cm^{-1} at 25 °C), then it decreases for the other compositions. The glass addition is not completely crystallized at low glass content. The addition might be located on the grain boundaries and provides good contact between particles of $Li_7La_3Zr_2O_{12}$. The glass partially or fully crystallizes in the sample with further addition and the crystallization leads to decrease in the total conductivity. The room temperature conductivity of LLZO–LYS composites increases with the addition of glass, attaining a maximum for composition with $x = 5$; then it slightly decreases for the other compositions. The conductivity reduction is likely to be due to the emerging impurity phases, $La_2Zr_2O_7$ and ZrO_2. The DSC, XRD, and local microanalysis data indicate that LYS interacts with LLZO, and the glass addition transforms into a crystalline state. The addition of 5 wt% of $40.2Li_2O{\cdot}5.7Y_2O_3{\cdot}54.1SiO_2$ glass increases the total conductivity of tetragonal LLZO to 2.5×10^{-5} S cm^{-1} at 25 °C.

Table 8.1 Total conductivity, activation energy, and density of the obtained composite electrolytes and tetragonal $Li_7La_3Zr_2O_{12}$

Solid electrolyte	σ_t at 25 °C, S cm^{-1}	ρ, g cm^{-3}	E_a, kJ mol^{-1}
$Li_7La_3Zr_2O_{12}$-5 wt% $LiPO_3$	2.5×10^{-6}	3.4	48.9 ± 0.8
$Li_7La_3Zr_2O_{12}$–3 wt% $65Li_2O \cdot 27B_2O_3 \cdot 8SiO_2$	1.5×10^{-5}	3.5	42.5 ± 0.5
$Li_7La_3Zr_2O_{12}$–5 wt% $40.2Li_2O \cdot 5.7Y_2O_3 \cdot 54.1SiO_2$	2.5×10^{-5}	4.2	37.5 ± 0.5
Tetragonal $Li_7La_3Zr_2O_{12}$ (this work, 1200 °C)	~5×10^{-7}	3.0	56.9 ± 0.7

Thus, one can conclude that a method for increasing the density and total conductivity of ceramic electrolytes based on tetragonal $Li_7La_3Zr_2O_{12}$ due to the $LiPO_3$, $65Li_2O \cdot 27B_2O_3 \cdot 8SiO_2$, and $40.2Li_2O \cdot 5.7Y_2O_3 \cdot 54.1SiO_2$ glass additives is described. The influence of these glasses on the Li^+ conductivity, phase composition, microstructure, and density of the composites based on the tetragonal LLZO was established. The density of the LLZO-based composites increases with the rise of the glass content due to the gluing grains by the glass during heat treatment. The increase in the density of the samples resulted in the conductivity rise. Then the drop in the total conductivity is caused by a decrease in the Li^+ mobility due to the appearance of the low-conductive crystalline phases and impurities at the grain boundaries.

Table 8.1 shows the measured sample density, not recalculated which is taking into account the introduced glass (the density of the glass is almost two times lower than the density of the LLZO). It can be concluded that the density affects the total conductivity of the samples, but there is still another effect from the glass addition. The using of the glass additives with high values of lithium-ion conductivity leads to obtaining the composite electrolytes with high total conductivity. Consequently, it can be assumed that lithium-ion transport occurs not only in LLZO ceramics but also through lithium-conducting glass or some high lithium-conductive phase, which can be formed at the ceramic grains. Besides the proper regime parameters should be found to avoid full crystallization in case of using glass additives. Also, a chemical interaction between glass and ceramic with impurity phase formation is not desirable.

The glass addition leads to increasing the density of the t-LLZO and its total conductivity by 1–2 orders of magnitude. Moreover, the thickness of the ceramic membrane can be reduced by increasing the mechanical strength of a solid electrolyte by a glassy component. Therefore, the obtained composite electrolytes can be used as solid electrolytes for all-solid-state batteries.

Acknowledgements The authors are grateful to Saetova N.S., Antonov B.D., Pankratov A.A., Vovkotrub E.G., Reznitskikh O.G., and Pryakhina V.I. The characterization of powder and composite materials has been carried out with the equipment of the Shared Access Center "Composition of Compounds" of the Institute of High Temperature Electrochemistry of Ural Branch of RAS and the equipment of the Ural Center for Shared Use "Modern Nanotechnologies" UrFU, Yekaterinburg, Russian Federation.

This work was carried out within the framework of the budgetary plans of the Institute of High Temperature Electrochemistry (Sects. 8.2.1 and 8.3.2) and was supported by the Russian Foundation for Basic Researches (under RFBR Project № 16-33-00354 mol_a, (Sects. 8.2.3 and 8.3.1) and the Russian Science Foundation (under RSF Project № 18-73-00099, Sects. 8.1.1, 8.1.2, and 8.2.2).

References

1. Yang M, Hou J (2012) Membranes in lithium ion batteries. Membranes 2:367–383. https://doi.org/10.3390/membranes2030367
2. Mizuno F, Yada C, Iba H (2014) Lithium-ion batteries: advances and applications. In: Pistoia G (ed) Solid-state lithium-ion batteries for electric vehicles. Elsevier, UK, pp 273–292
3. Chen K, Shen Y, Jiang J et al (2014) High capacity and rate performance of $LiNi_{0.5}Co_{0.2}Mn_{0.3}O_2$ composite cathode for bulk-type all-solid-state lithium battery. J Mater Chem A 2:13332–13337. https://doi.org/10.1039/C4TA02395A
4. Choi NS, Chen Z, Freunberger SA et al (2012) Challenges facing lithium batteries and electrical double-layer capacitors. Angew Chem Int Ed 51:9994–10024. https://doi.org/10.1002/anie.201201429
5. Goodenough JB, Kim Y (2010) Challenges for rechargeable Li batteries. Chem Mater 22:587–603. https://doi.org/10.1021/cm901452z
6. Oudenhoven JFM, Baggetto L, Notten PHL (2011) All solid state lithium ion microbatteries: a review of various three dimensional concepts. Adv Energy Mater 1:10–33. https://doi.org/10.1002/aenm.2010000027
7. Kondo S, Takada K, Yamamura Y (1992) New lithium ion conductors based on Li_2S–SiS_2 system. Solid State Ion 53–56:1183–1186. https://doi.org/10.1016/0167-2738(92)90310-L
8. Kim JG, Son B, Mukherjee S et al (2015) A review of lithium and non-lithium based solid-state batteries. J Power Sources 282:299–322. https://doi.org/10.1016/0167-2738(92)90310-L
9. Matsuyama T, Takano R, Tadanaga K et al (2016) Fabrication of all-solid-state lithium secondary batteries with amorphous TiS_4 positive electrodes and $Li_7La_3Zr_2O_{12}$ solid electrolytes. Solid State Ion 285:122–125. https://doi.org/10.1016/j.ssi.2015.05.025
10. Shin BR, Nam YJ, Oh DY et al (2014) Comparative study of TiS_2/Li-In all-solid-state lithium batteries using glass-ceramic Li_3PS_4 and $Li_{10}GeP_2S_{12}$ solid electrolytes. Electrochim Acta 146:395–402. https://doi.org/10.1016/j.electacta.2014.08.139
11. Kamaya N, Homma K, Yamakawa Y et al (2011) A lithium superionic conductor. Nat Mater 10:682–686. https://doi.org/10.1038/NMAT3066
12. Tatsumisago M, Nagao M, Hayashi A (2013) Recent development of sulfide solid electrolytes and interfacial modification for all-solid-state rechargeable lithium batteries (Review). J Asian Ceram Soc 1:17–25. https://doi.org/10.1016/j.jascer.2013.03.005
13. Han X, Gong Y, Fu K, He X et al (2017) Negating interfacial impedance in garnet-based solid-state Li metal batteries. Nat Mater 16:572–579. https://doi.org/10.1038/NMAT4821
14. Bai L, Xue W, Li Y et al (2018) The interfacial behaviours of all-solid-state lithium ion batteries. Ceram Int 44:7319–7328. https://doi.org/10.1016/j.ceramint.2018.01.190
15. Sun C, Liu J, Gong Y et al (2017) Recent advances in all-solid-state rechargeable lithium batteries. Nano Energy 33:363–386. https://doi.org/10.1016/j.nanoen.2017.01.028
16. Appetecchi GB, Hassoun J, Scrosati B et al (2003) Hot-pressed, solvent-free, nanocomposite, PEO-based electrolyte membranes: II. All solid-state Li/$LiFePO_4$ polymer batteries. J Power Sources 124:246–253. https://doi.org/10.1016/S0378-7753(03)00611-6
17. Bates JB, Dudney NJ, Gruzalski GR et al (1993) Fabrication and characterization of amorphous lithium electrolyte thin films and rechargeable thin-film batteries. J Power Sources 43:103–110. https://doi.org/10.1016/0378-7753(93)80106-Y

18. Kitaura H, Zhou H (2012) Electrochemical performance and reaction mechanism of all-solid-state lithium–air batteries composed of lithium, $Li_{1+x}Al_yGe_{2-y}(PO_4)_3$ solid electrolyte and carbon nanotube air electrode. Energy Environ Sci 5:9077–9084. https://doi.org/10.1039/C2EE22381C
19. Kobayashi E, Plashnitsa LS, Doi T et al (2010) Electrochemical properties of Li symmetric solid-state cell with NASICON-type solid electrolyte and electrodes. Electrochem Commun 12:894–896. https://doi.org/10.1016/j.elecom.2010.04.01420
20. Tatsumisago M, Hayashi A (2012) Superionic glasses and glass-ceramics in the $Li_2S–P_2S_5$ system for all-solid-state lithium secondary batteries. Solid State Ion 225:342–345. https://doi.org/10.1016/j.ssi.2012.03.01321
21. Tsai CL, Roddatis V, Chandran CV et al (2016) $Li_7La_3Zr_2O_{12}$ interface modification for Li dendrite prevention. ACS Appl Mater Interfaces 8:10617–10626. https://doi.org/10.1021/acsami.6b00831
22. Luo W, Gong Y, Zhu Y et al (2017) Reducing interfacial resistance between garnet-structured solid-state electrolyte and Li-metal anode by a Germanium layer. Adv Mater 29:1606042. https://doi.org/10.1002/adma.201606042
23. Xu B, Li W, Duan H et al (2017) Li_3PO_4-added garnet-type $Li_{6.5}La_3Zr_{1.5}Ta_{0.5}O_{12}$ for Li-dendrite suppression. J Power Sources 354:68–73. https://doi.org/10.1016/j.jpowsour.2017.04.026
24. Liu Q, Geng Z, Han C et al (2018) Challenges and perspectives of garnet solid electrolytes for all solid-state lithium batteries. J Power Sources 389:120–134. https://doi.org/10.1016/j.jpowsour.2018.04.019
25. Zheng F, Kotobuki M, Song S et al (2018) Review on solid electrolytes for all-solid-state lithium-ion batteries. J Power Sources 389:198–213. https://doi.org/10.1016/j.jpowsour.2018.04.022
26. Park K, Yu BC, Jung JW et al (2016) Electrochemical nature of the cathode interface for a solid-state lithium-ion battery: interface between $LiCoO_2$ and garnet-$Li_7La_3Zr_2O_{12}$. Chem Mater 28:8051–8059. https://doi.org/10.1021/acs.chemmater.6b03870
27. Ohta S, Komagata S, Seki J et al (2013) All-solid-state lithium ion battery using garnet-type oxide and Li_3BO_3 solid electrolytes fabricated by screen-printing. J Power Sources 238:53–56. https://doi.org/10.1016/j.jpowsour.2013.02.073
28. Wu JF, Pang WK, Peterson VK et al (2017) Garnet-type fast Li-ion conductors with high ionic conductivities for all-solid-state batteries. ACS Appl Mater Interfaces 9:12461–12468. https://doi.org/10.1021/acsami.7b00614
29. Kato T, Hamanaka T, Yamamoto K et al (2014) In-situ $Li_7La_3Zr_2O_{12}/LiCoO_2$ interface modification for advanced all-solid-state battery. J Power Sources 260:292–298. https://doi.org/10.1016/j.jpowsour.2014.02.102
30. Afyon S, Krumeich F, Mensing C et al (2014) New high capacity cathode materials for rechargeable Li-ion batteries: vanadate-borate glasses. Sci Rep 4:1–7. https://doi.org/10.1038/srep0711331
31. Pershina SV, Druzhinin KV, Shevelin PYu et al (2016) Cathode half-cell of all-solid-state battery, modified with $LiPO_3$ glass. Rus J Appl Chem 89:904–908. https://doi.org/10.1134/S1070427216060094
32. Kitaura H, Hayashi A, Ohtomo T et al (2011) Fabrication of electrode-electrolyte interfaces in all-solid-state rechargeable lithium batteries by using a supercooled liquid state of the glassy electrolytes. J Mater Chem 21:118–124. https://doi.org/10.1039/C0JM01090A
33. Hu S, Li YF, Yang R et al (2018) Structure and ionic conductivity of $Li_7La_3Zr_{2-x}Ge_xO_{12}$ garnet-like solid electrolyte for all solid state lithium ion batteries. Ceram Int 44:6614–6618. https://doi.org/10.1016/j.ceramint.2018.01.065
34. Kim S, Oguchi H, Toyama N et al (2019) A complex hydride lithium superionic conductor for high-energy-density all-solid-state lithium metal batteries. Nat Commun 1–9. https://doi.org/10.1038/s41467-019-09061-9
35. Ohta N, Takada K, Zhang L et al (2006) Enhancement of the high-rate capability of solid-state lithium batteries by nanoscale interfacial modification. Adv Mater 18:2226–2229. https://doi.org/10.1002/adma.200502604

36. Hirai K, Tatsumisago M, Minami T (1995) Thermal and electrical properties of rapidly quenched glasses in the systems $Li_2S–SiS_2–Li_xMO_y$ ($Li_xMO_y = Li_4SiO_4, Li_2SO_4$). Solid State Ion 78:269–273. https://doi.org/10.1016/0167-2738(95)00094-M
37. Raskovalov AA, Il'ina EA, Antonov BD (2013) Structure and transport properties of $Li_7La_3Zr_{2-0.75x}Al_xO_{12}$ superionic solid electrolytes. J Power Sources 238:48–52. https://doi.org/10.1016/j.jpowsour.2013.03.049
38. Li Y, Han J, Wang C (2012) Ionic distribution and conductivity in lithium garnet $Li_7La_3Zr_2O_{12}$. J Power Sources 209:278–281. https://doi.org/10.1016/j.jpowsour.2012.02.100
39. Murugan R, Thangadurai V, Weppner W (2007) Fast lithium ion conduction in garnet type $Li_7La_3Zr_2O_{12}$. Angew Chem Int Ed 46:7778–7781. https://doi.org/10.1002/anie.200701144
40. Aono H, Sugimoto E, Sadaoka Y et al (1989) Ionic conductivity of the lithium titanium phosphate ($Li_{1+x}M_xTi_{2-x}(PO_4)_3$, M = Al, Sc, Y, and La) systems. J Electrochem Soc 136:590–591. https://doi.org/10.1149/1.2096693
41. Duluard S, Paillassa A, Puech L et al (2013) Lithium conducting solid electrolyte $Li_{1.3}Al_{0.3}Ti_{1.7}(PO_4)_3$ obtained via solution chemistry. J Eur Ceram Soc 33:1145–1153. https://doi.org/10.1016/j.jeurceramsoc.2012.08.005
42. Inaguma Y, Chen LQ, Itoh M et al (1993) High ionic conductivity in lithium lanthanum titanate. Solid State Commun 86:689–693. https://doi.org/10.1016/0038-1098(93)90841-A
43. Aono H, Sugimoto E, Sadaoka Y et al (1990) Ionic conductivity of solid electrolytes based on lithium titanium phosphate. J Electrochem Soc 137:1023–1027. https://doi.org/10.1149/1.2086597
44. Yu XH, Bates JB, Jellison GE et al (1997) A stable thin film lithium electrolyte: lithium phosphorus oxynitride. J Electrochem Soc 144:524–532. https://doi.org/10.1149/1.1837443
45. Jones SD, Akridg JR (1994) Thin film rechargeable Li batteries. Solid State Ion 69:357–368. https://doi.org/10.1016/0167-2738(94)90423-5
46. Souquet JL, Duclot M (2002) Thin film lithium batteries. Solid State Ion 148:375–379. https://doi.org/10.1016/S0167-2738(02)00076-0
47. Kennedy JH (1989) Ionically conductive glasses based on SiS_2. Mater Chem Phys 23:29–50. https://doi.org/10.1016/0254-0584(89)90015-1
48. Ramakumar S, Deviannapoorani C, Dhivya L et al (2017) Lithium garnets: synthesis, structure, Li^+ conductivity, Li^+ dynamics and applications. Prog Mater Sci 88:325–411. https://doi.org/10.1016/j.pmatsci.2017.04.007
49. Minami T, Tatsumisago M, Minami T et al (ed) (2005) Solid state ionics for batteries. Springer, Japan, pp 1–269. https://doi.org/10.1007/4-431-27714-5
50. Bohnke O, Bohnke C, Fourquet JL (1996) Mechanism of ionic conduction and electrochemical intercalation of lithium into the perovskite lanthanum lithium titanate. Solid State Ion 91:21–31. https://doi.org/10.1016/S0167-2738(96)00434-1
51. Thangadurai V, Weppner W (2006) Effect of sintering on the ionic conductivity of garnet-related structure $Li_5La_3Nb_2O_{12}$ and In- and K-doped $Li_5La_3Nb_2O_{12}$. J Solid State Chem 179:974–984. https://doi.org/10.1016/j.jssc.2005.12.025
52. Yang J, Huang Z, Huang B et al (2015) Influence of phosphorus sources on lithium ion conducting performance in the system of $Li_2O–Al_2O_3–GeO_2–P_2O_5$ glass-ceramics. Solid State Ion 270:61–65. https://doi.org/10.1016/j.ssi.2014.12.013
53. Zhu Y, Hea X, Mo Y (2016) First principles study on electrochemical and chemical stability of solid electrolyte–electrode interfaces in all-solid-state Li-ion batteries. J Mater Chem A 4:3253–3266. https://doi.org/10.1039/C5TA08574H
54. Kammampata SP, Thangadurai V (2018) Cruising in ceramics-discovering new structures for all-solid-state batteries-fundamentals, materials, and performances (Review). Ionics 24:639–660. https://doi.org/10.1007/s11581-017-2372-7
55. Ramzy A, Thangadurai V et al (2010) Tailor-made development of fast Li-ion conducting garnet-like solid electrolytes. ACS Appl Mater Interfaces 2:385–390. https://doi.org/10.1021/am900643t

56. Chan CK, Yang T, Weller JM (2017) Nanostructured garnet-type $Li_7La_3Zr_2O_{12}$: synthesis, properties, and opportunities as electrolytes for Li-ion batteries. Electrochim Acta 253:268–280. https://doi.org/10.1016/j.electacta.2017.08.130
57. Cheng SHS, He KQ, Lio Y et al (2017) Electrochemical performance of all-solid-state lithium batteries using inorganic lithium garnets particulate reinforced PEO/$LiClO_4$ electrolyte. Electrochim Acta 253:430–438. https://doi.org/10.1016/j.electacta.2017.08.162
58. Il'ina EA, Andreev OL, Antonov BD et al (2012) Morphology and transport properties of the solid electrolyte $Li_7La_3Zr_2O_{12}$ prepared by the solid-state and citrate–nitrate methods. J Power Sources 201:169–173. https://doi.org/10.1016/j.jpowsour.2011.10.108
59. Awaka J, Kijima N, Hayakawa H (2009) Synthesis and structure analysis of tetragonal $Li_7La_3Zr_2O_{12}$ with the garnet-related type structure. J Solid State Chem 182:2046–2052. https://doi.org/10.1016/j.jssc.2009.05.020
60. Geiger CA, Alekseev E, Lazic B et al (2011) Crystal chemistry and stability of "$Li_7La_3Zr_2O_{12}$" garnet: A fast lithium-ion conductor. Inorg Chem 50:1089–1097. https://doi.org/10.1021/ic101914e
61. Ahn JH, Park S-Y, Lee J-M et al (2014) Local impedance spectroscopic and microstructural analyses of Al-in-diffused $Li_7La_3Zr_2O_{12}$. J Power Sources 254:287–292. https://doi.org/10.1016/j.jpowsour.2013.12.115
62. Bernuy-Lopez C, Manalastas JW, Lopez del Amo JM et al (2014) Atmosphere controlled processing of Ga-substituted garnets for high Li-ion conductivity ceramics. Chem Mater 26:3610–3617. https://doi.org/10.1021/cm5008069
63. Shinawi HE, Janek J (2013) Stabilization of cubic lithium-stuffed garnets of the type '$Li_7La_3Zr_2O_{12}$' by addition of gallium. J Power Sources 225:13–19. https://doi.org/10.1016/j.jpowsour.2012.09.111
64. Wolfenstine J, Ratchford J, Rangasamy E et al (2012) Synthesis and high Li-ion conductivity of Ga-stabilized cubic $Li_7La_3Zr_2O_{12}$. Mater Chem Phys 134:571–575. https://doi.org/10.1016/j.matchemphys.2012.03.054
65. Li Y, Han JT, Wang CA et al (2012) Optimizing. Li^+ conductivity in a garnet framework. J Mater Chem 22:15357–15361. https://doi.org/10.1039/C2JM31413D
66. Yi E, Wang W, Kieffer J et al (2017) Key parameters governing the densification of cubic-$Li_7La_3Zr_2O_{12}$ Li^+ conductors. J Power Sources 352:156–164. https://doi.org/10.1016/j.jpowsour.2017.03.126
67. Deviannapoorani C, Dhivya L, Ramakumar S et al (2013) Lithium ion transport properties of high conductive tellurium substituted $Li_7La_3Zr_2O_{12}$ cubic lithium garnets. J Power Sources 240:18–25. https://doi.org/10.1016/j.jpowsour.2013.03.166
68. Buschmann H, Berendts S, Mogwitz B et al (2012) Lithium metal electrode kinetics and ionic conductivity of the solid lithium ion conductors "$Li_7La_3Zr_2O_{12}$" and $Li_{7-x}La_3Zr_{2-x}Ta_xO_{12}$ with garnet-type structure. J Power Sources 206:236–244. https://doi.org/10.1016/j.jpowsour.2012.01.094
69. Murugan R, Ramakumar S, Janani N (2011) High conductive yttrium doped $Li_7La_3Zr_2O_{12}$ cubic lithium garnet. Electrochem Commun 13:1373–1375. https://doi.org/10.1016/j.elecom.2011.08.014
70. Song S, Yan B, Zheng F et al (2014) Crystal structure, migration mechanism and electrochemical performance of Cr-stabilized garnet. Solid State Ion 268:135–139. https://doi.org/10.1016/j.ssi.2014.10.009
71. Gai J, Zhao E, Ma F et al (2018) Improving the Li-ion conductivity and air stability of cubic $Li_7La_3Zr_2O_{12}$ by the co-doping of Nb, Y on the Zr site. J Europ Ceram Soc 38:1673–1678. https://doi.org/10.1016/j.jeurceramsoc.2017.12.002
72. Ohta S, Kobayashi T, Asaoka T (2011) High lithium ionic conductivity in the garnet-type oxide $Li_{7-X}La_3(Zr_{2-X},Nb_X)O_{12}$ ($X = 0$–2). J Power Sources 196:3342–3345 (2011). https://doi.org/10.1111/jace.14736
73. Kumazaki S, Iriyama Y, Kim K-H et al (2011) High lithium ion conductive $Li_7La_3Zr_2O_{12}$ by inclusion of both Al and Si. Electrochem Commun 13:509–512. https://doi.org/10.1016/j.elecom.2011.02.035

74. Huang M, Dumon A, Nan C-W (2012) Effect of Si, In and Ge doping on high ionic conductivity of $Li_7La_3Zr_2O_{12}$. Electrochem Commun 21:62–64. https://doi.org/10.1016/j.elecom.2012.04.032
75. Wolfenstine J, Rangasamy E, Allen JL et al (2012) High conductivity of dense tetragonal $Li_7La_3Zr_2O_{12}$. J Power Sources 208:193–196. https://doi.org/10.1016/j.jpowsour.2012.02.031
76. Kotobuki M, Yan B, Lu L et al (2016) Study on stabilization of cubic $Li_7La_3Zr_2O_{12}$ by Ge substitution in various atmospheres. Funct Mater Lett 9:1642005–1642008. https://doi.org/10.1142/S1793604716420054
77. Hu S, Li YF, Yang R et al (2018) Structure and ionic conductivity of $Li_7La_3Zr_{2-x}Ge_xO_{12}$ garnet-like solid electrolyte for all solid state lithium ion batteries. Ceram Intern 44:6614–6618. https://doi.org/10.1016/j.ceramint.2018.01.065
78. Takano R, Tadanaga K, Hayashi A et al (2014) Low temperature synthesis of Al-doped $Li_7La_3Zr_2O_{12}$ solid electrolyte by a sol–gel process. Solid State Ion 255:104–107. https://doi.org/10.1016/j.ssi.2013.12.006
79. Fu J (1997) Superionic conductivity of glass-ceramics in the system Li_2O–Al_2O_3–TiO_2–P_2O_5. Solid State Ion 96:195–200. https://doi.org/10.1016/S0167-2738(97)00018-0
80. Thokchom JS, Kumar B (2006) Microstructural effects on the superionic conductivity of a lithiated glass-ceramic. Solid State Ion 177:727–732. https://doi.org/10.1016/j.ssi.2006.01.027
81. Il'ina EA, Raskovalov AA, Saetova NS et al (2016) Composite electrolytes $Li_7La_3Zr_2O_{12}$–glassy Li_2O–B_2O_3–SiO_2. Solid State Ion 296:26–30. https://doi.org/10.1016/j.ssi.2016.09.003
82. Ohta S, Seki J, Yagi Y et al (2014) Co-sinterable lithium garnet-type oxide electrolyte with cathode for all-solid-state lithium ion battery. J Power Sources 265:40–44. https://doi.org/10.1016/j.jpowsour.2014.04.065
83. Hosono H, Abe Y (1991) Fast lithium conducting glass-ceramics in the Li_2O–CaO–TiO_2–Al_2O_3–P_2O_5 system. Solid State Ion 44:293–297. https://doi.org/10.1016/0167-2738(91)90021-3
84. Aono H, Sugimoto E, Sadaoka Y et al (1991) Electrical property and sinterability of $LiTi_2(PO_4)_3$ mixed with lithium salt (Li_3PO_4 or Li_3BO_3). Solid State Ion 47:257–264. https://doi.org/10.1016/0167-2738(91)90247-9
85. Kobayashi Y, Tabuchi M, Nakamura O (1997) Ionic conductivity enhancement in $LiTi_2(PO_4)_3$-based composite electrolyte by the addition of lithium nitrate. J Power Sources 68:407–411. https://doi.org/10.1016/S0378-7753(97)02520-2
86. Xiong L, Ren Z, Xu Y et al (2017) LiF assisted synthesis of $LiTi_2(PO_4)_3$ solid electrolyte with enhanced ionic conductivity. Solid State Ion 309:22–26. https://doi.org/10.1016/j.ssi.2017.06.018
87. Rosero-Navarro NC, Miura A, Higuchi M et al (2017) Optimization of Al_2O_3 and Li_3BO_3 content as sintering additives of $Li_{7-x}La_{2.95}Ca_{0.05}ZrTaO_{12}$ at low temperature. J Electron Mater 46:497–501. https://doi.org/10.1007/s11664-016-4924-4
88. Janani N, Deviannapoorani C, Dhivya L et al (2014) Influence of sintering additives on densification and Li^+ conductivity of Al doped $Li_7La_3Zr_2O_{12}$ lithium garnet. RSC Adv 4:51228–51238. https://doi.org/10.1039/C4RA08674K
89. Langer F, Bardenhagen I, Glenneberg J et al (2016) Microstructure and temperature dependent lithium ion transport of ceramic–polymer composite electrolyte for solid-state lithium ion batteries based on garnet-type $Li_7La_3Zr_2O_{12}$. Solid State Ion 291:8–13. https://doi.org/10.1016/j.ssi.2016.04.014
90. Chen L, Li Y, Li SP et al (2018) PEO/garnet composite electrolytes for solid-state lithium batteries: from "ceramic-in-polymer" to "polymer-in-ceramic". Nano Energy 46:176–184. https://doi.org/10.1016/j.nanoen.2017.12.037
91. Chen F, Yang D, Zha W et al (2017) Solid polymer electrolytes incorporating cubic $Li_7La_3Zr_2O_{12}$ for all-solid-state lithium rechargeable batteries. Electrochim Acta 258:1106–1114. https://doi.org/10.1016/j.electacta.2017.11.164

92. Yang T, Zheng J, Cheng Q et al (2017) Composite polymer electrolytes with $Li_7La_3Zr_2O_{12}$ garnet-type nanowires as ceramic fillers: mechanism of conductivity enhancement and role of doping and morphology. ACS Appl Mater Interfaces 9:21773–21780. https://doi.org/10.1021/acsami.7b03806
93. Zheng J, Hu YY (2018) New insights into the compositional dependence of Li-ion transport in polymer–ceramic composite electrolytes. ACS Appl Mater Interfaces 10:4113–4120. https://doi.org/10.1021/acsami.7b17301
94. Karthik K, Murugan R (2018) Lithium garnet based free-standing solid polymer composite membrane for rechargeable lithium battery. J Solid State Electrochem 22:2989–2998. https://doi.org/10.1007/s10008-018-4010-3
95. Yoshima K, Harada Y, Takami N (2016) Thin hybrid electrolyte based on garnet-type lithium-ion conductor $Li_7La_3Zr_2O_{12}$ for 12 V-class bipolar batteries. J Power Sources 302:283–290. https://doi.org/10.1016/j.jpowsour.2015.10.031
96. Keller M, Appetecchi GB, Kim GT et al (2017) Electrochemical performance of a solvent-free hybrid ceramic polymer electrolyte based on $Li_7La_3Zr_2O_{12}$ in P(EO)15LiTFSI. J Power Sources 353:287–297. https://doi.org/10.1016/j.jpowsour.2017.04.014
97. Pershina SV, Raskovalov AA, Antonov BD et al (2015) Extreme behavior of Li-ion conductivity in the Li_2O–Al_2O_3–P_2O_5 glass system. J Non-Cryst Solids 430:64–72. https://doi.org/10.1016/j.jnoncrysol.2015.10.003
98. Cruz AM, Ferreira EB, Rodrigues ACM (2009) Controlled crystallization and ionic conductivity of a nanostructured $LiAlGePO_4$ glass-ceramic. J Non-Cryst Solids 355:2295–2301. https://doi.org/10.1016/j.jnoncrysol.2009.07.012
99. Thokchom JS, Kumar B (2010) The effects of crystallization parameters on the ionic conductivity of a lithium aluminum germanium phosphate glass-ceramic. J Power Sources 195:2870–2876. https://doi.org/10.1016/j.jpowsour.2009.11.037
100. Mariappan CR, Yada C, Rosciano F et al (2011) Correlation between micro-structural properties and ionic conductivity of $Li_{1.5}Al_{0.5}Ge_{1.5}(PO_4)_3$ ceramics. J Power Sources 196:6456–6464. https://doi.org/10.1016/j.jpowsour.2011.03.065
101. Soman S, Iwai Y, Kawamura J et al (2012) Crystalline phase content and ionic conductivity correlation in LATP glass-ceramic. J Solid State Electrochem 16:1761–1766. https://doi.org/10.1007/s10008-011-1592-4
102. Chen H, Tao H, Wu QD et al (2012) Thermal behavior and lithium ion conductivity of L_2O–Al_2O_3–TiO_2–SiO_2–P_2O_5 glass-ceramics. J Wuhan Univ Technol 27:67–72. https://doi.org/10.1007/s11595-012-0409-3
103. Wang Y, He K, Zu CK et al (2012) Study on crystallization kinetics of Li_2O–Al_2O_3–TiO_2–P_2O_5 glass by non-isothermal technique. Adv Mater Res 535–537:1629–1633. https://doi.org/10.4028/www.scientific.net/AMR.535-537.1629
104. Thokchom JS, Kumar B (2008) Composite effect in superionically conducting lithium aluminium germanium phosphate based glass-ceramic. J Power Sources 185:480–485. https://doi.org/10.1016/j.jpowsour.2008.07.009
105. Meesala Y, Jena A, Chang H et al (2017) Recent advancements in Li-Ion conductors for all-solid-state Li-Ion batteries. ACS Energy Lett 2:2734–2751. https://doi.org/10.1021/acsenergylett.7b00849
106. Leo C, Rao GS, Chowdari B (2003) Effect of MgO addition on the ionic conductivity of $LiGe_2(PO_4)_3$ ceramics. Solid State Ion 159:357–367. https://doi.org/10.1016/S0167-2738(03)00032-8
107. Noi K, Suzuki K, Tanibata N et al (2018) Liquid-phase sintering of highly Na^+ ion conducting $Na_3Zr_2Si_2PO_{12}$ ceramics using Na_3BO_3 additive. J Am Ceram Soc 101:1255–1265. https://doi.org/10.1111/jace.15288
108. Suzuki K, Noi K, Hayashi A et al (2018) Low temperature sintering of $Na_{1+x}Zr_2Si_xP_{3-x}O_{12}$ by the addition of Na_3BO_3. Scripta Mater 145:67–70. https://doi.org/10.1016/j.scriptamat.2017.10.010
109. Tadanaga K, Takano R, Ichinose T et al (2013) Low temperature synthesis of highly ion conductive $Li_7La_3Zr_2O_{12}$–Li_3BO_3 composites. Electrochem Commun 33:51–54. https://doi.org/10.1016/j.elecom.2013.04.004

110. Shin RH, Son S, Han YS et al (2017) Sintering behavior of garnet-type $Li_7La_3Zr_2O_{12}$–Li_3BO_3 composite solid electrolytes for all-solid-state lithium batteries. Solid State Ion 301:10–14. https://doi.org/10.1016/j.ssi.2017.01.005
111. Rosero-Navarro NC, Yamashita T, Miura A et al (2016) Preparation of $Li_7La_3(Zr_{2-x},Nb_x)O_{12}$ ($x = 0$–1.5) and $Li_3BO_3/LiBO_2$ composites at low temperatures using a sol–gel process. Solid State Ion 285:6–12. https://doi.org/10.1016/j.ssi.2015.06.015
112. Yang C, Yi-Qiu L, Xiang-Xin G (2013) Densification and lithium ion conductivity of garnet-type $Li_{7-x}La_3Zr_{2-x}Ta_xO_{12}$ ($x = 0.25$) solid electrolytes. Chin Phys B 22:0782011–0782015. https://doi.org/10.1088/1674-1056/22/7/078201
113. Rosero-Navarro NC, Yamashita T, Miura A et al (2017) Effect of sintering additives on relative density and Li-ion conductivity of Nb-doped $Li_7La_3ZrO_{12}$ solid electrolyte. J Am Ceram Soc 100:276–285. https://doi.org/10.1111/jace.14572
114. Duan H, Zheng H, Zhou Y et al (2018) Stability of garnet-type Li ion conductors: an overview. Solid State Ion 318:45–53. https://doi.org/10.1016/j.ssi.2017.09.018
115. Hao S, Zhang H, Yao W et al (2018) Solid-state lithium battery chemistries achieving high cycle performance at room temperature by a new garnet-based composite electrolyte. J Power Sources 393:128–134. https://doi.org/10.1016/j.jpowsour.2018.05.028
116. Tian Y, Ding F, Zhong H et al (2018) $Li_{6.75}La_3Zr_{1.75}Ta_{0.25}O_{12}$@amorphous Li_3OCl composite electrolyte for solid state lithium-metal batteries. Energy Storage Materials 14:49–57. https://doi.org/10.1016/j.ensm.2018.02.015
117. Buschmann H, Dolle J, Berendts S et al (2011) Structure and dynamics of the fast lithium ion conductor "$Li_7La_3Zr_2O_{12}$". Phys Chem Phys 13:19378–19392. https://doi.org/10.1039/C1CP22108F
118. Il'ina EA, Aleksandrov AV, Raskovalov AA et al (2013) Optimization of the preparation conditions of $Li_7La_3Zr_2O_{12}$ ceramic electrolyte for lithium power cells. Rus. J Appl Chem 86:1225–1229. https://doi.org/10.1134/s1070427213080120
119. Tang Y, Zhang Q, Luo Z et al (2017) Effects of Li_2O–Al_2O_3–SiO_2 system glass on the microstructure and ionic conductivity of $Li_7La_3Zr_2O_{12}$ solid electrolyte. Mater Lett 193:251–254. https://doi.org/10.1016/j.matlet.2017.01.134
120. Zhao P, Wen Y, Cheng J et al (2017) A novel method for preparation of high dense tetragonal $Li_7La_3Zr_2O_{12}$. J Power Sources 344:56–61. https://doi.org/10.1016/j.jpowsour.2017.01.088
121. Ito Y, Miyauchi K, Oi T (1983) Ionic conductivity of Li_2O–B_2O_3 thin films. J Non-Cryst Solids 57:389–400. https://doi.org/10.1016/0022-3093(83)90426-X
122. Eddrief M, Dzwonkowski P, Julien C et al (1991) The ac conductivity in B_2O_3–Li_2O films. Solid State Ion 45:7–82. https://doi.org/10.1016/0167-2738(91)90105-K
123. Saetova NS, Raskovalov AA, Antonov BD et al (2016) The influence of lithium oxide concentration on the transport properties of glasses in the Li_2O–B_2O_3–SiO_2 system. J Non-Cryst Solids 443:75–81. https://doi.org/10.1016/j.jnoncrysol.2016.04.025
124. Alexander MG, Riley B (1986) Ion conducting glasses in the Na_2O–Y_2O_3–SiO_2 and Li_2O–Y_2O_3–SiO_2 systems. Solid State Ion 19:478–482. https://doi.org/10.1016/0167-2738(86)90163-3
125. Sokolov IA, Tarlakov YP, Ustinov NY et al (2005) Structure and electric properties of lithium phosphate glasses. Russ J Appl Chem 78:741–746. https://doi.org/10.1007/s11167-005-0383-z
126. Raskovalov AA, Pershina SV, Reznitskikh OG et al (2015) The transport properties of the composition glassy $LiPO_3$/crystalline SiO_2 system. Ionics 21:695–704. https://doi.org/10.1007/s11581-014-1203-3
127. Money BK, Hariharan K (2007) Lithium ion conduction in lithium metaphosphate based systems. Appl Phys A 88:647–652. https://doi.org/10.1007/s00339-007-4020-y
128. Pershina SV, Raskovalov AA, Antonov BD et al (2016) Structure and properties of (1–x)$LiPO_{3-x}SiO_2$ and $50Li_2O$–(50–x)$P_2O_{5-x}SiO_2$ glass-forming systems. Rus J Phys Chem A 90:2194–2201. https://doi.org/10.1134/S0036024416110200
129. Larraz G, Orera A, Sanjúan ML (2013) Cubic phases of garnet-type $Li_7La_3Zr_2O_{12}$: the role of hydration. J Mater Chem A 1:11419–11428. https://doi.org/10.1039/C3TA11996C

130. Tietz F, Wegener T, Gerhards MT et al (2013) Synthesis and Raman micro-spectroscopy investigation of $Li_7La_3Zr_2O_{12}$. Solid State Ion 230:77–82. https://doi.org/10.1016/j.ssi.2012.10.021
131. Thompson T, Wolfenstine J, Allen JL et al (2014) Tetragonal vs. cubic phase stability in Al–free Ta doped $Li_7La_3Zr_2O_{12}$ (LLZO). J Mater Chem A 2:13431–13436. https://doi.org/10.1039/C4TA02099E
132. Hitz GT, Wachsman ED, Thangadurai V (2013) Highly Li-Stuffed Garnet-type $Li_{7+x}La_3Zr_{2-x}Y_xO_{12}$. J Electrochem Soc 160:A1248–A1255. https://doi.org/10.1149/2.088308jes
133. Pershina SV, Il'ina EA, Reznitskikh OG (2017) Phase composition, density and ionic conductivity of the $Li_7La_3Zr_2O_{12}$-based composites with $LiPO_3$ glass addition. Inorgan Chem 56:9880–9891. https://doi.org/10.1021/acs.inorgchem.7b01379
134. Brethous J-C, Levasseur A, Villeneuve G et al (1981) Etudes par spectroscopic Raman et par RMN des verres du systhme B_2O_3–SiO_2–Li_2O. J Solid State Chem 39:199–208. https://doi.org/10.1016/0022-4596(81)90332-7
135. Cornaglia LM, Múnera J, Irusta S et al (2004) Raman studies of Rh and Pt on La_2O_3 catalysts used in a membrane reactor for hydrogen production. Appl Catal A 263:91–101. https://doi.org/10.1016/j.apcata.2003.12.003
136. Kruger H, Kahlenberg V, Kaindl R (2007) $Li_2Si_3O_7$: crystal structure and Raman spectroscopy. J Solid State Chem 180:922–928. https://doi.org/10.1016/j.jssc.2006.12.015
137. Il'ina EA, Raskovalov AA, Antonov BD et al (2017) Composite electrolytes ceramic $Li_7La_3Zr_2O_{12}$/glassy Li_2O–Y_2O_3–SiO_2. Mater Res Bull 93:157–161. https://doi.org/10.1016/j.materresbull.2017.04.050

Chapter 9
$Li_7La_3Zr_2O_{12}$ and Poly(Ethylene Oxide) Based Composite Electrolytes

Frederieke Langer, Robert Kun and Julian Schwenzel

Abstract Garnet type electrolytes based on the sum formula $Li_7La_3Zr_2O_{12}$ (LLZO) have been identified as promising materials for solid state electrolytes with high ionic conductivity and electrochemical stability. Since the large-scale processing of thin and brittle ceramic sheets remains challenging and possible volume changes in a solid state battery require flexibility from the electrolyte/separator, the concept of composite electrolytes consisting of LLZO and polymer electrolyte based on poly(ethylene oxide) (PEO) has been explored by multiple research groups. The chapter is dedicated to introduce the concept of polymer and composite electrolytes and gives an overview of LLZO/PEO electrolytes investigated so far.

9.1 Introduction

Ever since the discovery of lithium ion conducting garnets by Thangadurai and Weppner [1], this class of oxide ceramic material has attracted attention for its chemical stability—in air and against lithium metal—and high ionic conductivity. With Murugan's report on $Li_7La_3Zr_2O_{12}$ (LLZO) [2] intense research has been sparked to find the right composition with maximum ionic conductivity [3]. In recent years, lots

F. Langer (✉) · J. Schwenzel
Fraunhofer Institute for Manufacturing Technology and Advanced Materials IFAM, Electrical Energy Storage, Wiener Str. 12, 28359 Bremen, Germany
e-mail: frederieke.langer@ifam.fraunhofer.de

J. Schwenzel
e-mail: julian.schwenzel@ifam.fraunhofer.de

R. Kun
Institute of Materials and Environmental Chemistry, Research Centre for Natural Sciences, Magyar tudósok körútja 2., 1117 Budapest, Hungary
e-mail: kun.robert@ttk.hu

Faculty of Chemical Technology and Biotechnology, Department of Chemical and Environmental Process, Budapest University of Technology and Economics, Műegyetem rakpart 3., 1111 Budapest, Hungary
e-mail: robert.kun@mail.bme.hu

R. Murugan and W. Weppner (eds.), *Solid Electrolytes for Advanced Applications*,
https://doi.org/10.1007/978-3-030-31581-8_9

of studies have been devoted to crystallographic topics, such as determination of the right crystal structure (cubic [4] vs. tetragonal [5] vs. off centred cubic [6, 7]), lithium-ion and substitution ion positions in the lattice [8–12], as well as processing problems, such as densification [13], high temperature lithium loss [14] and others [15–17]. While progress has been made on these basic topics, with the development of knowledge, processing issues have been started to ask. While most studies use small (up to 20 mm) pellet-shaped LLZO samples with high density [13, 18–20], it is obvious that such dimensions are not feasible for large-scale production and application to mass-produced, commercial battery cells. Hence, methods are investigated to process the LLZO crystal structure in thin and dense layers that might suffice for industrial application. Sputtering processes yield thin layers, but still struggle with phase purity, crystallinity and scale-up [21]. Dip-coating [22] and tape casting have been successfully attempted [23, 24], however, large sheets of thin ceramics are usually brittle and challenging to handle.

At the same time, the polymer electrolyte community has (re-)discovered Li-ion conducting ceramics as filler materials for solid polymer electrolytes. Polymer electrolytes are competitors to ceramic electrolytes, such as LLZO, that have the advantage of facile processing and easy scale-up ability, but until recently they lacked high ionic conductivity at room temperature. Furthermore, polymer electrolytes are far softer than rigid and hard ceramic electrolytes and consequently offer the advantage to absorb volume changes of electrode layers during cycling of the battery cell. Now, a new wave of research is dedicated to composites or hybrids between LLZO and polymer electrolytes. This chapter aims to introduce briefly the concept of polymer electrolytes and composite electrolytes to the reader with the focus on polyethylene oxide-based electrolytes. The chapter will then elucidate the motivation of various researchers for LLZO/PEO composites and present the recent findings. In the end, ideas for further research are presented.

9.2 Concept of Polymer Electrolytes

The term 'polymer electrolytes' covers a wide range of electrolyte compositions usually comprising of a polymer, a Li salt and, in some cases, an organic liquid. Compositions range from gel electrolytes (liquid electrolytes with small amounts of polymer) to plasticised electrolytes (mixtures of polymer and Li salt with small liquid addition to improve the ionic conductivity) and, finally, dry solid polymer electrolytes (DSPE), which consist only of polymer and salt. The aim is usually to find an optimum between ionic conductivity, mechanical stability and safety [25]. Further groups are ionic rubbers, which are similar to plasticised electrolytes, though, instead of liquid electrolytes ionic liquids are added [26], and polyelectrolytes, where the Li salt anion is incorporated into the polymer chain backbone. The following sections will mainly focus on DSPEs that become ionically conductive by addition of metal salts [27].

In contrast to glass and ceramic Li^+ ion conductors, polymer electrolytes are usually mixed ionic conductors. In ceramics, only the Li^+ ion is mobile in a crystal structure while the other ions make up the crystal scaffold. Since the polymer chains dissolve the Li salt within DSPE [28], not only the Li^+ ions are mobile, but the anions as well. Hence, the ionic conductivity is the sum of cationic (Li^+) and anionic (i.e. counter anions) conductivity. Consequently, when a potential is applied, polarisation effects within the DSPE hinder Li^+ ion conduction. Therefore, the anionic conductivity is detrimental to the aptitude of polymers as electrolyte materials.

Nevertheless, DSPEs are considered for application, because of their relative softness and low-density compared to other solid electrolytes. Due to their softness, they are able to compensate the volume changes of anode and cathode materials during battery cycling. Their density, which is close to liquid electrolytes, results in higher gravimetric energy densities in solid state batteries when a polymer electrolyte is used compared to battery manufactured with a ceramic electrolyte with the same volume fraction. Furthermore, DSPEs can be processed using the already established processing routes for polymers and polymer matrix composites. The possibilities of solvent-based processing or post-process annealing ensure good contact between active material and electrolyte enabling three-dimensional, thick cathode designs.

9.2.1 Poly(Ethylene Oxide) Based Electrolytes

The most established DSPEs are based on poly(ethylene oxide) (PEO) [29]. The distance between the functional ether groups within the polymer chains as well as the absence of side groups results in ionic conductivity superior compared to polymer electrolytes with side groups (e.g. poly(propylene oxide)). In addition, PEO lacks groups that are potentially harmful to the chemical and electrochemical stability of the electrolyte, such as carbonyl or urethane groups [30]. However, its linearity causes the polymer to crystallize in its pure form.

In 1973, Fenton and co-workers described the ability of PEO to dissolve metal salts [31]. The ether oxygen within the polymer backbone acts as a Lewis base, which is responsible for the facile solvation of metal salts [32]. PEO is able to dissolve both Li^+ ions and other ions (such as Na^+, K^+, NH^{4+} and I^-), which makes it an interesting material for a wide range of ion conductors [33]. Bruce's group described the crystal structure of PEO LiX complexes with several different anions (X) [34]. In short, the polymer chains form helices with repeating units of seven monomer units and two turns of the helix. The Li^+ ions are coordinated to three ether oxygens within the helix [35, 36].

In theory, Li^+ ions move between coordination sites by a hopping mechanism strongly coupled to the segmental motion of the polymer chain [32, 37]. The Li^+ ions move preferably along the polymer chain. Movement between two polymer chains is the rate-limiting step [38]. Furthermore, ionic conduction predominantly takes place in the amorphous region of the polymer [39]. Consequently, the temperature dependency of ionic conductivity in PEO based electrolytes can often be fit using the Vogel–Tamman–Fulcher expression that applies to amorphous polymers at temperatures above the glass transition temperature [40]. Li^+ ion conductivity in

PEO highly depends on both specific polymer and specific Li salt properties. Polymer properties are the molecular weight of the polymer chains and the degree of crystallinity.

Often, high molecular weight PEO ($M_w > 10^5$ g/mol) is used for DSPE preparation, because low weight PEO with short-chain lengths ($M_w < 10^3$ g/mol, also called poly(ethylene glycol) (PEG)) is mechanically unstable or even liquid ($M_w <$ 400 g/mol). While at low molecular weight, Li^+ ion conductivity is high, at molecular weights above 10,000 g/mol, the conductivity is slightly lower but becomes nearly independent of the chain length [41]. At such high chain length, the polymer is mechanically stable.

As described above, amorphous polymers show high Li^+ ion conductivity [42]. Nevertheless, ion conduction also takes place in the crystalline regions and is sometimes considered superior to the amorphous conductivity [34]. For practical application, it is generally disregarded because random crystallinity reduces the Li^+ ion conductivity [42] and oriented crystal structures in polymers are only achieved with effort. Methods to suppress the crystallisation of PEO are discussed in the following paragraphs since they are related to additives, such as the Li salts or filler particles.

Li salt properties that influence the ionic conductivity are the choice of Li salt anion and the Li salt concentration.

The choice of Li salt affects both the crystallinity of the PEO and the conductivity. Large complex anions (e.g. perchlorate (ClO_4^-) [43], bis (trifluoromethane) sulfonimide (TFSI, $(CF_3SO_2)_2N$) [44] or triflate ($CF_3SO_3^-$) [45]) delocalise their negative charge. Hence, the salts possess low lattice energy resulting in facile solubility. The crystallinity of PEO is reduced by complex formation between Li^+ ions by the ether groups. In addition, anions, such as $TFSI^-$, work as plasticisers. Both effects result in increased ionic conductivity. Furthermore, reduced anion mobility due to the delocalised charge increases the cation transference number of the electrolyte.

The optimum concentration of Li salt to achieve maximum conductivity varies from Li salt to Li salt [46]. It is usually referred to as the ratio n of ethylene oxide monomer units to Li^+ ions ($P(EO)_nLiX$ with X as the anion). If the Li^+ ion concentration is low, ionic conductivity is low as well due to the small number of charge carriers. In contrast, at high concentrations, the number of Li^+ ions exceeds the number of coordination sites resulting in a decrease of conductivity. Instead, the high concentration of ions leads to ion pair formation. For example, the highest conductivity for a $P(EO)_nLiClO_4$ system is achieved with $n = 20$, whereas a completely amorphous electrolyte is achieved with $n = 8$ to $n = 10$ [45].

Since the addition of Li salt reduces the mechanical stability of PEO, simultaneously increasing conductivity and mechanical stability of polymer electrolytes is the aim of multiple studies. Both, co-polymerisation [47] (e.g. poly(styrene-block-ethylene oxide) [48]) and blending of polymers (e.g. thermoplastic polyurethane (TPU) and polyether modified polysiloxane (PEMPS) [49]) have been investigated. Furthermore, cross-linking of polymer chains [50], such as polyethers [51] or polyesters [52], and the formation of composite materials are described. Composites also include incorporation of plasticizers, ionic liquids or liquid electrolytes

[53]. However, although these liquid components increase conductivity, they have a negative effect on mechanical stability.

9.2.2 *Concept of Composite Electrolytes*

9.2.2.1 Composite Electrolytes with Non-conductive Fillers

The combination of polymer electrolytes with ceramic components is part of composite electrolytes. The majority of these composite electrolytes do not contain any liquid. Thus, the addition of ionically conductive (termed as "*active*") or insulating, non-conductive (termed "*inactive*") ceramic particles increases both ionic conductivity and mechanical stability. Weston and Steele [54] first introduced this concept several decades ago. They systematically investigated the influence of the addition of alumina particles to $P(EO)_8LiClO_4$. Mechanical stability increased with increasing filler content, while only a moderate decrease of ionic conductivity was noted. Since then, positive effects on conductivity upon addition of inactive ceramic fillers, such as Al_2O_3 [55], $BaTiO_3$ [56], SiO_2 [57], TiO_2 [58], ZrO_2 [59], zeolites [60] have been demonstrated.

The filler particles introduce defects in the polymer crystal structure. Therefore, the ceramic particles suppress the crystallisation of the polymer so that the ratio of amorphous/crystalline fraction increases. Furthermore, the polymer electrolyte forms an amorphous layer at the interface between ceramic and polymer that is supposed to be highly ion-conductive [61].

Parameters of the ceramic particles that affect the ionic conductivity include:

- weight fraction of filler in the composite material
- size of filler particles
- functional surface groups of filler particles.

For *inactive* fillers, the maximum improvement to ionic conductivity is often reached for an addition of about 10 wt% fillers, independent of the kind of ceramic used. Higher amounts of filler particles lead percolation of ceramic particles and dilution of polymer electrolytes and, consequently, lead to a decrease in ionic conductivity. If the concentration of filler exceeds the optimum, insulating agglomerates are formed (percolation) and the fraction of active material (polymer electrolyte) becomes too small (dilution), resulting in a decrease in ionic conductivity [54].

Often, nano-sized filler particles are dispersed in polymer electrolytes to increase ionic conductivity, because large and poorly dispersed particles can lead to a decreased ionic conductivity [54, 62]. However, for dielectric particles (i.e. $BaTiO_3$) also larger particles are considered because different effects are observed for different particle sizes. While small particles (0.06 μm) change the polymer's morphology (reduction of crystallinity) and increase the conductivity, larger particles (0.5 μm) increase the Li salt dissociation and, thereby, the ionic conductivity and cation transference number [56].

A higher cation transference number can also be realised by changing the functional surface groups of filler particles. The acidity of the particles' surface not only causes an increase in conductivity but also increases the cation transference number [63].

Finally, the addition of ceramic particles has a further positive effect on the Li metal/polymer interface [64, 65]. Filler particles cause a reduction of the charge transfer resistance and a stabilisation of the solid electrolyte interphase.

9.2.2.2 Composite Electrolytes with Ion-Conductive Fillers

Manufacturing composite electrolytes containing ionically conductive (*active*) fillers also approaches increasing conductivity and mechanical stability of polymer electrolytes at the same time. Furthermore, adding active fillers might also increase the Li transference number of the electrolyte system. In addition, ion-conducting ceramic powders can be processed using polymer processing routes with this approach. In 1986, Stevens and Mellander [66] dispersed $MAg4I_5$ (M = K, Ca) in a PEO matrix and investigated the influence on conductivity. At the same time, Wieczorek et al. [67] added Na conducting NaSICON material in high volume fraction to a PEO–NaI matrix. However, high interface transfer resistances inhibit the ion transport across the ceramic/polymer interface [68]. Skaarup et al. [69] used small amounts of PEO for hot pressing Li_3N. The conductivity of the composites was lower than sintered Li_3N but increased steadily with the reduction of PEO content. Nevertheless, the majority of early research regarding the sometimes so-called hybrid electrolytes focused on γ-$LiAlO_2$ in combination with PEO [60, 70].

Since then, the focus has shifted to materials investigated as bulk inorganic electrolytes today, such as $Li_xLa_yTiO_3$ (LLTO) [71], $Li_{1+x}Al_xTi_{2-x}(PO_4)_3$ (LATP) [72] and $Li_7La_3Zr_2O_{12}$ (LLZO) [73]. Compared to composite electrolytes with inactive fillers, the volume fraction of ceramics used within composites with active fillers is higher. Filler contents up to 90 wt% have been realised [74] and particle sizes range from submicron to micron sized. Furthermore, alternative processing routes (hot pressing [75]) or advanced three-dimensional structures (e.g. fibres and mats [76]) are investigated.

LLTO in form of particles as well as electrospun fibres and mats has been added to PEO based polymer electrolytes. Abe et al. [77] proposed a model for ideal ion transport pathways in a composite. Wang et al. described the increase in conductivity as a result of 'high-speed lithium pathways' [76] the LLTO fibres and the interface provided in a composite. In this case, 20 wt% LLTO fibres lead to a maximum conductivity of 2.4×10^{-4} S/cm at room temperature [78].

LATP and LAGP [79] powders have been mixed with PEO and various Li salts (triflate [80], $LiClO_4$ [81] and TFSI [82]) to form composites with up to 90 wt% ceramics. In all cases, room temperature conductivity improved compared to pure polymer electrolytes. Similarly, glass ceramics based on LATP were investigated in PEO matrices with $LiClO_4$ and LiTFSI [72]. With advanced processing, monolayers of Ta-doped LATP particles with polymer electrolytes have been reported [83]. LATP composites are not restricted to PEO matrices but expand to polyurethane [75] and polypropylene oxide [84] as well.

9.3 Composites Containing $Li_7La_3Zr_2O_{12}$

In the following paragraphs, an overview of the available literature on LLZO–PEO composite electrolytes is presented. In most cases, the preparation follows one of the two routes depicted in Fig. 9.1. Composites of LLZO particles embedded into a PEO based polymer electrolyte matrix are most often prepared using the tape casting procedure that has been established for composite electrolytes with inactive ceramic particles. In short, PEO and Li salt are dissolved in a polar solvent with high vapour pressure (most commonly acetonitrile). The ceramic particles are added to the solution forming a homogeneous dispersion that is cast onto a flat surface, such as a glass plate or polytetrafluoroethylene (PFTE) dish. The solvent is evaporated and the result is a composite electrolyte with dispersed LLZO particles. The second route is used to create a LLZO scaffold that is later infiltrated with polymer electrolyte. The scaffold is obtained from a two-step synthesis starting with a precursor solution containing the respective metal ions in form of dissolved salts. The solution may be electrospun or coated onto a sacrificial templating material. After a sintering step, a porous LLZO sample is obtained that is infiltrated with the PEO–salt solution. After a final drying step, the composite is prepared.

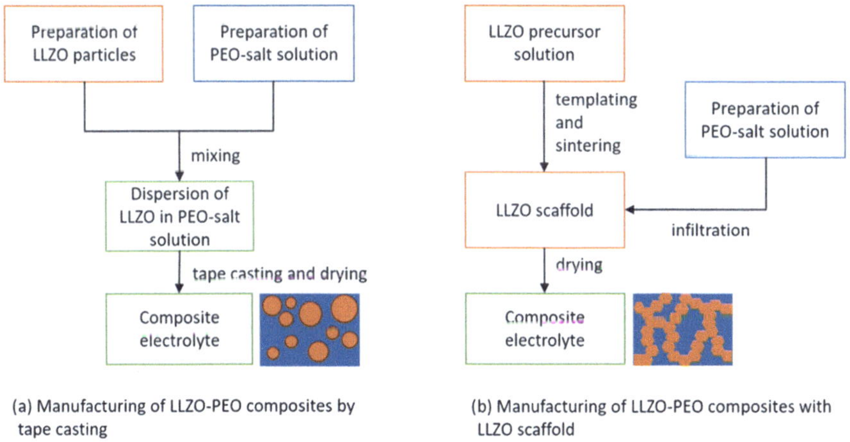

Fig. 9.1 Manufacturing routes commonly employed for LLZO–PEO composite electrolyte production, **a** tape casting technique to obtain composites with granular LLZO particles (orange) embedded in PEO matrix (blue), **b** scaffolding technique resulting in LLZO matrix (orange) with mechanical properties enhanced by PEO infiltration (blue)

9.3.1 Composites with Granular LLZO

The first laboratory experiment on such PEO/LLZO composites was executed by Choi et al. [73]. In 2015 they described a composite consisting of a $PEO_{15}LiClO_4$ matrix filled with tetragonal LLZO particles in the submicron range. The composites were prepared by tape casting a slurry of PEO, $LiClO_4$ and LLZO dissolved and dispersed in acetonitrile. The EO:Li ratio was kept constant at 15:1, while the LLZO content was increased from 0 wt% to a maximum of 85 wt%. The dispersion of 52.5 wt% LLZO was found to yield the maximum ionic conductivity of 4×10^{-4} S/cm [73]. Although the ionic conductivity of the LLZO–PEO composite is higher than the ionic conductivity of a Al_2O_3–PEO composite with the same amount of ceramic filler, unfortunately, no statement is made to interpret the effect other than the inhibition of crystallinity in the polymer matrix which is also observed for inactive fillers.

Similar ionic conductivities were obtained with lower amounts of cubic, Al-substituted LLZO [85]. The ceramic powder was prepared by field assisted sintering technique (FAST). The polymer electrolyte used was a PEO_8TFSI. Rather low amounts of LLZO (i.e. 2.5–10 wt%) were mixed into the polymer matrix and the composites were prepared by tape casting technique. A maximum in ionic conductivity with 5.5×10^{-4} S/cm was observed for LLZO content of 7.5 wt%. Furthermore, a positive effect of LLZO addition to the polymer electrolyte on the Li-ion transfer number (t_{Li+}) was noticed. The number increased to 0.207, which is higher than the transfer number for the same polymer electrolyte without LLZO addition. At last, the electrochemical stability window could be increased to 5.7 V versus Li.

Flexible and free standing composite electrolytes with LLZO particles with cubic crystal structure were also prepared by Langer et al. [86]. The LLZO particles were prepared using a combined co-precipitation and solid state lithiation method [87]. The cubic phase was stabilized by doping with Al^{3+}. Homogeneous membranes could be obtained by tape casting up to 40 vol% LLZO in a $PEO_{20}LiClO_4$ electrolyte. Higher ceramic content led to the segregation of phases during the drying process. In contrast to Choi's work, the addition of cubic LLZO particles did not improve the ionic conductivity, but instead it decreased with increasing LLZO content.

Similar experiments were carried out by Passerini's group [88]. They dispersed 70 wt% LLZO prepared by Schott in a $PEO_{15}LiTFSI$ electrolyte by milling and hot pressing, thus eliminating the risk of solvent residue in the resulting electrolyte. However, although they observed increased mechanical stability in the form of improved flexibility in comparison to rigid ceramic electrolytes, the ionic conductivity was reduced to about 10^{-6} S/cm at room temperature. However, the interface resistance between the electrolyte and Li metal could be decreased, reaching a plateau of about 200 Ω cm^2 after 30 days.

Although most of the composite membranes are prepared using a cubic LLZO stabilized with Al^{3+} ions, some studies also used Ta^{5+} substitution for their LLZO. Ta-doped LLZO was used in a study employing micron sized LLZO particles in a $PEO_{15}LiClO_4$ matrix [89]. The ceramic content was varied up to 56 wt%. LLZO was obtained from solid state reaction. Conventional tape casting was used to prepare the composite membranes. The highest ionic conductivity could be realised with 17 wt% LLZO. However, the ionic conductivity was still lower than the ionic conductivity of pellets prepared from the same LLZO material. Nevertheless, a solid state battery employing the composite electrolyte could be operated at 60 °C.

In another study, Ta-doped LLZO powder prepared by solid state reaction was used in a series of composites with LLZO content from 0 to 80 wt% [90]. Unlike the usual tape casting, the composites were prepared by hot pressing a mixture of LLZO and $PEO_8LiTFSI$ at 70 °C. Furthermore, poly(ethylene glycol), a short-chain version of PEO with a molecular weight of $M_w = 400$ g/mol, was added to increase the flexibility of the prepared membranes. A maximum ionic conductivity (1.17×10^{-4} S/cm) was observed for LLZO content of 10 wt% because the addition of ceramic in amounts below the percolation threshold increases the free volume in the polymer electrolyte and, thereby, facilitates the segmental motion necessary for the ion transport. At higher ceramic content (80 wt%), the ionic conductivity decreased because of the disadvantageous ceramic–polymer interface.

Instead of PEG, ionic liquid 1-butyl-3-methylimidazolium bis (trifluoromethylsulfonyl) imide ($[BMIM]TF_2N$) was used to improve the composite by Huo et al. [91]. They also used Ta-doped LLZO prepared from solid state reaction in a tape cast composite. The ionic liquid wets the LLZO/PEO interface and hence, improves the performance of the prepared electrolyte. The maximum ionic conductivity obtained with 15 vol% LLZO without ionic liquid is slightly lower than other values reported (1.7×10^{-5} S/cm). The addition of ionic liquid increases this value to 2.2×10^{-4} S/cm and, in addition, also increases the electrochemical stability window to 4.85 V versus Li.

These contradicting results lead to the question how Li^+ ions are transported in a composite electrolyte with granular dispersed LLZO. While in composites with LLZO scaffolds a continuous Li^+ ion pathway exists in the ceramic phase, in a composite with dispersed LLZO particles, the contact between LLZO particles is a physical contact at maximum with no sintered grain boundaries due to the lack of thermal treatment. Furthermore, the amount of LLZO and the particle size of the powder used is important. In order to achieve a continuous Li^+ ion pathway in composites with particular LLZO, a percolation threshold needs to be reached. This threshold also depends on the morphology of the particles and their size.

The concept of using the polymer as a binder only and forgo the addition of Li salt to the PEO phase, which was presented in the 1980s by Skaarup et al. for Li_3N [69], has been revisited by Zhang et al. [92]. They used a Ta-substituted LLZO and prepared composite membranes with 12 vol% LLZO in three different polymer matrices, pure PEO, PEO_8LiClO_4 and $PEO_8LiTFSI$. They found that the salt-free membrane outperformed the salt-containing samples. The ionic conductivity of the 40 μm thick membrane was 2.1×10^{-4} S/cm and the electrochemical stability window was greater than 4.7 V versus Li. Furthermore, in a plating/stripping experiment the salt-free membrane inhibited Li dendrite growth, while Li dendrites could be observed in salt-containing samples.

In summary, various compositions of LLZO–PEO based composite electrolytes have been reported so far. The achieved ionic conductivity varies in a broad range between 10^{-6} S/cm and 5.5×10^{-4} S/cm. To compare the different reported conductivities, in Fig. 9.2 room temperature ionic conductivities obtained from impedance spectroscopy are plotted as a function of the LLZO content. All displayed conductivities were measured at room temperature. In case of studies with multiple LLZO contents, only the highest ionic conductivity is shown. Figure. 9.3 shows that for similar LLZO contents, the ionic conductivity may vary drastically. This might be

Fig. 9.2 Photograph of a typical LLZO–PEO electrolyte membrane prepared by tape casting method. The homogeneous distribution of LLZO particles in the PEO matrix causes the opaque colour of the membrane. Usually, flexibility is demonstrated by showing photographs of bent samples. Reprinted from [86] with permission from Elsevier

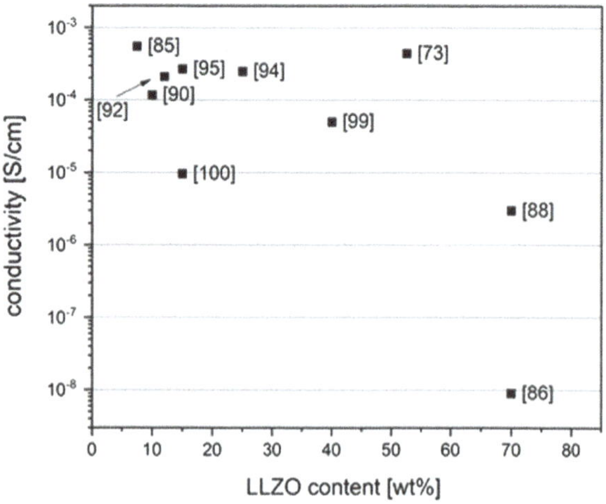

Fig. 9.3 Room temperature ionic conductivities of LLZO–PEO composite electrolytes as a function of LLZO content reported so far

due to the different preparation techniques (tape cast vs. infiltrated), different types of LLZO (e.g. substitution elements) or varying morphology of LLZO (e.g. particle size). However, it seems that the maximum possible ionic conductivity is in the region of 10^{-4} S/cm, a value commonly reached by dense LLZO pellets. The advantage of the composites in comparison to the pellets is the increased flexibility and facile processing.

9.3.2 *Composites with LLZO Scaffolds*

To our knowledge, the first mention of a composite consisting of a PEO based polymer electrolyte and LLZO is a computational simulation [93]. It studied the electrochemical and mechanical properties of a porous, sintered LLZO scaffold filled with $PEO_{10}LiTFSI$. It was found that, compared to the pure polymer electrolyte, the ionic conductivity increases with an increasing sintering neck size between LLZO particles.

Experimentally, this was realised by Wachsman's group in 2016. A three-dimensional scaffold of sintered LLZO was infiltrated with PEO–LiTFSI [94]. The LLZO scaffold was prepared by electrospinning a metal nitrate acidic precursor solution and sintering. The scaffold was infiltrated with the polymer electrolyte. The ceramic fraction was 25 wt% and offered the possibility of long-range Li^+ ion conduction. PEO was added for flexibility. The ionic conductivity was determined to 2.5×10^{-4} S/cm at room temperature. A similar approach was taken by the same group by changing the scaffolding technique from cost and time-consuming electrospinning to using cellulose textile as template [95]. The resulting composite electrolyte showed ionic conductivity comparable to the electrospun composite (2.7×10^{-4} S/cm at room temperature) and could be cycled in a Li–sulphur battery for 500 h. Although the ceramic phase only made up 15 vol% of the composite, the Li-ion conduction predominantly took place in the ceramic phase.

Another scaffolding technique proposed is freeze casting [96]. In this technique, an aqueous dispersion is frozen and the ice sublimed in a vacuum. By carefully choosing the freezing parameters, it is possible to align the resulting ceramic channels in the desired direction. Sintering of the scaffold and subsequent infiltration with a polymer electrolyte yields flexible composite films. Although the technique has been proposed to work with LLZO, so far, it has only been demonstrated with $Li_{1+x}Al_xTi_{2-x}(PO_4)_3$ (LATP).

As can be seen, the preparation of a suitable scaffold and the consequent infiltration with polymer electrolytes can be challenging. Hence, a lot of research focuses on the processing route similar to the processing of inactive fillers, namely tape casting of a ceramic–polymer slurry with finely dispersed LLZO particles in a PEO and salt solution.

9.3.3 Li-Ion Pathways in Composites and LLZO/PEO Interface

A few publications aim to identify the Li-ion pathways in LLZO–PEO composites and determine the interface resistance between the two phases—polymer and ceramic electrolyte. Mainly, two different approaches are used for investigation—electrochemical and chemical methods.

Theoretically, several Li-ion pathways are possible in composite electrolytes depending on the interface resistance between the polymer and LLZO phases and the LLZO content. The different pathways are shown in Fig. 9.4. In Fig. 9.4a–c, a composite electrolyte with low amount of LLZO is depicted. The ceramic phase does not form a percolating network. If the interface resistance between the PEO phase and the LLZO phase is high (Fig. 9.4a and b) the Li-ions do not cross the interface. Since LLZO provides no continuous pathway, Li-ions move through the

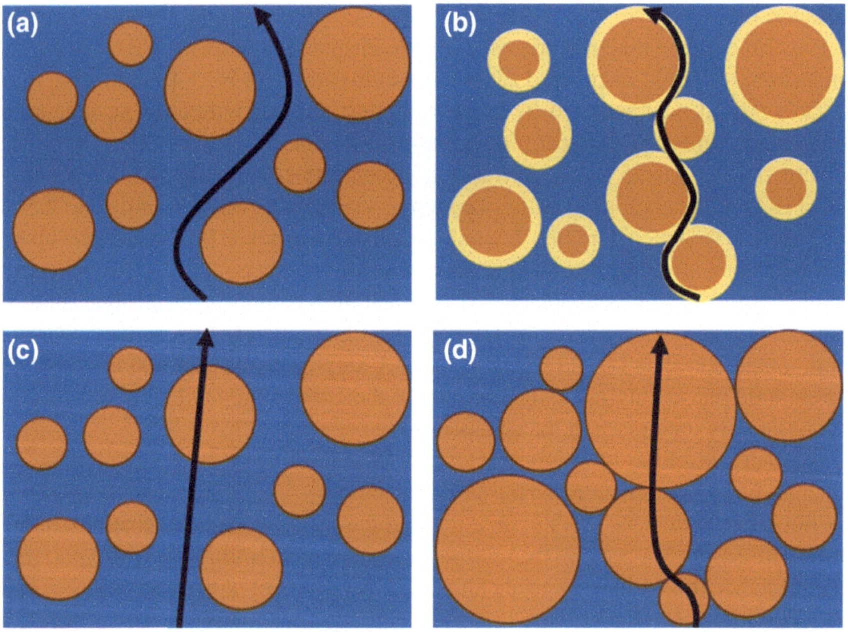

Fig. 9.4 Possible Li-ion pathways (black arrows) in composite electrolytes of LLZO (orange) and PEO (blue), depending on the composition of the composite and the interfaces, the Li-ion movement can take place by one or more of the mechanisms shown, **a** in a composite with non-percolating LLZO phase and with high interface resistance between LLZO and PEO may only move through the PEO bulk phase, **b** in a composite with non-percolating LLZO phase and highly conductive amorphous polymer phase surrounding the LLZO particles, Li-ion conduction in the polymer phase is possible **c** in a composite with low interface resistance between LLZO and PEO, the Li-ions may move across the interface, **d** with continuous LLZO network (percolating particles or scaffold) the LI-ions may move through the ceramic phase

polymer phase only, bypassing the LLZO particles. If the LLZO particles do not alter the polymer phase (Fig. 9.4a), ion conduction can only take place in the PEO bulk phase. In this case, the ionic conductivity of the composite should be dominated by the PEO phase. If the LLZO particles alter the PEO phase and, in the best case, the LLZO particles inhibit the crystallisation of PEO (Fig. 9.4b), the conductivity is enhanced. However, they do not take part in the Li-ion conduction actively. If the interface resistance between LLZO and PEO is low, then Li-ions can pass across this interface (Fig. 9.4c). In addition to the suppression of the poorly conductive crystalline PEO phase, the LLZO particles actively take part in the ionic conduction. As a result, the ionic conductivity might be increased, since the ionic conductivity of LLZO is usually higher than the conductivity of PEO. Furthermore, LLZO is a single ion conductor (Li) and consequently, the Li transference number could be increased in such composites. In Fig. 9.4d a highly filled composite is presented, where LLZO particles form a percolating network within the polymer matrix. Depending on the resistance between the individual particles, Li-ions may move through the composite in the LLZO phase exclusively, bypassing the PEO phase.

To determine the interface resistance between LLZO and PEO matrices, Langer et al. [97] prepared layers of pure LLZO and polymer electrolyte (i.e. $PEO_{20}LiClO_4$). LLZO pellets were prepared by conventional uniaxial pressing and sintering, while $PEO_{20}LiClO_4$ was prepared by tape casting technique. The two materials were brought into contact with well-defined interfaces by melting the polymer electrolyte onto the LLZO pellets' surfaces. The obtained samples were investigated using temperature dependent electrochemical impedance spectroscopy to obtain an interface resistance for the LLZO/PEO interface. While interpretation of the results was challenging due to the rough surface of the LLZO leading to the effect of porous electrodes, the interface resistance was found to be 9 kΩ cm^2 at 70 °C. This high value indicates that the interface transition of Li-ions across the PEO/LLZO interface is unlikely, which is in agreement with the results obtained for LLZO/PEO composites with high amounts of LLZO [86, 88]. It indicates that the addition of high degrees of LLZO dilutes the polymer electrolyte without contributing to the ionic conduction due to the high interface resistance and missing continuous Li-ion pathway.

However, a chemical isotope analysis revealed a Li-ion conduction preferably through the LLZO phase. Zheng et al. [98] prepared a composite electrolyte consisting of 20 vol% LLZO dispersed in a $PEO_{18}LiClO_4$ matrix by tape casting technique. A plating/stripping experiment employing Li metal electrodes with isotope ^{6}Li was used to replace ^{7}Li-ions within the electrolyte with ^{6}Li-ions from the electrodes. Since both isotopes are NMR active, a quantitative analysis of the isotope content could be used to identify the Li-ion pathway. The results show that the Li-ion preferably moves through the LLZO phase and not through the bulk polymer phase or the LLZO/PEO interface. The publication also stresses the importance of a percolated LLZO network for Li-ions to minimize the fraction of the pathway through the polymer electrolyte or the interface.

In their following publications, the same researches expanded their NMR method to a broader range of composites. First, they added triethylene glycol dimethyl ether (TEGDME) as a plasticizer to the LLZO–PEO composite. The investigated "ternary" system consists of 40 wt% Al-substituted LLZO prepared by sol–gel technique, 40 wt% $PEO_{18}LiClO_4$ and 20 wt% TEGDME [99]. The composite was prepared by doctor blading a slurry on a PTFE plate to obtain films with thickness ranging from 50 to 70 μm. It was found that, in contrast to their previous work, the Li-ions moved through the TEGDME-associated phases, bypassing the LLZO and PEO phase. The authors concluded that, thus, the conduction mechanism strongly depends on the composite composition. Hence, they further investigated the effect of LLZO content and TEGDME on the Li-ion pathway [89]. They once again prepared composites by tape casting method using Al-substituted LLZO from a sol–gel synthesis. However, for this publication, the Li salt $LiClO_4$ was exchanged in favour of LiTFSI, using a polymer electrolyte of PEO_{18}LiTFSI. The LLZO content was varied from 5 to 20 wt% and 50 wt%. Again, the Li-ion pathways were tracked by plating/stripping with ^{6}Li electrodes. It was found that for small amounts of LLZO with no percolating LLZO network, the Li-ion conduction takes place predominantly in the polymer electrolyte phase. The ionic conductivity was about 1×10^{-5} S/cm, however, no conclusive trend to the dependency of conductivity on LLZO content could be found. Only at 50 wt% LLZO, a percolating LLZO network forms and the Li-ion conduction pathway changes from the polymer phase to predominantly the LLZO phase. Furthermore, the addition of 20 wt% TEGDME as a plasticizer to the 50 wt% LLZO composite was investigated. With the addition of TEGDME, the ionic conductivity could be enhanced from below 10^{-5} S/cm to nearly 6×10^{-5} S/cm for the 50 wt% LLZO composite. However, the Li-ion pathway changed back from predominantly LLZO phase to predominantly polymer phase upon addition of TEGDME.

9.4 Summary and Outlook

In summary, while high ionic conductivities have been obtained in various cases [73, 85] and even application in solid state batteries have been demonstrated [100], as described in the previous sections, further investigations are necessary. A characterisation of the used LLZO with regard to crystallography (e.g. crystal structure, phase purity), chemistry (e.g. Li content and substitution elements, functional surface groups) and morphology—in case of tape cast composites with granular LLZO, e.g. particle size distribution, shape of particles; in case of scaffold LLZO, e.g. porosity and tortuosity—is useful. Full electrochemical characterisation comprising determination of ionic conductivity as well as Li transference numbers and the electrochemical stability window is necessary to judge the applicability of the produced composites. Furthermore, interface resistances, in general, need to be determined and optimised. This is equally true for the interface resistance between the two phases PEO and LLZO, but maybe even more necessary for the interface between active material (especially Li metal anodes) and the composite electrolyte.

To tailor the interfaces attempts with plasticizers have already been made. Further steps may include the chemical functionalisation of LLZO particle surfaces and chemical bonding between the polymer electrolyte and the ceramic. In this case, the polymer electrolyte has to be changed from already available PEO. In situ polymerisation of a precursor dispersion may be one way to ensure good interfaces between polymer and ceramic, regardless if the ceramic is an active material or LLZO.

In addition, while numerous publications claim increased mechanical properties of LLZO/PEO composite electrolytes in comparison to both LLZO pellets and PEO polymer electrolytes, systematic studies investigating mechanical properties and quantitative determination of parameters such as tensile strength, elastic and shear moduli as well as thermomechanical properties are still missing. Most authors demonstrate the flexibility of the prepared membranes in photographs with bent or rolled samples, showing that the composites are considerably less brittle compared to ceramic sheets. However, tensile strength tests or torsion tests quantifying the mechanical properties have not been reported yet. Furthermore, Monroe and Newman [101] calculated that the shear modulus of a solid state electrolyte should be 6 GPa or higher to inhibit the formation of lithium dendrites. It is unlikely that the addition of ceramic particles to a polymer electrolyte based on PEO will increase the shear modulus to such high values. However, it was recently shown that the shear modulus might not be sufficient as Li dendrite growth is possible in ceramic electrolytes with shear moduli higher than the proposed 6 GPa. In addition, Wachsman's group [102] recently showed that the wetting behaviour of the electrolyte material on Li metal might be more important than the mechanical properties.

To determine the functionality of the prepared composite electrolytes, cycling tests in full battery cells are necessary. This is occasionally done already. For example, LLZO was mixed with PVdF, and liquid electrolyte (1.5 mol $LiPF_6$ in EC/DEC) [74]. The focus of this study was not the electrolyte itself, but a pouch cell containing the composite electrolyte with up to 90 wt% LLZO particles (size 1–5 μm). A nail penetration test was performed to show the non-leaking of the pouch cell with the composite electrolyte. In this case, a liquid electrolyte was included in the cell. Eliminating the liquid components from the battery cell brings along the problem of cathode preparation. While in conventional Li-ion batteries with liquid electrolytes, the porous electrode is wetted by the electrolyte during the filling step, an electrode in solid state batteries needs to contain the solid electrolyte upon manufacturing, consequently being a composite of its own.

Finally, the manufacturing process itself presents a topic of further research. In the laboratory environment, the currently predominant tape casting technique is a facile method to produce composite electrolytes. However, the scale-up from laboratory to technical size and mass production may be unsuitable for organic solvent-based methods. In the enclosed environment of glove boxes, fumes of organic solvents may be considered a minor aspect, however, in industrial-scale application, the recovery of solvent and health security of workers could increase production cost significantly. Hence, solvent free production routes, such as already demonstrated hot pressing, or extrusion, should be considered.

In conclusion, the topic of composite electrolytes consisting of LLZO and PEO based polymer electrolytes is still a wide, interesting and not fully discovered research topic.

References

1. Thangadurai V, Weppner WJF (2005) $Li_6ALa_2Nb_2O_{12}$ (A = Ca, Sr, Ba): a new class of fast lithium ion conductors with garnet-like structure. J Am Chem Soc 88(2):411–418. https://doi.org/10.1111/j.1551-2916.2005.00060.x
2. Murugan R, Thangadurai V, Weppner WJF (2007) Schnelle lithiumionenleitung in granatartigem $Li_7La_3Zr_2O_{12}$. Angew Chem 119(41):7925–7928. https://doi.org/10.1002/ange.200701144
3. Zeier WG (2014) Structural limitations for optimizing garnet-type solid electrolytes: a perspective. Dalton Trans 43(43):16133–16138. https://doi.org/10.1039/c4dt02162b
4. Awaka J, Takashima A, Kataoka K et al (2011) Crystal structure of fast lithium-ion-conducting cubic $Li_7La_3Zr_2O_{12}$. Chem Lett 40(1):60–62. https://doi.org/10.1246/cl.2011.60
5. Awaka J, Kijima N, Hayakawa H et al (2009) Synthesis and structure analysis of tetragonal $Li_7La_3Zr_2O_{12}$ with the garnet-related type structure. J Solid State Chem 182(8):2046–2052. https://doi.org/10.1016/j.jssc.2009.05.020
6. Rettenwander D, Redhammer GJ, Preishuber-Pflugl F et al (2016) Structural and electrochemical consequences of Al and Ga cosubstitution in $Li_7La_3Zr_2O_{12}$ solid electrolytes. Chem Mater 28(7):2384–2392. https://doi.org/10.1021/acs.chemmater.6b00579
7. Robben L, Merzlyakova E, Heitjans P et al (2016) Symmetry reduction due to gallium substitution in the garnet $Li_{6.43(2)}Ga_{0.52(3)}La_{2.67(4)}Zr_2O_{12}$. Acta Crystallogr E Cryst Commun 72(3):287–289. https://doi.org/10.1107/s2056989016001924
8. Geiger CA, Alekseev E, Lazic B et al (2011) Crystal chemistry and stability of "$Li_7La_3Zr_2O_{12}$" garnet: a fast lithium-ion conductor. Inorg Chem 50(3):1089–1097. https://doi.org/10.1021/ic101914e
9. Rettenwander D, Geiger CA, Tribus M et al (2015) The solubility and site preference of $Li_{7-3x}Fe_xLa_3Zr_2O_{12}$ garnets. J Solid State Chem 230:266–271. https://doi.org/10.1016/j.jssc.2015.01.016
10. Rettenwander D, Langer J, Schmidt W et al (2015) Site occupation of Ga and Al in stabilized cubic $Li_{7-3(x+y)}Ga_xAl_yLa_3Zr_2O_{12}$ garnets as deduced from 27 Al and 71 Ga MAS NMR at ultrahigh magnetic fields. Chem Mater 27(8):3135–3142. https://doi.org/10.1021/acs.chemmater.5b00684
11. El Shinawi H, Janek J (2013) Stabilization of cubic lithium-stuffed garnets of the type "$Li_7La_3Zr_2O_{12}$" by addition of gallium. J Power Sources 225:13–19. https://doi.org/10.1016/j.jpowsour.2012.09.111
12. Buschmann H, Dölle J, Berendts S et al (2011) Structure and dynamics of the fast lithium ion conductor "$Li_7La_3Zr_2O_{12}$". Phys Chem Chem Phys 13(43):19378–19392. https://doi.org/10.1039/C1CP22108F
13. Wolfenstine J, Sakamoto J, Allen JL (2012) Electron microscopy characterization of hot-pressed Al substituted $Li_7La_3Zr_2O_{12}$. J Mater Sci 47(10):4428–4431. https://doi.org/10.1007/s10853-012-6300-y
14. Liu K, Ma J-T, Wang C-A (2014) Excess lithium salt functions more than compensating for lithium loss when synthesizing $Li_{6.5}La_3Ta_{0.5}Zr_{1.5}O_{12}$ in alumina crucible. J Power Sources 260:109–114. https://doi.org/10.1016/j.jpowsour.2014.02.065
15. Kang SG, Sholl DS (2014) First-principles study of chemical stability of the lithium oxide garnets $Li_7La_3M_2O_{12}$ (M = Zr, Sn, or Hf). J Phys Chem C 118(31):17402–17406. https://doi.org/10.1021/jp504314w
16. Jin Y, McGinn PJ (2013) $Li_7La_3Zr_2O_{12}$ electrolyte stability in air and fabrication of a $Li/Li_7La_3Zr_2O_{12}/Cu_{0.1}V_2O_5$ solid-state battery. J Power Sources 239:326–331. https://doi.org/10.1016/j.jpowsour.2013.03.155
17. Aguesse F, López del Amo, Juan Miguel, Roddatis V et al (2014) Enhancement of the grain boundary conductivity in ceramic $Li_{0.34}La_{0.55}TiO_3$ electrolytes in a moisture-free processing environment. Adv Mater Interfaces 1(7). https://doi.org/10.1002/admi.201300143

18. Kotobuki M, Kanamura K, Sato Y et al (2011) Fabrication of all-solid-state lithium battery with lithium metal anode using Al_2O_3-added $Li_7La_3Zr_2O_{12}$ solid electrolyte. J Power Sources 196(18):7750–7754. https://doi.org/10.1016/j.jpowsour.2011.04.047
19. Botros M, Djenadic R, Clemens O et al (2016) Field assisted sintering of fine-grained $Li_{7-3x}La_3Zr_2Al_xO_{12}$ solid electrolyte and the influence of the microstructure on the electrochemical performance. J Power Sources 309:108–115. https://doi.org/10.1016/j.jpowsour.2016.01.086
20. Tenhaeff WE, Rangasamy E, Wang Y et al (2014) Resolving the grain boundary and lattice impedance of hot-pressed $Li_7La_3Zr_2O_{12}$ garnet electrolytes. ChemElectroChem 1(2):375–378. https://doi.org/10.1002/celc.201300022
21. Lobe S, Dellen C, Finsterbusch M et al (2016) Radio frequency magnetron sputtering of $Li_7La_3Zr_2O_{12}$ thin films for solid-state batteries. J Power Sources 307:684–689. https://doi.org/10.1016/j.jpowsour.2015.12.054
22. Bitzer M, van Gestel T, Uhlenbruck S et al (2016) Sol-gel synthesis of thin solid $Li_7La_3Zr_2O_{12}$ electrolyte films for Li-ion batteries. Thin Solid Films 615:128–134. https://doi.org/10.1016/j.tsf.2016.07.010
23. Okubo K, Wang H, Hayashi K et al (2018) A dense NASICON sheet prepared by tape-casting and low temperature sintering. Electrochim Acta 278:176–181. https://doi.org/10.1016/j.electacta.2018.05.020
24. Jiménez R, del Campo A, Calzada ML et al (2018) Improved conductivity in tape casted Li-NASICON supported thick films: Effect of temperature treatments and lamination. J Eur Ceram Soc 38(4):1679–1687. https://doi.org/10.1016/j.jeurceramsoc.2017.12.017
25. Manuel Stephan A, Nahm KS (2006) Review on composite polymer electrolytes for lithium batteries. Polymer 47(16):5952–5964. https://doi.org/10.1016/j.polymer.2006.05.069
26. Sharova V, Kim G-T, Giffin GA et al (2016) Quaternary polymer electrolytes containing an ionic liquid and a ceramic filler. Macromol Rapid Commun 37(14):1188–1193. https://doi.org/10.1002/marc.201600025
27. Fergus JW (2010) Ceramic and polymeric solid electrolytes for lithium-ion batteries. J Power Sources 195(15):4554–4569. https://doi.org/10.1016/j.jpowsour.2010.01.076
28. Gray FM (1997) Polymer electrolytes. RSC materials monographs. The Royal Society of Chemistry, Cambridge
29. Xue Z, He D, Xie X (2015) Poly(ethylene oxide)-based electrolytes for lithium-ion batteries. J Mater Chem A 3(38):19218–19253. https://doi.org/10.1039/c5ta03471j
30. Daniel C, Besenhard JO (eds) (2011) Handbook of battery materials, second, completely revised and enlarged edition. Wiley-VCH Verlag GmbH & Co. KGaA, Weinheim, Germany
31. Fenton DE, Parker JM, Wright PV (1973) Complexes of alkali metal ions with poly(ethylene oxide). Polymer 14(11):589. https://doi.org/10.1016/0032-3861(73)90146-8
32. Zhang H, Liu C, Zheng L et al (2014) Lithium bis(fluorosulfonyl)imide/poly(ethylene oxide) polymer electrolyte. Electrochim Acta 133:529–538. https://doi.org/10.1016/j.electacta.2014.04.099
33. Armand MB, Chabagno JM, Duclot MJ (1979) Poly-ethers as solid electrolytes. In: Vashishta P, Mundy JN, Shenoy GK (eds) Fast ion transport in solids: electrodes and electrolytes. Proceedings of the international conference on fast ion transport in solids, electrodes, and electrolytes, Lake Geneva, Wisconsin, USA, May 21–25, 1979. Elsevier North Holland, New York, pp 131–136
34. Gadjourova Z, Andreev YG, Tunstall DP et al (2001) Ionic conductivity in crystalline polymer electrolytes. Nature 412(6846):520–523. https://doi.org/10.1038/35087538
35. Armand MB (1986) Polymer electrolytes. Annu Rev Mater Sci 16(1):245–261. https://doi.org/10.1146/annurev.ms.16.080186.001333
36. Golodnitsky D, Livshits E, Peled E (2003) Highly conductive oriented PEO-based polymer electrolytes. Macromol Symp 203(1):27–46. https://doi.org/10.1002/masy.200351303
37. Agapov AL, Sokolov AP (2011) Decoupling ionic conductivity from structural relaxation: a way to solid polymer electrolytes? Macromolecules 44(11):4410–4414. https://doi.org/10.1021/ma2001096

38. Blanga R, Golodnitsky D, Ardel G et al (2013) Quasi-solid polymer-in-ceramic membrane for Li-ion batteries. Electrochim Acta 114:325–333. https://doi.org/10.1016/j.electacta.2013.09.106
39. Berthier C, Gorecki W, Minier M et al (1983) Microscopic investigation of ionic conductivity in alkali metal salts-poly(ethylene oxide) adducts. Solid State Ion 11(1):91–95. https://doi.org/10.1016/0167-2738(83)90068-1
40. Dygas J, Misztal-Faraj B, Florjańczyk Z et al (2003) Effects of inhomogeneity on ionic conductivity and relaxations in PEO and PEO–salt complexes. Solid State Ion 157(1–4):249–256. https://doi.org/10.1016/S0167-2738(02)00217-5
41. Hallinan DT, Balsara NP (2013) Polymer electrolytes. Annu Rev Mater Res 43(1):503–525. https://doi.org/10.1146/annurev-matsci-071312-121705
42. Marzantowicz M, Dygas JR, Krok F (2008) Impedance of interface between PEO:LiTFSI polymer electrolyte and blocking electrodes. Electrochim Acta 53(25):7417–7425. https://doi.org/10.1016/j.electacta.2007.12.047
43. Ferloni P, Chiodelli G, Magistris A et al (1986) Ion transport and thermal properties of poly(ethylene oxide)–$LiClO_4$ polymer electrolytes. Solid State Ion 18–19(1):265–270. https://doi.org/10.1016/0167-2738(86)90124-4
44. Armand MB (1990) Polymers with ionic conductivity. Adv Mater Weinheim 2(6–7):278–286. https://doi.org/10.1002/adma.19900020603
45. Vallée A, Besner S, Prud'Homme J (1992) Comparative study of poly(ethylene oxide) electrolytes made with $LiN(CF_3SO_2)_2$, $LiCF_3SO_3$ and $LiClO_4$: Thermal properties and conductivity behaviour. Electrochim Acta 37(9):1579–1583. https://doi.org/10.1016/0013-4686(92)80115-3
46. Robitaille CD, Fauteux D (1986) Phase diagrams and conductivity characterization of some PEO–LiX electrolytes. J Electrochem Soc 133(2):315–325. https://doi.org/10.1149/1.2108569
47. Uno T, Kawaguchi S, Kubo M et al (2008) Ionic conductivity and thermal property of solid hybrid polymer electrolyte composed of oligo(ethylene oxide) unit and butyrolactone unit. J Power Sources 178(2):716–722. https://doi.org/10.1016/j.jpowsour.2007.11.088
48. Gomez ED, Panday A, Feng EH et al (2009) Effect of ion distribution on conductivity of block copolymer electrolytes. Nano Lett 9(3):1212–1216. https://doi.org/10.1021/nl900091n
49. Wang S, Min K (2010) Solid polymer electrolytes of blends of polyurethane and polyether modified polysiloxane and their ionic conductivity. Polymer 51(12):2621–2628. https://doi.org/10.1016/j.polymer.2010.04.038
50. Aldalur I, Zhang H, Piszcz M et al (2017) Jeffamine® based polymers as highly conductive polymer electrolytes and cathode binder materials for battery application. J Power Sources 347:37–46. https://doi.org/10.1016/j.jpowsour.2017.02.047
51. Alloin F, Sanchez J-Y (1995) New solvating polyether networks. Electrochim Acta 40(13–14):2269–2276. https://doi.org/10.1016/0013-4686(95)00175-E
52. Mindemark J, Imholt L, Montero J et al (2016) Allyl ethers as combined plasticizing and crosslinkable side groups in polycarbonate-based polymer electrolytes for solid-state Li batteries. J Polym Sci Part A: Polym Chem. https://doi.org/10.1002/pola.28080
53. Manuel Stephan A (2006) Review on gel polymer electrolytes for lithium batteries. Eur Polymer J 42(1):21–42. https://doi.org/10.1016/j.eurpolymj.2005.09.017
54. Weston J, Steele B (1982) Effects of inert fillers on the mechanical and electrochemical properties of lithium salt-poly(ethylene oxide) polymer electrolytes. Solid State Ion 7(1):75–79. https://doi.org/10.1016/0167-2738(82)90072-8
55. Malathi M, Tamilarasan K, Ganesan V (2015) Role of ceramic reinforcement in composite polymer electrolyte. Polym Compos 36(1):42–46. https://doi.org/10.1002/pc.22910
56. Capiglia C, Yang J, Imanishi N et al (2002) Composite polymer electrolyte: the role of filler grain size. Solid State Ion 154–155:7–14. https://doi.org/10.1016/S0167-2738(02)00448-4
57. Liu Y, Lee JY, Hong L (2002) Functionalized SiO_2 in poly(ethylene oxide)-based polymer electrolytes. J Power Sources 109(2):507–514. https://doi.org/10.1016/S0378-7753(02)00167-2

58. Syzdek JS, Armand MB, Marcinek MŁ et al (2010) Detailed studies on the fillers modification and their influence on composite, poly(oxyethylene)-based polymeric electrolytes: polymer electrolytes selection of papers from The 11th international symposium (ISPE-11) 31 August 5 September 2008, Ofir, Portugal. Electrochim Acta 55(4):1314–1322. https://doi.org/10.1016/j.electacta.2009.04.025
59. Derrien G, Hassoun J, Sacchetti S et al (2009) Nanocomposite PEO-based polymer electrolyte using a highly porous, super acid zirconia filler. Solid State Ion 180(23–25):1267–1271. https://doi.org/10.1016/j.ssi.2009.07.006
60. Scrosati B, Croce F (1993) Composite polymer ionics: advanced electrolyte materials for thin-film batteries. Polym Adv Technol 4(23):198–204. https://doi.org/10.1002/pat.1993.220040221
61. Zhou J, Fedkiw PS (2004) Ionic conductivity of composite electrolytes based on oligo(ethylene oxide) and fumed oxides. Solid State Ion 166(3–4):275–293. https://doi.org/10.1016/j.ssi.2003.11.017
62. Wieczorek WG, Such K, Wyciślik H et al (1989) Modifications of crystalline structure of peo polymer electrolytes with ceramic additives. Solid State Ion 36(3–4):255–257. https://doi.org/10.1016/0167-2738(89)90185-9
63. Croce F, Settimi L, Scrosati B (2006) Superacid ZrO_2-added, composite polymer electrolytes with improved transport properties. Electrochem Commun 8(2):364–368. https://doi.org/10.1016/j.elecom.2005.12.002
64. Liu S, Imanishi N, Zhang T et al (2010) Effect of nano-silica filler in polymer electrolyte on Li dendrite formation in Li/poly(ethylene oxide)–$Li(CF_3SO_2)_2N$/Li. J Power Sources 195(19):6847–6853. https://doi.org/10.1016/j.jpowsour.2010.04.027
65. Kumar J, Rodrigues SJ, Kumar B (2010) Interface-mediated electrochemical effects in lithium/polymer–ceramic cells. J Power Sources 195(1):327–334. https://doi.org/10.1016/j.jpowsour.2009.06.098
66. Stevens J, Mellander B-E (1986) Poly(ethylene oxide)-alkali metal–silver halide salt systems with high ionic conductivity at room temperature. Solid State Ion 21(3):203–206. https://doi.org/10.1016/0167-2738(86)90073-1
67. Płocharski J, Wieczorek WG (1988) PEO based composite solid electrolyte containing nasicon. Solid State Ion 28–30, Part 2: 979–982. https://doi.org/10.1016/0167-2738(88)90315-3
68. Przyłuski J, Wieczorek W (1989) Increasing the conductivity of polymer solid electrolytes: a review. Solid State Ion 36(3–4):165–169. https://doi.org/10.1016/0167-2738(89)90163-X
69. Skaarup S, West K, Zachau-Christiansen B (1988) Mixed phase solid electrolytes. Solid State Ion 28–30:975–978. https://doi.org/10.1016/0167-2738(88)90314-1
70. Appetecchi GB, Croce F, Dautzenberg G et al (1998) Composite polymer electrolytes with improved lithium metal electrode interfacial properties. J Electrochem Soc 145(12):4126. https://doi.org/10.1149/1.1838925
71. Wang Y-J, Pan Y (2005) $Li_{1.3}Al_{0.3}Ti_{1.7}(PO_4)_3$ filler effect on (PEO)$LiClO_4$ solid polymer electrolyte. J. Polym. Sci. B Polym. Phys. 43(6): 743–751. https://doi.org/10.1002/polb.20371
72. Zhang X-W, Wang C, Appleby AJ et al (2002) Characteristics of lithium-ion-conducting composite polymer-glass secondary cell electrolytes. J Power Sources 112(1):209–215. https://doi.org/10.1016/S0378-7753(02)00365-8
73. Choi J-H, Lee C-H, Yu J-H et al (2015) Enhancement of ionic conductivity of composite membranes for all-solid-state lithium rechargeable batteries incorporating tetragonal $Li_7La_3Zr_2O_{12}$ into a polyethylene oxide matrix. J Power Sources 274:458–463. https://doi.org/10.1016/j.jpowsour.2014.10.078
74. Jung Y-C, Kim S-K, Kim M-S et al (2015) Ceramic separators based on Li^+-conducting inorganic electrolyte for high-performance lithium-ion batteries with enhanced safety. J Power Sources 293:675–683. https://doi.org/10.1016/j.jpowsour.2015.06.001
75. Nairn K, Best AS, Newman PJ et al (1999) Ceramic–polymer interface in composite electrolytes of lithium aluminium titanium phosphate and polyetherurethane polymer electrolyte. Solid State Ion 121(1–4):115–119. https://doi.org/10.1016/S0167-2738(98)00537-2

76. Wang C, Zhang X-W, Appleby AJ (2005) Solvent-free composite PEO–ceramic fiber/mat electrolytes for lithium secondary cells. J Electrochem Soc 152(1):A205. https://doi.org/10.1149/1.1828952
77. Abe T, Sagane F, Ohtsuka M et al (2005) Lithium-ion transfer at the interface between lithium-ion conductive ceramic electrolyte and liquid electrolyte-a key to enhancing the rate capability of lithium-ion batteries. J Electrochem Soc 152(11):A2151–A2154. https://doi.org/10.1149/1.2042907
78. Liu W, Liu N, Sun J et al (2015) Ionic conductivity enhancement of polymer electrolytes with ceramic nanowire fillers. Nano Lett 15(4):2740–2745. https://doi.org/10.1021/acs.nanolett.5b00600
79. Zhao Y, Huang Z, Chen S et al (2016) A promising PEO/LAGP hybrid electrolyte prepared by a simple method for all-solid-state lithium batteries. Solid State Ion 295:65–71. https://doi.org/10.1016/j.ssi.2016.07.013
80. MacFarlane DR, Newman PJ, Nairn KM et al (1998) Lithium-ion conducting ceramic/polyether composites. Electrochim Acta 43(10–11):1333–1337. https://doi.org/10.1016/S0013-4686(97)10039-1
81. Jung Y-C, Lee S-M, Choi J-H et al (2015) All solid-state lithium batteries assembled with hybrid solid electrolytes. J Electrochem Soc 162(4):A704–A710. https://doi.org/10.1149/2.0731504jes
82. Inda Y, Katoh T, Baba M (2007) Development of all-solid lithium-ion battery using Li-ion conducting glass-ceramics. J Power Sources 174(2):741–744. https://doi.org/10.1016/j.jpowsour.2007.06.234
83. Aetukuri NB, Kitajima S, Jung E et al (2015) Flexible ion-conducting composite membranes for lithium batteries. Adv Energy Mater 5(14). https://doi.org/10.1002/aenm.201500265
84. Nairn K, Forsyth M, Every H et al (1996) Polymer–ceramic ion-conducting composites. Solid State Ion 86–88, Part 1: 589–593. https://doi.org/10.1016/0167-2738(96)00212-3
85. Chen F, Yang D, Zha W et al (2017) Solid polymer electrolytes incorporating cubic $Li_7La_3Zr_2O_{12}$ for all-solid-state lithium rechargeable batteries. Electrochim Acta 258:1106–1114. https://doi.org/10.1016/j.electacta.2017.11.164
86. Langer F, Bardenhagen I, Glenneberg J et al (2016) Microstructure and temperature dependent lithium ion transport of ceramic–polymer composite electrolyte for solid-state lithium ion batteries based on garnet-type $Li_7La_3Zr_2O_{12}$. Solid State Ion 291:8–13. https://doi.org/10.1016/j.ssi.2016.04.014
87. Langer F, Glenneberg J, Bardenhagen I et al (2015) Synthesis of single phase cubic Al-substituted $Li_7La_3Zr_2O_{12}$ by solid state lithiation of mixed hydroxides. J Alloy Compd 645:64–69. https://doi.org/10.1016/j.jallcom.2015.03.209
88. Keller M, Appetecchi GB, Kim G-T et al (2017) Electrochemical performance of a solvent-free hybrid ceramic-polymer electrolyte based on $Li_7La_3Zr_2O_{12}$ in P(EO) 15LiTFSI. J Power Sources 353:287–297. https://doi.org/10.1016/j.jpowsour.2017.04.014
89. Zheng J, Hu Y-Y (2018) New insights into the compositional dependence of Li-ion transport in polymer–ceramic composite electrolytes. ACS Appl Mater Interfaces 10(4):4113–4120. https://doi.org/10.1021/acsami.7b17301
90. Chen L, Li Y, Li S-P et al (2018) PEO/garnet composite electrolytes for solid-state lithium batteries: from "ceramic-in-polymer" to "polymer-in-ceramic". Nano Energy 46:176–184. https://doi.org/10.1016/j.nanoen.2017.12.037
91. Huo H, Zhao N, Sun J et al (2017) Composite electrolytes of polyethylene oxides/garnets interfacially wetted by ionic liquid for room-temperature solid-state lithium battery. J Power Sources 372:1–7. https://doi.org/10.1016/j.jpowsour.2017.10.059
92. Zhang J, Zhao N, Zhang M et al (2016) Flexible and ion-conducting membrane electrolytes for solid-state lithium batteries: dispersion of garnet nanoparticles in insulating polyethylene oxide. Nano Energy 28:447–454. https://doi.org/10.1016/j.nanoen.2016.09.002
93. Kalnaus S, Tenhaeff WE, Sakamoto J et al (2013) Analysis of composite electrolytes with sintered reinforcement structure for energy storage applications. J Power Sources 241:178–185. https://doi.org/10.1016/j.jpowsour.2013.04.096

94. Fu KK, Gong Y, Dai J et al (2016) Flexible, solid-state, ion-conducting membrane with 3D garnet nanofiber networks for lithium batteries. Proc Natl Acad Sci USA 113(26):7094–7099. https://doi.org/10.1073/pnas.1600422113
95. Gong Y, Fu K, Xu S et al (2018) Lithium-ion conductive ceramic textile: a new architecture for flexible solid-state lithium metal batteries. Mater Today. https://doi.org/10.1016/j.mattod.2018.01.001
96. Zhai H, Xu P, Ning M et al (2017) A flexible solid composite electrolyte with vertically aligned and connected ion-conducting nanoparticles for lithium batteries. Nano Lett 17(5):3182–3187. https://doi.org/10.1021/acs.nanolett.7b00715
97. Langer F, Palagonia MS, Bardenhagen I et al (2017) impedance spectroscopy analysis of the lithium Ion transport through the $Li_7La_3Zr_2O_{12}/P(EO)_{20}Li$ interface. J Electrochem Soc 164(12):A2298–A2303. https://doi.org/10.1149/2.0381712jes
98. Zheng J, Tang M, Hu Y-Y (2016) Lithium ion pathway within $Li_7La_3Zr_2O_{12}$–polyethylene oxide composite electrolytes. Angew Chem Int Ed Engl 55(40):12538–12542. https://doi.org/10.1002/anie.201607539
99. Zheng J, Dang H, Feng X et al (2017) Li-ion transport in a representative ceramic–polymer–plasticizer composite electrolyte: $Li_7La_3Zr_2O_{12}$–polyethylene oxide–tetraethylene glycol dimethyl ether. J Mater Chem A 5(35):18457–18463. https://doi.org/10.1039/C7TA05832B
100. Tao X, Liu Y, Liu W et al (2017) Solid-state lithium-sulfur batteries operated at 37 °C with composites of nanostructured $Li_7La_3Zr_2O_{12}$/carbon foam and polymer. Nano Lett 17(5):2967–2972. https://doi.org/10.1021/acs.nanolett.7b00221
101. Monroe C, Newman J (2005) The Impact of elastic deformation on deposition kinetics at lithium/polymer interfaces. J Electrochem Soc 152(2):A396. https://doi.org/10.1149/1.1850854
102. Han X, Gong Y, Fu KK et al (2017) Negating interfacial impedance in garnet-based solid-state Li metal batteries. Nat Mater 16(5):572–579. https://doi.org/10.1038/nmat4821

Part II
Electrodes and Interfaces with Solid Electrolytes

Chapter 10
Zero-Strain Insertion Materials for All-Solid-State Li-Ion Batteries

Kazuhiko Mukai

Abstract Lithium insertion materials for lithium-ion batteries generally alter their lattice dimensions during the electrochemical reactions. However, some lithium insertion materials, which are called as zero-strain insertion materials, maintain the lattice dimensions during the whole electrochemical reaction. This zero-strain characteristic is ideal for electrode materials for all-solid-state lithium-ion batteries. The next chapter overviews characteristic of zero-strain insertion materials regarding macro- and microscopic structural change during the electrochemical reaction, and thermal stability when applying all-solid-state lithium-ion batteries. Furthermore, pseudo-zero-strain insertion materials and zero-strain insertion materials for sodium-ion batteries are briefly discussed.

10.1 What Are Zero-Strain Insertion Materials?

Lithium insertion materials, which are employed in lithium-ion batteries (LIBs), generally alter their lattice parameters and lattice volumes, possibly owing to phase transitions within Li content, changes in ionic radii (r)s of redox species, and Coulombic repulsions of Li and/or O^{2-} ions. Figure 10.1 shows several examples of changes in the lattice volume (ΔV) during charge and discharge reactions. Positive electrode materials such as $LiCoO_2$ [1–3], $LiMn_2O_4$ [4, 5], and $LiFePO_4$ [6], exhibit $+2$ or -6% ΔV, while negative electrode materials such as graphite [7] and silicon [8, 9], indicate large ΔV values above 10%. In contrast, lithium titanium oxide, $Li[Li_{1/3}Ti_{5/3}]O_4$ (LTO) indicates a virtually 0% ΔV leading to the recognition of a "zero-strain" insertion material [10, 11].

K. Mukai (✉)
Toyota Central Research & Development Laboratories, Inc., 41-1 Yokomichi, Nagakute 480-1192, Japan
e-mail: e1089@mosk.tytlabs.co.jp

R. Murugan and W. Weppner (eds.), *Solid Electrolytes for Advanced Applications*,
https://doi.org/10.1007/978-3-030-31581-8_10

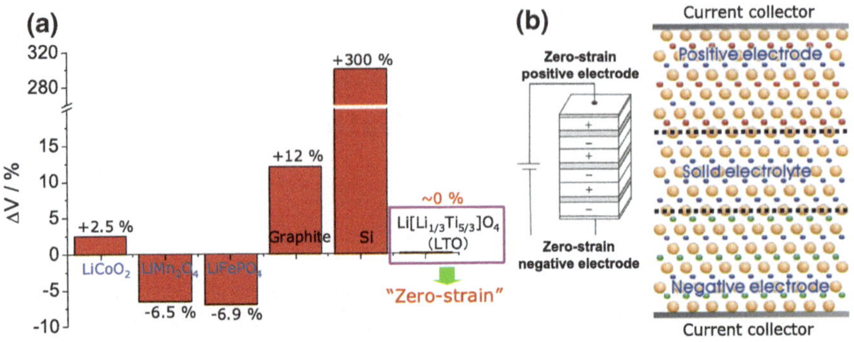

Fig. 10.1 **a** ΔV during charge and discharge reactions for various lithium insertion materials. LTO is recognized as a zero-strain insertion material because $\Delta V \simeq 0\%$. **b** Conceptual design for an ideal ALIB using two zero-strain insertion materials. Connecting several ALIBs in series increase the operating voltage of the ALIB. Each solid–solid interface are constructed by sharing the oxygen ions. Reproduced with permission [12]. Copyright 2017, Springer

Zero-strain insertion materials are ideal for all-solid-state LIBs (ALIBs) because robust interfaces between electrodes and solid electrolytes are required to maintain stable cycles and high-rate performances. Figure 10.1b shows the schematics of an ideal architecture for an ALIB, in which two zero-strain insertion materials are connected with a solid-electrolyte via the O^{2-} ions [12]. The operating voltage can be increased to the limit of the voltage window of solid electrolytes, by connecting several ALIBs in series.

Zero-strain insertions materials can be classified into two categories: the materials that indicate neither changes in lattice parameters nor ΔV like LTO, and the materials that indicate changes in lattice parameters, but do not show ΔV. Positive electrode materials like $LiNi_{1/2}Co_{1/2}O_2$ (NC11) [12, 13] and $LiNi_{1/3}Co_{1/3}Mn_{1/3}O_2$ (NCM111) [12, 14] are included in the latter category, and these compounds could be regarded as *pseudo* zero-strain insertion materials. The next four sections introduce the characteristics of zero-strain or pseudo zero-strain insertion materials. Finally, strategies for designing future zero-strain insertion materials are presented.

10.2 Characteristics of LTO

The most popular and useful zero-strain insertion material is LTO, which has a spinel structure with $Fd\bar{3}m$ space group as shown in the Fig. 10.2a. In the crystal lattice of LTO, Li^+ ions occupy tetrahedral 8*a* and octahedral 16*d* sites, while Ti^{4+} ions occupy the 16*d* site. The cubic lattice parameter (a_c) is ~8.356 Å. The relatively large O^{2-} ions occupy the tetrahedral 32*e* site, and form a cubic close-packed (ccp) array. The electrochemical reaction of LTO is represented as follow:

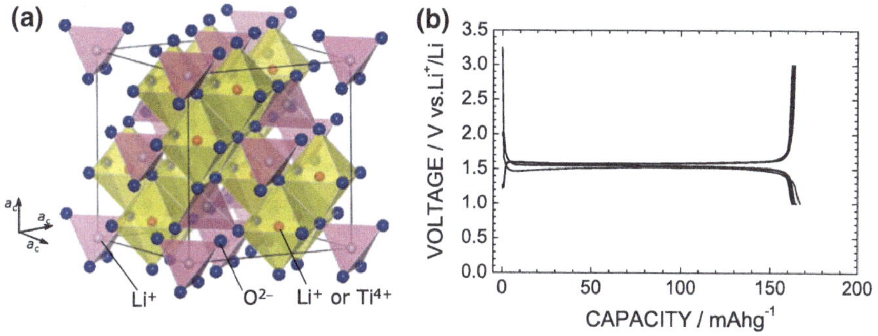

Fig. 10.2 **a** Schematics of the crystal structure of the LTO spinel with the $Fd\bar{3}m$ space group: Li^+ ions occupy both tetrahedral 8*a* and octahedral 16*d* sites, Ti^{4+} ions occupy the 16*d* site, and O^{2-} ions occupy the tetrahedral 32*e* site. Oxygen positional parameter (u) is approximately 0.265, leading to the deformation of the LiO_6 and TiO_6 octahedra. **b** Discharge and charge curves of the LTO/Li cell operated at a current density of $0.15\,mA \cdot cm^{-2}$ at room temperature. Reproduced with permission [17]. Copyright 2015, American Chemical Society

$$Li[Li_{1/3}Ti_{5/3}]O_4 + Li^+ + e^- \leftrightarrow Li_2[Li_{1/3}Ti_{5/3}]O_4, \quad (10.1)$$

where the Li^+ ions are inserted into the initially vacant 16*c* site. Figure 10.2b shows the typical discharge and charge curves of the LTO/Li cell [15–19]. The cell voltage rapidly drops to ~1.5 V from ~3.3 V at the beginning of the discharge reaction, and then maintains an almost constant voltage at approximately 1.55 V. The discharge capacity (Q_{dis}) and the subsequent charge capacity (Q_{cha}) are 165 and $162\,mAh \cdot g^{-1}$, respectively. These capacities correspond to ~94% of the theoretical capacity (Q_{theo}) ($=175\,mAh \cdot g^{-1}$) based on (10.1). The flat operating voltage indicates that the reaction of LTO proceeds in a two-phase reaction; however, it is still a debatable subject. Ohzuku et al. [11] claimed that the definitive description of the reaction mechanism cannot be given by ex situ X-ray diffraction (XRD) measurements in their early report. Then, Scharner et al. [20] found two different a_c values of $a_{c1} = 8.3595$ Å for $Li[Li_{1/3}Ti_{5/3}]O_4$ and $a_{c2} = 8.3538$ Å for $Li_2[Li_{1/3}Ti_{5/3}]O_4$, by high-angle XRD measurements up to $2\theta \simeq 150°$. These results provide evidence of the two-phase reaction of LTO [20]. However, Scorosati and co-workers [21, 22] reported that the a_c value monotonically decreases from 8.360 to 8.353 Å as the discharge reaction proceeds, by means of energy dispersive in situ XRD measurements. This indicates the one-phase reaction of LTO [21, 22].

In contrast to XRD measurements, Raman spectroscopy provides structural information on an atomic scale, leading to better understanding of the relationship between crystal structures and electrochemical properties [23]. Raman spectroscopy is based on the inelastic scattering of phonons, and Raman active modes can be predicted by a factor group analysis [24]. The site symmetries of the 8*a*, 16*d*, and 32*e* sites in the $Fd\bar{3}m$ space group are T_d, D_{3d}, and C_{3v}, respectively. Since the 16*d* site in LTO is randomly occupied by both Li^+ and Ti^{4+} ions [10, 11], the factor group analysis on LTO provides five Raman active modes of $A_{1g} + E_g + 3F_{2g}$. Note that in all Raman

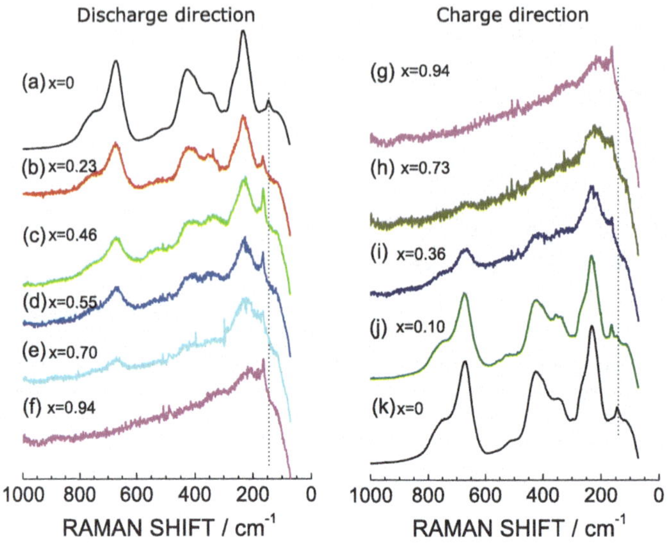

Fig. 10.3 Raman spectra of the $Li_{1+x}[Li_{1/3}Ti_{5/3}]O_4$ samples during both discharge and charge directions. Raman spectra of **a** $x = 0$ (initial sample), **b** $x = 0.23$, **c** $x = 0.46$, **d** $x = 0.55$, **e** $x = 0.70$, and **f** $x = 0.94$ are those in the discharge direction, while Raman spectra of **g** $x = 0.94$, **h** $x = 0.73$, **i** $x = 0.36$, **j** $x = 0.10$, and **k** $x = 0$ (initial sample) are those in the charge direction. The samples of parts (**h**), (**i**), and (**j**) were prepared by charging the fully discharged cell, i.e., the cell was discharged to 1.0 V, and then charged to the designed x value. The Raman spectra of (**a**) and (**f**) are the same as those of parts (**k**) and (**g**), respectively. The dotted lines indicate the Raman band for TiO_2 anatase (=141 cm^{-1}). Reproduced with permission [16]. Copyright 2014, American Chemical Society

active modes, the Li^+/Ti^{4+} ions at the 16d site are motionless, while only the O^{2-} ions change their positions. Therefore, the Raman spectrum of LTO offers insight into the local structural changes of the O^{2-} ions.

Figure 10.3 shows the Raman spectra for the LTO samples during both discharge and charge directions. [16] The Raman spectra of (a) $x = 0$, (b) $x = 0.23$, (c) $x = 0.46$, (d) $x = 0.55$, (e) $x = 0.70$, and (f) $x = 0.94$ are those in the discharge direction, while the Raman spectra of (h) $x = 0.73$, (i) $x = 0.36$, and (j) $x = 0.10$ are those in the charge direction. Note that x value of each sample was calculated by comparing the observed Q_{dis} and Q_{theo}. Furthermore, the $x = 0.73$, 0.36, and 0.10 samples were prepared by the fully discharged reaction, before charging to the designed x value. Three major changes appear in the discharge direction. That is, (i) the intensity of the whole Raman spectrum decreases with increasing x, (ii) the intensities of the Raman bands at 671 and 426 cm^{-1} gradually weaken with increasing x, and (iii) the Raman band at 671 cm^{-1} maintains a constant frequency up to $x = 0.94$, while the Raman bands at 426 and 230 cm^{-1} exhibit a blue shift with x. The decrease in the intensity is caused by the electronic conductive behavior of $x = 1$ [18].

As seen from Fig. 10.3g–k, the three major changes in the Raman spectra are reversible. Although the Raman spectrum of $x = 0.73$ is almost similar to that of $x = 0.94$, in the Raman spectrum of $x = 0.36$, at least three Raman bands appear at approximately 670, 420, and 230 cm^{-1}. The Raman spectrum of $x = 0.10$ resembles with that of $x = 0$; namely, there are three major Raman bands at 672, 423, and 234 cm^{-1} and four minor Raman bands at 751, 524, 347, and 268 cm^{-1}. The increase of the intensity of the whole Raman spectrum is consistent with the fact that the Ti^{3+} ions with the d^1 ($S = 1/2$the Ti^{4+} ions with the d^0 ($S = 0$) state with decreasing x. There is a discrepancy between the $x = 0.10$ and $x = 0$ samples in the Raman band below 160 cm^{-1}, which arises from the TiO_2 anatase. Since the $x = 0.10$ sample indicates a small Raman band at 164 cm^{-1}, the Li^+ ions still exist in the TiO_2 lattice even in the fully charged state. Hence, changes in the Raman spectrum of LTO are reversible.

Figure 10.4 shows the x dependence of the Raman bands at (a) 670 cm^{-1}, (b) 420 cm^{-1}, and (c) 230 cm^{-1}, which are assigned as A_{1g}, E_g, and F_{2g} modes, respectively, based on previous experimental [23, 25] and theoretical [26] Raman studies on $LiMn_2O_4$. Moreover, according to the character table of O_h symmetry, the A_{1g}, E_g, and F_{2g} modes correspond to a symmetric stretching vibration (ν_{sym}), an asymmetric stretching vibration (ν_{asym}), and a bending vibration (δ), respectively.

The frequency of A_{1g} is independent of x in both discharge and charge directions. This indicates that the bond length between the Ti and O atoms (d_{Ti-O}) is constant with x. This is surprising because the mean d_{Ti-O} value is expected to increase with increasing x, i.e., $d_{Ti-O} = 2.005$ Å for Ti^{4+}–O^{2-} and $d_{Ti-O} = 2.07$ Å for Ti^{3+}–O^{2-}, using the r values for Ti^{4+} ions [$r = 0.605$ Å (C.N. = 6)], Ti^{3+} ions [$r = 0.670$ Å (C.N. = 6)], and O^{2-} ions [$r = 1.40$ Å (C.N. = 6)], where C.N. is the coordination number [27]. The frequencies of the E_g and F_{2g} modes monotonically decrease with increasing x. The decrease in the frequency of E_g means a decrease in the bond length between the Li and O atoms (d_{Li-O}), because the reduced mass for the LiO_6 octahedron is constant during charge and discharge reactions. [28] On the other hand, the decrease in the frequency of F_{2g} suggests a change in the bond angle (α) between the O–Ti–O atoms in the TiO_6 octahedron. Ariyoshi et al. [29] reported that the average α varies with x induced by a decrease in the oxygen parameter (u). Here, the present u value changes from 0.262(1) for $Li[Li_{1/3}Ti_{5/3}]O_4$ to 0.258(1) for $Li_{1.94}[Li_{1/3}Ti_{5/3}]O_4$. The decrease in u generates changes in two bond angles: $\alpha = 84$ and 96° for $Li[Li_{1/3}Ti_{5/3}]O_4$ and $\alpha = 87$ and 93° for $Li_{1.94}[Li_{1/3}Ti_{5/3}]O_4$. Therefore, the decrease in the frequency of F_{2g} correlates with the change in α with x.

Figure 10.4d, e show the schematics of the zero-strain reaction scheme. The initial a_c value ($=a_{c1}$) was 8.357 Å, while the fully discharged a_c value ($=a_{c2}$) was 8.353 Å. Due to the deformation from the regular octahedron, the local structures of both LiO_6 and TiO_6 octahedra changed associated with the Li insertion. Furthermore, α in the TiO_6 octahedron was changed with x; $\alpha = 84$ and 96° for $x = 0$ and $\alpha = 87$ and 93° for $x = 0.94$. The d_{Li-O} value decreased with increasing x to maintain the spinel structure, although the d_{Ti-O} value was constant with x. These changes were reversible with x as for the electrochemical reaction. Since all the structural changes

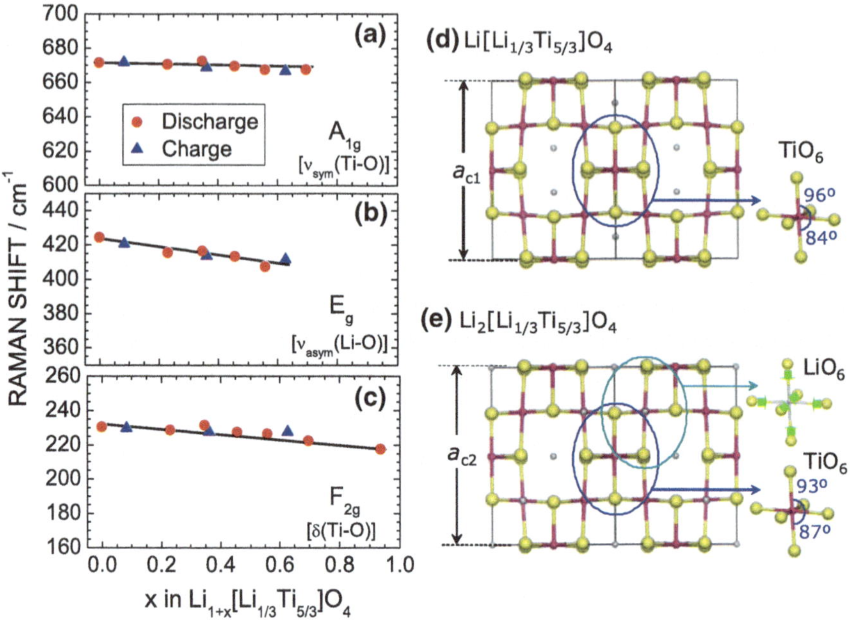

Fig. 10.4 x dependence of Raman bands of **a** A_{1g}, **b** E_g, and **c** F_{2g}, which correspond to the modes of ν_{sym}(Ti-O), ν_{asym}(Ti-O), and δ(Ti-O), respectively. The closed blue circles and red triangles are the Raman bands during the discharge and charge reactions, respectively. Schematics of the zero-strain reaction scheme between the **d** initial and **e** fully discharged states. The a_{c1} value (=8.357 Å) is almost the same with the a_{c2} value (=8.353 Å) owing to the zero-strain characteristic. In the initial state, the average bond angles (α) between the O-Ti-O atoms are 84 and 96° as a result of the deviation from the ideal oxygen parameter ($u = 0.25$). In the fully discharged state, the deformed TiO_6 and LiO_6 octahedra approach a regular octahedron form, resulting in $\alpha = 87$ and 93°. The bond length between the Li and O atoms decreases with x to adjust with the change in the TiO_6 octahedra. These changes are reversible with x, as for the electrochemical reaction. Reproduced with permission [16]. Copyright 2014, American Chemical Society

were monotonic with x, LTO proceeded in the one-phase reaction from the Raman spectroscopy point of view.

The Raman spectrum of $LiTi_2O_4$, which undergoes a superconducting transition at $T_C \simeq 13$ K, indicates four Raman bands at approximately 628, 494, 429, and 339 cm^{-1} [30]. The ΔV value of $LiTi_2O_4$ (=−1.3%) was slightly larger than that of LTO (≃0%). [10] Meanwhile, TiO_2 anatase indicates the structural transformation from tetragonal ($I4_1/amd$) to orthorhombic ($Imma$) phases with the Li insertion. The ΔV value of the reaction was +3.7% [31]. These results support that the Li ions at the 16d site act as "absorbers" to adjust the changes in the local structures in the LTO lattice.

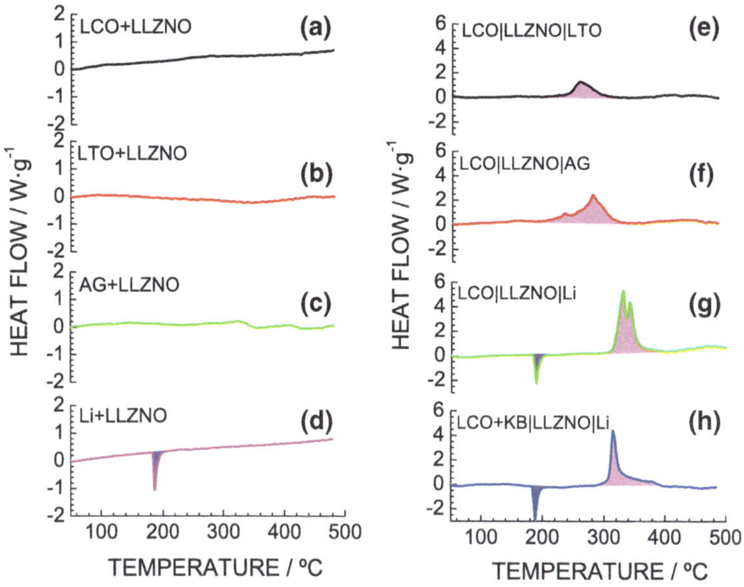

Fig. 10.5 DSC profiles of delithated and lithiated materials with a garnet structured $Li_{6.75}La_3Zr_{1.75}Nb_{0.25}O_{12}$ (LLZNO): **a** $Li_{0.43}CoO_2$ (LCO) with LLZNO, **b** $Li_{1.95}[Li_{1/3}Ti_{5/3}]O_4$ (LTO) with LLZNO, **c** C_6Li (AG) with LLZNO, and **d** Li metal with LLZNO. The endothermic peak at 180°C corresponds to the heat of fusion of the Li metal. DSC profiles of various ALIB-AIMs: **e** LCO|LLZNO|LTO, **f** LCO|LLZNO|AG, **g** LCO|LLZNO|Li, and LCO + Ketjenblack (KB)|LLZNO|Li. The exothermic peaks in (**e**–**h**) above 250 °C are due to the reaction between the released oxygen from LCO and Li metal. Reproduced with permission [32]. Copyright 2017, American Chemical Society

10.3 Thermal Stability of Virtual ALIBs Using LTO

One of the advantages of ALIBs over conventional LIBs is better thermal stability at elevated temperatures, since organic solutions are used in conventional LIBs as electrolytes. The thermal stability of *virtual* ALIBs is investigated by an all-inclusive microcell (AIM) method [32, 33], where the AIM comprises all the battery components with an equivalent capacity ratio of positive to negative electrodes.

Figure 10.5 shows the DSC profiles of delithiated and lithiated electrodes materials with a garnet structured $Li_{6.75}La_3Zr_{1.75}Nb_{0.25}O_{12}$ (LLZNO): (a) $Li_{0.47}CoO_2$ (LCO) with LLZNO, (b) $Li_{1.95}[Li_{1/3}Ti_{5/3}]O_4$ (LTO) with LLZNO, (c) C_6Li (AG) with LLZNO, and (d) Li metal with LLZNO. All the delithiated and lithiated samples were prepared by an electrochemical reaction using a conventional liquid electrolyte, and were rinsed with diethyl carbonate (DEC). The value of heat flow ($W \cdot g^{-1}$) is based on the weight of the LCO, LTO, AG, or Li metal sample. Neither exothermic nor endothermic peaks are observed up to 480 °C in the DSC profiles of LCO, LTO, and AG. A sharp endothermic peak is seen at approximately 185 °C in the DSC profile of the Li metal, but this endothermic peak ($=-3.0\ kJ \cdot mol^{-1}$) is attributed to the heat of fusion of the Li metal. These DSC profiles confirm the excellent thermal stability of ALIBs up to 480 °C, unless the usage of the Li metal.

However, the actual thermal stability should be examined using all the battery components. Figure 10.5 shows the DSC profiles of various types of ALIB-AIM: (e) LCO|LLZNO|LTO, (f) LCO|LLZNO|AG, (g) LCO|LLZNO|Li, and (h) LCO + Ketjenblack (KB)|LLZNO|Li. The samples of LCO, LTO, and AG are essentially same with those in Fig. 10.5a–c, and the value of heat flow ($W \cdot g^{-1}$) is normalized by the weight of the LCO sample. The LCO|LLZNO|LTO cell exhibits a broad exothermic peak at 260 °C with a change in enthalpy (ΔH) of 610 $J \cdot g^{-1}$ (57 $kJ \cdot mol^{-1}$). The LCO|LLZNO|AG cell also shows two broad exothermic peaks at around 235 and 280 °C, where the total ΔH value is calculated to be 1455 $J \cdot g^{-1}$ (137.1 $kJ \cdot mol^{-1}$). The LCO|LLZNO|Li cell indicates two sharp exothermic peaks at around 330 and 342 °C, together with an endothermic peak (=185 °C) due to the fusion of the Li metal. The temperatures of these two exothermic peaks are approximately 100 °C higher than the reported temperatures of LCO with the $LiPF_6$-based electrolyte [34]. The total ΔH value in LCO|LLZNO|Li is calculated to be 2200 $J \cdot g^{-1}$ (207.2 $kJ \cdot mol^{-1}$) ignoring the contribution of the endothermic peak at 185 °C. Comparing the ΔH values in the three ALIB systems, the ALIB using LTO exhibits the best thermal stability.

The DSC profiles shown in Fig. 10.5e–h demonstrated the heat generation even in ALIBs. According to the XRD measurements after the DSC measurements, Li_xCoO_2, Co_3O_4, and LLZNO phases were observed in the positive electrode, while Li_2O, Li metal, and LLZNO phases were observed in the negative electrode. The lattice parameters of Li_xCoO_2 were calculated to be $a_h = 2.817(1)$ Å and $c_h = 14.032(1)$ Å, indicating that $x \simeq 1$ based on the previous reports on LCO [2, 3]. Therefore, the exothermic peaks in LCO|LLZNO|Li are caused by the oxidation reaction of the Li metal. This reaction proceeds in the following two steps: oxygen was first released from the $Li_{0.47}CoO_2$ lattice, forming the $LiCoO_2$ and Co_3O_4 phases,

$$Li_{0.47}CoO_2 \rightarrow 0.47LiCoO_2 + 0.177Co_3O_4 + 0.176O_2. \quad (10.2)$$

Then the released oxygen reacted with Li metal in the reaction of

$$0.704Li + 0.176O_2 \rightarrow 0.352Li_2O. \quad (10.3)$$

The formation enthalpies ($\Delta_f H$s) of $LiCoO_2$ and $Li_{0.5}CoO_2$ are currently scattered depending on the method employed, [35–37] while $\Delta_f H$ of Co_3O_4 is known to be $-910\ kJ \cdot mol^{-1}$). [38] Hence, the ΔH value of (10.2) ranges between −138 and 67.98 $kJ \cdot mol^{-1}$. However, the ΔH value of (10.3) can be neglected if we consider the heat generation in LCO|LLZNO|Li. The DSC analysis on $Li_{0.47}CoO_2$ without the $LiPF_6$-based electrolyte showed no exothermic/endothermic peaks up to 480 °C. Moreover, the TG analysis revealed the formation of Co_3O_4 phase, not CoO phase. The $\Delta_f H$ value of Li_2O is known to be $-598.73\ kJ \cdot mol^{-1}$ [38]. Hence, the ΔH value for (10.4) is calculated to be 210.75 $kJ \cdot mol^{-1}$, which is similar to the observed ΔH (=207.2 $kJ \cdot mol^{-1}$).

The heat generations in LCO|LLZNO|LTO and LCO|LLZNO|AG are also attributed to the reaction between the released oxygen and the Li metal, but for both cases, Li metal is first required to be removed from the LTO or AG lattice. $\Delta_f H$ of $Li_2[Li_{1/3}Ti_{5/3}]O_4$ is known to be -165 kJ · mol^{-1}. [39] $\Delta_f H$ of AG is reoprted to be -2.3 [40] or -60 kJ · mol^{-1} [41]. Thus, the ΔH value of LTO is calculated to be 94.59 kJ · mol^{-1}, while the ΔH value of AG is calculated to be 208.65 or 168.13 kJ · mol^{-1}]. Since these calculated ΔH values are slightly different from observed ΔH values, the process of the extraction of Li^+ ions from the LTO or AG lattice strongly affects the heat generations in LCO|LLZNO|LTO and LCO|LLZNO|AG.

A trial for decreasing ΔH was performed in the LCO|LLZNO|Li cell. We mixed 28 wt% of Ketjenblack (KB) into the positive electrode as an O_2 scavenger. As shown in Fig. 10.5h, one exothermic peak was disappeared in the DSC profile of LCO + KB|LLZNO|Li. The ΔH value up to 420 °C was calculated to 79 kJ · mol^{-1}, which was half of the ΔH for LCO|LLZNO|Li (Fig. 10.5g). Hence, O_2 scavenger additives are found to be effective for decreasing ΔH in ALIBs.

10.4 Zero-Strain Insertion Materials with Non-topotactic Reaction

Except for LTO, zero-strain characteristics are seen in other spinel oxides such as $Li[CrTi]O_4$ [42], $LiRh_2O_4$ [43, 44], and $Li_{1/2+x/2}Fe_{5/2-3x/2}Ti_xO_4$ (LFTO) with $0.875 \leq x < 5/3$ [45, 46]. In this section, the reaction scheme of LFTO is presented as it proceeds in a *non-topotactic* manner.

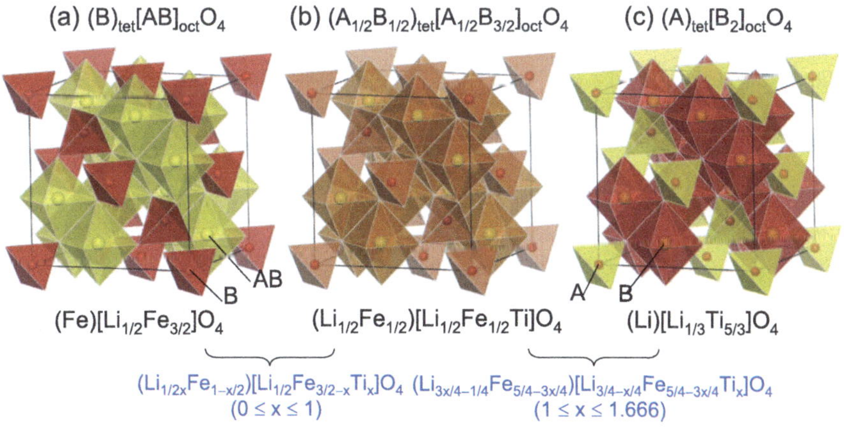

Fig. 10.6 Crystal structure of **a** $Fe[Li_{1/2}Fe_{3/2}]O_4$, **b** $Li_{1/2}Fe_{1/2}[Li_{1/2}Fe_{1/2}Ti]O_4$, and **c** $Li[Li_{1/3}Ti_{5/3}]O_4$ with spinel structures. The space group of (**a**) is transformed into $P4_332$ (or $P4_132$) due to the cation order between the Li and Fe ions, while those of (**b**) and (**c**) are $Fd\bar{3}m$. Reproduced with permission [45]. Copyright 2017, Elsevier

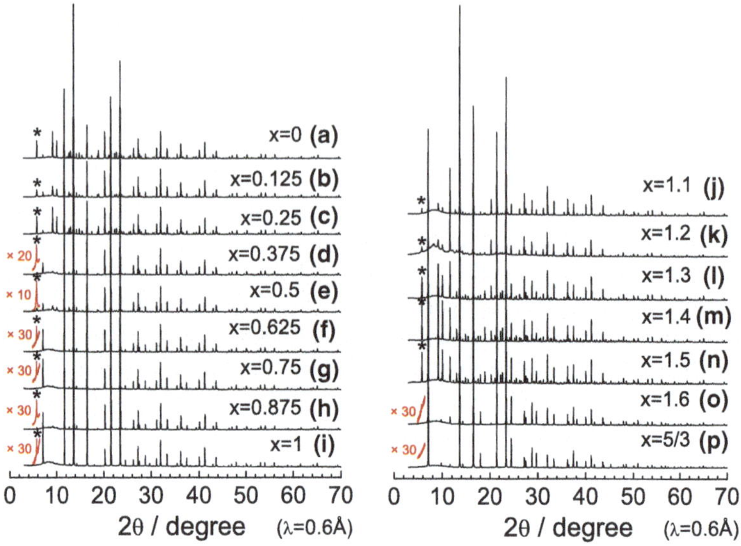

Fig. 10.7 XRD patterns of the LFTO samples with **a** $x = 0$, **b** $x = 0.125$, **c** $x = 0.25$, **d** $x = 0.375$, **e** $x = 0.5$, **f** $x = 0.625$, **g** $x = 0.75$, **h** $x = 0.875$, **i** $x = 1$, **j** $x = 1.1$, **k** $x = 1.2$, **l** $x = 1.3$, **m** $x = 1.4$, **n** $x = 1.5$, **o** $x = 1.6$, and **p** $x = 5/3$. Superlattice diffraction lines, which are caused by the 1:3 cation order at the octahedral site, are shown by ⋆. Reproduced with permission [45]. Copyright 2017, Elsevier

The LFTO compounds are the solid solution of $Fe[Li_{1/2}Fe_{3/2}]O_4$–$LiFeTiO_4$–$Li[Li_{1/3}Ti_{5/3}]O_4$ (LTO), as shown in Fig. 10.6. These three compounds are categorized into an inverse spinel with $(B)_{tet}[AB]_{oct}O_4$, an intermediate spinel with $(A_{1/2}B_{1/2})_{tet}[A_{1/2}B_{3/2}]_{oct}O_4$, and a derivative of a normal spinel with $(A)_{tet}[B_2]_{oct}O_4$, respectively. $(Fe)_{tet}[Li_{1/2}Fe_{3/2}]_{oct}O_4$ and $(Li_{1/2}Fe_{1/2})_{tet}[Li_{1/2}Fe_{1/2}Ti]_{oct}O_4$ forms a solid solution of $(Li_{x/2}Fe_{1-x/2})_{tet}[Li_{1/2}Fe_{3/2-x}Ti_x]_{oct}O_4$ in the x range between 0 and 1. In contrast, $(Li_{1/2}Fe_{1/2})_{tet}[Li_{1/2}Fe_{1/2}Ti]_{oct}O_4$ and $(Li)_{tet}[Li_{1/3}Ti_{5/3}]_{oct}O_4$ compounds form a solid solution of $(Li_{3x/4-1/4}Fe_{5/4-3x/4})_{tet}[Li_{3x/4-x/4}Fe_{5/4-3x/4}Ti_x]_{oct}$ O_4 in the x range between 1 and 5/3.

Figure 10.7a–i show the XRD patterns of the LFTO samples with $0 \leq x \leq 1$, and Fig. 10.7j–k show those of the LFTO samples with $1.1 \leq x \leq 5/3$ [45]. All the samples were prepared by heating the mixture of $Li(OH) \cdot H_2O$, α-FeOOH, and TiO_2 anatase at 750 °C for 12 h. The XRD pattern of $x = 0$ is assigned as the inverse spinel with the cation distribution of $[Fe]_{tet}[Li_{1/2}Fe_{3/2}]_{oct}$. Superlattice diffraction lines due to the cation order at the octahedral site are clearly observed as indicated by ⋆. Hence, the space group of $x = 0$ is $P4_332$, as reported previously [47–49]. The intensity of the superlattice diffraction line decreases with increasing x up to $x = 1$, and then increases again further increasing x. This supports that the space group of the samples with $0.125 \leq x \leq 1.5$ is also $P4_332$. The superlattice diffraction line is disappeared in the $x = 1.6$ and 5/3 samples, suggesting that these samples are assigned as a $Fd\bar{3}m$ space group.

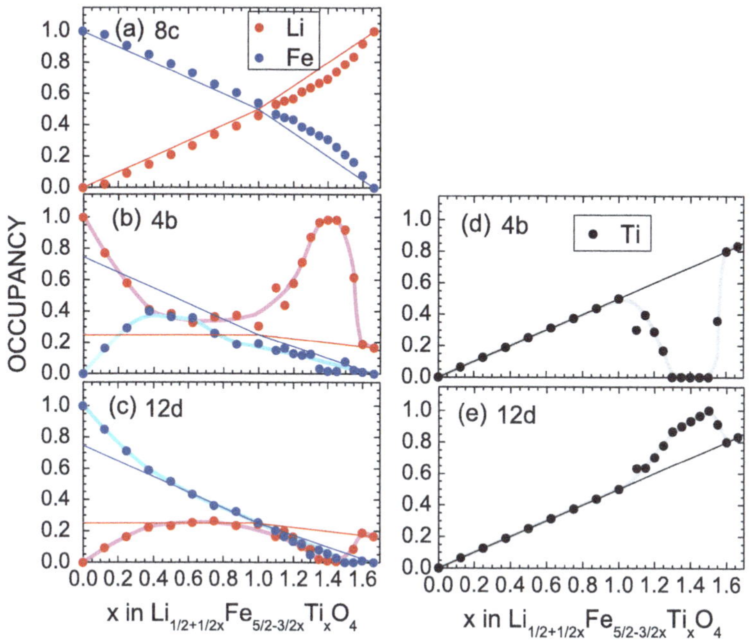

Fig. 10.8 x dependence of g at the **a** $8c$, **b** $4b$, **c** $12d$ sites for the Li and Fe ions, and **d** $4b$ and **e** $12d$ sites for the Ti ions. The solid lines in (**a**), (**b**), and (**c**) are the calculated g values using the chemical formula of $(Li_{x/2}Fe_{1-x/2})_{tet}[Li_{1/2}Fe_{3/2-x}Ti_x]_{oct}O_4$, while those in **d** and **e** are the calculated values using the chemical formula of $(Li_{3x/4-1/4}Fe_{5/4-3x/4})_{tet}[Li_{3x/4-x/4}Fe_{5/4-3x/4}Ti_x]_{oct}O_4$. Reproduced with permission [45]. Copyright 2017, Elsevier

Figure 10.8 shows the results of Rietveld analyses on the LFTO samples with $0 \leq x \leq 5/3$. For the LFTO samples with the $P4_332$ space group, the assumption that Ti^{4+} ions only occupy the octahedral $4b$ and $12d$ sites was employed. This is because Ti^{4+} ions usually occupy an octahedral coordination with O^{2-} ions, such as Li_2TiO_3 and TiO_2 anatase [16, 18].

As seen in Fig. 10.8a, the occupancy (g) of Li ions at the $8b$ site increases with x, whereas g of Fe ions at the $8b$ site decreases with x. The blue and red solid lines are the calculated g values based on the chemical formula of $(Li_{x/2}Fe_{1-x/2})_{tet}[Li_{1/2}Fe_{3/2-x}Ti_x]_{oct}O_4$ with $0 \leq x \leq 1$ or $(Li_{3x/4-1/4}Fe_{5/4-3x/4})_{tet}$ $[Li_{3x/4-x/4}Fe_{5/4-3x/4}Ti_x]_{oct}O_4$ with $1 \leq x \leq 5/3$. The g values at the octahedral $4b$ and $12d$ sites exhibit a drastic change with x, especially at the $4b$ sites for Li, Fe, and Ti ions (see Fig. 10.8b–e). In the x range between 0 and ~0.4, gs at the $4b$ and $12d$ sites indicate a linear decrease or increase with x. At $\sim 0.4 \leq x \leq 1.0$, the g value of Li ions at the $4b$ site is slightly greater than the calculated value. The similar trend is obtained in the x dependence in g of Fe ions at the $4b$ and $12d$ sites. At $1 \leq x \leq 5/3$, the g value of Li ions at the $4b$ site indicates the maximum at around $x = 1.4$, while that of Ti ions at the $4b$ site approaches ~0.

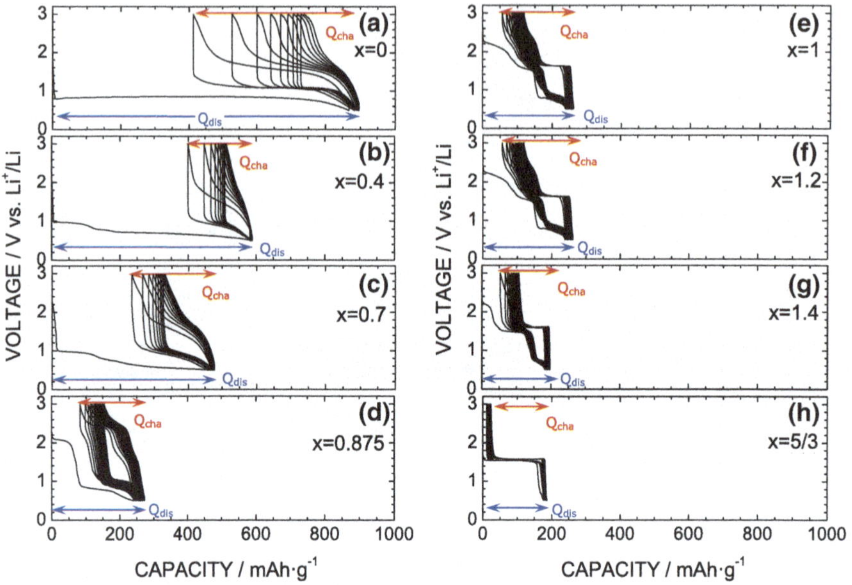

Fig. 10.9 Discharge and charge curves of the lithium cells operated at current of 0.5 mA in the voltage range between 0.5 and 3.0 V: **a** $x = 0$, **b** $x = 0.4$, **c** $x = 0.7$, **d** $x = 0.875$, **e** $x = 1$, **f** $x = 1.2$, **g** $x = 1.4$, and **h** $x = 5/3$. Reproduced with permission [46]. Copyright 2018, Elsevier

Due to the x dependence of gs of Li, Fe, and Ti ions, the 1:3 cation order in LFTO can be categorized into two types: (i) the cation order between Li and Fe ions at $0 \leq x \leq 0.375$ and (ii) the cation order between Li and Ti ions at $1 < x \leq 1.55$. Except at $x = 0$, the combination of Li, Fe, and Ti ions, more specifically, the combination of 1× ($Li^{+}_{1-x1-y1}Fe^{3+}_{x1}Ti^{4+}_{y1}$) and 3× ($Li^{+}_{1-x2-y2}Fe^{3+}_{x2}Ti^{4+}_{y2}$), participates in this cation order.

Figure 10.9 shows the discharge and charge curves of the LFTO/Li cells operated in the voltage range between 0.5 and 3.0 V: (a) $x = 0$, (b) $x = 0.4$, (c) $x = 0.74$, (d) $x = 0.875$, (e) $x = 1$, (f) $x = 1.2$, (g) $x = 1.4$, and (h) $x = 5/3$. The cell voltage of $x = 0$ drops to ~0.8 V at the beginning of the discharge reaction, and then maintains a constant voltage at ~0.8 V up to Q_{dis} of ~800 $\text{mAh} \cdot \text{g}^{-1}$. The subsequent charge curve differs from the previous discharge curve with a large irreversible capacity. That is, the Q_{dis} and Q_{cha} values at the initial cycle were 864 and 452 $\text{mAh} \cdot \text{g}^{-1}$, respectively. After the second cycle, there are large polarization of ~0.5 V between the discharge and charge curves. The Q_{dis} and Q_{cha} values decrease cycle by cycle. These results are reminiscent of the electrochemical properties of nano-sized metal oxides, such as FeO and NiO, which are characterized by a conversion reaction [50, 51]. The conversion reaction typically produces the reduced metal and Li_2O phases [50, 51]. Indeed, the Li_2O phase was observed in the fully discharged state of $x = 0$. Hence, the electrochemical reaction of $x = 0$ is represented as

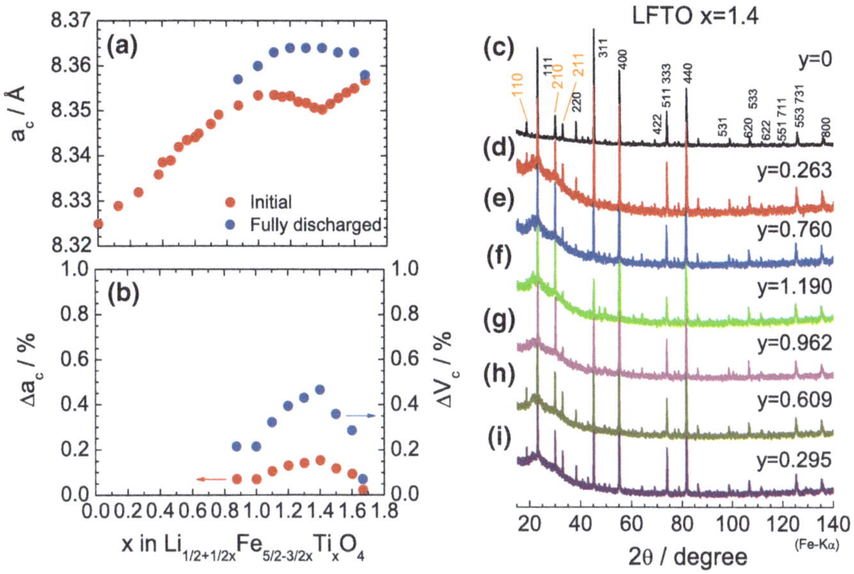

Fig. 10.10 **a** a_cs of the initial and fully discharged states, **b** Δa_c and ΔV of the LFTO samples. Ex situ XRD patterns of the LFTO sample with $x = 1.4$: **c** $y = 0$, **d** $y = 0.263$, **e** $y = 0.760$, **f** $y = 1.190$, **g** $y = 0.962$, **h** $y = 0.602$, and **i** $y = 0.295$. The $y > 0$ samples were prepared using the electrochemical reaction. Reproduced with permission [46]. Copyright 2018, Elsevier

$$Fe^{3+}[Li_{1/2}Fe^{3+}_{3/2}]O_4 + 15/2Li^+ + 15/2e^- \leftrightarrow 5/2Fe^0 + 4Li_2O, \qquad (10.4)$$

where the Q_{theo} is calculated to be 970. 5 mAh · g^{-1}. This Q_{theo} value was comparable with the Q_{dis} at the initial cycle. The conversion reaction is also observed in the $x = 0.4$ and 0.7 samples, but Q_{dis} and Q_{cha} values decrease with x. Since the Q_{dis} value indicated a relationship of $(5/2 - 3x/2)/(5/2) \times$ 970. 5 mAh · g^{-1}, Ti^{4+} ions inhibits the conversion reaction and stabilizes the spinel structure. The electrochemical properties of the $x \leq 0.875$ samples are different from those of the $x \leq 0.7$ samples. The cell voltage of $x = 0.875$ maintains at ~2.1 V at the initial discharge reaction, then decreases to ~0.9 V, and finally remains a constant voltage at ~0.8 V up to Q_{dis} of ~250 mAh · g^{-1}. The large polarization of ~1.5 V is observed in the subsequent discharge and charge curves. The plateau at ~1.6 V is observed in the charge curve of the $1 \leq x \leq 1.4$ samples. The Q_{dis} and Q_{cha} values of the $0.875 \leq x \leq 1.6$ samples range ~250 and ~160 mAh · g^{-1}, independent of x.

To clarify change in cubic lattice parameter (Δa_c) and ΔV of the LFTO samples with $x \geq 0.875$, XRD measurements were performed at the fully discharged state. Figure 10.10a shows the a_c values both initial and fully discharge states, and Fig. 10.10b shows the Δa_c and ΔV values of the $x \geq 0.875$ samples. The initial a_c value increases almost linearly from 8.325(1) Å at $x = 0$ to 8.354(1) Å at $x = 1$, then slightly decreases to 8.350(1) Å at $x = 1.4$, and finally increases with further increasing x. On the other hand, the a_c value at the fully discharged state indicates

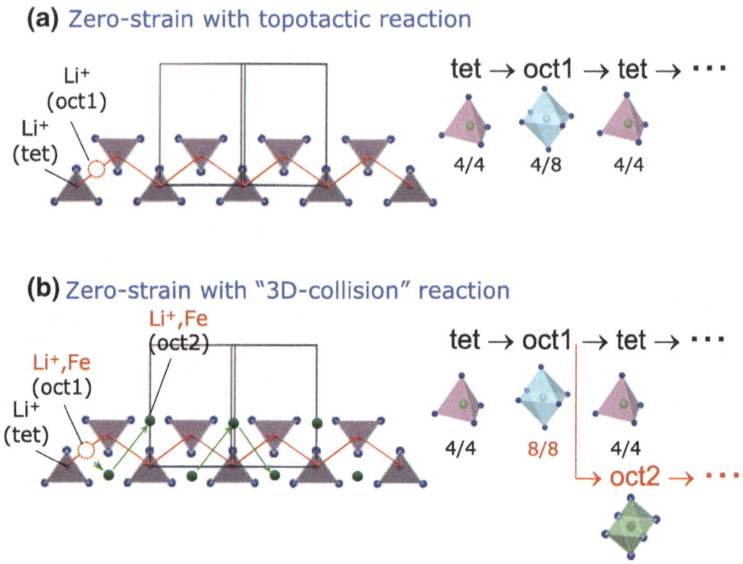

Fig. 10.11 Zero-strain reaction mechanisms with **a** topotactic and **b** non-topotactic reactions. Both lithium insertion materials adopt the spinel structure. Topotactic reaction exhibits a zigzag conduction pathway for Li^+ ions, while non-topotactic indicates an omnidirectional conduction pathway for Li^+ ions. The effective planes for the Li-conduction are different from each other: in the case for the non-topotactic, 4/4 for the tet → oct1 direction and 8/8 for the oct1 → tet and oct1 → oct2 directions. Reproduced with permission [46]. Copyright 2018, Elsevier

a broad maximum at $x = 1.4$. Hence, the Δa_c (ΔV) value exhibit the maximum at $x = 1.4$ with +40.16% (+0.47%). Since these Δa_c and ΔV values are smaller than those for $Li[CrTi]O_4$ ($\Delta V = +0.72\%$) [15, 42], the $x \geq 0.875$ samples can be regarded as one of zero-strain lithium insertion materials.

Further XRD patterns were recorded for the $x = 1.4$ sample, to clarify its zero-strain reaction scheme. Figure 10.10c–i show ex situ XRD patterns of $x = 1.4$ during the discharge and charge reactions: (c) $y = 0$, (d) $y = 0.263$, (e) $y = 0.760$, (f) $y = 1.190$, (g) $y = 0.962$, (h) $y = 0.602$, and (i) $y = 0.295$. Note that y was calculated by the observed Q_{dis} (or Q_{cha}) value and the Q_{theo}. In the XRD pattern of $y = 0$, the 110, 210, and 211 diffraction lines, which are attributed to the superlattice diffraction lines [45, 49], are clearly observed at $2\theta < 35°$. As y increases from 0 to 1.190, the intensity of the 311 diffraction line decreases, while that of the 111 diffraction line increases. This indicates that the Fe ions located at the tetrahedral 8*c* site move to the octahedral 4*a* and 12*d* sites, to construct a conduction pathway for Li^+ ions. Besides this change, the intensity of the 210 superlattice diffraction line increases with y, whereas the intensities of the 110 and 211 superlattice diffraction lines decrease with y. In addition, the 110 and 211 superlattice diffraction lines almost disappear at $y = 1.190$, but these diffraction lines are clearly observable at $y = 0.295$. These changes confirm that the Li, Fe, and Ti ions located at the octahedral 4*b* and 12*d* sites participate in the electrochemical reaction. This reaction scheme is different from conventional topotactic reaction, in which only Li^+ move back and forth between two lithium insertion materials while transition metal ions maintain their positions.

Figure 10.11 shows the reaction mechanism of the (a) topotactic and (b) non-topotactic reactions with the spinel structure. The conventional topotactic reaction indicates a zigzag conduction pathway for Li^+ ions from tet ($8a$) → oct1 ($16c$) → tet ($8a$) direction. The effective planes for the Li^+ conduction are limited to 4/8 for the oct1 → tet direction. In contrast, the non-topotactic reaction exhibits an omnidirectional conduction pathway for Li^+ ions. The effective planes for the Li^+ are 8/8 for the oct1 → tet and oct1 → oct2 directions. Since the superlattice diffraction lines were reversibly disappeared and appeared during the discharge and charge reactions, Li- and Fe-conductions obey certain rules. Further experiments, such as electron diffraction analyses and Möossbauer spectroscopy, could clarify the details of the non-topotactic reaction.

10.5 A Nearly Zero-Strain Insertion Material for Positive Electrodes

Not a perfect, but a nearly zero-strain insertion material for positive electrodes can be seen in spinel oxide $LiCo_xMn_{2-x}O_4$ (LCMO) with $x = 1$ [52]. Figure 10.12 shows the charge and discharge curves of the lithium cells of (a) $x = 0$, (b) $x = 0.25$, (c) $x = 0.5$, (d) $x = 0.75$, and (e) $x = 1$. The cells were operated in the voltage range between 3.0 and 5.4 V at a current density of ~0.3 $mA \cdot cm^{-2}$. The cell voltage of $x = 0$ rapidly increases to ~4.0 V at the beginning of the charge reaction, then maintains an almost constant voltage at approximately 4.2 V up to Q_{cha} of ~80 $mAh \cdot g^{-1}$. Finally, the cell indicates another voltage plateau at approximately 5.2 V, which was also observed in $Li[Li_xMn_{2-x}]O_4$ with $0 \leq x \leq 1/3$ [53]. The Q_{cha} values of $x = 0.25$ and 0.5 below ~4.0 V decrease, but those above ~5.0 V increase compared with that for $x = 0$. Based on the results of the X-ray absorption near-edge structure analyses [54] and electron paramagnetic resonance measurements [55, 56], the electrochemical reaction at 4.0 V is attributed to the redox reaction of $Mn^{3+} \leftrightarrow Mn^{4+}$, whereas that at 5.0 V is attributed to the redox reaction of $Co^{3+} \leftrightarrow Co^{4+}$. As x increases from 0.5, the Q_{cha} value below 4.0 V decreases indicating the disappearance of Mn^{3+} ions. However, the total Q_{cha} (or Q_{dis}) value also decreases with x: for instance, the Q_{cha} and Q_{dis} values of $x = 1$ were 112 and 108 $mAh \cdot g^{-1}$, respectively, at the initial cycle. Since the Q_{dis} value is slightly observed below 4.0 V, small amount of the Mn^{3+} ions still exists in the $x = 1$ sample.

Figure 10.12 shows the (f) Q_{dis} at the initial cycle and corresponding the (g) average volatge (E_{ave}) and (h) energy density (W) as a function of x in LCMO. The Q_{dis} value reaches a broad maximum at $x = 0.5$, whereas the E_{ave} value monotonically increases from 4.166 V at $x = 0$ to 4.996 V at $x = 1$. Thus, the maximum W value is obtained at $x = 0.5$ (=627 $mWh \cdot g^{-1}$). This W value is slightly greater than

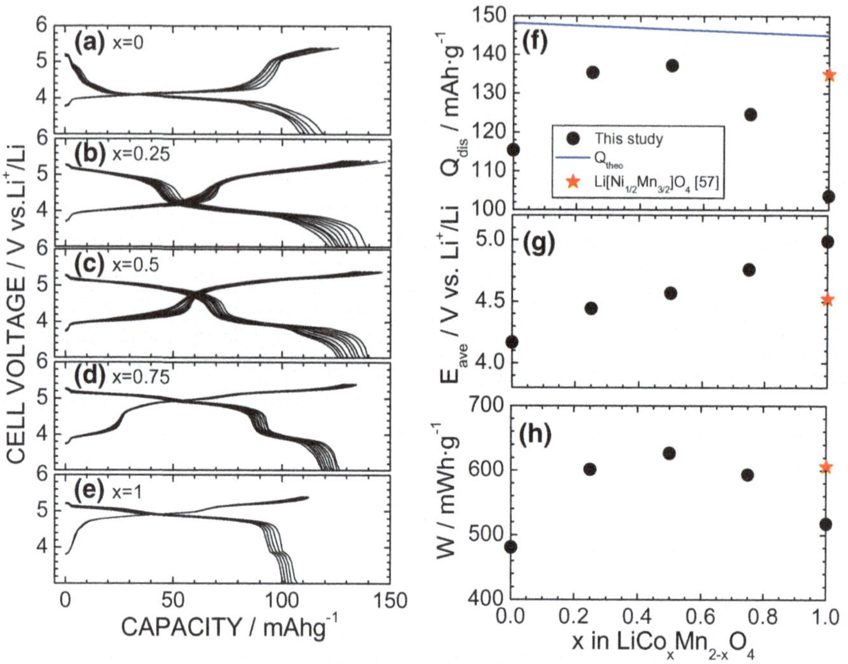

Fig. 10.12 Charge and discharge curves of the lithium cells with the LCMO samples: **a** $x = 0$, **b** $x = 0.25$, **c** $x = 0.5$, **d** $x = 0.75$, and **e** $x = 1$. x dependence of **d** Q_{dis} at the first cycle, **e** corresponding E_{ave} and **f** W. The solid line in (**f**) and ⋆ in (**f**)–(**h**) are the Q_{theo} of LCMO and the Q_{dis}, E_{ave}, and W values of $Li[Ni_{1/2}Mn_{3/2}]O_4$. [57]. Reproduced with permission [52]. Copyright 2017, American Chemical Society

that for $Li[Ni_{1/2}Mn_{3/2}]O_4$ (=607 mWh $\cdot$ g^{-1}), which has been regarded as a positive electrode material with high operating voltages [57].

Figure 10.13a shows a_cs of pristine and fully charged LCMO samples. The pristine a_c monotonically decreases from 8.2225(1) Å at $x = 0$ to 8.0589(1) Å at $x = 1$, being consistent with the change in r: $r_{Mn3+} = 0.58$ Å with C.N. = 6 and $r_{Co3+} = 0.55$ Å with C.N. = 6 [27]. The fully charged a_c slightly decreases from 8.0581(1) Å at $x = 0$ to 8.0083(1) Å at $x = 1$. Figure 10.13b shows Δa_c and ΔV of LCMO, where both values decrease with increasing x. That is, $\Delta a_c = -1.99\%$ and $\Delta V = -5.88\%$ for $x = 0$ and $\Delta a_c = -0.63\%$ and $\Delta V = -1.87\%$ for $x = 1$. $Li[Ni_{1/2}Mn_{3/2}]O_4$ indicated that $\Delta a_c = -1.98\%$ and $\Delta V = -5.83\%$ [57]. The Δa_c and ΔV values of $x = 0.5$ are still smaller than those of $Li[Ni_{1/2}Mn_{3/2}]O_4$, although the W value for $x = 0.5$ is greater than that for $Li[Ni_{1/2}Mn_{3/2}]O_4$ (Fig. 10.12h).

The Δa_c and ΔV values of $x = 1$ were significantly smaller than those of other positive electrode materials, such as $LiCoO_2$ with $\Delta V = +2.5\%$ [3] and $LiFePO_4$ with $\Delta V = -6.9\%$ [6] (see Fig. 10.1a). Furthermore, $Li[CrTi]O_4$, which indicates

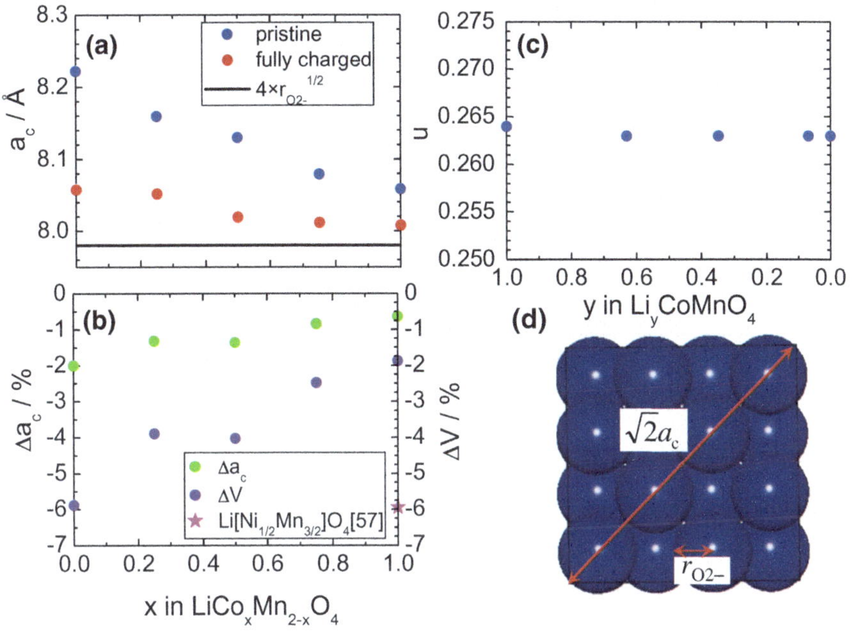

Fig. 10.13 **a** a_c values of the pristine and fully charged LCMO samples and **b** corresponding Δa_c and ΔV values. The solid line in (**a**) represents the calculated value of $4\sqrt{2}r_{O2-}$. The ΔV value of $Li[Ni_{1/2}Mn_{3/2}]O_4$ [57] is also shown in (**b**) for comparison. **c** y dependence of u in Li_yCoMnO_4 and **d** schematics of zero-strain reaction scheme in Li_yCoMnO_4. The minimum a_c is calculated to be ~8.0 Å based on the relation of $4\sqrt{2}r_{O2-} \simeq 7.8$ Å, which was close to a_c value of Li_yCoMnO_4. Reproduced with permission [52]. Copyright 2017, American Chemical Society

a ΔV of +0.72%, is recognized as one of the zero-strain lithium insertion materials [15, 42]. Therefore, the $x = 1$ sample can be regarded as a nearly zero-strain lithium insertion material.

To clarify the mechanism of the nearly zero-strain, Fig. 10.13c shows the change in u as a function of y in Li_yCoMnO_4. Since u maintains a constant value of ~0.263, local structure in MnO_6 and CoO_6 does not drastically change with y. As seen in Fig. 10.13d, the minimum a_c value is calculated to be ~7.8 Å for the ccp consisting of only O^{2-} ions. Thus, the nearly zero-strain characteristic of $x = 1$ would come from a rigid framework structure of O^{2-} with ccp packing. In other words, negative electrode materials with zero-strain are due to reversible changes in the local structures, whereas positive electrode materials with (nearly) zero-strain are due to invariance in the local structures.

Fig. 10.14 Cross-sectional views of the **a** initial NC11 and **b** charged NC11, **c** initial NCM111, **d** charged NCM111, **e** initial NCA, and **f** charged NCA. The charged NC11, NCM111, and NCA samples were prepared using the electrochemicalreactions. Reproduced with permission [12]. Copyright 2017, Springer

10.6 Pseudo Zero-Strain Insertion Materials

There are several pseudo zero-strain insertion materials for positive electrodes, although their lattice parameters change as a function of Li content. Examples are seen in NC11 [12, 13] and NCM111 [12, 14]. Figure 10.14 shows cross-sectional images of the (a) initial NC11, (b) charged NC11, (c) initial NCM111, (d) charged

NCM111, (e) initial $LiNi_{0.8}Co_{0.15}Al_{0.05}O_2$ (NCA), and (f) charged NCA samples. The Li compositions of the charged samples were 0.43 for NC11, 0.33 for NCM111, and 0.33 for NCA. In the initial state for all the samples, primary particles with an average size of 1 μm aggregate together, forming a large secondary particle with a diameter of ~10 μm. Any clearances nor, fractures are observed between primary particles. In the charged state, the primary particles of NC11 and NCM111 adhere to one another, whereas those of NCA separate each other.

According to the XRD measurements, the lattice parameters in the hexagonal setting, a_h and c_h, were determined to be 2.794(1) Å and 14.496(1) Å for the charged NC11, 2.826(1) Å and 14.255(1) Å for the charged NCM111, and 2.806(1) Å and 14.433(1) Å for the charged NCA. Hence, the ΔV values of the charged NC11, NCM111, and NCA samples were calculated to be −0.02, −0.44, and −2.2%, respectively. Note that the lattice parameters of both NC11 and NCM111 change, i.e., $\Delta a_h = -1.3\%$ and $\Delta c_h = +2.7\%$ for the charged NC11 sample. These results indicate that the clearances between primary particles are caused by ΔV, not by Δa_h or Δc_h. This means that the clearances are canceled as long as $\Delta V = 0\%$.

The electronic and ionic conductions occur only at the interfaces between secondary particles of the active material and the solid electrolyte in the bulk-type ALIB. Particle fractures inevitably isolate the electrode from the solid electrolyte leading to degrade of the battery performance. The pseudo zero-strain materials such as NC11 and NCM111 are suitable for the bulk-type ALIB, but further experiments would be required to clarify roles of zero-strain characteristics on the electrochemical performances of ALIBs.

10.7 Summary of Zero-Strain and Pseudo Zero-Strain insertion Materials

Table 10.1 summarizes the characteristics of zero-strain and pseudo zero-strain insertion materials. Zero-strain insertion materials LTO [11], $Li[CrTi]O_4$ [42], $LiRh_2O_4$ [43], and LFTO [45] adopt the cubic spinel structure for negative electrodes. LCMO also has the cubic spinel structure; however, it is for positive electrodes and is a nearly zero-strain insertion material [52]. Although perfectly zero-strain insertion materials for positive electrodes are currently unavailable, pseudo zero-strain insertion materials such as NC11 and NCM111 [12] are seen in the trigonal structure. Pseudo zero-strain insertion materials for negative electrodes are also reported in $LiCaFeF_6$ with $\Delta V = +0.20\%$ [58].

If we explore sodium insertion materials, zero-strain insertion materials $Na_3V(PO_3)_3N$ with $\Delta V = \sim -0.2\%$ [59, 60] and Na_2FeSiO_4 [61] with $\Delta V = -0.9\%$ are reported, where the former is for positive electrodes and the latter is for negative electrodes. As for NC11 and NCM111 [12], a pseudo zero-strain material for positive electrodes is also seen in the trigonal structure of $Na_{0.66}[Li_{0.22}Ti_{0.78}]O_2$ [62]. XRD densities of sodium zero-strain and pseudo zero-strain insertion materials

Table 10.1 Characteristic of zero-strain insertion materials with the cubic spinel structures

Compound	Crystal system	Lattice parameter (Å)	Change in lattice parameter (%)	Change in lattice volume (%)	References
$Li[Li_{1/3}Ti_{5/3}]O_4$	Cubic	8.356	~0	~0	[11]
$Li[CrTi]O_4$	Cubic	8.318	+0.24	+0.72	[42]
$LiRh_2O_4$	Cubic	8.460	+0.18	+0.53	[43]
Li-Fe-Ti-O system	Cubic	8.350–8.356	0 ± 0.18	0 ± 0.42	[45]
$LiCoMnO_4$	Cubic	8.059	−0.63	−1.87	[52]
NC11	Trigonal	$a_h = 2.832$	$\Delta a_h = -1.3$	−0.02	[12]
		$c_h = 14.113$	$\Delta c_h = +2.7$		
NCM111	Trigonal	$a_h = 2.862$	$\Delta a_h = -1.3$	−0.44	[12]
		$c_h = 14.251$	$\Delta c_h = +2.1$		
$LiCaFeF_6$	Trigonal	$a_h = 5.129$	$\Delta a_h = -1.3$	+0.20	[58]
		$c_h = 9.777$	$\Delta c_h = +2.1$		
$Na_3V(PO_3)_3N$	Cubic	9.440	−0.08	−0.24	[59]
		9.445	−0.06	−0.16	[60]
Na_2FeSiO_4	Cubic	7.330	−0.3	−0.9	[61]
$Na_{0.66}[Li_{0.22}Ti_{0.78}]O_2$	Trigonal	$a_h = 2.993$	$\Delta a_h = -0.96$	+0.21	[62]
		$c_h = 10.954$	$\Delta c_h = +1.66$		

are typically ~3.0 $g \cdot cm^{-3}$, which is much lower than those of lithium zero-strain and pseudo zero-strain insertion materials. For sodium insertion materials, zero-strain materials with higher densities are required to increase the energy density of sodium-ion batteries.

References

1. Mizushima K, Jones PC, Wiseman PJ, Goodenough JB (1980) Mater Res Bull 15:783–789
2. Remiers JN, Dahn JR (1992) J Electrochem Soc 139:2091–2097
3. Ohzuku T, Ueda A (1994) J Electrochem Soc 141:2972–2977
4. Huner JC (1981) J Solid State Chem 39:142–147
5. Ohzuku T, Kitagawa M, Hirai T (1990) J Electrochem Soc 137:769–775
6. Padhi AK, Nanjundaswamy KS, Goodenough JB (1997) J Electrochem Soc 144:1188–1194
7. Ohzuku T, Iwakoshi Y, Sawai K (1993) J Electrochem Soc 140:2490–2498
8. Beaulieu LY, Hatchard TD, Bonakdarpour A, Fleichauer MD, Dahn JR (2003) J Electrochem Soc 150:A1457–A1464
9. McDowell MT, Lee SW, Nix WD, Cui Y (2013) Adv Mater 25:4966–4985
10. Colbow KM, Dahn JR, Haering RR (1989) J Power Sources 26:397–402
11. Ohzuku T, Ueda A, Yamamoto N (1995) J Electrochem Soc 142:1431–1435
12. Mukai K (2017) Ionics. https://doi.org/10.1007/s11581-017-2385-2
13. Ueda A, Ohzuku T (1994) J Electrochem Soc 141:2010–2014
14. Yabuuchi N, Makimura Y, Ohzuku T (2007) J Electrochem Soc 154:A314–A321

15. Mukai K, Ariyoshi K, Ohzuku T (2005) J Power Sources 146:213–216
16. Mukai K, Kato Y, Nakano H (2014) J Phys Chem C 118:2992–2999
17. Mukai K, Kato Y (2015) J Phys Chem C 119:10273–10281
18. Mukai K, Sugiyama J (2015) Phys Chem Chem Phys 17:22652–22658
19. Mukai K, Nunotani N, Moriyasu R (2015) ACS Appl Mater Interf 7:20314–20321
20. Scharner S, Weppner W, Scmid-Beurmann P (1999) J Electrochem Soc 146:857–861
21. Panero S, Reale P, Ronci F, Scrosati B, Perfetti P, Albertini VR (2001) Phys Chem Chem Phys 3:845–847
22. Ronci F, Reale P, Scrosati B, Panero S, Albertini VR, Perfetti P, di Michiel M, Merino JM (2002) J Phys Chem B 106:3082–3086
23. Baddour-Hadjean R, Pereira-Ramos JP (2010) Chem Rev 110:1278–1319
24. Bhagavantam S, Venkatarayudu T (1969) Theory of groups and its application to physical problems. Academic Press, New York
25. Julien C, Massot M (2003) Mater Sci Eng B 97:217–230
26. Ammundsen B, Burns GR, Islam MS, Kanoh H (1999) Rozière. J J Phys Chem B 103:5175–5180
27. Shannon RD, Prewitt CT (1969) Acta Crystallogr B 25:925–946
28. Smirnov M, Baddour-Hadjean R (2004) J Chem Phys 121:2348–2355
29. Ariyoshi K, Yamato R, Ohzuku T (2005) Electrochim Acta 51:1125–1129
30. Liu DZ, Hayes W, Kurmoo M, Dalton M, Chen C (1994) Physica C 235–240:1203–1204
31. Ohzuku T, Kodama T, Hirai T (1985) J Power Sources 14:153–166
32. Inoue T, Mukai K (2017) ACS Appl Mater Interf 9:1507–1515
33. Inoue T, Mukai K (2017) Electrochem Commun 77:28–31
34. MacNeil DD, Dahn JR (2002) J Electrochem Soc 149:A912–A919
35. Reynier Y, Graetz J, Swan-Wood T, Rez P, Yazami R, Fultz B (2004) Phys Rev B 70:174304-1–174304-7
36. Wang M, Navrotsky A, Venkatraman S, Manthiram A (2005) J Electrochem Soc 152:J82–J84
37. Wang L, Maxisch T, Ceder G (2007) Chem Mater 19:543–552
38. Chase MW Jr, Davies CA, Downey JR Jr, Frurip DJ, McDonald RA, Syverud AN (1985) NIST-JANAF Thermochemical Tables, 3rd ed. JPCRD 14:1–1856
39. Yi TF, Xie Y, Zhu YR, Zhu RS, Shen H (2013) J Power Sources 222:448–454
40. Reynier YF, Yazami R, Fultz B (2004) J Electrochem Soc 151:A422–A426
41. Avdeev VV, Savchenkova AP, Monyakina LA, Nikolskaya IV, Khvostov AV (1996) J Phys Chem Solids 57:947–949
42. Ohzuku T, Tatsumi K, Matoba N, Sawai K (2000) J Electrochem Soc 147:3592–3597
43. Gu Y, Taniguchi K, Tajima R, Nishimura S, Hashimuze D, Yamada A, Takagi H (2013) J Mater Chem A:6550–6552
44. Nakamura A, Gu Y, Taniguchi K, Shibata N, Takagi H, Ikuhara Y (2015) Chem Mater 27:938–943
45. Mukai K, Kato Y, Nakano H (2017) J Solid State Chem 247:67–76
46. Mukai K (2018) Electrochim Acta 263:508–514
47. Braun PB (1952) Nature 170:1123–1123
48. Blasse G (1964) Philips Res Rep Suppl 3:1–138
49. Scharner S, Weppner W, Schmid-Beurmann P (1997) J Solid State Chem 134:170–181
50. Poizot P, Laruelle S, Grugeon S, Dupont L, Tarascon JM (2000) Nature 407:496–499
51. Obrovac MN, Dunlap RA, Sanderson RJ, Dahn JR (2001) J Electrochem Soc 148:A576–A588
52. Mukai K, Uyama T (2017) ACS Omega 2:5142–5149
53. Mukai K, Sugiyama J, Ikedo Y, Nozaki H, Kamazawa K, Andreica D, Amato A, Månsson M, Brewer JH, Ansaldo EJ, Chow KH (2010) J Phys Chem C 114:11320–11327
54. Kawai H, Nagata M, Kageyama H, Tukamoto H, West AR (1999) Electrochim Acta 45:315–327
55. Zhecheva E, Stoyanova R (2006) Alcaántara R, Lavela P, Tirado. J J J Power Sources 159:1389–1394
56. Stoyanova RK, Zhecheva EN, Gorova MY (2000) J Mater Chem 10:1377–1381
57. Mukai K, Sugiyama J (2010) J Electrochem Soc 157:A672–A676

58. de Biasi L, Lieser G (2017) Dräger C, Indris S, Rana J, Shumacher G, Mönig R, Ehrenberg H, Binder JR, Geßwein H. J Power Sources 362:192–201
59. Kim J, Yoon G, Lee MH, Kim H, Lee S, Kang K (2017) Chem Mater 29:7826–7832
60. Zhang H, Buchholz D, Passerini S (2017) Energies 10:889. https://doi.org/10.3390/en10070889
61. Li S, Guo J, Ye Z, Zhao X, Wu S, Mi JX, Wang CZ, Gong Z, McDonald MJ, Zhu Z, Ho KM, Yang Y (2016) ACS Appl Mater Interf 8:17233–17238
62. Wang Y, Yu X, Xu S, Bai J, Xiao R, Hu YS, Li H, Yang XQ, Chen L, Huang X (2013) Nat Commun 4:2365. https://doi.org/10.1038/ncomms3365

Chapter 11
Interfacial Engineering for Lithium Metal Batteries Based on Garnet Structured Solid Fast Lithium-Ion Conductors

Mir Mehraj Ud Din, George V. Alexander and Ramaswamy Murugan

Abstract Dense garnet structured solid electrolytes exhibit great promise for lithium metal batteries owing to its high lithium-ion conductivity, good mechanical and electrochemical properties. However, the rigid interfacial contact between garnet structured electrolyte and electrode stymie their practical application. In this chapter, salient features of various strategies utilized to enhance the room temperature performance of lithium metal battery based on garnet structured solid electrolyte are discussed. In particular, a detailed electrochemical investigation on lithium | garnet interface and cathode | garnet interface with buffer layers for solid-state battery. The application of garnet structured solid electrolyte in suppressing polysulfide shuttling in lithium-sulfur (Li-S) battery is also discussed. It is shown that the cycling performance of lithium metal and Li-S batteries can be greatly improved with the incorporation of garnet structured solid electrolyte.

11.1 Introduction

Lithium-ion batteries (LiBs) have attained great importance in recent times owing to their wide application in portable electronic devices [1–3]. The electrochemical reaction mechanism in conventional LiBs involves the intercalation and de-intercalation of lithium ions (Li^+) during respective charge and discharge cycles. During charge, Li^+ are released from the layered cathode via oxidation process and migrate to other end through electrolyte to intercalate into the layered graphitic anode. During discharge, a reverse process takes place in which lithium (Li) is oxidized at the anode and upon migration to other end it gets intercalated into the cathode. The electrons released on oxidation of Li flow through the external circuit. Figure 11.1a presents the schematics of a traditional LiB with layered graphitic anode. Although the graphitic anode helps in preventing the Li dendrite formation by allowing the Li^+ intercalation into its layered structure, there are several challenges that need to be well addressed.

M. M. U. Din · G. V. Alexander · R. Murugan (✉)
High Energy Density Batteries Research Laboratory, Department of Physics, Pondicherry University, Puducherry 605014, India
e-mail: moranamurugan.phy@pondiuni.edu.in

R. Murugan and W. Weppner (eds.), *Solid Electrolytes for Advanced Applications*,
https://doi.org/10.1007/978-3-030-31581-8_11

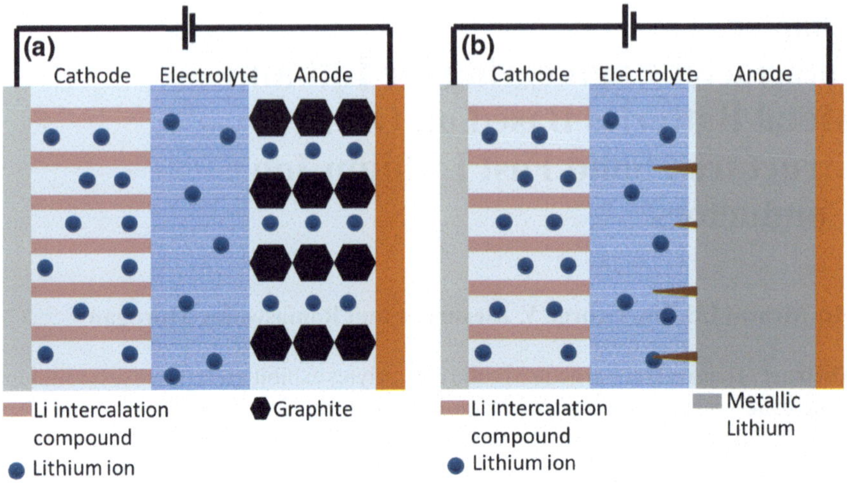

Fig. 11.1 Schematic of **a** traditional Li^+ battery and **b** Li metal battery with liquid electrolyte system

The primary factor in lithium batteries is the safety. The organic liquid electrolytes used in present LiBs are flammable, have narrow electrochemical window and tend to leak [4, 5]. In addition, the usage of graphite-based anodes in traditional LiBs limits the energy density to just 300 W h kg^{-1}.

Metallic Li is considered as an ideal anode for secondary batteries due to its lower negative electrochemical potential (−3.041 V vs. standard hydrogen electrode), high theoretical capacity and low density. Figure 11.1b presents the schematics of a Li metal battery with metallic Li as anode. Unlike traditional LiBs, Li metal batteries consist of a pure metallic Li as an anode which allows the replacement of conventional insertion-type cathodes by sulfur (Li–S) and air electrode (Li–Air) for high energy densities. However, there are technical challenges in developing the Li metal batteries. The major drawback is the growth of Li dendrites which could lead to safety hazard because of internal short-circuits in the battery [6].

11.1.1 All-Solid-State Lithium Metal Batteries

In recent times all-solid-state lithium batteries have received enormous attention as a replacement for present LiBs. Similar to liquid electrolyte-type Li^+ battery, during charge process, Li^+ come out of the cathode lattice, pass through the cathode/solid electrolyte interface, enter into the solid electrolyte structure, move across the anode/solid electrolyte interface and finally deposit onto the Li metal anode.

However, for a kinetically effective Li^+ diffusion process to happen all over the system, one of the important requirements is the suitability of solid electrolyte with electrode components.

For an efficient all-solid-state battery system, the proposed solid Li^+ conductors should possess the following features; (i) high Li^+ conductivity (~10^{-3} S/cm) at room temperature, (ii) insignificant electronic conductivity, (iii) good chemical stability with both the electrodes, (iv) larger electrochemical voltage window (greater than 5 V vs. Li), cost-effective and simple preparation technique [6–8]. Although high Li^+ conduction has been reported for crystalline-metal oxides and sulfides, oxides are assumed to be superior due to their high thermal and chemical stabilities [9]. However, many of the reported inorganic Li^+ conductors have had either high Li^+ conduction or high electrochemical stability, but not both. Among various inorganic solid electrolytes reported so far, garnet-structured solid electrolyte has received great attention due to its high Li^+ conductivity and excellent stability with Li metal [10–12].

Although this electrolyte possesses superior properties in comparison to other solid electrolyte, integration of this electrolyte with the present electrodes is still challenging because of large polarization and poor wettability with lithium metal [13]. Hence, it is of utmost importance to introduce effective strategies in improving the electrode–electrolyte interface. This chapter focuses on the overview of the interfacial behavior between garnet-structured solid electrolyte and electrode material. The highlight of this chapter is the emphases on the serious issues related to solid electrode/electrolyte interface and possible solutions toward their mitigation for practical realization of the all-solid-state batteries.

11.1.1.1 Interfacial Engineering at Garnet-Structured Solid Electrolyte and Metallic Lithium Interface

To develop an all-solid-state battery with Li metal as anode, it is of utmost importance that the chosen solid electrolyte should have extraordinary chemical stability against Li metal, and it should suppress Li dendrite formation. Earlier studies by Monroe et al. predicted that to suppress Li dendrite initiation, the solid electrolyte should possess double the shear modulus than that of metallic lithium (4.2 GPa) [14]. Ni et al. showed that the shear modulus of $Li_7La_3Zr_2O_{12}$ (LLZ) garnet structured electrolyte was 61 GPa, which is adequate to suppress Li dendrite initiation [15]. In addition, La and Zr in LLZ form a stable atomic framework which shows higher compatibility when in contact with metallic Li. Though garnet-structured solid electrolyte is electrochemically stable with Li metal, the poor wettability prevents intimate physical contact with the electrodes, which results in high interfacial resistance and an inhomogeneous current distribution [13, 16–21]. In recent times several strategies have been adopted to reduce the interface resistance between garnet-structured solid electrolyte and Li metal, which includes heat treatment, applying high external pressure, removal of Li_2CO_3 passivation layer, control of microstructure and surface modifications [22, 23].

Sharafi et al. demonstrated that LLZ|Li interface resistance decreases appreciably by heating Li|LLZ|Li symmetric cells to 175 °C and cooling down to room temperature [22]. The reduction in interface resistance is ascribed to the improved wettability between Li and LLZ, thus improving the effective contact area. However, Li dendrite propagation through polycrystalline LLZ was observed at rather low current density leading us to conclude that high shear modulus is necessary but not sufficient for blocking the Li dendrites.

The exact mechanism for dendrite growth between garnet-structured solid electrolyte and Li metal is not known, but the Li diffusion kinetics at the grain boundary could be an important factor. Suzuki et al. studied the Li dendrite formation with a transparent 1 wt% Al_2O_3-doped LLZ prepared by hot isostatic process (HIP) [24]. They observed an improvement in the relative density of 1 wt% Al_2O_3-doped LLZ from 91.5 to 99.1% through HIP treatment. The cell fabricated with HIP treated transparent 1 wt% Al_2O_3-doped LLZ showed no rapid voltage drop which suggested that the grain boundary-free LLZ may help in suppressing the Li dendrite growth through the solid electrolyte [25].

Metal and metal oxide coating over garnet-structured electrolyte has been found very effective in reducing the interfacial resistance at metallic lithium side and to barricade Li dendrite growth [16–21]. Metal coating such as Al, Si, Ge and metal oxide coatings like Al_2O_3 have been found effective in improving interface wettability [16–19]. However, the poor ductility and compressibility of reported interfacial layers may not tolerate for the long cycle life. It was proposed by Wang et al. that soft substrate as interlayer at the anode side helps to release the compressive stress during Li plating and prevent the dendrite growth [26]. Hence, it is important to introduce a soft interlayer on garnet surface to reduce interfacial resistance and to mitigate dendrite growth.

11.1.1.2 Interfacial Engineering at Garnet-Structured Solid Electrolyte and Cathode Interface

Although research is more focused on interface between lithium garnet and Li metal, the reduction of interfacial resistance at lithium garnet and cathode is much required because it contributes a lot to the total resistance of an all-solid-state lithium battery [9]. Lithium ion movement across the interface becomes very much difficult when the electrodes experience large volume changes. In order to overcome these issues, there should be favorable physical contact between cathode and garnet-structured solid electrolyte [27–30]. Studies on the interfacial resistance between the lithium garnet electrolyte and cathode are scanty. At the outset, pulsed laser thin film deposition technique was used for depositing cathode ($LiCoO_2$) over LLZ [31]. However, during high temperature sintering, diffusion of Co from $LiCoO_2$ to LLZ was detected along with La/Zr migration from LLZ to $LiCoO_2$. It has been reported that by coating $LiNbO_3$, $Li_4Ti_5O_{12}$, TaO_3 and so on over the solid electrolyte, the formation of space-charge layer could be suppressed [32–34]. Later on, the effect of Li_3BO_3 at the cathode|LLZ interface was studied [27]. The $LiCoO_2/Li_3BO_3$ composite as positive electrode deposited on LLZ solid electrolyte provided an initial discharge capacity of ~80 mA h/g. However, the performance was still limited as the addition of Li_3BO_3 to cathode causes reaction products at the interface amid sintering and charge/discharge process. To overcome the issue related to positive electrode side, one possible approach is the complete separation of cathode and LLZ by a buffer layer that has reasonable Li^+ conductivity, better wettability, chemically and electrochemically stable with LLZ and cathode [35].

Here, an emphasis will be placed on effective interfacial modification at (i) the Li|$Li_{6.28}La_3Zr_2Al_{0.24}O_{12}$ (LLZA) and (ii) $LiNi_{0.33}Mn_{0.33}Co_{0.33}O_2$ (NMC)|LLZA interface. Investigation was carried out with metallic lithium strip and thermal evaporated (TE) thin layer of lithium over pristine LLZA surface and Au deposited LLZA surface. Along with these studies, systematic investigation of NMC|LLZA interface was carried out with a thin amorphous lithium silicate (Li_2SiO_3) interlayer. With this developed interfaces, an all-solid-state lithium metal battery with LLZA as solid electrolyte, NMC as cathode and metallic lithium as anode was fabricated and studied.

11.1.2 High Energy Density Lithium–Sulfur (Li–S) Batteries

Li–S batteries are believed to be the possible replacement for conventional insertion-type Li^+ batteries in future due to higher theoretical capacity of elemental sulfur. Sulfur is environmentally friendly compared with toxic cobalt oxide, which is used as cathode in conventional Li^+ batteries. Also sulfur is chemically active and accepts up to two electrons per atom at ~2.1 V versus Li/Li^+. Li–S batteries therefore display high theoretical energy density of 2600 W h kg^{-1} [36, 37]. However, there are challenges in realizing the practical application of Li–S batteries system. Sulfur is non-conducting in nature and undergoes reduction reaction on reacting with Li [38], forming a series of intermediate reaction products (Li_2S_x, $4 \leq x \leq 8$) during electrochemical process. These reaction byproducts formed during cell discharging are highly soluble in ether-based organic liquid electrolytes. Due to the dissolution process the polysulfides diffuse through the separator membrane and reach at the anode side where they react with excess Li and get reduced into lower order intermediate products. The dissolution and the diffusion of intermediate reaction products lead to two serious problems in Li–S batteries. The primary challenge is the volume expansion in the cell due to the accumulation of insoluble lower order polysulfides (Li_2S_2 and Li_2S), and the second challenge is the powdering of Li metal due to dendrite formation related to secondary inimical reactions of Li anode with intermediate reaction products and the electrolyte exhaustion, which often leads to the early collapsing of the Li–S cell [39]. Attempts have been made to investigate different porous carbons with large surface area such as carbon black, hollow carbon, mesoporous (pore diameter of 2–50 nm) and microporous (pore diameter of <2 nm) carbons, carbon nano fibers, carbon nano rods, graphene oxide, conductive polymers, metal oxides and so on, where maximum sulfur loading has been achieved [40–47]. To mitigate the shuttling effect, bi-functional interlayers have been reported as polysulfide-shuttle inhibitors. The bi-functional interlayers or advanced separators can confine the polysulfides within the cathode area of the battery, thus resulting in high utilization of active material and enhanced cycling life of the battery [48–50]. In order to increase the active mass loading of sulfur, pre-lithiated materials such as Li_2S is proposed as a potential cathode in Li–S batteries. However, Li_2S has its own challenges, which include its high resistive nature and incomplete conversion to sulfur during charging, resulting in low Li^+ diffusivity.

11.1.2.1 Solid Electrolytes as Polysulfide Shuttle Inhibitors for Li–S Batteries

In recent times, extensive research work has been reported on application of various solid electrolytic systems to suppress the shuttle phenomenon in Li–S system. The unprecedented properties, like non-flammability, no leakage, high safety, reliability and prevention of polysulfide diffusion, make the ion conducting solids as potential electrolyte system for Li–S batteries. Solid-state batteries are advantageous over conventional batteries (using organic liquid electrolytes) in terms of both energy density and safety. Rechargeable solid-state batteries could offer excellent properties and could be safely used in long-distance electric vehicles (EVs) and hybrid electric vehicles (HEVs). Sulfur as a cathode material for Li metal batteries thus bears a promise of high theoretical capacity. However, the major challenges of intermediate reaction product formation and their dissolution in electrolytic media pose serious issues leading to the poor performance of the Li–S batteries. Various inorganic solid electrolytes have been reported [51]. However, the intrinsic instability of these materials with metallic lithium, relatively lower ionic conductivities and tendency to form inimical reaction products render them unsafe for Li–S battery applications [51].

Lithium garnet as a solid electrolyte has been proved to be effective in suppressing the polysulfide diffusion in Li–S batteries. The stability of lithium polysulfides with lithium garnet has been recently reported for hybrid Li–S system [52]. It is revealed that garnet solid electrolyte is stable with polysulfides due to the formation of self-inhibiting interlayer at the garnet surface.

Fu et al. reported a three-dimensional (3D) solid electrolyte framework for hybrid Li–S batteries [53]. This 3D framework consists of laminated bilayer comprising a thin dense garnet layer and a relatively thick porous garnet layer, which acts as host for higher mass loading of active sulfur cathode. Maximum loading of 2.5 mg sulfur has been reported in this study. This bilayer framework design is proved to be effective in blocking the polysulfide diffusion and suppressing the lithium dendrite penetration, thus improving the capacity retention in a Li–S battery. Liquid electrolyte modified garnet interface for Li–S has been reported for battery application where ether-based liquid electrolyte keeps the fast Li^+ transport kinetics and the solid-state electrolyte serves as a barricade to prevent the diffusion of polysulfides and suppresses the Li dendrite penetration [52].

Here, we discuss the application of Ta-doped lithium garnet solid electrolyte with nominal composition $Li_{6.4}La_3Zr_{1.4}Ta_{0.6}O_{12}$ (LLZT) for Li–S battery. Surface-engineered thin-walled LLZT membrane as a solid electrolyte and LLZT decorated polypropylene interlayer (PP-D(LLZT)) were utilized to reduce the electrode/electrolyte interface resistance and to effectively suppress the polysulfide shuttling phenomenon in Li–S battery.

11.2 Experimental

11.2.1 Preparation of All-Solid-State Lithium Metal Battery Based on Lithium Garnet as Electrolyte

11.2.1.1 Preparation of $Li_{6.28}Al_{0.24}La_3Zr_2O_{12}$ (LLZA)

Lithium garnets with composition $Li_{6.28}Al_{0.24}La_3Zr_2O_{12}$ (LLZA) were prepared by simple solid-state reaction. $LiOH{\cdot}H_2O$ (Alfa Aesar 98%; 10 wt% excess was added to compensate the lithium loss at higher temperature), La_2O_3 (Alfa Aesar 99.9%, pre-dried at 950 °C for 24 h), Al_2O_3 (Sigma Aldrich 99%) and ZrO_2 (Sigma Aldrich 99%) in stoichiometric amounts were mixed in a ball mill for 6 h in 2-propanol. The dried powder was later annealed at 950 °C for 12 h. The obtained powder was ball milled for 12 h. After removing the residual solvent, the powder was pressed into pellets. The LLZA pellets were placed in alumina crucible and covered with mother powder to circumvent lithium loss. The pellets were sintered at 1200 °C for 15 h. After sintering process the recovered LLZA pellets were mirror polished using different grit size sand papers.

11.2.1.2 Fabrication of $LiCoO_2$ Electrode

$LiCoO_2$ (LCO) powder (Sigma Aldrich), active carbon and polyvinylidene fluoride (PVDF) were taken in the ratio 80:15:5, respectively, and mixed in *N*-methyl-2-pyrrolidinone (NMP) to prepare the slurry. Later on, the prepared slurry was coated over aluminum foil. After room drying, the coated slurry was heat treated at 120 °C in vacuum overnight to remove the traces of solvent. For battery fabrication the prepared LCO electrodes were punched into discs of nearly 1 cm diameter with an active material content of 4.5 mg/cm^2.

11.2.1.3 Synthesis of Lithium Silicate

Li_2SiO_3 sol was prepared through sol–gel process by adding 2:1 (Li:Si) atomic ratios of lithium ethoxide [Li(OEt), high purity chemicals, 99.9%] and tetraethoxysilane

[TEOS, Shin Etsu, 99%] precursors [54]. Stoichiometric amount of Li(OEt) and TEOS were stirred for 2 h in ethanol solution in separate beakers. After receiving a complete transparent solution of both the precursors, TEOS solution was added gently to Li(OEt) solution under stirring. Later on, hydrolysis was carried out by adding the required amount of 0.1 M HCl to the solution.

11.2.1.4 Preparation of NMC Composite Cathode

The as-received NMC (pristine) powder [NMC(P)] and the prepared Li_2SiO_3 sol were taken in the ratio 70:30 and uniformly dispersed in NMP solvent. The finely dispersed particles were then coated over Li_2SiO_3-coated LLZA (LS–LLZA) surface. The amount of active material loading was ~2 mg cm^{-2}. After room drying, the sample was heat treated at 300 °C to achieve NMC composite|LLZA structure with Li_2SiO_3 buffer as an interlayer.

11.2.2 Fabrication of Symmetric Cells

The following symmetric cells were fabricated to investigate the interface between LLZA|Li and LLZA|NMC.

(i) Li(strip)|LLZA|Li(strip): For the fabrication of Li(strip)|LLZA|Li(strip) symmetric cell, lithium strips were punched into circular discs and were kept on both the sides of the LLZA pellet.
(ii) Li(strip)|LLZA|Li(strip) heat treated: The stacked configuration of Li(strip)|LLZA|Li(strip) was heat treated for 1 h at 180 °C and was slowly cooled down to room temperature. This was done in sequence on either side.
(iii) Li(strip)|Au|LLZA|Au|Li(strip): Li(strip)|Au|LLZA|Au|Li(strip) symmetric cells were fabricated by initially coating gold (Au) on either sides of LLZA pellet using a table top DC sputter coater. Later on, metallic lithium strip was pressed over Au-coated LLZA surface.
(iv) Li(strip)|Au|LLZA|Au|Li(strip) heat treated: Li(strip)|Au|LLZA|Au|Li (strip) symmetric cell after heat treatment at 180 °C (to melt the lithium) was slowly cooled to attain room temperature.
(v) Li(TE)|LLZA|Li(TE): Li(TE)|LLZA|Li(TE) symmetric cell was fabricated by depositing lithium (~2 μm) on either sides of LLZA pellet through thermal evaporation technique.
(vi) Li(TE)|LLZA|Li(TE) heat treated: The stacked Li(TE)|LLZA|Li(TE) cell was heated at 180 °C and was then slowly cooled to room temperature.
(vii) Li(TE)|Au|LLZA|Au|Li(TE): To the Au sputtered surface of LLZA, Li was deposited on both sides.

(viii) Li(TE)|Au|LLZA|Au|Li(TE): The stacked Li(TE)|Au|LLZA|Au|Li(TE) cell was heated at 180 °C on either sides in sequence and was slowly cooled to room temperature for electrochemical measurements.

(ix) NMC(P)|LLZA|NMC(P): NMC particles [NMC(P)] after dispersing in NMP were drop-coated on either sides of LLZA pellet. The symmetric cell configuration was later heated at 300 °C to evaporate the solvent.

(x) NMC|LLZA|NMC: NMC:LS composite cathode (70:30) [composite cathode is denoted as NMC] was coated on either sides of LLZA pellet and heated at 300 °C.

(xi) NMC|LS-LLZA-LS|NMC: The cells were fabricated by coating Li_2SiO_3 over LLZA surface prior to NMC coating. Li_2SiO_3 coating was performed by dropping 10 μL of as-prepared Li_2SiO_3 sol on both sides of LLZA. The pellet was then dried at 300 °C to completely evaporate the solvent to achieve LS-LLZA-LS. Later on, NMC was coated on either sides of LS–LLZA to form NMC|LS-LLZA-LS|NMC structure.

11.2.3 Fabrication of Li Metal Batteries

11.2.3.1 Li Metal Battery with Polypropylene (PP) Interlayer at Cathode

A quasi-solid-state battery was fabricated using LCO cathode, LLZA solid electrolyte and Li metal as anode. For smooth Li-ion diffusion through the interface, a thin layer of Au was deposited at the Li side. To improve the wettability at the LCO side, a PP separator with a minimal amount of liquid electrolyte (10 μL of 1.0 M $LiPF_6$ in DMC) was placed between LLZA and LCO. The cell was assembled in CR 2032 coin-type cell configuration.

11.2.3.2 All-Solid-State Battery with Li_2SiO_3 Interlayer

For the fabrication of an all-solid-state battery, Li metal was used as the anode, LLZA as solid electrolyte and NMC as cathode. A thin layer of Au was deposited at the anode side of LLZA and a thin amorphous Li_2SiO_3 was drop-casted at the NMC side to construct NMC||LS-LLZA||Au|Li structured solid-state cell. The cell was fabricated in swagelok and was heat treated at 180 °C inside glove box for melting Li.

11.2.4 Fabrication of Li–S Battery Based on $Li_{6.4}La_3Zr_{1.4}Ta_{0.6}O_{12}$ (LLZT) Solid Electrolyte

11.2.4.1 Preparation of Thin-Walled LLZT Membrane and LLZT Decorated PP Interlayer Membrane

Lithium garnet with composition $Li_{6.4}La_3Zr_{1.4}Ta_{0.6}O_{12}$ (LLZT) was synthesized by adopting a similar procedure as described in Sect. 11.2.1.1. A dense and high Li^+ conducting LLZT pellet of nearly 1.2 cm in diameter was mirror polished using sand papers to obtain a thin-walled (300 μm) LLZT pellet. The thin pellet was later cleaned and dried before shifting to glove box for cell assembling.

PP interlayer decorated with LLZT (PP-D(LLZT)) was utilized as separator membrane to prevent the shuttling of polysulfides in Li–S battery. High Li^+ conducting thick LLZT pellet was initially crushed into powder using planetary ball mill and later sieved to obtain the LLZT particles of uniform size. For the preparation of PP-D(LLZT) membrane, the finely ground LLZT powder was mixed with poly(vinylidene fluoride) (PVDF) dissolved in *N*-methyl-2 pyrrolidone (NMP) in a weight ratio of 9:1, respectively, to prepare the slurry. The slurry was blade-coated onto the polypropylene separator. The surface engineered PP separator was shifted to glove box for battery fabrication after drying for 24 h in vacuum oven at 50 °C.

11.2.4.2 Preparation of Sulfur–Carbon Composite and Sulfur Cathode

Composite sulfur was prepared by melt diffusion technique. Active carbon was dried for 10 h at 300 °C to remove the residual moisture. Active carbon and elemental sulfur were then hand mixed in agate mortar and sealed in a glass tube under Ar atmosphere and heat treated in vacuum at 160 °C for 12 h to diffuse the sulfur into porous carbon.

To prepare the sulfur cathode, a composition consisting of an active material, PVDF (dissolved in NMP) and acetylene black in 8:1:1 weight ratio, respectively, was mixed to form slurry. The prepared slurry was blade-cast onto the aluminum foil. The thickness of cathode during tape casting was set to 100 μm. After coating, the aluminum foil was dried for 24 h at 60 °C in vacuum. The coated aluminum foil was later pressed between two stainless steel dies using hydraulic press and then punched into small discs of nearly 1 cm^2 in area. The cathode discs were again dried under vacuum at 50 °C to remove the residual traces of solvent and transferred into glove box for cell fabrication.

11.2.4.3 Cell Assembling

(a) Symmetric cells were assembled to study the interface between LLZT and lithium electrode and the electrochemical test were performed in a Swagelok type cell inside Ar-filled glove box.

(i) Li|LLZT|Li: Symmetric cell with configuration Li|LLZT|Li was fabricated by sandwiching the LLZT membrane between two lithium strips.
(ii) Li|PP-LE-LLZT-PP-LE|Li: Symmetric cell of Li|PP-LE-LLZT-PP-LE|Li configuration was fabricated by staking the surface-modified LLZT membrane with soaked PP interlayer between two lithium strips. 10 μL of 0.25 M lithium bis (trifluoromethane) sulfonimide (LiTFSI) in a mixed solvent of 1,3-dioxolane (DOL) and 1,2-dimethoxyethane (DME) (V = 1:1) was used as liquid electrolyte (LE) to soak the PP interlayer.

(b) Li–S batteries based on thin-walled LLZT membranes and LLZT decorated PP separator were assembled inside glove box.

(i) Hybrid Li–S battery based on LLZT membrane: CR2032 coin cell with Li||LE-PP-LLZT-LE||S configuration was fabricated inside glove box by stacking lithium strip, PP–LE, LLZT pellet (thickness ~ 300 μm) and sulfur cathode with mass loading of 2.27 mg cm^{-2}. 10 μL of LE containing 0.25 M LiTFSI in mixed solvents of DME:DOL (V = 1:1) was utilized to soak PP interlayer and the sulfur cathode.
(ii) Hybrid Li–S battery based on LLZT decorated interlayer: Coin cell (CR 2032) was constructed with a configuration of Li||LE-PP-D(LLZT)||S consisting lithium strip as anode, engineered PP interlayer as separator and sulfur as cathode (active mass of sulfur = 2.34 mg cm^{-2}). The cathode was placed over the LLZT decorated side of PP interlayer and the anode was placed on the other side of the interlayer wetted with LE. 20 μL of 0.25 M LiTFSI in DME:DOL (V = 1:1) was utilized to wet the PP-D(LLZT) interlayer.

11.2.5 Characterization

Phase analysis of LLZA and LLZT pellets were carried out using X-ray diffraction (X'pert PRO PANalytical) with a Cu Kα radiation of 1.5418 Å wavelength in 2θ ranging from 10 to 80°. The Raman spectrum (Renishaw) of LLZA and LLZT was recorded with 50 mW internal Ar^+ laser source with excitation wavelength of 514 nm. The morphologies of LLZA, PP, LLZT decorated PP surface and the LLZA/electrode interfaces were investigated through scanning electron microscope (Hitachi S-4800).

11.2.6 Electrochemical Measurements

Cyclic voltammetry (CV), electrochemical impedance spectroscopy (EIS) and charge–discharge tests were performed using Bio-Logic VMP 3 battery tester. The conductivity of LLZA and LLZT pellets was determined by coating two blocking electrodes on either side of the pellets. EIS was measured at room temperature (25 °C) in the frequency range of 7 MHz to 1 Hz with 10 mV of perturbation amplitude.

11.3 Discussion

11.3.1 Characterization of LLZA Electrolyte

Dense Li^+ conducting solid electrolytes are crucial for smooth Li transportation and to suppress the dendrite growth. Lithium garnets can exist in three distinct polymorphs, namely high Li^+ conductive cubic ($Ia\bar{3}d$) phase, low Li^+ conductive cubic phase and tetragonal ($I4_1/acd$) phase [55, 56].

Figure 11.2a displays the XRD pattern of sintered LLZA ceramic pellet along with the reported cubic phase LLZ [12]. The XRD pattern of sintered LLZA pellet is in well match with the reported system indicating that the LLZA pellet is having a cubic phase ($Ia\bar{3}d$). The Raman spectrum of the synthesized LLZA sample (Fig. 11.2b) exhibits broad and fairly overlapping Raman bands in the 200–500 cm^{-1} region, which corresponds to the internal modes of LiO_4 and LiO_6 [55]. The Raman spectrum of LLZA further confirms the formation of cubic phase ($Ia\bar{3}d$). Li^+ conductivity of LLZA was determined from its corresponding Nyquist plot and was estimated to be 3×10^{-4} S cm^{-1} (Fig. 11.2c) at 25 °C.

11.3.2 All-Solid-State Lithium Battery Based on Lithium Garnet Electrolyte

11.3.2.1 Evaluation of Metallic Lithium–LLZA Interface

A qualitative assessment of thermal treatment and the effect of Au interlayer between metallic Li and LLZA was performed through cross-sectional SEM image. Figure 11.3 displays the cross-sectional SEM images of Li(TE)|LLZA|Li(TE), Li(TE)|Au|LLZA|Au|Li(TE) cell before and after heat treatment. On comparing the SEM image of Li(TE)|LLZA|Li(TE) (Fig. 11.3a) and heat treated (180 °C) Li(TE)|LLZA|Li(TE) (Fig. 11.3b), the latter shows an improved physical contact between metallic Li and LLZA, suggesting molten Li has better wettability over LLZA surface. The cross-sectional SEM image of heat treated (Fig. 11.3d) Li(TE)|Au|LLZA|Au|Li(TE) cell undoubtedly indicates the formation of a thin buffer layer in comparison with its untreated (Fig. 11.3c) sample. When the Li(TE)|Au|LLZA|Au|Li(TE) cell was heat treated at 180 °C, the Au coating over LLZA might diffuse inside molten Li, thus forming a Li–Au alloy, which might have brought LLZA in close and homogenous contact with metallic Li, as seen in Fig. 11.3d [21].

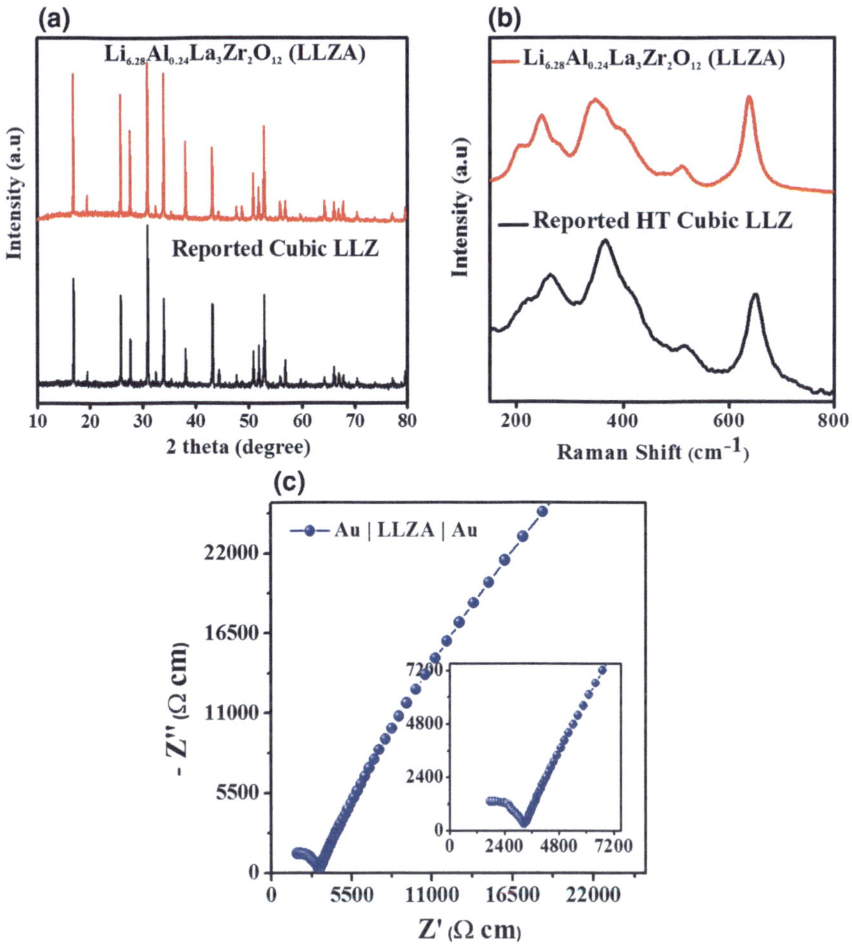

Fig. 11.2 **a** XRD pattern of LLZA (red) sintered pellet at 1200 °C and reported cubic LLZ (black) [12] (Ia$\bar{3}$d). **b** Raman spectra of LLZA (red) sintered pellet at 1200 °C and reported cubic LLZ (black) [56] in 150–800 cm^{-1} range. **c** Nyquist plot of LLZA pellet measured at room temperature (25 °C). Adapted from [21] with permission from Elsevier

A systematic quantitative analysis on the thermal treatment and the application of Au buffer layer between (i) thermal evaporated Li, (ii) metallic Li strip, and LLZA were evaluated using EIS. Figure 11.4 displays the impedance plot of Li(strip)|LLZA|Li(strip) symmetric cell before and after heat treatment. Figure 11.4a shows the impedance plot of Li(strip)|LLZA|Li(strip) symmetric cell before heat treatment. The Nyquist plot has two semicircles (as shown in inset) with the large semicircle at lower frequency corresponding to interface resistance between Li (strip)|LLZA. The reason for high interfacial resistance is mainly due to poor wettability of metallic lithium over LLZA [21]. Figure 11.4b displays the impedance

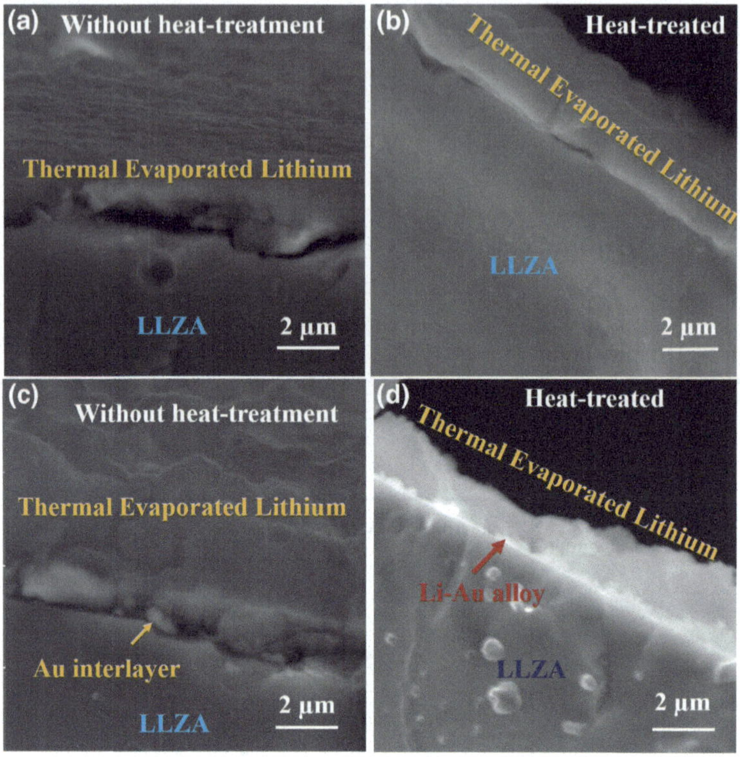

Fig. 11.3 Cross-sectional SEM images of Li(TE)|LLZA|Li(TE) symmetric cells before **a** and after **b** heat treatment at 180 °C. Cross-sectional SEM images of Li(TE)|Au|LLZA|Au|Li(TE) symmetric cells before **c** and after **d** heat treatment at 180 °C. Adapted from [21] with permission from Elsevier

spectrum of Li(strip)|LLZA|Li(strip) symmetric cell after heat treatment. It was noted that the Li(strip)|LLZA|Li(strip) symmetric cell after heat treatment exhibits less interface resistance in comparison with the untreated sample, which suggests that molten Li has improved wettability over LLZA [21].

For further understanding of the effect of thermal treatment of metallic Li on LLZA surface, a thin layer (~2 μm) of Li was deposited through thermal evaporation over LLZA surface. Figure 11.5a displays the impedance plot of Li(TE)|LLZA|Li(TE) before heat treatment. The impedance plot had two semicircles and the interfacial resistance was calculated to be 1.61 kΩ cm^2 (for each side) [21]. It was noted that the interfacial resistance between Li (TE) and LLZA has been reduced appreciably even compared to that of melting Li over LLZA (4.524 kΩ cm^2). From Fig. 11.5b, it was observed that the interfacial resistance of Li(TE)|LLZA|Li(TE) after heat treatment decreased significantly to 0.133 kΩ cm^2. From these results, it is evident that the uneven surface of LLZA causes microscopic gaps at the interface, thus leading to high interfacial resistance. However, by melting Li on the surface of LLZA, molten Li can access the microscopic gaps, further improving the interfacial contact [21].

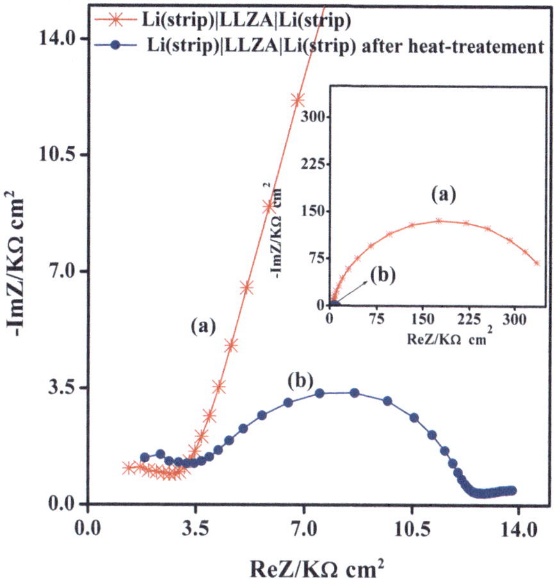

Fig. 11.4 Impedance plot of: **a** Li(strip)|LLZA|Li(strip) and **b** Li(strip)|LLZA|Li(strip) after heat treatment at 180 °C. The inset shows the impedance plot in the entire measured frequency range. Adapted from [21] with permission from Elsevier

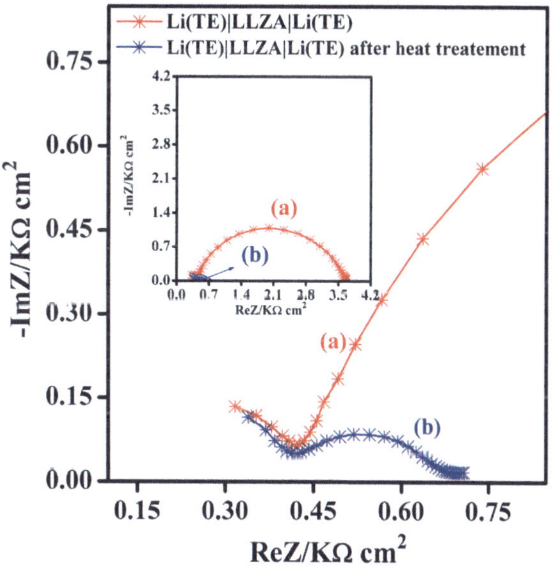

Fig. 11.5 Impedance plot of Li(TE)|LLZA|Li(TE) **a** before and **b** after heat treatment. The inset shows the impedance plot in the entire measured frequency range. Adapted from [21] with permission from Elsevier

Further to study the effect of buffer layer at the interface, a thin layer of Au was deposited over LLZA with different architectures of symmetric cell. Initially, metallic Li(strip) was placed above Au deposited LLZA surface and its impedance was measured. Figure 11.6a shows the impedance plot of Li(strip)|Au|LLZA|Au|Li(strip) before heat treatment. The interfacial resistance from each side of the symmetric

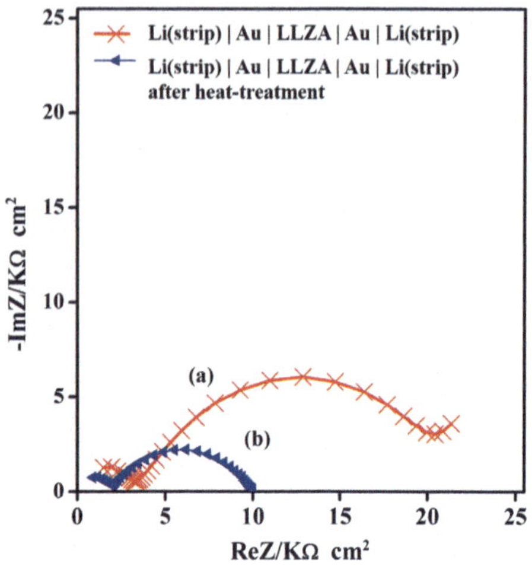

Fig. 11.6 Impedance plot of Li(strip)|Au|LLZA|Au|Li(strip) **a** before and **b** after heat treatment

cell was calculated to be 8.656 kΩ cm^2. The reduction in interfacial resistance when compared with Li(strip)|LLZA|Li(strip) could be because of the modified LLZA surface containing Au layer with higher wettability toward lithium leading to homogeneous contact [21]. Figure 11.6b displays the impedance plot of Li(strip)|Au|LLZA|Au|Li(strip) after heat treatment, which revealed an interfacial resistance of 3.978 kΩ cm^2. The reduction in interface resistance may be attributed to the enhanced wettability of molten Li with Au-coated LLZA [21].

In addition, the effect of Au interlayer between thermal evaporated Li and LLZA was studied. The Li(TE)|Au|LLZA|Au|Li(TE) cell exhibited an interfacial resistance of 1.587 kΩ cm^2 (Fig. 11.7a). Figure 11.7b shows the impedance plot of Li(TE)|Au|LLZA|Au|Li(TE) cell after heat treated at 180 °C. The interface resistance for each side was calculated to be 0.054 kΩ cm^2 [21]. This reduction in interfacial resistance may be because Au fills the microscopic gaps on the surface of LLZA, thereby reducing the interfacial flaws. When Li(TE)|Au|LLZA|Au|Li(TE) symmetric cell is heat treated at 180 °C, Li melts and diffuses into the metal coating, forming a favorable Li-ion conducting pathway between metallic Li and LLZA. Hence, it is evident that a metal coating between Li and LLZA is effective in reducing the resistance at electrode–electrolyte interface.

Galvanostatic cycling measurements were carried out to assess the lithium-ion transport capability across the Li|LLZA interface with a current density of 0.03 mA cm^{-2}. Figure 11.8a shows the DC Li stripping and platting in Li(strip)|LLZA|Li(strip) cell at 25 °C. The cell displayed a noisy potential with a fluctuated potential rise and fall, indicating non-uniform Li^+ transport through the interface. High polarization could be because of poor contact between metallic Li and LLZA [21]. Similarly, in case of Li(TE)|LLZA|Li(TE) symmetric cell voltage spikes were observed as shown in Fig. 11.8b. The voltage drops down to almost 0 V, suggesting possible dendrite growth through the LLZA solid electrolyte. Hence it can be concluded that Li deposition by thermal evaporation helps to reduce the interface resistance

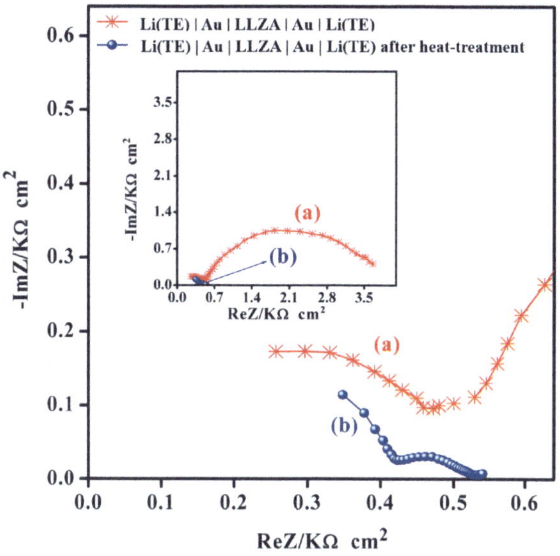

Fig. 11.7 Impedance plot of Li(TE)|Au|LLZA|Au|Li(TE) **a** before and **b** after heat treatment. The inset shows the impedance plot in the entire measured frequency. Adapted from [21] with permission from Elsevier

but fails to barricade the dendrite growth. Whereas Li(TE)|Au|LLZA|Au|Li(TE) cell after heat treatment at 180 °C stabilized at ~15 mV at room temperature, as shown in Fig. 11.8c [21]. The excellent cycling with stable voltage response could be because of homogeneous contact and lower interfacial resistance. The stable cycling confirms no short circuit, suggesting metal coating over LLZA is a promising strategy to enhance the interface contact between metallic Li and LLZA.

A summary on Au buffer layer utilization at the Li|LLZA interface is presented in Fig. 11.9. The uneven surface of LLZA leads to non-homogeneous contact between LLZA solid electrolyte and Li metal anode, as shown in Fig. 11.9a. The reason for the high interface resistance was due to poor interface contact between LLZA and Li. In addition, the increase in voltage under constant current density implies an inhomogeneous dissolution and deposition of Li metal during cycling which could later promote the Li dendrite growth through the point contact. Upon using thin layer of Au as a buffer between LLZA and Li electrode, the contact area between LLZA and Li electrode was improved as shown in Fig. 11.9b. Additionally, a homogeneous dissolution and deposition of Li at the interface could support the uniform Li^+ transport. It is therefore concluded that the contact between Li electrodes and LLZA electrolyte is critical for preventing the Li dendrite growth. Thus, proper microstructure engineering is required to reduce interface resistance and to prevent dendrite growth for the development of high performance all-solid-state Li batteries.

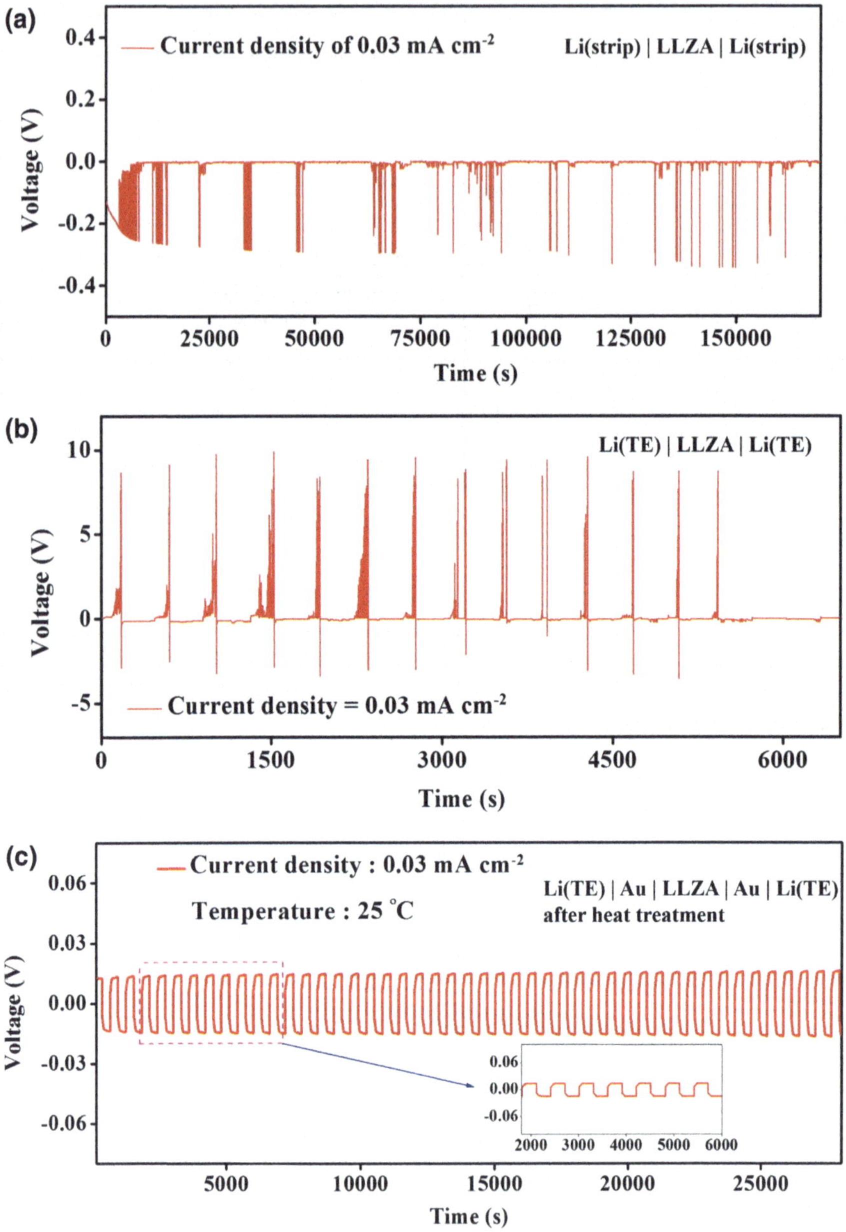

Fig. 11.8 D.C. cycling of **a** Li(strip)|LLZA|Li(strip) cell, **b** Li(TE)|LLZA|Li(TE), **c** Li-stripping/platting profile of Li(TE)|Au|LLZA|Au|Li(TE), with current density of 0.03 mA cm^{-2}. Adapted from [21] with permission from Elsevier

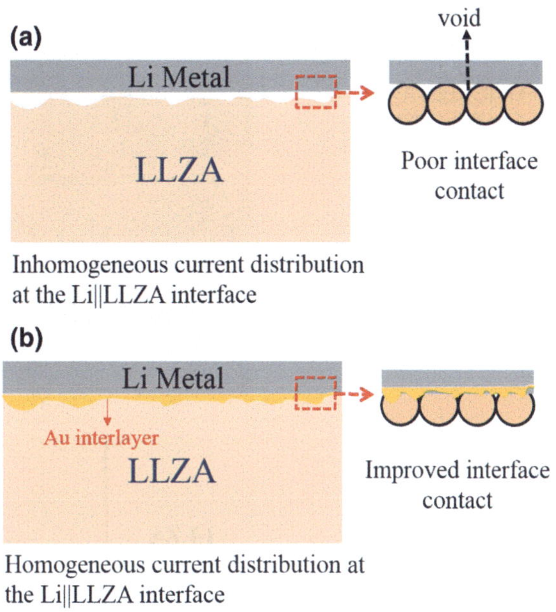

Fig. 11.9 Schematic diagram of **a** Li|LLZA with poor interface contact and **b** Li|Au|LLZA with improved interface contact

11.3.2.2 Evaluation of NMC–LLZA Interface

To study about the cathode|LLZA interface, we have chosen NMC as the cathode, and Li_2SiO_3 as intermediate buffer layer. A qualitative assessment of Li_2SiO_3 (LS) buffer layer was done from cross-sectional SEM images. Figure 11.10a displays the cross-sectional SEM image of NMC|LLZA interface. From the SEM image interfacial gaps between LLZA and NMC can be observed which is due to poor wettability between NMC and LLZA. Figure 11.10b shows cross-sectional SEM image of NMC|LS-LLZA-LS|NMC cell, with a Li_2SiO_3 buffer layer in between LLZA and NMC, filling the voids and improving the contact.

Systematic investigations were carried out, to understand the interfacial resistance between LLZA and NMC. Initially, pristine NMC(P) was coated over both the sides of LLZA and the EIS measurements were taken at room temperature. The impedance plot of NMC(P)|LLZA|NMC(P) shows high charge transfer resistance of 65 kΩ cm^2, as shown in Fig. 11.11a [57]. The high charge transfer resistance could be attributed to poor interfacial contact between NMC(P) and LLZA. Later on, composite cathode (NMC) was uniformly coated on both sides of LLZA. Figure 11.11b shows the impedance plot of NMC|LLZA|NMC. An interfacial resistance of ~7.2 kΩ cm^2 was estimated for each side. The reduction in interfacial resistance in comparison with NMC(P)|LLZA|NMC(P) can be attributed to the improved contact of composite cathode with LLZA surface [57]. Although composite cathode helped in reducing the interface resistance, the effective contact area between NMC and LLZA was still poor [57].

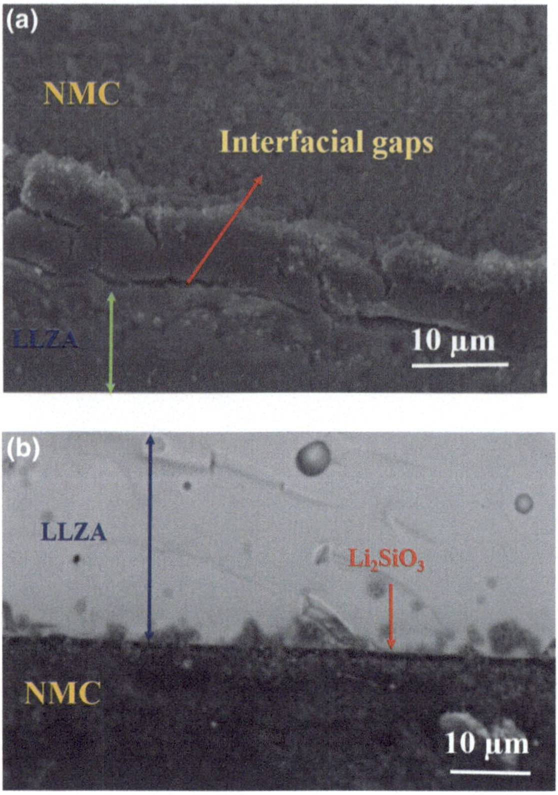

Fig. 11.10 Cross-sectional SEM images of **a** NMC|LLZA|NMC and **b** NMC|LS-LLZA-LS|NMC. Adapted from [57] with permission from Elsevier

The impedance plot of symmetric cell with NMC|LS-LLZA-LS|NMC configuration, as shown in Fig. 11.11c, describes the interface resistance between LLZA and NMC with Li_2SiO_3 buffer layer. The plot consists of a clear semicircle at high frequency region corresponding to the resistance of LLZA, a semicircle at medium frequency region related to the charge transfer resistance and a sloping line corresponding to the diffusion impedance. For NMC|LS-LLZA-LS|NMC cell, a charge transfer resistance of 0.675 kΩ cm^2 for each side was estimated [57]. This reduction in interface resistance could be due to the following: (i) the addition of Li_2SiO_3 fills the gaps at the interface, thus creating a uniform contact and (ii) Li_2SiO_3 coating favors Li^+ transport at the interface to the active material [57]. In addition, it has been observed that for the NMC|LS-LLZA-LS|NMC cell the interfacial resistance was reduced from 0.675 to 0.351 kΩ cm^2 at higher operating temperature of 100 °C, as shown in Fig. 11.11d [57]. The impedance plot at 100 °C exhibited distinct semicircles and a sloping line. On comparing the impedance plot at 25 °C, a significant reduction in interface resistance was observed, indicating that the fabricated cell exhibits improved performance at elevated temperatures.

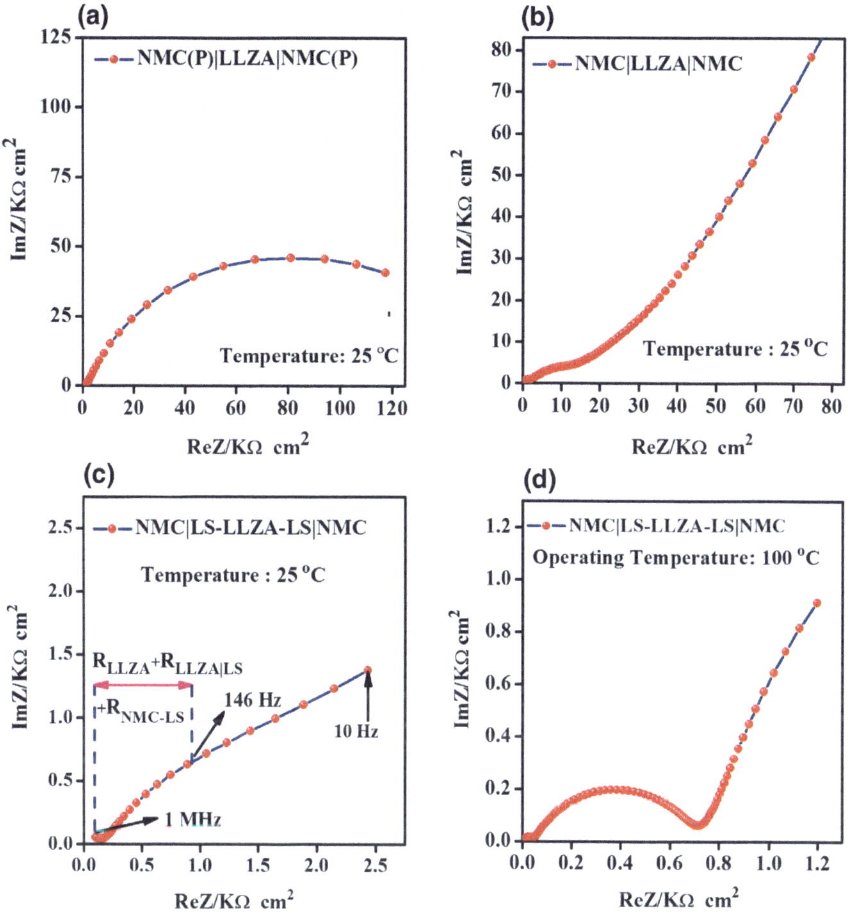

Fig. 11.11 Impedance plot of **a** symmetric cell with NMC(P)|LLZA|NMC(P) configuration at 25 °C. **b** Symmetric cell with NMC|LLZA|NMC configuration at 25 °C. **c** Symmetric cell with NMC|LS-LLZA-LS|NMC configuration at 25 °C. **d** Symmetric cell with NMC|LS-LLZA-LS|NMC configuration at 100 °C. Adapted from [57] with permission from Elsevier

11.3.2.3 Evaluation of Solid-State Lithium Metal Battery

(i) Electrochemical Characterization of Quasi Solid-State Battery

Figure 11.12a shows the impedance plot of Li(TE)|Au||LLZA-PP-LE||LCO fabricated cell at 25 °C. The impedance plot has two clearly distinct semicircles and a diffusion tail. The smaller semicircle near high frequency corresponds to the contribution from LLZA, and the semicircle at medium and low frequency relates to the charge transfer resistance between LLZA and electrodes. It was observed that the cell showed a total resistance of 1830 Ω cm^2, which could be the sum of LLZA||Li and LLZA||LCO interfaces. Further, galvanostatic charge–discharge studies were

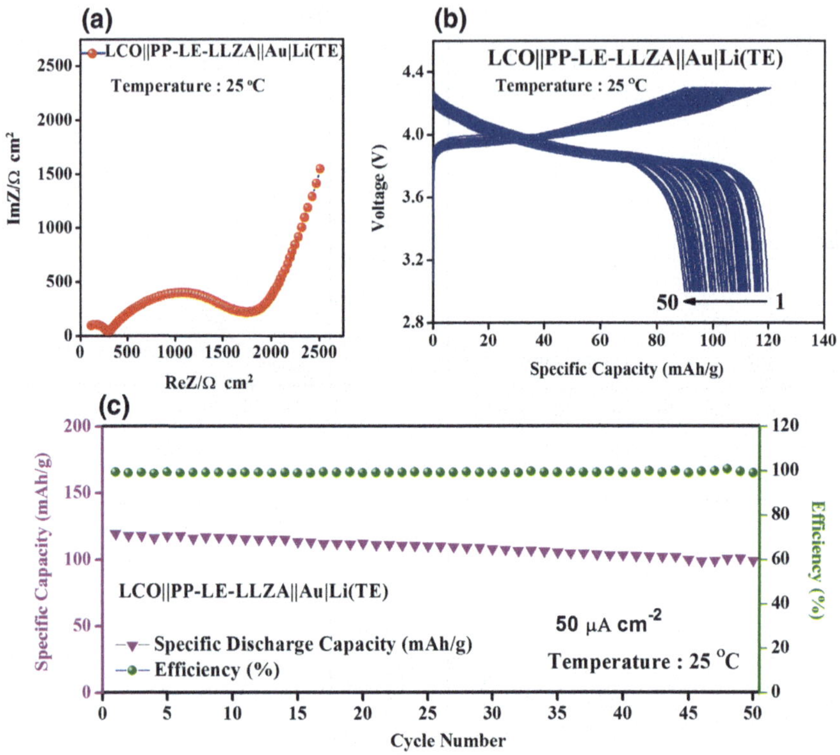

Fig. 11.12 **a** Impedance plot of Li(TE)|Au||LLZA-PP-LE||LCO at 25 °C. **b** Charge–discharge profile of Li(TE)|Au||LLZA-PP-LE||LCO for 50 cycles. Inset represents the voltage versus time profile of Li(TE)|Au||LLZA-PP-LE||LCO cell. **c** Discharge capacity and cycling efficiency of Li(TE)|Au||LLZA-PP-LE||LCO at 50 μA cm^{-2} for 50 cycles. Adapted from [21] with permission from Elsevier

performed on the fabricated cell at a current density of 50 μA cm^{-2}, as shown in Fig. 11.12b. Inset in Fig. 11.12b shows voltage versus time profile of the cell. The cell showed exactly the same behavior throughout the cycles, suggesting a stable interface. The cell delivered a discharge capacity of 121 mA h g^{-1} after initial cycle with good capacity retention over 50 cycles. Figure 11.12c portrays the discharge capacity and coulombic efficiency of Li(TE)|Au||LLZA-PP-LE||LCO cell. The fabricated cell displayed good capacity retention over 50 cycles with coulombic efficiency nearly 100%.

(ii) Electrochemical Characterization of All-Solid-State Lithium Metal Battery

Figure 11.13a shows the schematic of an all-solid-state Li metal battery utilizing a thin layer of Li_2SiO_3 at the cathode side as a buffer and Au as an interlayer at the anode side. Figure 11.13b shows the impedance plot of NMC||LS-LLZA||Au|Li cell at 25 and 100 °C. It was observed that the cell showed higher total resistance at 25 °C in comparison with the 100 °C. This reduction in interface resistance at high

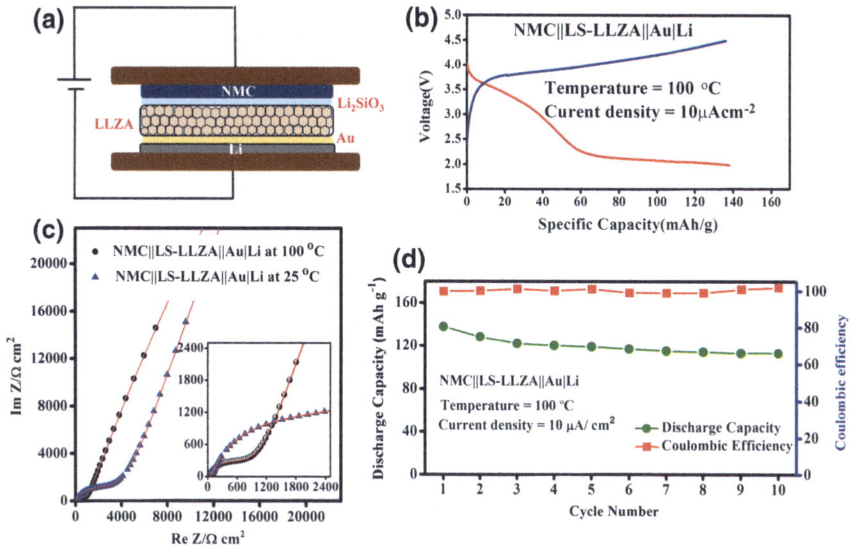

Fig. 11.13 **a** Schematic diagram of the fabricated cell with modified interface. **b** Impedance plot of NMC||LS-LLZA||Au|Li cell, measured at 25 and 100 °C. **c** First cycle charge/discharge profile of fabricated cell. **d** Discharge capacity and coulombic efficiency of fabricated cell for 10 cycles. Adapted from [57] with permission from Elsevier

temperature may be attributed to the improved wettability of NMC and Li with LLZA solid electrolyte. Figure 11.13c shows the charge–discharge profile of the fabricated all-solid-state cell (first cycle) in the voltage range of 4.5–2 V at a current density of 10 μA cm^{-2}. The cell demonstrated an initial discharge capacity of 138 mA h g^{-1} at 100 °C. It should be noted that a low cut-off voltage (2 V) was used to attain this capacity because of the lower ionic conductivity of Li_2SiO_3 layer, as it may slow down the Li^+ diffusion through the NMC|LLZA interface. Figure 11.13d shows the cycling stability and coulombic efficiency of the fabricated cell. The cell exhibited good capacity retention for 10 cycles with a coulombic efficiency of nearly 100%.

11.3.3 Electrochemical Evaluation of LLZT Based Lithium–Sulfur Battery

11.3.3.1 Characterization of LLZT Electrolyte

The X-ray diffraction pattern of LLZT pellet is in well agreement with that of reported cubic phase LLZ [12], as shown in Fig. 11.14a, indicating that the sintered LLZT pellet is purely a high temperature cubic phase ($Ia\bar{3}d$). The Raman spectrum of LLZT sample (Fig. 11.14b) displays broad Raman modes in the region ranging from 200

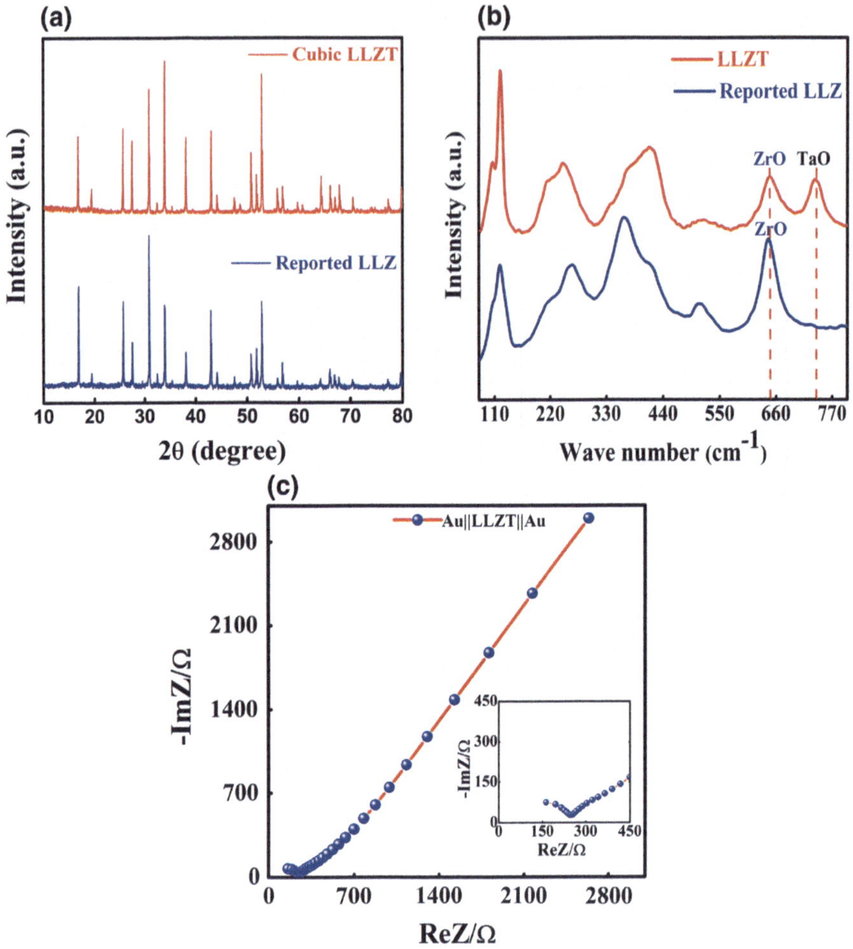

Fig. 11.14 **a** XRD pattern of the reported high temperature cubic LLZ (blue) (Ia$\bar{3}$d), LLZT (red). **b** Raman modes of the reported HT cubic LLZ (blue), LLZT (red) sintered at 1200 °C. **c** Impedance plot of LLZT at 25 °C

to 500 cm^{-1} representing the vibrational modes of LiO_4 and LiO_6 [55]. The Raman mode at above 740 cm^{-1} corresponds to the Ta–O bond stretching [58]. The ionic conductivity LLZT was determined from the Nyquist impedance plot. The total Li^+ conductivity of LLZT at 25 °C was calculated to be 2.8×10^{-4} S cm^{-1} (Fig. 11.14c).

11.3.3.2 Characterization of Li|LLZT|Li and Li|PP-LE-LLZT-PP-LE|Li Symmetric Cells

Impedance plots of the assembled Li|LLZT symmetric cells are shown in Fig. 11.15a and b, respectively.

The insets in Fig. 11.15a and b are the schematic representation of Li|LLZT|Li and Li|PP-LE-LLZT-PP-LE|Li symmetric cells, respectively, displaying a rough surface of LLZT in the former and the engineered surface in the latter case. The Li|LLZT|Li symmetric cell exhibited the total interface resistance of magnitude 260 kΩ cm^2 as depicted in Fig. 11.15a. Note that the total interface resistance was obtained by dividing the diameter of low-frequency half-circle by two, as the total interface resistance corresponds to two identical interfaces. The large interfacial resistance in Li|LLZT|Li symmetric cell can be ascribed to the improper contact between metallic lithium and uneven LLZT surface. The symmetric cell with configuration Li|PP-LE-LLZT-PP-LE|Li exhibited a low interfacial resistance of 0.435 kΩ cm^2 due to enhanced interface contact area, as shown in Fig. 11.15b. The improved contact between LLZT and metallic lithium is ascribed to the application of PP–LE, resulting in smooth interface and reduced interfacial resistance [59].

Lithium stripping and platting in the symmetric cells was galvanostatically carried out in voltage window of −5 to 5 V. Time-dependent voltage profiles of cells with cycle duration of 10 min at a steady current density of 0.025 mA cm^{-2} are depicted in Fig. 11.16a and b, respectively. The Li|LLZT|Li symmetric cell displayed a large voltage polarization, indicating a non-uniform Li^+ transport across the interface as depicted in Fig. 11.16a. It is believed that the poor contact area between metallic lithium and LLZT membrane could be the possible cause for charges to accumulate

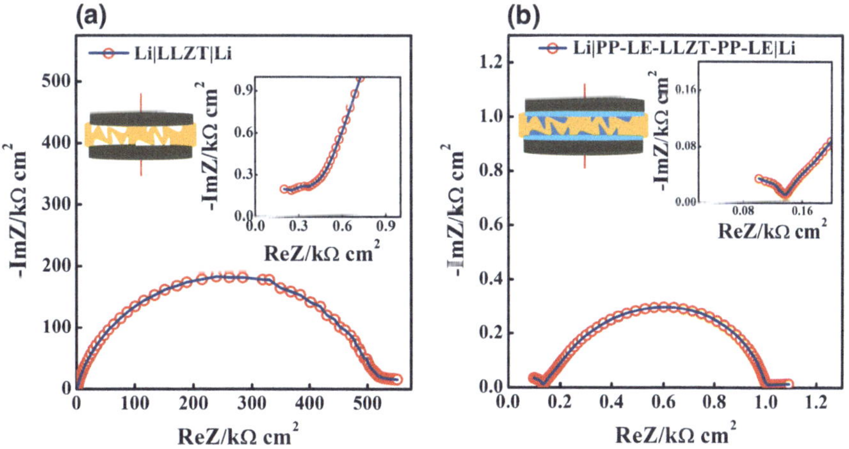

Fig. 11.15 Impedance plots of **a** Li|LLZT|Li and **b** Li|PP-LE-LLZT-PP-LE|Li symmetric cells at 25 °C. Insets in **a** and **b** represent the schematics and impedance data (in high frequency region) of respective symmetric cells. Adapted from [59] with permission from Elsevier

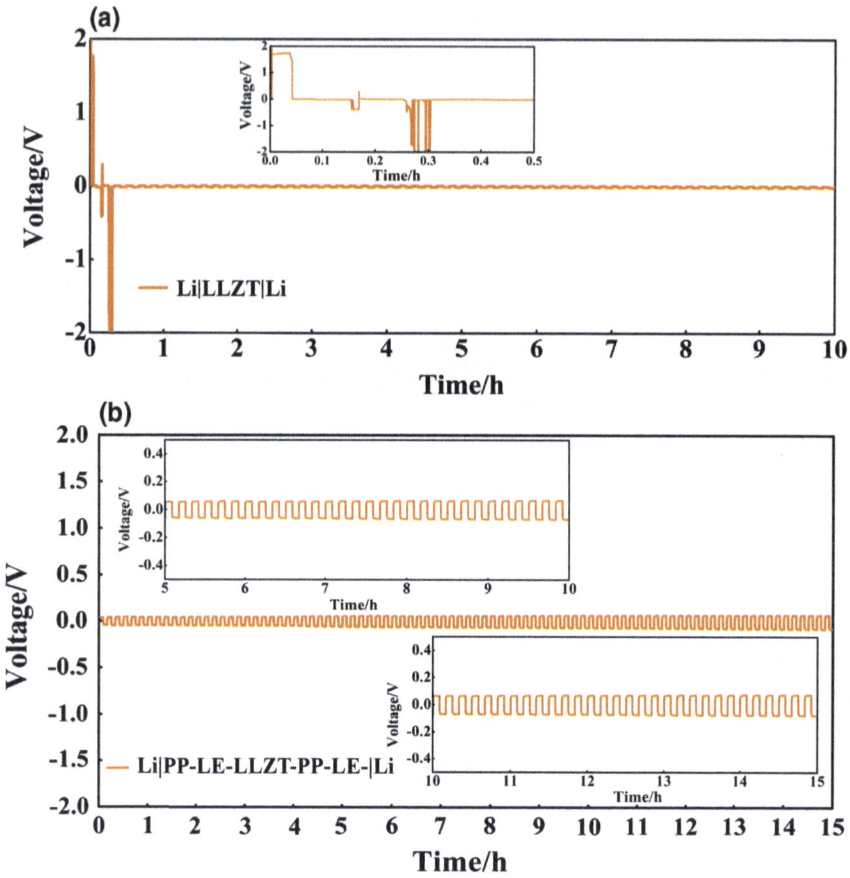

Fig. 11.16 Li-stripping and platting profiles in **a** Li|LLZT|Li and **b** Li|PP-LE-LLZT-PP-LE|Li symmetric cells at 0.025 mA cm^{-2}. Insets in **a** and **b** represent the enlarged profile of specific cycles. Adapted from [59] with permission from Elsevier

at the interface, resulting in huge cell polarization [59]. A sudden potential drop confirms the short circuiting in the cell, which is possibly due to the lithium dendrite penetration in Li|LLZT|Li symmetric cell [22]. On the other hand, Li|PP-LE-LLZT-PP-LE|Li cell displayed a stable Li stripping and plating voltage profiles when cycled for 15 h at a constant current density of 0.025 mA cm^{-2}, as depicted in Fig. 11.16b. The smooth and even voltage profile observed in Li|PP-LE-LLZT-PP-LE|Li cell is ascribed to the lower interface resistance and uniform contact provided by the interlayer membrane [59].

11.3.3.3 Electrochemical Performance of Li–S Battery with Surface-Modified LLZT Membrane

Electrochemical impedance measurements on Li||LE-PP-LLZT-LE||S coin cell were carried out in AC frequency of 7 MHz to 0.1 Hz at excitation amplitude of 10 mV. Impedance plot of the assembled Li||LE-PP-LLZT-LE||S battery displays the total interface resistance of 3.19 kΩ, which on cycling decreased to 1.79 kΩ, as depicted in Fig. 11.17a and b, respectively. The reduction in interface resistance indicated the formation of stable electrode–electrolyte interface. The charge–discharge cycling measurements of Li||LE-PP-LLZT-LE||S cell were carried out in a voltage range of 1.5–3 V with a current density of 0.037 mA cm^{-2}.

The Li||LE-PP-LLZT-LE||S cell exhibited an initial specific discharge capacity of nearly 1100 mA h g^{-1} and a stabilized discharge capacity of 783 mA h g^{-1} for the 20th cycle with 100% efficiency, as depicted in Fig. 11.17c and d, respectively. The good capacity retention indicates that the high Li^{+} conducting LLZT membrane

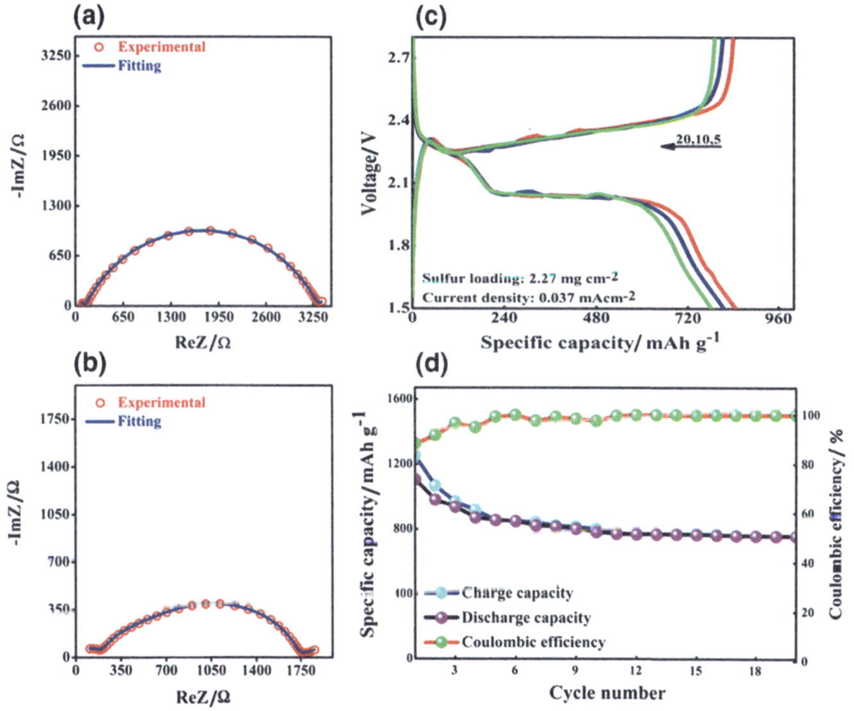

Fig. 11.17 **a** Impedance plot of Li||LE-PP-LLZT-LE||S before cell cycling. **b** Impedance plot of Li||LE-PP-LLZT-LE||S after first three charge–discharge cycles. **c** Charge–discharge cycles of Li||LE-PP-LLZT-LE||S at room temperature (25 °C). **d** Charge–discharge capacity and coulombic efficiency of Li||LE-PP-LLZT-LE||S for 20 cycles. Adapted from [59] with permission from Elsevier

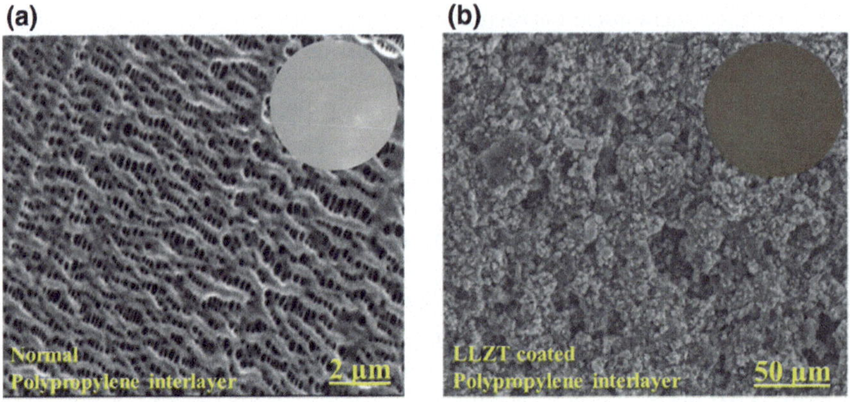

Fig. 11.18 SEM images of **a** routine PP separator and **b** LLZT decorated PP separator. Insets represent the digital images of **a** routine PP separator and **b** LLZT decorated PP separator

can effectively improve the cyclic efficiency by restricting the movement of reaction products on the cathode region of Li–S cell.

11.3.3.4 Electrochemical Performance of Li–S Battery with LLZT Decorated PP Separator

The microstructure of routine PP separator and the LLZT decorated PP separator is shown in Fig. 11.18.

The routine polypropylene with highly porous structure (Fig. 11.18a) could be responsible for the polysulfide shuttling, which results in capacity loss and early cell collapse. Thin layer of LLZT coated on PP separator (PP-D(LLZT)) can block the pores and help in enhancing the cell performance by restricting the flow of polysulfides on the cathode region during electrochemical process (Fig. 11.18b). Electrochemical impedance measurements of Li||LE–PP-D(LLZT)||S cell were carried out in AC frequency of 1 MHz to 0.1 Hz at excitation amplitude of 10 mV s^{-1}. Before cell cycling the impedance plot near high frequency area consists of a compressed half-circle and a sloping line near low frequency area, as shown in Fig. 11.19a. After cycling, two depressed half-circles in the impedance spectrum near the high and medium frequency regions are evolved, as depicted in Fig. 11.19b.

A depressed semicircle observed near high frequency region may be attributed to the interface contact resistance and the region near low frequency is related to the charge transfer resistance. The Li||LE–PP-D(LLZT)||S cell was galavnostatically cycled in a voltage range of 1.5–3 V by applying a current density of 0.39 mA cm^{-2}. The cell displayed an initial discharge capacity of 855 mA h g^{-1} and a discharge capacity of nearly 680 mA h g^{-1} after 20 cycles (Fig. 11.19c) [59]. A coulombic efficiency of above 97% was achieved after 20 cycles (Fig. 11.19d). It can be observed from the galvanostatic cycling that no prolonged charge plateaus are seen

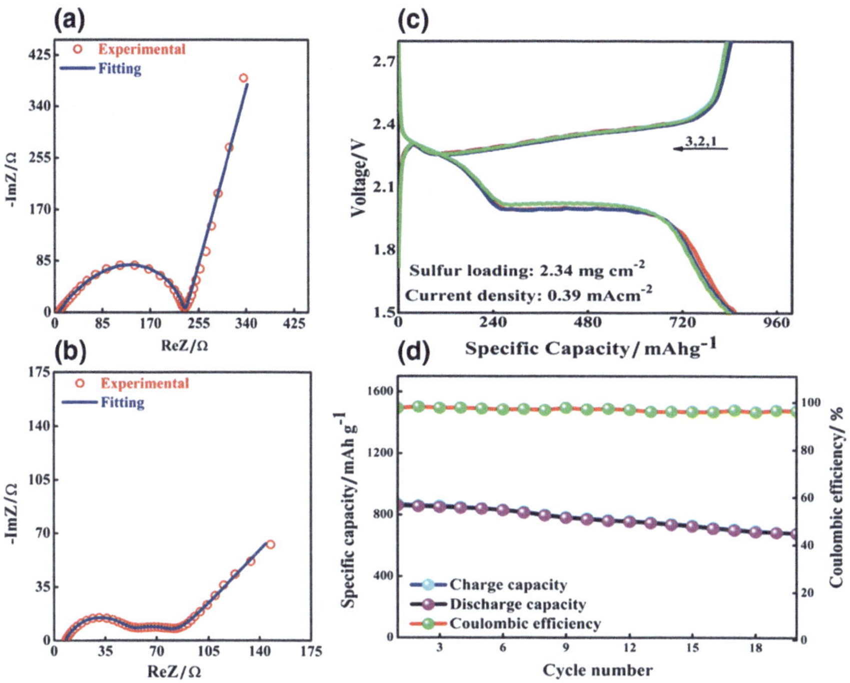

Fig. 11.19 **a** Impedance plot of Li–S cell with Li||LE-PP-LLZT-LE||S configuration before galvanostatic cycling. **b** Impedance plot of Li–S cell with Li||LE-PP-LLZT-LE||S configuration after first 3 galvanostatic charge/discharge cycles. **c** Charge/discharge cycles of Li||LE-PP-LLZT-LE||S at room temperature (25 °C). **d** Specific capacity and efficiency of Li||LE-PP-LLZT-LE||S for 20 cycles. Adapted from [59] with permission from Elsevier

during the initial cycles, indicating that the shuttling of polysulfides could be effectively prevented by the application of LLZT layer on PP separator, as depicted in Fig. 11.20. The LE–PP-D(LLZT) interlayer, therefore, plays a crucial role in localizing the movement of polysulfides and allows smooth Li^+ transport by improving the area of contact with metallic lithium anode, thereby enhancing the electrochemical performance of Li–S cell [59].

11.4 Conclusions

Effective interfacial engineering was performed by deposition of thin layer of Li via thermal evaporation and introduction of Au interface via DC sputtering to mitigate high interface resistance between metallic Li and LLZA solid electrolyte. For the cathode|LLZA interface, an effective strategy was demonstrated to reduce the encountered high interfacial resistance. The Li_2SiO_3 as buffer layer in liquid phase possesses high affinity to completely wet the surfaces of both NMC and LLZA, which

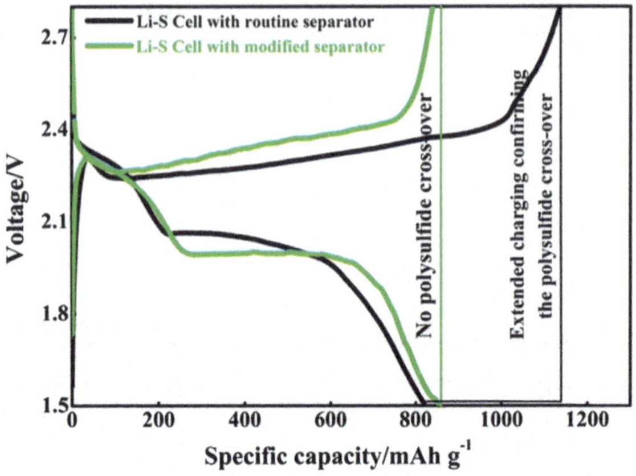

Fig. 11.20 Charge/discharge profiles of conventional Li–S coin cell with normal polypropylene as separator and Li–S coin cell of Li|LE–PP-D(LLZT)|S configuration with LLZT decorated polypropylene as separator

in turn could provide a lithium conduction pathway between NMC and LLZA, thus reducing the interfacial resistance. A *quasi*-solid-state battery with new interface chemistry delivered high capacity and improved coulombic efficiency. An all-solid-state battery constructed with NMC cathode, LLZA pellet as electrolyte and Li metal anode delivered a discharge capacity of 138 mA h g^{-1} for the first galvanostatic cycle and stable cycling performance at 100 °C.

To mitigate the polysulfide diffusion and lithium dendrite penetration, thin-walled LLZT pellet and PP coated with LLZT powder slurry as separators were utilized to fabricate Li–S cell. The thin LLZT membrane restricts the polysulfide diffusion and does not allow the polysulfides to come in contact with lithium anodes, hence mitigating lithium corrosion, while the soft PP interlayer can form an intimate contact with lithium anode and promote the homogeneous Li^+ transport resulting in enhanced capacity and good electrochemical stability of Li–S cell. The hybrid Li–S cell containing high areal loading of active material delivered high initial discharge capacity and displayed improved coulombic efficiency. Thus garnet-structured solid electrolytes with proper interface modification are promising in developing safe high energy density all-solid-state lithium metal and lithium-sulfur batteries. However, further research work is essential to reduce the high interfacial resistance at lithium garnet–cathode interface for the realization of room temperature all-solid-state lithium metal batteries.

References

1. Armand M, Tarascon JM (2008) Building better batteries. Nature 451:652–657
2. Goodenough JB, Kim Y (2009) Challenges for rechargeable Li batteries. Chem Mater 22(3):587–603
3. Kato Y, Hori S, Saito T, Suzuki K, Hirayama M, Mitsui A, Yonemura M, Iba H, Kanno R (2016) High-power all-solid-state batteries using sulfide superionic conductors. Nat Energy 1(4):16030

4. Xu W, Wang J, Ding F, Chen X, Nasybulin E, Zhang Y, Zhang JG (2014) Lithium metal anodes for rechargeable batteries. Energy Environ Sci 7:513–537
5. Kim H, Jeong G, Kim YU, Kim JH, Park CM, Sohn HJ (2013) Metallic anodes for next generation secondary batteries. Chem Soc Rev 42:9011–9034
6. Ramakumar S, Deviannapoorani C, Dhivya L, Shankar L, Murugan R (2017) Lithium garnets: synthesis, structure Li^+ conductivity, Li^+ dynamics and applications. Prog Mater Sci 88:325–411
7. Wu B, Wang S, Evans IV, Deng DZ, Yang J, Xiao J (2016) Interfacial behaviours between lithium ion conductors and electrode materials in various battery systems. J Mater Chem A 4(40):15266–15280
8. Liu Z, Fu W, Payzant EA, Yu X, Wu Z, Dudney NJ, Kiggans J, Hong K, Rondinone AJ, Liang C (2013) Anomalous high ionic conductivity of nanoporous β-Li_3PS_4. J Am Chem Soc 135(3):975–978
9. Luntz AC, Johannes V, Karsten R (2015) Interfacial challenges in solid-state Li ion batteries. J Phys Chem Lett 6(22):4599–4604
10. Thangadurai V, Kaack H, Weppner WJ (2003) Novel fast lithium ion conduction in garnet-type $Li_5La_3M_2O_{12}$ (M = Nb, Ta). J Am Ceramic Soc 86(3):437–440
11. Fergus JW (2010) Ceramic and polymeric solid electrolytes for lithium-ion batteries. J Power Sources 195(15):4554–4569
12. Murugan R, Thangadurai V, Weppner W (2007) Fast lithium ion conduction in garnet-type $Li_7La_3Zr_2O_{12}$. Angew Chem 46(41):7778–7781
13. Han X, Gong Y, Fu KK, He X, Hitz GT, Dai J, Pearse A, Liu B, Wang H, Rubloff G, Mo Y, Thangadurai V, Wachsman ED, Hu L (2017) Negating interfacial impedance in garnet-based solid-state Li metal batteries. Nat Mater 16(5):572
14. Monroe C, Newman J (2005) The impact of elastic deformation on deposition kinetics at lithium/polymer interfaces. J Electrochem Soc 152(2):396–404
15. Ni JE, Case ED, Sakamoto JS, Rangasamy E, Wolfenstine JB (2012) Room temperature elastic moduli and Vickers hardness of hot-pressed LLZO cubic garnet. J Mater Sci 47(23):7978–7985
16. Fu KK, Gong Y, Liu B, Zhu Y, Xu S, Yao Y, Luo W, Wang C, Lacey SD, Dai J, Chen Y, Mo Y, Wachsman ED, Hu L (2017) Toward garnet electrolyte-based Li metal batteries: An ultrathin, highly effective, artificial solid-state electrolyte/metallic Li interface. Sci Adv 3(4):e1601659
17. Luo W, Gong Y, Zhu Y, Li Y, Yao Y, Zhang Y, Fu KK, Pastel G, Lin CF, Mo Y, Wachsman ED, Hu L (2017) Reducing interfacial resistance between garnet-structured solid-state electrolyte and Li-metal anode by a germanium layer. Adv Mater 29:1606042
18. Luo W, Gong Y, Zhu Y, Fu KK, Dai J, Lacey SD, Wang C, Liu B, Han X, Mo Y, Wachsman ED, Hu L (2016) Transition from superlithiophobicity to superlithiophilicity of garnet solid-state electrolyte. J Am Chem Soc 138:12258–12262
19. Fu KK, Gong Y, Fu Z, Xie H, Yao Y, Liu B, Carter M, Wachsman ED, Hu L (2017) Transient behavior of the metal interface in lithium metal–garnet batteries. Angew Chem Int Ed 56:14942–14947
20. Tsai CL, Roddatis V, Chandran CV, Ma Q, Uhlenbruck S, Bram M, Heitjans P, Guillon O (2016) $Li_7La_3Zr_2O_{12}$ interface modification for Li dendrite prevention. ACS Appl Mater Interfaces 8:10617–10626
21. Alexander GV, Patra S, Raj SVS, Sugumar MK, Din MMU, Murugan R (2018) Electrodes–electrolyte interfacial engineering for realizing room temperature lithium metal battery based on garnet structured solid fast Li^+ conductors. J Power Sources 396:764–773
22. Sharafi A, Meyer HM, Nanda J, Wolfenstine J, Sakamoto J (2016) Characterizing the Li–$Li_7La_3Zr_2O_{12}$ interface stability and kinetics as a function of temperature and current density. J Power Sources 302:135–139
23. Cheng L, Crumlin EJ, Chen W, Qiao R, Hou H, Lux SF, Zorba V, Russo R, Kostecki R, Liu Z, Persson K (2014) The origin of high electrolyte–electrode interfacial resistances in lithium cells containing garnet type solid electrolytes. Phys Chem Chem Phys 16(34):18294–18300
24. Suzuki Y, Kami K, Watanabe K, Watanabe A, Saito N, Ohnishi T, Takada R, Sudo R, Imanishi N (2015) Transparent cubic garnet-type solid electrolyte of Al_2O_3-doped $Li_7La_3Zr_2O_{12}$. Solid State Ion 278:172–176

25. Takeda Y, Yamamoto O, Imanishi N (2016) Lithium dendrite formation on a lithium metal anode from liquid, polymer and solid electrolytes. Electrochemistry 84(4):210–218
26. Wang X, Zeng W, Hong L, Xu W, Yang H, Wang F, Duan H, Tang M, Jiang H (2018) Stress-driven lithium dendrite growth mechanism and dendrite mitigation by electroplating on soft substrates. Nat Energy 3:227
27. Ohta S, Komagata S, Seki J, Saeki T, Morishita S, Asaoka T (2013) All-solid-state lithium ion battery using garnet-type oxide and Li_3BO_3 solid electrolytes fabricated by screen-printing. J Power Sources 238:53–56
28. Liu T, Zhang Y, Zhang X, Wang L, Zhao SX, Lin YH, Shen Y, Luo J, Li L, Nan CW (2018) Enhanced electrochemical performance of bulk type oxide ceramic lithium batteries enabled by interface modification. J Mater Chem A 6:4649–4657
29. Shoji M, Munakata H, Kanamura K (2016) Fabrication of all-solid-state lithium-ion cells using three-dimensionally structured solid electrolyte $Li_7La_3Zr_2O_{12}$ pellets. Front Energy Res 4:32
30. Liu T, Ren Y, Shen Y, Zhao SX, Lin Y, Nan CW (2016) Achieving high capacity in bulk-type solid-state lithium ion battery based on $Li_{6.75}La_3Zr_{1.75}Ta_{0.25}O_{12}$ electrolyte: interfacial resistance. J Power Sources 324:349–357
31. Kato T, Hamanaka T, Yamamoto K, Hirayama T, Sagane F, Motoyama M, Iriyama Y (2014) In-situ $Li_7La_3Zr_2O_{12}/LiCoO_2$ interface modification for advanced all-solid-state battery. J Power Sources 260:292–298
32. Ohta N, Takada K, Sakaguchi I, Zhang L, Ma R, Fukuda K, Osada M, Sasaki T (2007) $LiNbO_3$-coated $LiCoO_2$ as cathode material for all solid-state lithium secondary batteries. Electrochem Commun 9(7):1486–1490
33. Ohta N, Takada K, Zhang L, Ma R, Osada M, Sasaki T (2006) Enhancement of the high-rate capability of solid-state lithium batteries by nanoscale interfacial modification. Adv Mater 18(17):2226–2229
34. Xu X, Takada K, Fukuda K, Ohnishi T, Akatsuka K, Osada M, Hang BT, Kumagai K, Sekiguchi T, Sasaki T (2011) Tantalum oxide nanomesh as self-standing one nanometer thick electrolyte. Energy Environ Sci 4(9):3509–3512
35. Shao Y, Wang H, Gong Z, Wang D, Zheng B, Zhu J, Lu Y, Hu YS, Guo X, Li H, Huang X, Yang Y, Nan CW, Chen L (2018) Drawing a soft interface: an effective interfacial modification strategy for garnet-type solid-state Li batteries. ACS Energy Lett 3(6):1212–1218
36. Pang Q, Kundu D, Cuisinier M, Nazar LF (2014) Surface-enhanced redox chemistry of polysulphides on a metallic and polar host for lithium–sulphur batteries. Nat Commun 5:4759
37. See KA, Jun YS, Gerbec JA, Sprafke JK, Wudl F, Stucky GD, Seshadri R (2014) Sulfur-functionalized mesoporous carbons as sulfur hosts in Li–S batteries: increasing the affinity of polysulfide intermediates to enhance performance. ACS Appl Mater Interfaces 6(14):10908–10916
38. Liang L, Nazar LF (2016) In situ reactive assembly of scalable core–shell sulfur–MnO_2 composite cathodes. ACS Nano 10(4):4192–4198
39. Jin B, Kim JU, Gu HB (2003) Electrochemical properties of lithium–sulphur batteries. J Power Sources 117(1–2):148–152
40. Wang H, Yang Y, Liang Y, Robinson JT, Li Y, Jackson A, Cui Y, Dai H (2011) Graphene-wrapped sulphur particles as a rechargeable lithium–sulphur-battery cathode material with high capacity and cycling stability. Nano Lett 11(7):2644–2647
41. Zhou W, Yu Y, Chen H, DiSalvo FJ, Abruña HD (2013) Yolk–shell structure of polyaniline-coated sulphur for lithium–sulphur batteries. J Am Chem Soc 135(44):16736–16743
42. Wang C, Chen H, Dong W, Ge J, Lu W, Wu X, Guo L, Chen L (2014) Sulfur–amine chemistry-based synthesis of multi-walled carbon nanotube–sulfur composites for high performance Li–S batteries. Chem Commun 50(10):1202–1204
43. Fu Y, Manthiram A (2012) Core–shell structured sulphur–polypyrrole composite cathodes for lithium–sulphur batteries. RSC Adv 2(14):5927–5929
44. Liang C, Dudney NJ, Howe JY (2009) Hierarchically structured sulphur/carbon nanocomposite material for high-energy lithium battery. Chem Mater 21(19):4724–4730

45. Jayaprakash N, Shen J, Moganty SS, Corona A, Archer LA (2011) Porous hollow carbon@sulphur composites for high-power lithium–sulphur batteries. Angew Chem 50(26):5904–5908
46. Zhang B, Qin X, Li GR, Gao XP (2010) Enhancement of long stability of sulphur cathode by encapsulating sulphur into micropores of carbon spheres. Energy Environ Sci 3(10):1531–1537
47. Zeng T, Hu X, Ji P, Shang B, Peng Q, Zhang Y, Song R (2017) Promotional role of $Li_4Ti_5O_{12}$ as polysulfide adsorbent and fast Li^+ conductor on electrochemical performances of sulfur cathode. J Power Sources 359:250–261
48. Li GC, Jing HK, Su Z, Lai C, Chen L, Yuan CC, Li HH, Liu L (2015) A hydrophilic separator for high performance lithium–sulfur batteries. J Mater Chem A 3(20):11014–11020
49. Manthiram A, Fu Y, Chung SH, Zu C, Su YS (2014) Rechargeable lithium–sulphur batteries. Chem Rev 114(23):11751–11787
50. Zhang Z, Lai Y, Zhang Z, Zhang K, Li J (2014) Al_2O_3-coated porous separator for enhanced electrochemical performance of lithium sulfur batteries. Electrochim Acta 129:55–61
51. Yu X, Bi Z, Zhao F, Manthiram A (2015) Hybrid lithium–sulfur batteries with a solid electrolyte membrane and lithium polysulfide catholyte. ACS Appl Mater Interfaces 7(30):16625–16631
52. Fu KK, Gong Y, Xu S, Zhu Y, Li Y, Dai J, Wang C, Liu B, Pastel G, Xie H, Yao Y (2017) Stabilizing the garnet solid-electrolyte/polysulfide interface in Li–S batteries. Chem Mater 29(19):8037–8041
53. Fu KK, Gong Y, Li Y, Xu S, Wen Y, Zhang L, Wang C, Pastel G, Dai J, Liu B, Xie H (2017) Three-dimensional bilayer garnet solid electrolyte based high energy density lithium metal–sulfur batteries. Energy Environ Sci 10(7):1568–1575
54. Furusawa SI, Kasahara T, Kamiyama A (2009) Fabrication and ionic conductivity of Li_2SiO_3 thin film. Solid State Ionics 180(6–8):649–653
55. Tietz F, Wegener T, Gerhards M, Giarola M, Mariotto G (2013) Synthesis and raman microspectroscopy investigation of $Li_7La_3Zr_2O_{12}$. Solid State Ion 230:77–82
56. Dhivya L, Murugan R (2014) Effect of simultaneous substitution of Y and Ta on the stabilization of cubic phase, microstructure, and Li^+ conductivity of $Li_7La_3Zr_2O_{12}$ lithium garnet. ACS Appl Mater Interfaces 6(20):17606–17615
57. Alexander GV, Rosero-Navarro NC, Miura A, Tadanaga K, Murugan R (2018) Electro chemical performance of a garnet solid electrolyte based lithium metal battery with interface modification. J Mater Chem A 6(42):21018–21028
58. Janani N, Ramakumar S, Kannan S, Murugan R (2015) Optimization of lithium content and sintering aid for maximized Li^+ conductivity and density in Ta-doped $Li_7La_3Zr_2O_{12}$. J Am Ceramic Soc 98(7):2039–2046
59. Din MMU, Murugan R (2018) Garnet structured solid fast Li^+ conductor as polysulfide shuttle inhibitor in Li–S battery. Electrochem Commun 93:109–113

Part III
Solid-State Batteries

Chapter 12
Grain Boundary Engineering for High Short-Circuit Tolerance

Rajendra Hongahally Basappa and Hirotoshi Yamada

Abstract Dendritic lithium plating is one of the critical issues that hinder practical application of lithium metal anode on secondary batteries. Even dense ceramic solid electrolytes, which have been expected to show high resistance on lithium penetration, suffer from short-circuit. In this work, to suppress the lithium penetration through voids along grain boundaries, garnet-type solid electrolytes of $Li_{6.5}La_3Zr_{1.5}Ta_{0.5}O_{12}$ (LLZT) were sintered with grain boundaries modified with Li_2CO_3 or LiOH. Although all prepared solid electrolyte pellets showed relative density higher than 96% and high ionic conductivity of 7×10^{-4} S cm^{-1}, critical current densities (CCD) depended on grain boundary phases. While CCD of LLZT without modification was 0.2 mA cm^{-2}, Li_2CO_3 and LiOH modified LLZT pellets exhibited CCD of 0.4 and 0.6 mA cm^{-2}, respectively. This is possibly caused by the two effects of modification phases as supported by microstructure analyses using SEM, STEM, and EIS. First, Li_2CO_3 and LiOH remaining on the grain boundary plug voids and protect lithium growth. Secondly, Li_2CO_3 and LiOH played roles of sintering agent and facilitated densification of grain boundaries. The obtained results indicate a strategy toward practical application of lithium metal batteries.

12.1 Introduction

High performance energy storage devices are one of the key technologies to reduce dependence on fossil fuels. Among various types of energy storage devices, rechargeable batteries using lithium-metal anode are attractive due to its high theoretical capacity, and thus, by coupling with high-capacity cathodes such as sulfur or oxygen, high energy density can be achieved [1–4]. However, it is well known that during lithium plating in liquid electrolytes, lithium metal is dendritically deposited. After repetitive dendritic growth of lithium metal on charging processes, finally lithium reaches to cathode, resulting in internal short-circuit. This issue hinders practical application of lithium metal anode in rechargeable batteries. Toward lithium metal

R. Hongahally Basappa · H. Yamada (✉)
Graduate School of Engineering, Nagasaki University, Nagasaki, Japan
e-mail: h-yama@nagasaki-u.ac.jp

R. Murugan and W. Weppner (eds.), *Solid Electrolytes for Advanced Applications*,
https://doi.org/10.1007/978-3-030-31581-8_12

batteries, inorganic solid electrolytes (ISEs) are anticipated to effectively suppress dendritic growth of lithium and improve the safety of next-generation energy storage devices [5]. On the development of practical lithium metal batteries, various properties are required to ISEs: high lithium ionic conductivity, high electrochemical stability against lithium metal, and high mechanical strength that protect lithium penetration through the solid electrolytes. According to a model proposed by Monroe and Newman, high shear modulus of ISEs (typically more than two times higher than lithium metal) would decrease the possibility of lithium growth in ISEs [6].

After a long history of development of solid electrolytes, a new class of lithium ion conductors with garnet-type framework emerged which clears the above mentioned criteria. The garnet-type ISE with a standard composition of $Li_7La_3Zr_2O_{12}$ (LLZ) shows high ionic conductivity of 3×10^{-4} S cm^{-1} at room temperature (RT) and electrochemical stability against lithium metal [7, 8]. In addition, the shear modulus of the garnet-type ISE is reported to be as high as ca. 55 GPa [9, 10].

The emergence of the garnet-type solid electrolytes attracted the interests of many researchers, but they soon found that short-circuit occurs due to lithium penetration even when they use the garnet-type ISE. Imanishi et al. reported short-circuit of symmetric cells consisting of Li|doped-LLZ|Li during direct current (DC) [11–13]. The solid electrolytes that were used were prepared by a conventional solid-state sintering process and exhibited rather high relative density of 92.8–96.7%. In spite of high relative density, they short-circuited in 1000 s at a current density of 0.5 mA cm^{-2}. They also confirmed positive correlation of relative density to lithium penetration through Ta-doped LLZ (LLZT) [14], which suggests that lithium grows through open pores on grain boundaries in LLZT.

To suppress the lithium growth, several efforts have been made. Suzuki et al. prepared a highly densified Al-doped LLZ (Al-LLZ) pellet with a relative density of 99.1% by conventional solid-state sintering (1180 °C, 36 h) followed by hot isostatic press (1160 °C, 127 MPa, 2 h) [15]. Some parts of the pellet were transparent, which exhibited high tolerance against lithium growth. Cell voltage did not drop at 0.5 mA cm^{-2} up to 1000 s, after which cell voltage increased because of all lithium was stripped on one side. Cheng et al. investigated the influence of microstructure on critical current density (current density at which short-circuit occurred) [16]. The Al-LLZ with smaller grains showed higher critical current density than that with larger grains. Tsai et al. reported the effect of interfacial layers. They inserted Au thin layers between LLZT and Li metal electrodes [17], which suppressed the lithium growth. Au layers homogeneously deposited by sputtering on LLZT surface forms alloy with Li, and thus, contact between lithium and LLZT is more homogeneous than interfaces without Au layers. Therefore, it is supposed that the homogeneity of current across the interfaces of electrodes–solid electrolyte is one of the key factors to improve the critical current density.

Critical current density also depends on preparation methods of the garnet-type ISEs. Although hot-pressing of Al-LLZ results in high relative density of 97%, its critical current density was rather low (0.05 mA cm^{-2} at 30 °C) [18]. On the contrary, in our previous works, LLZT pellets sintered by spark plasma sintering (SPS, also known as field-assisted sintering technique: FAST) exhibited rather high critical current density of 0.2 mA cm^{-2} [19], which were further enhanced to 0.4 mA cm^{-2} by employing vacuum-evaporation deposited lithium electrodes [20]. The rather high critical current densities for the SPS-processed pellets may be caused by the plastic deformation of grains during sintering [19].

Although the critical current density has been gradually improved, even the best value cannot satisfy the required values for the practical application. In the case of conventional lithium-ion batteries, area-specific capacity is at least 2–3 mAh cm^{-2} (this corresponds to a composite cathode (~50 μm in thickness, containing ~80 vol% of $LiCoO_2$). To charge/discharge these batteries at 1 C-rate, current density of 2–3 mA cm^{-2} is applied. ASSBs are also required to be operated at least up to this current density. Therefore, it is critical to further improve the critical current density of ISE for the practical lithium-metal batteries.

Taking into account the works reported so far, it is supposed that the short-circuit is mainly caused by lithium growth through interconnected open pores, which is facilitated by the inhomogeneous distribution of local current density. However, the reported pellets exhibit rather high relative density of 92–99% and it would be difficult to further densify. Therefore, another approach is necessary to completely eliminate the interconnected open pores. In this study, we prepared LLZT pellets with modified grain boundaries [20]. Such grain boundary structure was obtained by SPS using LLZT powders with surface coating layers of Li_2CO_3, LiOH or nothing. In conventional solid-state sintering operated at high temperature for long time, these surface layers are decomposed and do not remain. On the other hand, SPS is operated at relatively low temperature and in short duration time. Therefore, these surface phases remain on the grain boundary of the pellets.

12.2 Experimental Section*

12.2.1 Materials and Preparation of LLZT Pellets

LLZT powders with a composition of $Li_{6.5}La_3Zr_{1.5}Ta_{0.5}O_{12}$ were prepared by a solid-state reaction and subsequent SPS, as reported in our previous works [19–21]. Stoichiometric amounts of $LiOH{\cdot}H_2O$ (Kishida Chemical Co. Ltd., 99.0%), La_2O_3 (Kishida Chemical Co. Ltd., 99.99%), ZrO_2 (Kishida Chemical Co. Ltd., 99.9%) and Ta_2O_5 (Wako Pure Chemical Industries Ltd., 99.0%) were used as starting materials to synthesize LLZT. Excess lithium by 10% was added in order to compensate potential loss during a high temperature process. $LiOH{\cdot}H_2O$ and La_2O_3 were heated at 200 and 800 °C in advance, respectively, to remove H_2O and CO_2 from those materials. The mixture of the starting materials was calcined at 900 °C for 12 h. The synthesized powders were pelletized with a graphite die and punches by the SPS technique at 950 °C for 10 min with uniaxial pressure of 25 MPa under reduced

*Most of Experimental section is reproduced with permission from [20]. Copyright 2017 Elsevier B.V.

chamber pressure (<5 Pa). Graphite sheets were inserted between the die and the LLZT powder. After naturally cooling down to room temperature, the pellets were taken out of the die and the graphite sheets were removed by grinding the surface of the pellets. The obtained pellets were typically 10 mm and 0.8–0.9 mm in diameter and thickness, respectively.

In this work, three types of LLZT pellets were prepared with the same procedures and differently modified powders. The first specimen was handled in air except for the SPS, which is called as LLZT-air. Owing to the strong reactivity of LLZT with moisture and CO_2, the surface of LLZT particles is covered with Li_2CO_3. The second specimen was prepared without exposure to air, which is named as LLZT-w/o-air. The calcination was conducted in pure O_2 using a furnace directly connected to an Ar-filled glovebox (O_2: 0.1 ppm, $H_2O < 1$ ppm). The calcined powder was directly transferred to the glovebox and was put into the graphite die for the SPS. Subsequent processes (removal of graphite sheets and polishing) were also carried out in the glovebox. The last one, named as LLZT-2calc, was prepared using LLZT powder calcined twice. After the first calcination, LiOH that corresponds to 0.65 Li in LLZT using a planetary ball-mill with a ZrO_2 pot and balls (3 mm in diameter, ball-powder ratio was 1:1 by volume) at 300 rpm for 1 h. Then, the mixture was calcined again with the same condition.

12.2.2 Structural Characterization

The relative density of the pellets was calculated as the ratio of experimental density to the theoretical density. The experimental density was estimated from weight and geometric volume (i.e. thickness × area) of the pellets, while the theoretical density was obtained from the lattice parameters determined by X-ray diffraction (XRD) and the mole weight for the composition $Li_{6.5}La_3Zr_{1.5}Ta_{0.5}O_{12}$. The SPS-processed pellets exhibited a relative density of 96.6, 95.8, and 96.4% for LLZT-air, LLZT-w/o-air, and LLZT-2calc, respectively. These relative densities are averaged values of three different pellets for each specimen. Crystalline phases of the LLZT powders and pellets were investigated by XRD on a SmartLab (Rigaku Corp.) equipped with an X-ray source of Cu Kα. The sintered pellets were characterized (45 kV, 200 mA) using Ar-filled sample holders covered with a polyimide film. Microstructure of the powders and the pellets were studied by using a scanning electron microscope (SEM-5100 M, JEOL Ltd.) and a scanning transmission electron microscope (STEM; JEM-ARM200F, JEOL Ltd.). To observe the microstructure of inner parts of the LLZT pellets, the pellets were thinned by dimple-grinding and subsequent Ar-ion milling. Element distribution on grain boundaries of pellets was studied by energy dispersive X-ray spectroscopy (EDS) on a STEM. For the STEM-EDS observation, thinned specimens were prepared from the LLZT pellets using focused ion beam (FIB; JIB-4501, JEOL Ltd.). Fourier transform infrared spectroscopy (FT-IR) was carried out using an IR-Prestage21 (Shimadzu Corp.) to obtain information of amorphous or poorly crystalline phases in the LLZT powders. The spectra were recorded and were in the range from 400 to 4000 cm^{-1}. Elemental analysis of LLZT powders and pellets was conducted by inductively coupled plasma—optical emission spectroscopy (ICP-OES, ULTIMA2, HORIBA Ltd.).

12.2.3 Electrochemical Analysis

Ionic conductivity of the LLZT pellets was obtained from AC impedance measurement using HP4192A (Keysight Technologies) in the frequency range from 10 MHz to 10 Hz with an oscillation voltage of 10 mV. Au layers with a thickness of ca. 300 nm were deposited by sputtering on both sides of the pellets. The pellets were sandwiched by Au-covered Cu electrodes and placed in a glass tube in which Ar flowed. Temperature of the glass tube was controlled by an oven from 110 to −15 °C with a slow cooling rate of 0.25 °C/min, which was slow enough to reach quasi-equilibrated state. The obtained impedance data were analyzed using the Z-View software (Scribner Associates Inc.).

To investigate the short-circuit tolerance, DC polarization was applied to symmetric cells of Li|LLZT|Li. First, surfaces of the pellets were ground to remove inert layers such as Li_2CO_3 using emery paper (grit number #400). Next, lithium metal layer (5 mm and 2 μm in diameter and thickness, respectively) was deposited by a vacuum evaporation method on both sides of the pellets. Next, lithium foils with a diameter of 5 mm were attached to the lithium thin layers. Finally, the symmetric cells were put in gas-tight stainless-steel vessels, in which a constant pressure of 1.4 MPa was applied to the Li electrodes by a spring. All of these processes were carried out in the Ar-filled glovebox. Further details of the symmetric cell of Li|LLZT|Li was described in our previous work [21]. Before the DC polarization, the vessels were heated at 175 °C for 2 h to ensure good contact between Li foils and LLZT. Charge transfer resistance (R_{CT}) at room temperature was analyzed using a potentio-galvanostat (Solartron Analytical, S1470E) equipped with a frequency response analyzer (Solarton Analytical, S1255B) in the frequency range from 500 kHz to 0.1 Hz with an oscillation voltage of 10 mV. DC polarization was conducted on the symmetric cells to obtain the critical current density at which short-circuit occurs, in the following procedure. First, DC with a constant current density of 0.05 mA cm^{-2} was applied for 30 min. Then, the same current density with opposite polarity was applied for 30 min between each galvanostatic process, 10-min rest, and subsequent AC impedance was inserted. The set of DC polarization, rest and impedance were repeated five times, and the same procedure was conducted with progressively increasing current density until cell voltage dropped, that is, short-circuit happened. After the DC cycling, the cells were disassembled in the glovebox, and the lithium foils were removed from the short-circuited LLZT pellets.

12.3 Results and Discussion

12.3.1 Phase and Morphology Analysis of Solid Electrolytes

As shown in Fig. 12.1, all diffraction peaks of the LLZT powders and pellets were attributed to the garnet-type structure. Amorphous phases and/or poorly crystallized phases in the LLZT powders, FT-IR was carried out (Fig. 12.2). For LLZT-air, strong absorption peaks at 1429 and 1495 cm^{-1} are ascribed to ν_3 mode vibration of trigonal planar CO_3^{2-}. The other two peaks at 866 and 1088 cm^{-1} are ν_2 and ν_1 modes of CO_3^{2-}, respectively. In addition, OH bond in the garnet-type structure is also indicated by a broad middle peak at 3537 cm^{-1} [22]. These results indicate H^+/Li^+ ion exchange and following reaction with CO_2 to form Li_2CO_3 for LLZT-air [23, 24]. On the contrary, Li_2CO_3 in LLZT-w/o-air is drastically suppressed. The formation of Li_2CO_3 on LLZT-w/o-air may indicate the strong sensitivity of LLZT to H_2O and CO_2. LLZT-2calc also contained less Li_2CO_3, while it exhibited sharp vibration peaks of OH at 3564 cm^{-1} (free OH) and 3674 cm^{-1} (hydrogen bonded OH) [22]. In addition, a small peak at 1572 cm^{-1} is attributed to ν_2 mode of a water molecule. Therefore, LLZT-2calc is supposed to be covered with dominantly $LiOH{\cdot}H_2O$. The XRD profiles did not confirm Li_2CO_3 nor $LiOH{\cdot}H_2O$, which suggests that these phases are amorphous and/or poorly crystallized.

Figure 12.3 shows the composition of all LLZT powders and pellets estimated by ICP-OES analysis. The compositions of all the specimens were the same. The same compositions for LLZT-air and LLZT-w/o-air are because of the same compositions of the starting materials. On the contrary, for LLZT-2calc, extra LiOH was added before the second calcination. However, the Li composition of LLZT-2calc was almost the same as those of others. It is supposed that most of the excess Li component

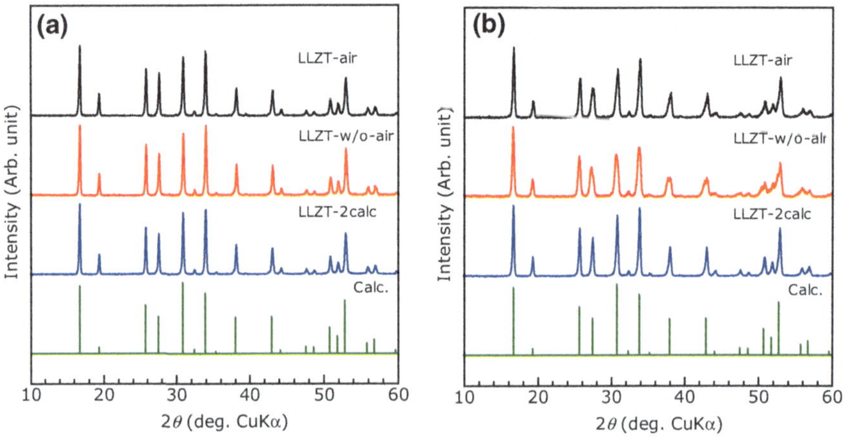

Fig. 12.1 XRD profiles of **a** powders and **b** pellets of LLZT-air, LLZT-w/o-air, and LLZT-2calc. Reproduced with permission from [20]. Copyright 2017 Elsevier B.V.

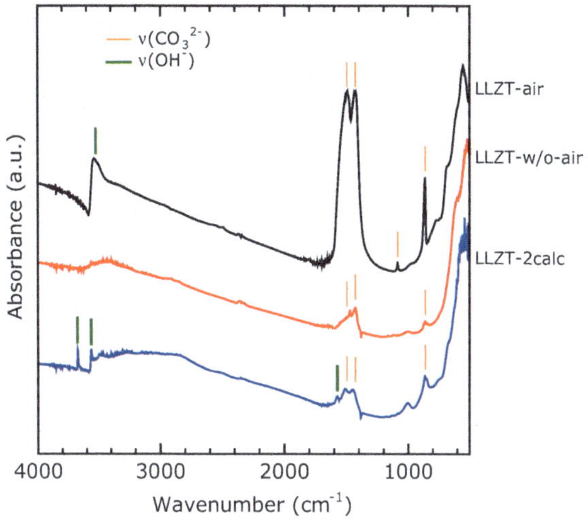

Fig. 12.2 FT-IR spectra of LLZT powders prepared in different conditions. Reproduced with permission from [20]. Copyright 2017 Elsevier B.V.

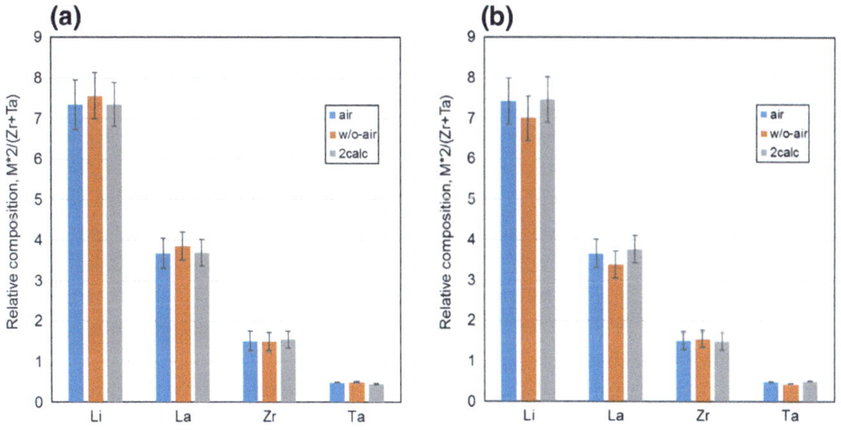

Fig. 12.3 Composition of **a** LLZT powders and **b** LLZT pellets, obtained by ICP-OES. Relative composition was obtained by assuming that the sum of Zr and Ta is equal to 2. Reproduced with permission from [20]. Copyright 2017 Elsevier B.V.

on the LLZT-2calc evaporated and decomposed during the second calcination, and only little amount of $LiOH{\cdot}H_2O$ remains on the LLZT particles. From the thermal gravimetric analysis of $LiOH{\cdot}H_2O$ (see Appendix I), the following process on the second calcination is proposed. First, LiOH is decomposed to Li_2O. Next, Li_2O (and remaining LiOH) reacts with CO_2 in atmosphere to form Li_2CO_3, but most of Li_2CO_3 is decomposed to Li_2O above Li_2O above 700 °C. On the other hand, Li_2CO_3

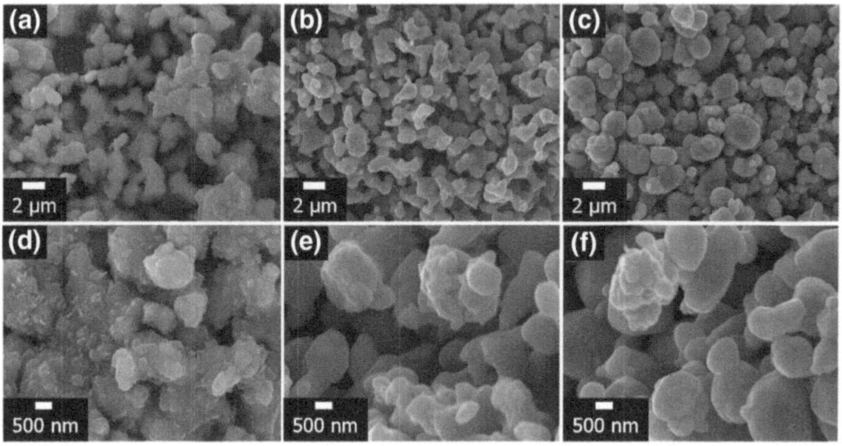

Fig. 12.4 SEM images of LLZT powders: **a** and **d** LLZT-air, **b** and **e** LLZT-w/o-air, and **c** and **f** LLZT-2calc. Reproduced with permission from [20]. Copyright 2017 Elsevier B.V.

evaporates, which includes reverse reaction of the equilibrium: $Li_2CO_3 \rightleftarrows Li_2O + CO_2$. Through these reactions, extra Li is decreased from the system, and only trace of Li_2O remains on the LLZT particles. Finally, Li_2O absorbs H_2O to produce LiOH and $LiOH{\cdot}H_2O$ again below 100 °C.

The morphology of the LLZT powders is shown in Fig. 12.4. Particle size of LLZT-air and LLZT-w/o-air is almost monodispersed around ca. 1 μm, while LLZT-2calc contained slightly larger particles (ca. 2 μm in diameter) as well as 1-μm particles. From the density of LLZT and LiOH, the volume fraction of additional LiOH is ca. 7.3%, which cannot account for the large grains. Therefore, it is supposed that LiOH partly facilitates the grain growth of LLZT particles during the second calcination. The surface of LLZT-air particles is covered with small plates, which would be Li_2CO_3 crystalline. On the other hand, the other powders consist of round-shaped smooth particles.

As for the pellets, in spite of the almost same relative density for the three specimens (95.8–96.6%), the pellets look differently (see Fig. 12.5): LLZT-air and LLZT-w/o-air are opaque and white, while LLZT-2calc is translucent. The translucence indicates less scattering in the pellets.

Figure 12.6 demonstrates SEM images of surface of the LLZT pellets after the surface was grinded by at least 0.1 mm and then mirror-polished. In the images, several voids are confirmed as indicated by arrows. Although it is impossible to know whether each void is interconnected or not, the images clearly confirm that LLZT-w/o-air contains more voids than the others, which cause higher possibility of lithium growth through the voids. In the cross-section images of the pellets (Fig. 12.7), not only grains but also continuous regions are confirmed in all specimens.

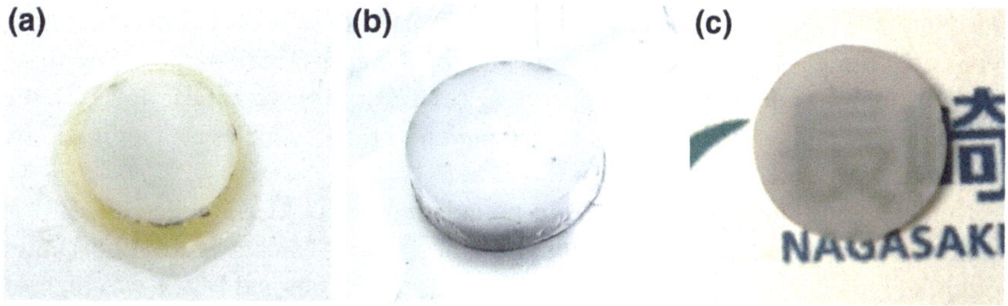

Fig. 12.5 Optical images of LLZT pellets. **a** LLZT-air, **b** LLZT-w/o-air, and **c** LLZT-2calc. Reproduced with permission from [20]. Copyright 2017 Elsevier B.V.

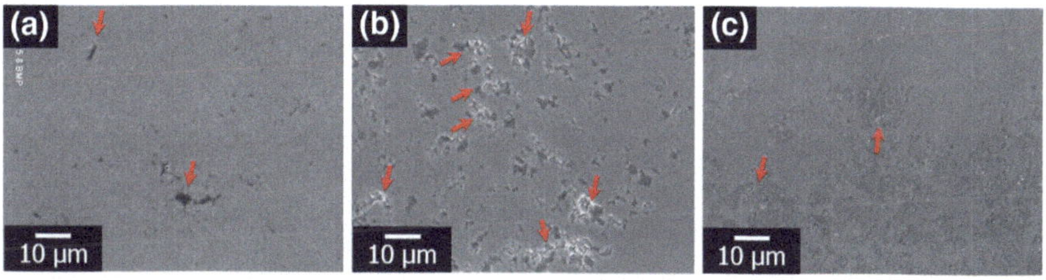

Fig. 12.6 SEM images of mirror-polished surface of LLZT pellets: **a** LLZT-air, **b** LLZT-w/o-air, and **c** LLZT-2calc. Reproduced with permission from [20]. Copyright 2017 Elsevier B.V.

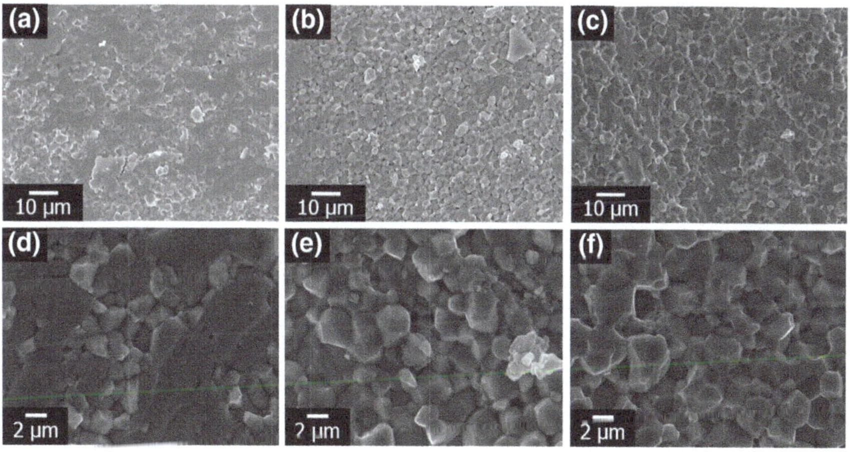

Fig. 12.7 SEM images of fractured cross-section of LLZT pellets: **a** and **d** LLZT-air, **b** and **e** LLZT-w/o-air, and **c** and **f** LLZT-2calc. Reproduced with permission from [20]. Copyright 2017 Elsevier B.V.

The continuous regions were wider for LLZT-air. The continuous region is supposed to be Li_2CO_3 and/or LiOH, which initially covered the surface of particles. As these materials melt at relatively low temperature (Li_2CO_3: 700 °C, LiOH: 462 °C), the surface layers melt and aggregate at voids among grains, which was facilitated by the high temperature and pressure of the SPS.

To understand the structure of grain boundaries of the LLZT pellets, STEM is conducted to observe three-dimensional connection of voids. Figure 12.8 shows bright field images of inner parts of LLZT pellets (LLZT-air and LLZT-w/o-air). At the low magnification (Fig. 12.8a, c), bright lines along open grain boundaries are observed, which is probably formed during the thinning processes (dimple-grinding and Ar-ion milling). Except for the open grain boundaries, the STEM images of LLZT-air (Fig. 12.8a, b) consist of two regions with different brightness: dark regions and translucent (bright) regions. As the brightness of the STEM bright field images

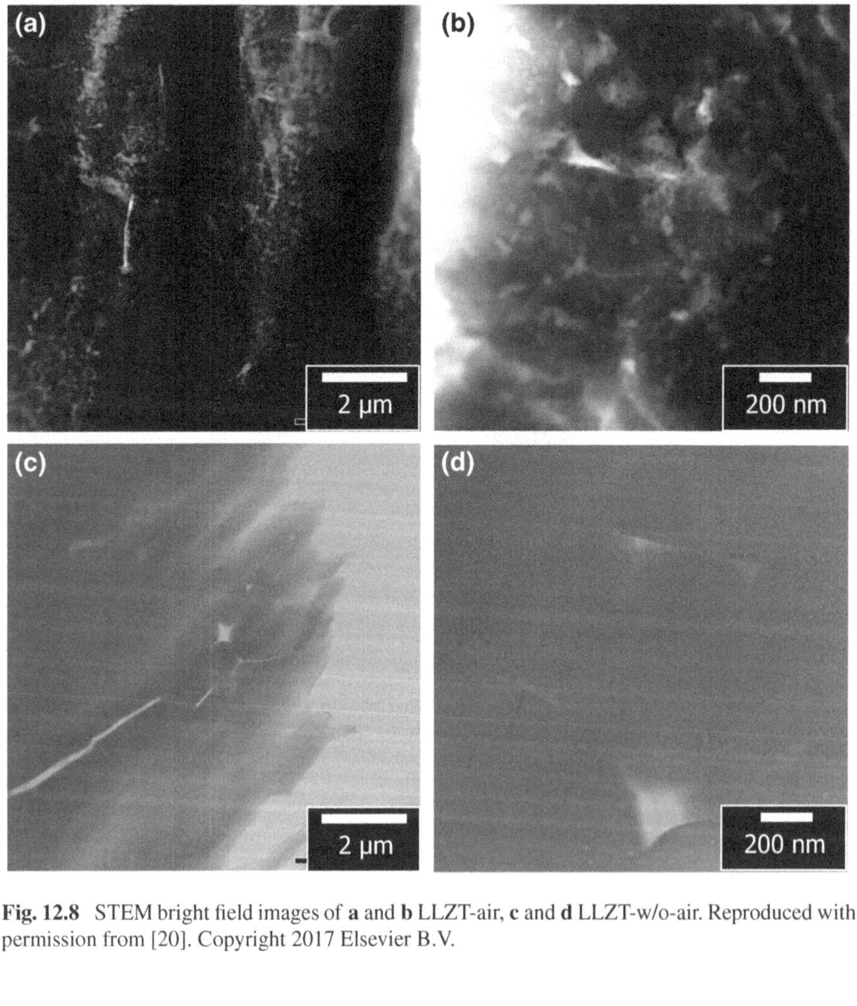

Fig. 12.8 STEM bright field images of **a** and **b** LLZT-air, **c** and **d** LLZT-w/o-air. Reproduced with permission from [20]. Copyright 2017 Elsevier B.V.

is related to the number of electrons of materials, the dark regions are attributed to LLZT grains, and the translucent regions consist of lighter elements. The translucent regions seem to fill the grain boundaries, which is possibly Li_2CO_3. On the other hand, for LLZT-w/o-air, several triangular or square voids are observed but no translucent regions were confirmed (Fig. 12.8c, d). It is obvious that these polygonal voids are on the grain boundaries of three or four grains, and some of the voids are interconnected (see Fig. 12.8d). These voids are possibly thermal etching grooves formed on the grain boundaries [25]. On thermal equilibrium, dihedral angle (ϕ) on a grain boundary is co-related to a ratio of interfacial energy on the grain boundary (γ_{gb}) to interfacial energy between solid and vapor phase (γ_{SV}). These parameters satisfy the following equation:

$$\gamma_{gb} = 2\gamma_{SV} \cos \frac{\phi}{2} \tag{12.1}$$

Small ϕ is given when γ_{gb} is large and/or γ_{SV} is small. In this case, solid–vapor interface gets wider while grain boundary contact area becomes smaller. This results in interconnected pores along triple junction grain boundaries. The critical dihedral angle at which the interconnected pores are formed is 60°. When ϕ is smaller than 60°, voids are interconnected along triple junction grain boundaries [25]. For LLZT-air, it is supposed that molten Li_2CO_3 fills the voids and (12.1) can be applied by replacing γ_{SV} with interfacial energy between solid (LLZT) and liquid phases (Li_2CO_3). In Fig. 12.8b, Li_2CO_3 (the liquid phase above 700 °C) seems to be interconnected with smaller ϕ. However, the voids are filled with Li_2CO_3, which would probably hinder the lithium penetration.

Figure 12.9 demonstrates the STEM-EDS mapping images around grain boundaries of LLZT-air. The STEM dark-field image (top left) shows grain boundaries among three grains. One grain boundary is obviously open (from center to top left of the image), which may be opened during the thinning process of the specimen (polishing and/or FIB). Here, another grain boundary running from the center to the bottom is focused. Note that the STEM image clearly demonstrates an interphase between the two grains, while EDS map images show constant element distribution for Zr, La, and Ta around this grain boundary. This suggests that LLZT is newly deposited on the grain boundary. On the other hand, carbon is intensely detected around the grain boundary. These results suggest that LLZT-air is sintered via a process with liquid phase [25, 26].

Around another grain boundary horizontally running from the center to the right of the image, La, Zr, and Ta seem relatively enriched. A pyrochlore-type $La_2Zr_2O_7$ crystalline (with Ta doping) is possibly produced. This phase is probably produced during sintering of the LLZT-air particles, on which Li-poor region exists due to the ion exchange of Li and H.

12.3.2 Ionic Conductivity

Ionic conductivity of each pellet was obtained from AC impedance measurements. The Nyquist plots recorded at 25 and −15 °C are shown in Fig. 12.10. At 25 °C, the plots consist of part of a semi-circle above 10^6 Hz and a straight line for a low

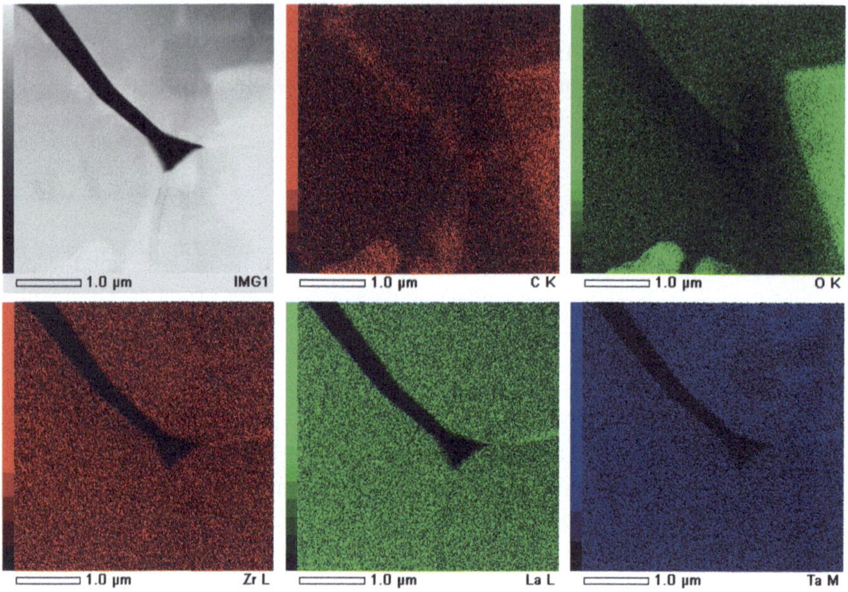

Fig. 12.9 STEM dark-field image and EDS map images of LLZT-air pellet. Reproduced with permission from [20]. Copyright 2017 Elsevier B.V.

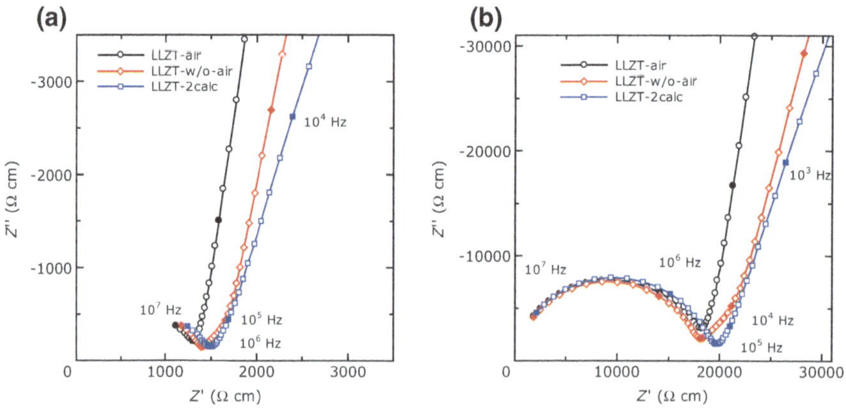

Fig. 12.10 Nyquist plots of LLZT pellets recorded at **a** 25 °C and **b** −15 °C. Reproduced with permission from [20]. Copyright 2017 Elsevier B.V.

frequency range. Table 12.1 summarizes total ionic conductivity and activation energy of the LLZT pellets. The three specimens showed similar results at 25 °C, while decreasing temperature made the difference clear. At −15 °C, a trace of another semi-circle is confirmed as change in the slope of the straight line in the frequency range from 10^5 to 10^3 Hz (Fig. 12.10b). For detailed analysis, the Nyquist plots below −5 °C are fitted using an equivalent circuit shown in Fig. 12.11a. With decreasing temperature, the semi-circle became obvious. In addition, the change in the slope of the straight line can be seen between 10^5 and 10^3 Hz. This possibly indicates additional semi-circle (Fig. 12.10b). To investigate the origin of this semi-circle, the Nyquist plots recorded below −5 °C are fitted using an equivalent

Table 12.1 Relative density and ionic conductivity of LLZT pellets (averaged value of three pellets)

	Relative density (%)	Total ionic conductivity @25 °C (S cm^{-1})	Activation energy (eV)
LLZT-air	96.6	7.2×10^{-4}	0.458
LLZT-w/o-air	95.8	6.6×10^{-4}	0.452
LLZT-2calc	96.4	7.6×10^{-4}	0.459

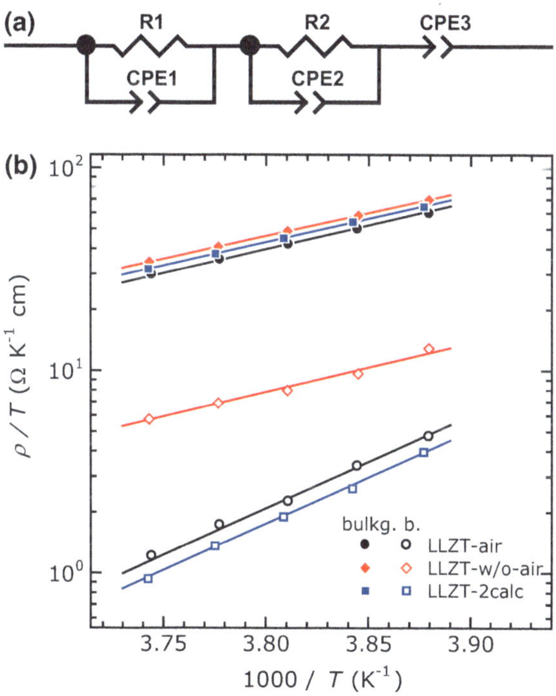

Fig. 12.11 **a** Equivalent circuit used for fitting Nyquist plots of LLZT pellets recorded below −5 °C. **b** Arrhenius plots of resistivity of bulk and grain boundary of LLZT pellets. Reproduced with permission from [20]. Copyright 2017 Elsevier B.V.

circuit (Fig. 12.11a). Specific capacitance is obtained from the resistance (R) and parameters (P, T) of a constant phase element (CPE) using (12.2) [27, 28]:

$$C = R^{(1-P)/P} T^{1/P} \tag{12.2}$$

Table 12.2 lists resistance and capacitance of two electrochemical processes included in the Nyquist plot recorded at −15 °C. Characteristic capacitance for the first process (C_1, large clear semi-circle) is ~10^{-11} F, which can be attributed to the bulk resistance of solid electrolytes. On the other hand, the characteristic capacitance for the second process (C_2) is 10^{-8} F, which is a typical value for the grain boundary. Figure 12.11b plots Arrhenius plot of the inverse resistivity of the bulk and the grain boundary. Although bulk resistivity is almost similar for all pellets, grain boundary resistivity is higher for LLZT-w/o-air than the others. This is due to the less sintered grain boundary of LLZT-w/o-air, which agrees with the SEM images showing more pores in LLZT-w/o-air. As indicated by STEM-EDS for LLZT-air and LLZT-2calc (Fig. 12.9), it is supposed that Li_2CO_3 and/or LiOH would not only fill voids but also play a role of sintering additives. Note that activation energy of the grain boundary of these two pellets is higher than LLZT-w/o-air. This is probably due to the residual Li_2CO_3 and/or LiOH on the grain boundary.

Compared with LLZT-air, LLZT-2calc exhibits higher bulk resistivity and lower grain boundary resistivity (Fig. 12.11b). The higher bulk resistivity is explained by resistance of extra LiOH, while LLZT-air contains only Li_2CO_3 that originates from LLZT itself. The lower grain boundary resistivity suggests the better-sintered grain boundary for LLZT-2calc (Fig. 12.11 and Table 12.2).

Table 12.2 Fitting parameters of Nyquist of LLZT pellets recorded at −15 °C

	R_1 (Ω)	R_2 (Ω)	C_1 (F)	C_2 (F)	χ^2
LLZT-air	1577	125.2	3.7×10^{-11}	3.9×10^{-8}	2.0×10^{-5}
LLZT-w/o-air	2241	413.2	3.0×10^{-11}	2.1×10^{-8}	2.6×10^{-5}
LLZT-2calc	1822	115.4	3.4×10^{-11}	1.5×10^{-8}	6.1×10^{-5}

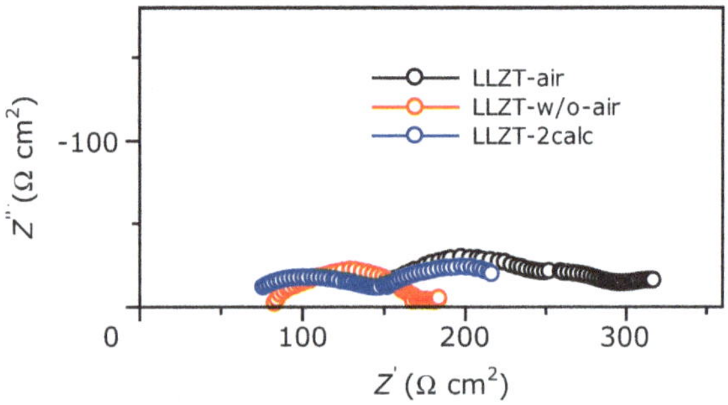

Fig. 12.12 Nyquist plots for Li|LLZT|Li cells measured at room temperature after heated at 175 °C. **a** LLZT-air, **b** LLZT-w/o-air, and **c** LLZT-2calc. Reproduced with permission from [20]. Copyright 2017 Elsevier B.V.

12.3.3 *Charge Transfer Resistance and Short-Circuit Tolerance*

Symmetric cells of Li|LLZT|Li are used to evaluate both charge transfer resistance (R_{ct}) and short-circuit tolerance, which were investigated by AC impedance and DC polarization of the cell. In the Nyquist plots, one or two semi-circle is observed. The semi-circles at high and low frequency ranges are attributed to the Rct and Warburg impedance, respectively. R_{ct} of all pellets were as small as 47 to 69 Ω cm^2, showing homogeneous contacts between vacuum-evaporated Li layers and LLZT pellets (Fig. 12.12).

DC polarization curves of the symmetric cells are demonstrated in Fig. 12.13. All the cells showed low and stable voltage at 0.05 mA cm^{-2}. As the current density increases, the cell voltage starts to fluctuate, resulting in decreased effective contact area during lithium dissolution process. After the fluctuation, the cell voltage dropped suddenly, at which current density short-circuit occurred due to lithium growth between two lithium electrodes. Nyquist plots after the voltage drop show small resistance (Fig. 12.14), which also supports the short-circuit. A similar phenomenon has been observed in previous reports [15, 16, 18, 21]. Critical current

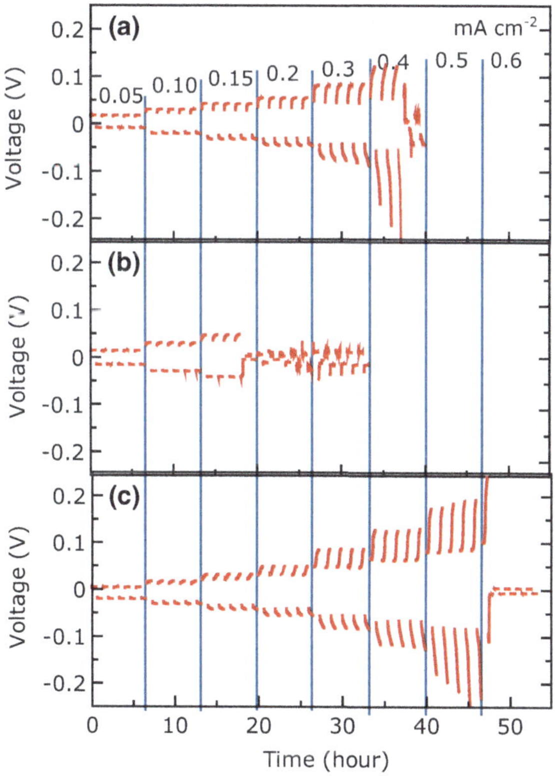

Fig. 12.13 DC polarization curves of Li|LLZT|Li cells measured at 25 °C using LLZT pellets prepared in different conditions: **a** LLZT-air, **b** LLZT-w/o-air, and **c** LLZT-2calc. Reproduced with permission from [20]. Copyright 2017 Elsevier B.V.

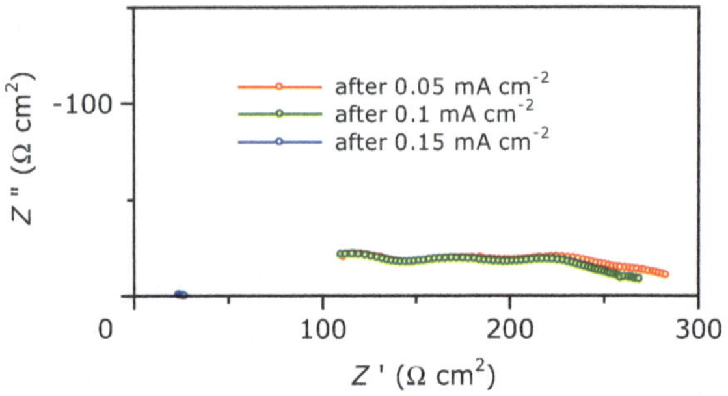

Fig. 12.14 Nyquist plots for Li|LLZT-w/o-air|Li after DC polarization at current densities of 0.05, 0.1, 0.15 mA cm^{-2}. Reproduced with permission from [20]. Copyright 2017 Elsevier B.V.

density, a current density at which the cell voltage dropped was the highest for LLZT-2calc (0.6 mA cm^{-2}) and the lowest for LLZT-w/o-air (0.15 mA cm^{-2}). LLZT-air exhibited critical current density between the other two pellets. Each pellet showed reproducible critical current density. As the relative density of the pellets is almost the same, it is supposed that the critical current density depends on the microstructure of the LLZT pellets. As shown in Fig. 12.10 and Table 12.1, LLZT-w/o-air exhibits larger grain boundary resistance and grain boundaries of this pellet is supposed to be less sintered than the others. In addition, SEM and STEM images demonstrated more voids in the pellets (Figs. 12.6b and 12.8c, d), and thus, interconnected open voids would be highly included. On the other hand, voids are filled with Li_2CO_3 and/or LiOH in LLZT-air and LLZT-2calc and less voids were observed. The highest critical current density of LLZT-2calc can be attributed to the well-sintered grain boundaries and less voids.

The critical current density of 0.6 mA cm^{-2} of LLZT-2calc is placed around the top. However, it should be noted that this value is still far below the required value of 2–3 mA cm^{-2} for practical application to ASSBs. Lithium growth cannot be perfectly suppressed even for the pellets that excluded interconnected open pores by highly densified grain boundaries and plugging grain boundary phases. Therefore, another mechanism of lithium growth may exist. One of the possible mechanisms is a "wedge mechanism". As reported by Porz et al., short-circuit happens even for single crystalline garnet-type solid electrolyte, which is explained by crack propagation on lithium plating [29]. In this mechanism, tensile stress is applied to the solid electrolyte at the bottom of cracks by emerging lithium in the crack on lithium plating. When the tensile stress exceeds the limit of the mechanical strength of the solid electrolyte, the cracks propagate. In the newly widened and deepened voids, lithium grows and tensile stress is applied to the solid electrolyte again. And finally, the cracks reach the other side of the pellet, resulting in short-circuit. According to this mechanism, a key factor is fracture toughness of solid electrolytes. Figure 12.15 compares fracture toughness (K_{1C}) of the LLZT pellets and other materials. Obviously, LLZT-w/o-air, in which grain boundaries are not modified, exhibited relatively high fracture toughness. On the other hand, grain boundary modified LLZT, LLZT-air and LLZT-2calc, exhibits low fracture toughness, which is comparable to or even lower than graphite and Pyrex glass. This means cracks would easily propagate through these

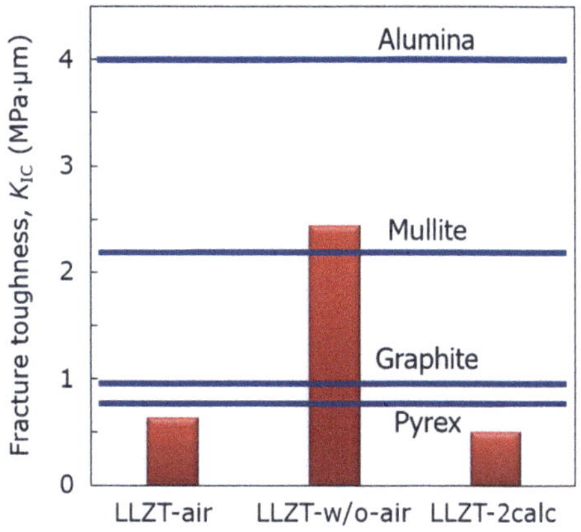

Fig. 12.15 Fracture toughness of the LLZT pellets estimated by microindentation method

pellets. The mechanically weak grain boundary is also suggested by the fractured cross-section SEM images (Fig. 12.7). When pellets were broken, the mechanically weak points would be broken. Therefore, the fractured cross-section images show mechanically weak points. LLZT-air and LLZT-2calc exhibit continuous region, which is attributed to Li_2CO_3 and/or LiOH. Obviously, the amount of these region is much more than expected from the included Li_2CO_3 and/or LiOH (typically 2 wt%). The discrepancy indicates that the grain boundaries with Li_2CO_3 and/or LiOH are mechanically weak, which agree with the fracture toughness.

Therefore, the grain boundary modification effectively plugged the interconnected open voids to improve the critical current density. However, on the contrary, those modification layers deteriorated the mechanical strength of the grain boundary to facilitate the crack propagation. Therefore, it may be important to employ other grain boundary phases that exhibit higher mechanical properties. For higher critical current density, it is important to optimize the amount of the modifier as well as sintering condition.

12.4 Conclusions

In conclusion, we prepared garnet-type solid electrolytes with various microstructures by different synthesis conditions. In this study, it is demonstrated that lithium growth through the solid electrolytes strongly depends on the microstructure of the pellets. Compared with LLZT without any grain boundary modification, LLZT-air and LLZT-2calc with Li_2CO_3 and LiOH on the grain boundaries exhibit higher critical current density of 0.4 and 0.6 mA cm^{-2}, respectively. The STEM images and EDS map images indicate that surface modification layers of Li_2CO_3 and/or LiOH not only fill voids but also facilitate sintering to exhibit high critical current density. Furthermore, the SEM images and EIS analysis also supported the influence of the surface modification. The modification of the microstructure improves the interconnection among grains and decreasing voids, which would be an effective way to suppress the lithium growth along the grain boundaries.

Acknowledgments A part of this work was financially supported by a research project, "Advanced Low Carbon Technology Research and Development Program for Specially Promoted Research for Innovative Next Generation Batteries" of Japan Science and Technology Agency (JST-ALCA SPRING), which is gratefully acknowledged. The authors are also thankful to Ms. T. Ito (Nagasaki University) for her synthesis and preparation of specimens, and to Dr. T. Morimura (Nagasaki University) and Dr. R. Bekarevich (National Institute for Materials Science) and Dr. K. Mitsuishi (National Institute for Materials Science) for advanced skills and helpful discussions on the STEM observation and EDS-ED analyses.

Appendix I**

To confirm the thermal stability of LiOH during the second calcination, thermal gravimetric analysis of $LiOH{\cdot}H_2O$ was conducted in air (TG/DTA7300, Hitachi High-Tech Science Corp.). The weight changes in Fig. 12.16 can be attributed to the following reactions.

$$< 100\ ^{\circ}\mathrm{C}: \quad \mathrm{LiOH \cdot H_2O(s) \rightarrow LiOH(s) + H_2O(g)} \tag{12.3}$$

$$> 100\ ^{\circ}\mathrm{C}: \mathrm{LiOH(s) + 1/2CO_2(g) \rightarrow 1/2\,Li_2CO_3(s) + 1/2H_2O(g)} \tag{12.4}$$

$$> 400\ ^{\circ}\mathrm{C}: \mathrm{LiOH(s) \rightarrow LiOH(l) \rightarrow 1/2\,Li_2O(s) + H_2O(g)} \tag{12.5}$$

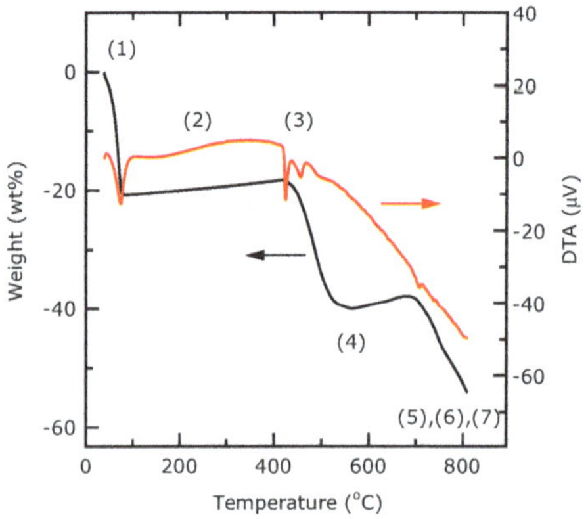

Fig. 12.16 TG-DTA curves of $LiOH{\cdot}H_2O$ in air. Reproduced with permission from [20]. Copyright 2017 Elsevier B.V.

**Reproduced with permission from [20]. Copyright 2017 Elsevier B.V.

$$500\text{–}700\ ^\circ\mathrm{C}: \quad \mathrm{Li_2O(s) + CO_2(g) \rightarrow Li_2CO_3(s)} \tag{12.6}$$

$$> 700\ ^\circ\mathrm{C}: \mathrm{Li_2CO_3(s) \rightarrow Li_2CO_3(l)} \tag{12.7}$$

$$\mathrm{Li_2CO_3(s, l, g) \rightleftarrows Li_2O(s) + CO_2(g)} \tag{12.8}$$

$$\mathrm{Li_2CO_3(s, l) \rightleftarrows Li_2CO_3(g)} \tag{12.9}$$

Note that materials in the system on the heating process are not only dependent on the thermodynamics but also influenced by their kinetics. Therefore, for instance, at 700 °C, Li_2O and Li_2CO_3 coexist. Taking into consideration the above reactions and kinetics, it is supposed that the surface $LiOH{\cdot}H_2O$ (and LiOH) on LLZT-2calc is produced as follows. First, on heating, LiOH melts around 400–450 °C, and is decomposed to Li_2O (12.5). Next, Li_2CO_3 is produced from LiOH and Li_2O with atmospheric CO_2 by (12.4) and (12.6), but most of Li_2CO_3 is decomposed to Li_2O above 700 °C (12.8) [30]. However, due to the equilibrium reaction of (12.8), small amount of Li_2CO_3 exists on LLZT surface at 900 °C. Then, Li_2CO_3 evaporates (12.9), which results in loss of Li-component from the reaction system [31]. As the forward reaction of (12.9) proceeds, it induces the reverse reaction of (12.8). As a result, excess Li decreases gradually. After the calcination, remaining Li_2O on LLZT particles reacts with moisture to form LiOH and $LiOH{\cdot}H_2O$.

References

1. Tarascon JM, Armand M (2001) Issues and challenges facing rechargeable lithium batteries. Nature 414:359–367
2. Xu W, Wang J, Ding F, Chen X, Nasybulin E, Zhang Y, Zhang JG (2014) Lithium metal anodes for rechargeable batteries. Energy Environ Sci 7:513–537
3. Manthiram A, Fu Y, Chung S, Zu C, Su Y (2014) Rechargeable lithium-sulfur batteries. Chem Rev 114:11751–11787
4. Grande L, Paillard E, Hassoun J, Park JB, Lee YJ, Sun YK, Passerini S, Scrosati B (2015) The lithium/air battery: still an emerging system or a practical reality? Adv Mater 27:784–800
5. Takada K (2013) Progress and prospective of solid-state lithium batteries. Acta Mater 61:759–770
6. Monroe C, Newman J (2005) The impact of elastic deformation on deposition kinetics at lithium/polymer interfaces. J Electrochem Soc 152:A396–A404
7. Murugan R, Thangadurai V, Weppner W (2007) Fast lithium ion conduction in garnet-type $Li_7La_3Zr_2O_{12}$. Angew Chemie Int Ed 46:7778–7781
8. Kotobuki M, Munakata H, Kanamura K, Sato Y, Yoshida T (2010) Compatibility of $Li_7La_3Zr_2O_{12}$ solid electrolyte to all-solid-state battery using Li metal anode. J Electrochem Soc 157:A1076–A1079
9. Ni JE, Case ED, Sakamoto JS, Rangasamy E, Wolfenstine JB (2012) Room temperature elastic moduli and Vickers hardness of hot-pressed LLZO cubic garnet. J Mater Sci 47:7978–7985
10. Yu S, Schmidt RD, Garcia-Mendez R, Herbert E, Dudney NJ, Wolfenstine JB, Sakamoto J, Siegel DJ (2016) Elastic properties of the solid electrolyte $Li_7La_3Zr_2O_{12}$ (LLZO). Chem Mater 28:197–206

11. Ishiguro K, Nakata Y, Matsui M, Uechi I, Takeda Y, Yamamoto O, Imanishi N (2013) Stability of Nb-doped cubic $Li_7La_3Zr_2O_{12}$ with lithium metal. J Electrochem Soc 160:A1690–A1693
12. Ishiguro K, Nemori H, Sunahiro S, Nakata Y, Sudo R, Matsui M, Takeda Y, Yamamoto O, Imanishi N (2014) Ta-doped $Li_7La_3Zr_2O_{12}$ for water-stable lithium electrode of lithium-air batteries. J Electrochem Soc 161:A668–A674
13. Sudo R, Nakata Y, Ishiguro K, Matsui M, Hirano A, Takeda Y, Yamamoto O, Imanishi N (2014) Interface behavior between garnet-type lithium-conducting solid electrolyte and lithium metal. Solid State Ion 262:151–154
14. Ren Y, Shen Y, Lin Y, Nan CW (2015) Direct observation of lithium dendrites inside garnet-type lithium-ion solid electrolyte. Electrochem Commun 57:27–30
15. Suzuki Y, Kami K, Watanabe K, Watanabe A, Saito N, Ohnishi T, Takada K, Sudo R, Imanishi N (2015) Transparent cubic garnet-type solid electrolyte of Al_2O_3-doped $Li_7La_3Zr_2O_{12}$. Solid State Ion 278:172–176
16. Cheng L, Chen W, Kunz M, Persson K, Tamura N, Chen G, Doeff M (2015) Effect of surface microstructure on electrochemical performance of garnet solid electrolytes. ACS Appl Mater Interfaces 7:2073–2081
17. Tsai CL, Roddatis V, Chandran CV, Ma Q, Uhlenbruck S, Bram M, Heitjans P, Guillon O (2016) $Li_7La_3Zr_2O_{12}$ interface modification for Li dendrite prevention. ACS Appl Mater Interfaces 8:10617–10626
18. Sharafi A, Meyer HM, Nanda J, Wolfenstine J, Sakamoto J (2016) Characterizing the Li–$Li_7La_3Zr_2O_{12}$ interface stability and kinetics as a function of temperature and current density. J Power Sources 302:135–139
19. Yamada H, Ito T, Hongahally Basappa R (2016) Sintering mechanisms of high-performance garnet-type solid electrolyte densified by spark plasma sintering. Electrochim Acta 12:648–656
20. Hongahally Basappa R, Ito T, Morimura T, Bekarevich R, Mitsuishi K, Yamada H (2017) Grain boundary modification to suppress lithium penetration through garnet-type solid electrolyte. J Power Sources 363:145–152
21. Hongahally Basappa R, Ito T, Yamada H (2017) Contact between garnet-type solid electrolyte and lithium metal anode: influence on charge transfer resistance and short circuit prevention. J Electrochem Soc 164:A666–A671
22. Liu C, Rui K, Shen C, Badding ME, Zhang G, Wen Z (2015) Reversible ion exchange and structural stability of garnet-type Nb-doped $Li_7La_3Zr_2O_{12}$ in water for applications in lithium batteries. J Power Sources 282:286–293
23. Cheng L, Wu CH, Jarry A, Chen W, Ye Y, Zhu J, Kostecki R, Persson K, Guo J, Salmeron M, Chen G, Doeff M (2015) Interrelationships among grain size, surface composition, air stability, and interfacial resistance of Al-substituted $Li_7La_3Zr_2O_{12}$ solid electrolytes. ACS Appl Mater Interfaces 7:17649–17655
24. Truong L, Howard M, Clemens O, Knight KS, Slater PR, Thangadurai V (2013) Facile proton conduction in H^+/Li^+ ion-exchanged garnet type fast Li-ion conducting $Li_5La_3Nb_2O_{12}$. J Mater Chem A 1:13469–13475
25. German RM, Suri P, Park SJ (2009) Review: liquid phase sintering. J Mater Sci 44:1–39
26. Ren Y, Deng H, Chen R, Shen Y, Lin Y, Nan CW (2015) Effects of Li source on microstructure and ionic conductivity of Al-contained $Li_{6.75}La_3Zr_{1.75}Ta_{0.25}O_{12}$ ceramics. J Eur Ceram Soc 35:561–572
27. Fleig J (2002) The grain boundary impedance of random microstructures: numerical simulations and implications for the analysis of experimental data. Solid State Ion 150:181–193
28. Li Q, Thangadurai V (2009) Synthesis, structure and electrical properties of mo-doped CeO_2–materials for SOFCs. Fuel Cells 9:684–698
29. Porz L, Swamy T, Sheldon BW, Rettenwander D, Till Frömling, Thaman HL, Berendts S, Uecker R, Carter WC, Chiang YM (2017) Mechanism of lithium metal penetration through inorganic solid electrolytes. Adv Energy Mater 7:1701003
30. West AR (2014) Solid state chemistry and its applications. Wiley, Chichester. Chapter 1

Chapter 13
All-Solid-State Batteries Based on Glass-Ceramic Lithium Vanadate

Anton A. Raskovalov and Nailya S. Saetova

Abstract In this chapter, challenges in the development of all-solid-state batteries designed to solve the problem of safety of the chemical power sources are discussed. Difficulties in the development of such batteries are poor adhesion and electrical conductivity of solids and change in the volume of materials during charge/discharge processes. To solve these problems, the use of glassy or glass-ceramic materials as the electrode is suggested. Based on vanadates glasses and glass-ceramics have attracted the most significant attention among glassy electrode materials since glassy vanadates have rather high electrical conductivity in comparison with other oxide glasses (about 10^{-5} S·cm^{-1} at room temperature). In addition, the electrical conductivity of vanadate glasses can be significantly improved by obtaining of glass-ceramic based on them. Further, results of the test of all-solid-state battery with vanadate cathode are presented. It is shown that the voltage of Li–Ga | glassy vanadate single cell achieves the value of 3.3 V.

Abbreviations and Contractions

ASSB	ALL-SOLID-STATE BATTERY
CI	Current interrupt
CPE	Constant phase element
CV	Cyclic voltammetry
DSC	Differential scanning calorimetry
EIS	Electrochemical impedance spectroscopy
EMF	Electromotive force
EPR	Electron paramagnetic resonance
h	An hour

A. A. Raskovalov (✉) · N. S. Saetova
Institute of High Temperature Electrochemistry, Ural Branch of RAS, Yekaterinburg, Russian Federation
e-mail: other@e1.ru

N. S. Saetova
e-mail: n.saetova@yandex.ru

R. Murugan and W. Weppner (eds.), *Solid Electrolytes for Advanced Applications*,
https://doi.org/10.1007/978-3-030-31581-8_13

LAGP $Li_{1.3}Al_{0.3}Ge_{1.7}(PO_4)_3$
LATP $Li_{1.3}Al_{0.3}Ti_{1.7}(PO_4)_3$
LLTO $Li_{0.5}La_{0.5}TiO_3$
LLZO $Li_7La_3Zr_2O_{12}$
NaS Sodium–sulfur battery
OCV Open-circuit voltage
S Siemens (=Ohm^{-1})
SHE Standard hydrogen electrode
TFB Thin-film battery
T_g Glass transition temperature
XRD X-ray diffraction analysis

13.1 Introduction

13.1.1 Why All-Solid-State?

Almost all commercially available chemical power sources are realized using liquid electrodes or electrolytes. The use of liquid electrodes is typical for sodium-sulfur (NaS) [1, 2] and sodium-metal halide (Zebra) [3–8] batteries. In this case, the most significant drawback is that liquid sodium and sulfur are extremely dangerous and chemically active. Therefore, it is very difficult to select sustainable materials for shell and electrolyte. Sodium beta-alumina (Na–β–Al_2O_3), which requires robust technologies for synthesis, is usually used as the electrolyte. It is known that any leakage of liquid electrodes to a surrounding media or their interaction with each other can lead to an accident. For example, in September 2011, NaS batteries installed at the Tsukuba plant (Japan) caught fire [9]. In addition, it is necessary to maintain a working temperature at which electrodes are in the liquid state (in the case of the NaS battery this temperature is about 300–350 °C).

Batteries with liquid electrodes are less common than batteries with a liquid electrolyte. However, liquid electrolytes are usually dangerous because of aggressive substances like sulfuric acid in lead-acid batteries and alkali in primary alkaline and secondary metal-hydride batteries. Widely used in lithium power sources, liquid electrolytes are flammable and explosive organic solutions of the lithium salts. Such electrolytes can react with electrode materials and the organic solvent can evaporate. Moreover, some of lithium salts containing in the electrolyte composition decompose due to thermodynamic instability [10]. All the mentioned facts can lead to gas liberation, a volume expansion and, as a result, an explosion of a battery. Some accidents with electric car batteries are well-known [11–14]. Owing to insufficient safety of lithium metal batteries, Federal Aviation Administration has forbidden carrying such batteries as bulk cargo shipments on passenger aircrafts [15].

Thus, all liquid-containing power sources have problems with their safety. The next ensuring problem is a narrow operating temperature range, which is determined by the temperature of electrode melting and electrolyte decomposition in the case of batteries with liquid electrodes and electrolyte, respectively. In addition, liquid electrolytes are susceptible to a dendrite formation [16]. To avoid these drawbacks, a concept of an all-solid-state battery (ASSB) was developed, which means that all components of such battery (anode, cathode and electrolyte) are solids. The design of ASSB can potentially extend the region of applicability of power sources, improve their safety and make them able to operate under extreme conditions.

13.1.2 Challenges of All-Solid-State Battery Development

To understand challenges connected with the construction of all-solid-state-batteries, let us, at first, consider the arrangement of a convenient lithium-ion battery (Fig. 13.1a). The electrode material is in a finely dispersed state and its surface is moistened by the liquid electrolyte to support high enough current densities (low ohmic resistance). One of the important properties of the electrolyte is the ability to moisten the electrode material to provide a satisfactory ionic contact. Electronic conduction of the system is provided by a corresponding additive. The liquid covers all grains of the solid electrolyte. The particles of the electrode material have a definite depth of utilization due to limited diffusion coefficients in solids. In addition, electronic conductivity of some electrode materials is rather small. Thus, recently popular $LiFePO_4$ has very low lithium diffusion coefficient and electronic conductivity [17]. Lithium diffusion coefficients of most popular cathode materials, namely, Li_2FeSiO_4, $LiFePO_4$ and $LiNi_{0.5}Co_{0.2}Mn_{0.3}O_2$ are 10^{-15}, $6{\cdot}10^{-15}$ and $3{\cdot}10^{-11}$ $cm^2{\cdot}s^{-1}$ at 25 °C, correspondingly [18]. To provide a maximal degree of the material utilization, it is necessary to increase the near-surface share of the material. In order to tighten an electrical contact and remove adsorbed bubble of gases, evacuation is used during a battery manufacturing.

Solids have specific mechanical properties, namely, they hardly envelop to each other, have a poor adhesion, and, in most cases, do not form a tight contact. Therefore,

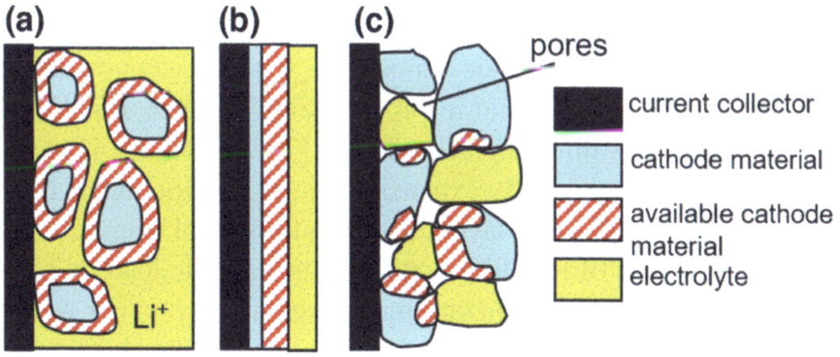

Fig. 13.1 The schematic representation of a cathodic half-cell of **a** a conventional lithium power source with a liquid electrolyte, **b** a thin-film battery and **c** a result of naive attempt to apply convenient technology of lithium-ions batteries to solid-state ones. Electron-conducting additives were omitted for clarity

it is difficult to apply the abovementioned techniques for solids. It is hard to obtain a good contact even between two flat solid surfaces. The creation of a developed surface in this case is nearly impossible. The use of abovementioned convenient ways to create ASSB can lead to a situation depicted in Fig. 13.1c. Single-point contacts between electrode and electrolyte materials are formed and there are a lot of pores and, as a consequence of these factors, the cell has extremely high resistance. The formation of point contacts results in poor kinetics of ion-exchanging processes and limits the usefulness of an all-solid-state battery. In principle, it is possible to obtain the situation depicted in Fig. 13.1a in the case of solids, for example, using solid polymer electrolytes. Another problem is the low conductivity of solid electrolytes. Even if the particles of the cathode material would be enveloped by the solid electrolyte, any benefits, most likely, are not obtained because solid electrolytes, except some sulfide-based [19–21], have conductivity about two orders lower than liquid ones. In fact, the situation is more difficult. According to literature data, liquid electrolytes based on $LiPF_6$ solutions in organic carbonates can undergo decrease in the dissociation degree of a lithium salt. This process result in an increase in electrolyte resistance by one or two orders [22, 23] and the conductivity of such electrolytes is comparable with a bulk conductivity of some solid electrolytes. On the other hand, the conductivity of solid electrolytes is generally limited by a lower conductivity of grain boundaries. In any case, requirements for electrode materials for ASSB are stronger; their electronic and ionic conductivity should be as high as possible to overcome low ionic conductivity of solid electrolytes. The conductivity of pure and MgO-doped lithium cobaltite is 10^{-3} and 10^{-1} $S{\cdot}cm^{-1}$ [24], respectively; therefore it is frequently used in ASSB [25–38]. Glenneberg et al. [39] noticed that major share of thin film batteries (TFB) use lithium cobaltite as the cathode material.

Some ways to create a tight contact between solid electrodes and electrolytes can be highlighted:

- Thin film technologies;
- Soft materials (lithium metal, Li–In alloy, polymer electrolytes);
- Adhesive layers (low-melting glasses);
- Glassy electrode materials.

Thin film technologies include different kinds of sputtering (magnetron sputtering, vacuum evaporation, pulsed laser deposition, spark plasma sintering etc.). A listing of these methods can be found in [40]. A large number of published works deal with ASSB obtained by thin film technologies [26–29, 34–39, 41–43]. Unfortunately, these technologies are usually expensive, the deposition rate is typically low and the thickness of the deposited layers is strongly limited. Figure 13.1b clearly demonstrates the limitation of thin film batteries: the cathode load of TFB is determined by the depth of a material utilization; all excess width is not working. Such microscopic batteries operate well and reach very high specific characteristic as per the electrode weight, however, only in a few works it is mentioned that such characteristics can be achieved only for limited layers. For example, the thicknesses of cathode and electrolyte layer of ASSB are about 25 and 50 μm, respectively [44].

In addition, diffusion in thin films can be lower than in bulk material, for example, in the case of MoS_2 [45].

The next way is the application of soft materials with a good adhesion. For example, polymer electrolytes were successfully used in a number of works [33, 44, 46–48]. Typically, a few problems appear on the anode side because metallic lithium and its alloys are soft enough and can be deposited on the solid electrolyte directly. Indium is used in the series of works [25, 31, 32, 36, 49]. There are some research that demonstrated contact between the solid electrolyte and metallic lithium can be improved using a gold [50] or Al_2O_3 [51] interlayer.

Another trick for providing a tight connection on the solid–solid interface is the use of adhesive additives such as low melting electrolyte glasses [52, 53]. These glasses soften with heating and glue surfaces of solids to support a tight contact between them. The evident development of adhesive additives methodology is the creation of electrode materials with properties of adhesive additives. These may be glassy or glass-ceramic electrode materials [54]. Oxide glasses can be heat-treated at a temperature higher than a melting temperature to fill pores on the electrode/electrolyte interface which allows one to provide a good contact with a high specific area. Moreover, glasses have slightly higher interatomic distances that can promote an ion intercalation, but their electronic conductivity is, typically, lower than for crystalline analogs for the same reason. However, it is possible to transform glasses into glass-ceramic with combined properties. In addition to abovementioned problems, it should be noted that the volume of electrode materials changes during intercalation process. It is not crucial when an electrode is in a contact with a liquid electrolyte but can lead to a mechanical destruction of ASSB. It can be expected that greater interatomic distances in glassy electrodes allow the degree of chemical expansion of the material be reduced. At present, there are a small amount of works studying glassy or glass-ceramic electrode materials [45, 54–68]. The most part of cited works deals with a convenient lithium-ion technology with a powdered material but not with ASSB.

13.2 Cathode Materials

Let us consider some materials for ASSB that can be designed by the glass technology. A review of more commonly used cathode materials for lithium batteries can be found in [69]. Vanadate system attracts some attention due to the following advantages:

- High redox potential of V^{5+}/V^{4+} pair (~1 V against standard hydrogen electrode);
- Relativity high electronic conductivity in both reduced (V^{4+}) and oxidized (V^{5+}) forms;
- The possibility of lithium intercalation;
- Low cost;
- Ability of glass formation in some cases.

For these reasons cathode materials based on vanadate systems have been thoroughly studied [70]. It is important to note that one of the first ASSB was made based on lithium vanadate (LiV_3O_8) in 1988 [46]. A main disadvantage of vanadates as cathode materials for lithium batteries is their poor cycleability due to irreversible phase transformation during the process of lithium intercalation/deintercalation [71, 72]. This is an additional reason to make an attempt to use glassy electrodes without crystal lattice and, probably, phase transitions. Another electrochemically active oxide able to glass forming is lead oxide (PbO) [73–76]. However, lead is too heavy and toxic; therefore, lead-containing materials are not discussed in the present work.

In some cases vanadium oxide can act as the glass former, however, such behavior is usually possible in the presence of any traditional glass former (SiO_2, B_2O_3, P_2O_5, etc.); pure V_2O_5 cannot be obtained in the vitreous state. According to literature data, vanadium pentoxide can be the glass former when its concentration is rather high [77–79]. The most attractive among traditional glass former oxides is boron oxide with the least molecular mass because specific capacity is one of the most important characteristics of a power source. The specific capacity of a cathode material can be evaluated by the formula:

$$Q = \frac{nzF}{M}, \tag{13.1}$$

where n is the number of compensation ions that can be intercalated/deintercalated during discharge/charge, z is their charge, F is the Faraday constant, ~96484 $C \cdot mol^{-1}$, M is a molar mass. According to (13.1), the resulting unit is $C \cdot g^{-1}$ and it can be easily converted into $mA \cdot h \cdot g^{-1}$ by the division of the obtained number into 3.6. It should be noted that n is not always equal to the total number of mobile ions in the formula unit of completely intercalated material. Intercalation/deintercalation process is accompanied by a change in oxidation degree of transition metal ion; therefore, the deintercalation of Li^+ ions is impossible when all ions of a transition metal are in maximal oxidation degree. According to formula (13.1), it is desirable that the electrode material consists of mainly electroactive and compensating ions. Table 13.1 shows several cathode materials in their charged and discharged states with corresponding specific capacities (some of the tabulated states are hypothetic, maximal charged/discharged states are often unreachable).

If the glassy electrode material is applied on electrolyte by heating and premelting, the electrolyte must be stable in the short-time contact with the melt. Thus, polymer electrolytes are unsuitable for such way of battery processing. In this case, oxide ceramics are the most suitable electrolyte materials. Sulfides can also react with molten vanadates with the liberation of sulfur dioxide. The best Li^+-conducting oxide electrolytes are $Li_{0.5}La_{0.5}TiO_3$ (LLTO) [80], $Li_{1.3}Al_{0.3}Ti_{1.7}(PO_4)_3$ (LATP) [81], $Li_{1.3}Al_{0.3}Ge_{1.7}(PO_4)_3$ (LAGP) [82, 83] and $Li_7La_3Zr_2O_{12}$ (LLZO) [84]. LLZO is interesting due to the fact that it combines both relatively high ionic conductivity and good chemical stability against metallic lithium [84].

Table 13.1 Theoretical values of specific capacity (*Q*) of some electrode materials. *N* is a number of intercalated/deintercalated ions

Discharged state	Charged state	N	Q, $mA{\cdot}h{\cdot}g^{-1}$
$LiCoO_2$	CoO_2	1	274
$LiNiO_2$	NiO_2	1	274
LiV_2O_5	V_2O_5	1	141
$Li_2V_2O_5$	V_2O_5	2	274
$LiFePO_4$	$FePO_4$	1	170
$Li_4V_3O_8$	LiV_3O_8	3	260
$LiMn_2O_4$	2 MnO_2	1	148
$LiNi_{1/3}Mn_{1/3}Co_{1/3}O_2$	$Ni_{1/3}Mn_{1/3}Co_{1/3}O_2$	1	279
$0.3Li_2O{\cdot}0.3V_2O_4{\cdot}0.15V_2O_5{\cdot}0.25B_2O_3$	$0.45V_2O_5{\cdot}0.25B_2O_3$	0.6	155

Since individual LLZO usually exists in the form of low-conducting tetragonal phase it is rarely applied as the electrolyte. However, a high-conducting cubic modification can be obtained by doping with Al^{3+}, Y^{3+}, Ga^{3+}, Nb^{5+} and other cations [85–97]. According to literature data, the conductivity and density of both tetragonal and cubic LLZO phases can be improved by the use of low-melting glasses as sintering additives. Recently, many papers with glass-containing composite solid electrolytes have appeared [97–101]. It should be noted that, at the present time, the glass additives to LLZO are used for decreasing sintering time and temperature [91, 96–99].

13.3 Vanadate Glass and Glass-Ceramic

In this part, some published data about vanadate glass and glass-ceramics are considered. Electron conducting glasses of V_2O_5–P_2O_5 and V_2O_5–TeO_2 systems are the most widely studied glasses which can be used as a cathode material of Li and Na-ion rechargeable batteries. The first glasses in this system containing about 89% V_2O_5 were, likely, obtained by Tammann et al. in 1929 [102]. At first, these glasses attracted the attention of scientists due to the ability of vanadium oxide to be a glass former; however, in 1952 Stanworth mentioned low electrical resistance (~$6.3{\cdot}10^6$ $\Omega{\cdot}cm^{-1}$ at room temperature) of vanadium oxide-based glasses with $15BaO$–$31.4TeO_2$–$53.6V_2O_5$ composition [103]. This finding significantly increased the interest in the investigation of electrical properties of vanadium glasses. The group of Stanworth carried out a series of research devoted to the investigation of the properties of BaO–V_2O_5, V_2O_5–TeO_2, BaO–V_2O_5–P_2O_5 and V_2O_5–PbO glasses [104, 105]. A detailed investigation of glasses of V_2O_5–PbO–P_2O_5 and V_2O_5–WO_3–P_2O_5 systems was carried out by Kitaigirodskii et al. [106]. It was found that the replacement of vanadium oxide by P_2O_5, PbO, BaO, WO_3 and MoO_3 leads to a considerable growth of the electrical resistance, while the replacement of phosphorus oxide by PbO, BaO, WO_3 and MoO_3 at the constant concentration of V_2O_5 leads to an increase in conductivity. The composition dependence of electrical conductivity of glasses with high vanadium oxide concentration was studied in detail by Munakata [107]. This investigation allowed the factors affecting electrical conductivity of vanadium glasses to be determined, namely, vanadium concentration, ratio of V^{4+} to the total amount of vanadium and the kind of basic cation. The lowest electrical resistance (about 10^5 $\Omega \cdot cm^{-1}$) has the glass in which phosphorus oxide is partially substituted by cesium oxide. Basing on his previous studies [108, 109], in 1968 Mott presented the theory of polaron hopping in glasses [110], which is commonly used for the description of electron conductivity of glasses nowadays. Glasses of binary V_2O_5–P_2O_5 system with vanadium oxide content of 50, 60, 70, 80 and 90 mol% were investigated in [111]; electrical conductivity of these glasses was determined in a wide temperature range from −200 to ~200 °C, electrical conductivity of the glass containing 90 mol% V_2O_5 is ~10^{-11} $S \cdot cm^{-1}$ at the temperature of −163 °C. The structural model of vanadium phosphate glasses was proposed by Frazier and France in 1977 [112]; this model was based on the calculation of n value (n is "the number of uncoupled nearest sites surrounding the final hopping site of the small polaron" [112]). It was found that the n value can be non-integer and the equation taking into account non-integer n values was developed. In 1980, Feltz studied the influence of polarizability of the oxide ions on the electrical conductivity of vanadium glasses [113]. According to the obtained data, the polarizability of ions surrounding a polaron state is more important for the electrical conductivity than the site spacing or the presence of ions in different oxidation state.

Electrical properties of binary glasses with the constant concentration of V_2O_5 of 55 mol% and CeO_2, TeO_2, PbO, BaO and As_2O_3 as the second oxide were investigated by Chung [114]. As a result of the investigation, structural model of electron hopping in such glass was proposed and the Mott's equation for the calculation of activation energy was extended to take into consideration existence of inhomogeneity in the glass structure. Electrochemical behavior of Li_2O–V_2O_5 glass with LiV_3O_8 composition was investigated by Nassau [115] and it was found that despite the higher lithium storage capacity of glassy LiV_3O_8 in comparison with the crystalline one during first 1–5 cycles, the capacity of glass rapidly decreases after 14 cycles. The electrical conductivity of vanadium glasses containing TeO_2, ZnO, Bi_2O_3 was investigated in [116–119]. Glasses containing from 50 to 80 mol% V_2O_5 were obtained and its electrical conductivity was measured in the temperature range of 77–400 K; composition with vanadium oxide concentration of 80 mol% has the highest value of conductivity [116]. Levy et al. [119] showed the possibility of reversible lithium intercalation in the glass with the composition of $60V_2O_5$–$40TeO_2$ under the voltage higher than 4 V and, in the further work, made the composites with graphite to overcome the low conductivity of such glasses [120]. According to [117], in the case of ZnO–V_2O_5 system the charge transport obeys a non-adiabatic mechanism; the highest value of conductivity was observed for the glass with 75 mol% of V_2O_5 and it was $1.4 \cdot 10^{-5}$ $S \cdot cm^{-1}$ at room temperature. Glass with the electrical conductivity of about 10^{-5} $S \cdot cm^{-1}$ at room temperature had the composition of $90V_2O_5$–$10TeO_2$ [121]. In the case of M_2O_3–V_2O_5–P_2O_5 (M_2O_3 = Bi_2O_3, Sb_2O_3) ternary system, the partial replacement of phosphorus oxide by bismuth oxide leads to an increase in conductivity of Bi_2O_3–V_2O_5–P_2O_5 system in comparison with the binary V_2O_5–P_2O_5 system and the replacement of phosphorus oxide by antimony oxide decreases the conductivity in comparison with the binary V_2O_5–P_2O_5 system [118]. It should also be noted that there is an interesting effect on the concentration dependence of conductivity in both Bi_2O_3 and Sb_2O_3 cases, namely, a minimum of conductivity, which can be explained by the phase separation. The influence of small (from 2 to 15 wt%) additions of nickel oxide to the V_2O_5–P_2O_5 system was investigated in [122]; it was found that the 2 wt% of NiO leads to a slight increase in the conductivity of the glass and a further increase in the NiO concentration decreases the conductivity in comparison with the binary V_2O_5–P_2O_5 system. Electrochemical insertion of lithium ions in binary V_2O_5–M_mO_n glasses, where M_mO_n is Ag_2O, CuO, ZnO, Fe_2O_3, Co_3O_4, MoO_3 or WO_3, was studied by Machida [56]; it was shown that the glass containing Fe_2O_3 had high specific capacity and the best chemical diffusion coefficient among all studied glasses. A detailed electrochemical investigation of $20Fe_2O_3$–$80V_2O_5$ glass is presented in [57].

Electrical properties of $0.4TeO_2$–$0.6(xV_2O_5$–$(1-x)MoO_3)$ glasses with $x = 0.2$–1 were investigated by Lebrun et al. [123]. They found that the electrical conductivity significantly decreases with an increase in molybdenum oxide concentration. The influence of the lithiation process on the structure of $50V_2O_5$–$50TeO_2$ glass was investigated in [124]. This study showed that the lithiation of such glasses is inhomogeneous and the existence of some phase separation is possible; however, the investigated glass was stable under electrochemical lithiation and the crystallization was not observed. Electrical conductivity of V_2O_5–Sb_2O_3–TeO_2 and V_2O_5–SnO–TeO_2 glasses was investigated by Mori [125, 126]. It was found that the conductivity is provided by small polaron hopping mechanism and it is adiabatic for glasses with V_2O_5 concentration more than 50 mol% and non-adiabatic when the amount of V_2O_5 is less than 50 mol%; the highest conductivity is $1.46 \cdot 10^{-3}$ $S \cdot cm^{-1}$ at 473 K for the composition with V_2O_5 concentration of 62 mol% [125]. The conductivity of V_2O_5–PbO–TeO_2 glasses synthesized in a wide concentration range was studied in [127]; the highest value of conductivity was $1.2 \cdot 10^{-4}$ $S \cdot cm^{-1}$ at 408 K for the glass containing 70 mol% of V_2O_5. The influence of the concentration of transition metal oxides on the conductivity mechanism of glasses with both ionic and electronic conductivity was investigated by Jayasinghe et al. [128, 129]. It was found for glasses of xNa_2O–$(1-x)V_2O_5$–$2TeO_2$ system that there are two regions, namely, V_2O_5 rich region ($0 \leq x < 0.5$) and Na_2O ($0.5 < x \leq 1$) rich region having predominantly electronic and ionic conductivity, respectively [128]. In the case of xLi_2O–$(1-x)V_2O_5$–$3TeO_2$ system, there is a minimum concentration dependence of conductivity at $x = 0.5$ that can be explained by the inability of percolation of both ionic and electronic paths [129]; the concentration ranges of glasses with predominantly electronic and ionic conductivity are the same for the abovementioned xNa_2O–$(1-x)V_2O_5$–$2TeO_2$ system. The conductivity of P_2O_5–$[(1-x)V_2O_5$–$xNa_2O]$ glasses was studied by Levy et al. [130] and the behavior of the conductivity of this system during the transition from binary V_2O_5–P_2O_5 to binary Na_2O–P_2O_5 glass was similar to the result observed by Jayasinghe. The minimum conductivity is located at the concentration of Na_2O of 0.7.

Electrical conductivity of V_2O_5–Sb–TeO_2 glasses was investigated in [131]; the glass with the composition of $70V_2O_5$–20Sb–$10TeO_2$ was found to be the highest conducting glass of the studied system and its conductivity was $7.13 \cdot 10^{-3}$ S·cm^{-1} at 473 K. The concentration ranges of adiabatic and non-adiabatic small polaron hopping were found (despite authors wrote the composition of glasses using elementary antimony, it can be assumed that Sb was oxidized by vanadium pentoxide). The effect of lithium ions on the electrical properties of $x$$Li_3PO_4$–(1–$x$)($76V_2O_5$–$24P_2O_5$) glasses was studied by Salman et al. [132]. It was found that electrical conductivity decreases with an increase in lithium concentration and predominantly electronic conductivity becomes predominantly ionic at Li_3PO_4 concentration more than 50 mol%. Glasses of $2TeO_2-xK_2O-(1-x)V_2O_5$ system (x from 0 to 1 mol%) were investigated by El-Desoky [133]. The conductivity of such glasses achieved minimum at $x = 0.5$ mol%, and in all cases the value of conductivity is less than the conductivity of initial $2TeO_2-V_2O_5$ glass. El-Desoky also investigated the glasses of V_2O_5–Fe_2O_3–TeO_2 system [134]; it was found that the conductivity increases with an increase in vanadium oxide concentration and the highest value of conductivity was $2.88 \cdot 10^{-4}$ S·cm^{-1} at 400 K for $65V_2O_5$–$5Fe_2O_3$–$30TeO_2$ composition. In the work [135] it was shown that the electrical conductivity of CuO–V_2O_5–TeO_2 glasses is primarily determined by the hopping carrier mobility rather than carriers concentration. The influence of Ag_2O and the concentration of transition metal oxides on the properties of vanadium tellurite [136, 137] and vanadium–molibdenium tellurite [138] glasses was investigated by Montany et al. According to the results of [136], there is a minimum of the conductivity at the Ag_2O concentration of 50 mol% and at this point the conductivity changes from predominantly electronic to predominantly ionic. In the work [137], glasses with general composition of $0.27Ag_2O-0.73[yV_2O_5-(1-y)TeO_2]$ were investigated and the minimum of conductivity was found at $y = 0.27$. In both cases the minimum was explained in terms of the assumption which was offered by Jayasinghe [129] that there are two types of pathways (ionic and electronic) which are blocking each other. To confirm this theory, the investigation of glasses of $xAg_2O-(1-x)[0.5V_2O_5 \cdot 0.5MoO_3]2TeO_2$ and $xLi_2O \cdot (1-x)[0.5V_2O_5 \cdot 0.5MoO_3]2TeO_2$ systems was carried out [138, 139]. It was established that the addition of network modifier (Ag_2O, Li_2O) leads to the blocking of electronic conductivity pathways and at the crucial point (minimum of conductivity) the conductivity mechanism changes from electronic to ionic due to the formation ionic pathways. Electrical and structural properties of $65TeO_2$–$(35-x)V_2O_5$–xSm_2O_3 glasses were studied in [140], and an increase in samarium oxide concentration leads to a decrease in conductivity at $x \leq 2$ and the following increase in the Sm_2O_3 content leads to an increase in conductivity due to the incorporation of Sm_2O_3 into the glass network. A series of research devoted to the investigation of the influence of the effect of mixed transition-metal ions on the glass properties was conducted by Dutta et al. [141–143]. It was found that for V_2O_5–Fe_2O_3–P_2O_5 system (V/(V + Fe) ratio was varied), there is a mixed transition-ion effect which is similar to the mixed alkali effect in glasses [141]. This phenomenon was explained in terms of the enhancement of Anderson localization of the polarons when two transition ions are mixed in a glass. In further works [142, 143], it was shown that the manganese ions have no significant influence on the electrical properties of glasses of Fe_2O_3–MnO–P_2O_5 and V_2O_5–MnO–P_2O_5 systems. Electrical properties of BaO–V_2O_5–TeO_2 glasses were investigated in [144]; according to the results of the investigation, an increase in vanadium oxide concentration leads to an increase in conductivity and an increase in barium oxide concentration results in a decrease in the conductivity, which is connected with the blocking of polaron hopping due to the formation of non-bringing oxygen caused by Ba^{2+} ions. Structural and electrical properties of xLi$_2$O–$(1-x)$[0.3V_2O_5–0.7TeO_2] glasses with $0 \leq x \leq 0.4$ were investigated in [145]. It was found that an increase in Li_2O concentration leads to an increase in the distance between vanadium ion hopping sites and, consequently, a decrease in electronic conductivity with an increase in ionic one. Electrical properties of $(50–x)$Li$_2$O–xV$_2$O$_5$–50P$_2$O$_5$ (x from 10 to 40 mol%) glasses were investigated by Behzard et al. [146]. The highest conductivity had the composition with $x = 40$ mol%. Pant et al. studied the properties of 10BaO–xAg$_2$O–(85-x)V$_2$O$_5$–5TeO$_2$ glasses [147]. The maximal value of conductivity of ~$1.6 \cdot 10^{-6}$ S·cm^{-1} had the composition with $x = 55$ mol%. The transition from electronic to ionic conductivity occurs at $x = 30$ mol%.

Electrical conductivity of xA$_2$O–50V$_2$O$_5$–$(50-x)$P$_2$O$_5$ (A = Li, Na, K) glasses was investigated by Barczynski [78]; a decrease in electrical conductivity with an increase in alkali ion concentration was explained by the change in structural units of vanadium atoms from square to tetrahedral pyramids; as a consequence, the large part of vanadium ions was excluded from the conductivity process due to the small activity of tetrahedral pyramids as hopping centers. The conductivity of $(60-x)$V$_2$O$_5$–40TeO$_2$–xSb$_2$O$_3$ glasses was studied in [148]; the conductivity occurs through the non-adiabatic small polaron hopping mechanism and increases with an increase in vanadium oxide concentration. The influence of heat treatment on the 40V$_2$O$_5$–20Fe$_2$O$_3$–40P$_2$O$_5$ glass was investigated in [149]; the growth of room temperature conductivity from 1.4 to $4.1 \cdot 10^{-4}$ S·cm^{-1} with an increase in the treatment time to 4 h was observed; however, a further increase in the treatment time led to the decrease in conductivity. It was shown in [150] that the introduction of La_2O_3 in vanadium phosphate glasses leads to a slight increase in conductivity, and these glasses have mixed ionic–electronic conductivity which is provided by La^{3+} ions (in authors opinion) and polarons. Electrical conductivity of $(1–x)$SrO–xV$_2$O$_5$ ($0.5 \leq x \leq 0.8$) glasses was studied and analyzed by Siddiqui et al. in terms of different hopping models [151]; the conductivity of these glasses rises with an increase in V_2O_5 concentration and the polaron hopping is non-adiabatical. Mirzayi et al. investigated the properties of xZnO–$(75-x)$V$_2$O$_5$–25P$_2$O$_5$ (x = 10, 20, 30, 40, 50, 60, 70 mol%) glasses and they found that the conductivity increases with the growth of vanadium oxide concentration [152]. The investigation of properties of xBaO–$(100-x)$V$_2$O$_5$ (x = 30, 40 and 45 mol%) glasses is presented in [153]; it was found that despite all samples are amorphous according to XRD data, in glasses with V_2O_5 concentration more than 55 mol% there are some nanostructures. Electrical conductivity of such glasses decreases with the growth of barium oxide concentration. Electrochemical properties of glassy and glassy-crystalline materials of Li_2O–V_2O_5–P_2O_5–Fe_2O_3 system were investigated by Aoyagi et al. [65]; both glass and glassy-crystalline material showed good cycleability: the capacity retention was 66 and 68% after 100 cycles for glass and crystallized glass, respectively. Properties of xK_2O–$(50\text{-}x)V_2O_5$–$50TeO_2$ glasses with $x = 0$, 2.5, 5, 7.5, 10 and 15 mol% were investigated in [154]. The replacement of V_2O_5 by K_2O leads to the appearance of the conductivity maximum at $x = 7.5$ mol%, which is connected with the change in the type of charge carriers from ions to electrons. The influence of annealing in H_2 atmosphere on the properties of $70V_2O_5{\cdot}30TeO_2$ glasses was studied by Mori [155]. The glass pellets were annealed at 473 K for 1 and 2 h and the highest value of conductivity of $5.78{\cdot}10^{-5}$ $S{\cdot}cm^{-1}$ was achieved for the sample annealed in H_2 atmosphere for 2 h. It should be noted that this value of conductivity is higher than obtained by Krins [145] for the same composition (~$3.2{\cdot}10^{-6}$ $S{\cdot}cm^{-1}$).

Although boron oxide is a widely used glass former, vanadium glasses with B_2O_3 having pure electronic or mixed ionic–electronic conductivity received significantly less attention [79]. Dixit et al. [156] studied the electrical properties of xV_2O_5–$50B_2O_3$–$(1\text{-}x)TeO_2$ glasses in the range of x from 10 to 50. The glass with $x = 40$ has the value of conductivity of $7.8{\cdot}10^{-4}$ $S{\cdot}cm^{-1}$ at 310 K, that is, 1.5 times higher than the conductivity of $50V_2O_5$–$50B_2O_3$ glass. A further decrease in x leads to the fall of conductivity to $4.2{\cdot}10^{-9}$ $S{\cdot}cm^{-1}$ at 310 K for the composition of $10V_2O_5$–$50B_2O_3$–$40TeO_2$. Properties of binary V_2O_5–B_2O_3 glasses in the wide range of concentration up to 90 mol% of V_2O_5 were investigated by Sharma et al. [157]. It was found that glasses with a small vanadium oxide concentration are hydroscopic that leads to the crystallization of B_2O_3; therefore, only samples with V_2O_5 concentration of 80 and 90 mol% were obtained in fully amorphous state. Despite electrical conductivity of such glasses is smaller than binary V_2O_5–P_2O_5 and V_2O_5–TeO_2 glasses, its value is rather high (~10^{-5} $S{\cdot}cm^{-1}$ at 300 K). Glasses of V_2O_5–B_2O_3 system in the V_2O_5 concentration range from 60 to 90 mol%, and the influence of heat treatment on its electrical conductivity was investigated by Culea et al. [158]. According to the obtained data, the annealing of $90V_2O_5$–$10B_2O_3$ glass at 200–300 °C for 1 h increases the conductivity from $6.3{\cdot}10^{-5}$ $S{\cdot}cm^{-1}$ to $1.0{\cdot}10^{-3}$ $S{\cdot}cm^{-1}$ at room temperature. Properties of SrO–V_2O_5–B_2O_3 glasses were studied by Ichinose et al. [159]; it was found that an increase in B_2O_3 concentration leads to the growth of conductivity, while the rise of SrO content results in a decrease in the conductivity; the highest value of conductivity was ~$5.7{\cdot}10^{-5}$ $S{\cdot}cm^{-1}$ at 350 K for glass $13SrO$–$84V_2O_5$–$3B_2O_3$. In [160] the physical parameters of small polaron hopping in V_2O_5–B_2O_3 glasses were evaluated and the relaxation processes were studied. Research devoted to the investigation of electrical properties of BaO–V_2O_5–B_2O_3 glasses was carried out by Sanghi et al. [161]. Among glasses of BaO–V_2O_5–B_2O_3 system the highest value of conductivity is $5.68{\cdot}10^{-5}$ $S{\cdot}cm^{-1}$ at 473 K for the composition of $34BaO$–$16V_2O_5$–$40B_2O_3$. The investigation of glass with $80V_2O_5$–$20LiBO_3$ wt% composition and a test of its suitability as a cathode material were carried out by Afion et al. [68]. The influence of crystallization on the properties of $75V_2O_5$–$15B_2O_3$–$10P_2O_5$ glass was studied by Choi et al. [162]. Electrical conductivity of the glass was increased by the heat treatment at 275 °C for 5 h from $2.8 \cdot 10^{-8}$ to $7.5 \cdot 10^{-7}$ $S \cdot cm^{-1}$ at room temperature. Glasses of $xV_2O_5 \cdot 20Li_2O \cdot (80-x)[0.6B_2O_3{:}0.4ZnO]$ ($10 \leq x \leq 50$) system obtained by microwave synthesis were studied by Reddy et al. [163]; it was found that electrical conductivity changes nonlinearly that is connected with the present both transition metal and alkali ions. The character of conductivity changes from predominantly ionic to electronic with an increase in V_2O_5 concentration. Rajashekara et al. [164] studied the influence of the mixing of transition metal oxides on the electrical properties of $xCo_3O_4-(50-x)(V_2O_5)-50B_2O_3$ glasses where $x = 0.05$–0.45 and found that the concentration dependence of the conductivity passes through the maximum at $x = 0.3$ and reaches a value of $8.9 \cdot 10^{-5}$ $S \cdot cm^{-1}$ at 553 K.

A large contribution to the investigation of Li_2O–P_2O_5–V_2O_5 glasses and glass-ceramics was made by Garbarczyk et al. Investigation of the mixed-conducting glasses of Li_2O–V_2O_5–P_2O_5 system was presented in [165, 166]; the ratio of $[Li_2O]/[V_2O_5 + P_2O_5]$ was changed. The minimum of conductivity was observed at the ratio of 0.67 and the explanation of such conductivity behavior was similar to Montani [136–138]. Further, in [167, 168], the electrical conductivity of xLi_2O–$(100\text{-}2x)V_2O_5$–xP_2O_5 glasses for $x = 15, 25, 35, 40, 45$ was improvement by nanocrystallization at the temperature close to the crystallization temperature. The investigation was focused on the glass with $15Li_2O$–$70V_2O_5$–$15P_2O_5$ composition since it has rather high contribution of electronic conductivity to the total one and rather low crystallization temperature. The glass was annealed at 311 °C for 7 h and its conductivity was significantly increased; however, the annealing at 398 °C led to the decrease in conductivity. The influence of nanocrystallization on $xV_2O_5-(100-x)P_2O_5$ glasses with $60 < x < 90$ was studied in [168, 169]; it was shown that the conductivity and thermal stability of glasses were enhanced as a result of the growth of V_2O_5 nanocrystals within the glass matrix. The enhancement of conductivity can be connected with an increase in the amount of V^{4+}/V^{5+} couples or a formation of well-conducting regions. In further work [170] it was postulated that the formation of interfacial region between glass and crystallites promotes the formation of conduction paths which contain more V^{4+}/V^{5+} pairs than glassy matrix or crystalline by itself. The effect of microstructure and annealing time and temperature on the electrical conductivity of $90V_2O_5-10P_2O_5$ glasses was investigated in detail [171]. It was established that annealing time has the most influence on the conductivity in the range of 1–5 h and then the growth of conductivity significantly reduces; an increase in annealing temperature led to the growth of conductivity from $8 \cdot 10^{-5}$ $S \cdot cm^{-1}$ (without annealing) to $2 \cdot 10^{-3}$ $S \cdot cm^{-1}$ (annealing at 420 °C) at room temperature. Li_2O–V_2O_5–P_2O_5 glassy and nanocrystallized samples in a wide concentration range were investigated in [172]; the highest increase in conductivity has the composition of $50Li_2O$–$25V_2O_5$–$25P_2O_5$ and, despite this composition has predominantly ionic conductivity in a glassy state, drastic increase in electronic conductivity was observed after crystallization. The investigation of conductivity in the low-temperature range to –170 °C was carried out in [173]; the "core-shell" theory of nanocrystal formation was proposed. "Core-shell" theory was applied to the partly crystallized samples in which the cores are orthorhombic V_2O_5 nanocrystallites and the shells around them consist of strongly defective V_2O_5-rich phase. In the work [174] some theoretical basis of the proposed previous work "core-shell" theory is presented. The influence of synthesis and annealing temperature on the electrical conductivity of $90V_2O_5 \cdot 10P_2O_5$ glass was studied [175]; it was found that there is a strong dependence of electrical properties on thermal history of glass. The highest value of conductivity (0.07 $S \cdot cm^{-1}$ at room temperature) was obtained for the glasses melted at 1300 °C and nanocrystallized at 550 °C. The improvement of conductivity of $Li_3Me_2(PO_4)_2F_3$ glass-ceramics (Me = V, Fe, Ti) by nanocrystallization was described in [176]; the conductivity of glass with the composition of $Li_3V_2(PO_4)_2F_3$ was enhanced from $2 \cdot 10^{-7}$ to $1.7 \cdot 10^{-3}$ $S \cdot cm^{-1}$ at room temperature due to annealing at 500 °C. The properties of $Li_3V_2(PO_4)_3$ glass-ceramic nanocomposites were investigated by El-Desoky [177–179]. The effect of grain size on the transport in $Li_3V_2(PO_4)_3$ glass-ceramics annealed at different temperature was investigated [177, 178]; it was found that the conductivity had a maximum value ($1.7 \cdot 10^{-3}$ $S \cdot cm^{-1}$ at 400 K) when the glass was annealed at 450 °C for 4 h. An increase in the annealing time led to a decrease in conductivity.

As can be seen from the literature review, electrical conductivity of vanadate glasses is provided by polaron hopping mechanism [180] between V^{4+} and V^{5+} ions. The Arrhenius growth of the conductivity with temperature confirms this conclusion. Some values of electrical conductivity for vanadate glasses are summarized in Table 13.2 and visualized in Fig. 13.2. The same data for glass-ceramic are given in Table 13.3. According to Table 13.2, the conductivity of vanadate glasses at room temperature has an order of 10^{-5} $S \cdot cm^{-1}$. It is interesting to note that the conductivity can vary even for the same composition cited in different papers. Probably, this fact is due to different fraction of V^{4+} in the glass, since the amount of V^{4+} is difficult to control. This suggestion is confirmed by the mentioned research of the influence of annealing in H_2 atmosphere. The conductivity of vanadate-tellurite glass increased after annealing in H_2 [155]; probably because of reduction of the part of vanadium to tetravalent state.

Conductivity of vanadate glass-ceramics can reach values of order of 10^{-3} and even almost 10^{-1} $S \cdot cm^{-1}$ [175] at room temperature (Table 13.3). Probably, crystalline phases have a good conductivity due to lesser interatomic distances, a formation of percolation pathways and reduction vanadium to V^{4+} state. In any case transport characteristics of vanadate glasses can be sufficiently improved by heat treatment and preparation of glass-ceramics.

13.4 Experimental Techniques

13.4.1 Preparation Methods

The cheapest, fastest and more convenient way to obtain glass is a melt quenching technique. Precursors are mixed in desired ratios, melted and quenched using a special mold. The homogeneity of the melt can be achieved by long-time exposure and stirring. After quenching samples are usually annealed at a temperature below

Table 13.2 Electrical conductivity (σ) at temperature (T) of some glassy vanadates

No	The best composition in a studied system	σ, $S \cdot cm^{-1}$ (T, K)	References
1	$25ZnO \cdot 75V_2O_5$	$1.4 \cdot 10^{-5}$ (300 K)	[117]
2	$5TeO_2 \cdot 95V_2O_5$	~10^{-3} (300 K)	[118]
3	$(0.6V_2O_5 \cdot 0.4P_2O_5) + 2wt\%$ NiO	~10^{-6} (350 K)	[122]
4	$0.4TeO_2 \cdot 0.6V_2O_5$	~$1.6 \cdot 10^{-5}$ (298 K)	[123]
5	$70V_2O_5 \cdot 10PbO \cdot 20TeO_2$	$1.2 \cdot 10^{-4}$ (408 K)	[127]
6	$70V_2O_5 \cdot 20Sb \cdot 10TeO_2$	$7.13 \cdot 10^{-3}$ (473 K)	[131]
7	$65V_2O_5 \cdot 5Fe_2O_3 \cdot 30TeO_2$	$2.88 \cdot 10^{-4}$ (400 K)	[134]
8	$10CuO \cdot 57.5V_2O_5 \cdot 32.5TeO_2$	~10^{-3} (300 K) ~10^{-2} (400 K)	[135]
9	$81.56V_2O_5 \cdot 18.44P_2O_5$	~$2.2 \cdot 10^{-3}$ (575 K)	[143]
10	$10BaO \cdot 55Ag_2O \cdot 30V_2O_5 \cdot 5TeO_2$	~$1.6 \cdot 10^{-6}$ (room)	[147]
11	$60V_2O_5 \cdot 40TeO_2$	$4.35 \cdot 10^{-5}$ (room)	[148]
12	$10La_2O_3 \cdot 30V_2O_5 \cdot 60P_2O_5$	~$7.2 \cdot 10^{-6}$ (475 K)	[150]
13	$20SrO \cdot 80V_2O_5$	~$4.5 \cdot 10^{-4}$ (400 K)	[151]
14	$30BaO \cdot 70V_2O_5$	$2.4 \cdot 10^{-6}$ (303 K)	[153]
15	$50TeO_2 \cdot 42.5V_2O_5 \cdot 7.5K_2O$	$6.6 \cdot 10^{-5}$ (300 K)	[154]
16	$90V_2O_5 \cdot 10B_2O_3$	$6.3 \cdot 10^{-5}$ (room)	[158]
17	$75V_2O_5 \cdot 15B_2O_3 \cdot 10P_2O_5$	$2.8 \cdot 10^{-8}$ (room)	[162]
18	$35Co_3O_4 \cdot 15V_2O_5 \cdot 50B_2O_3$	$8.9 \cdot 10^{-5}$ (553 K)	[164]

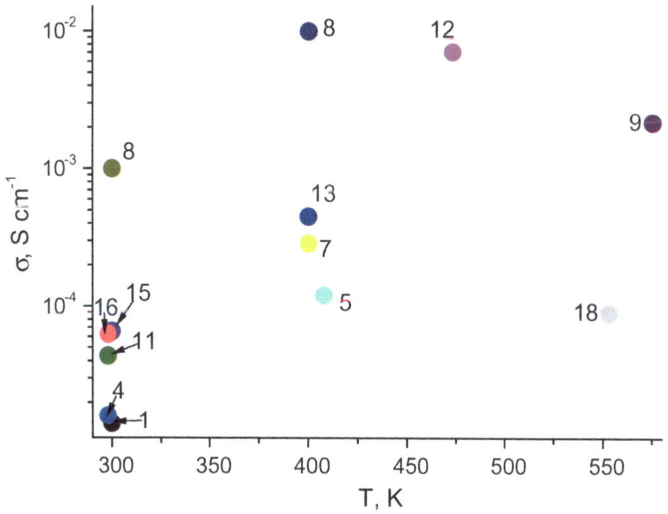

Fig. 13.2 Electrical conductivity (σ) of some glassy vanadates as a function of temperature. The glass compositions decoded by the numbers in Table 13.2

Table 13.3 The conductivity (σ) at room temperature of initial glasses and glass-ceramics obtained by annealing of glasses

Composition	Annealing conditions	σ glass, $S\cdot cm^{-1}$	σ glass-ceramics, $S\cdot cm^{-1}$	References
$90V_2O_5\cdot 10B_2O_3$	200–300 °C for 1 h	$6.3\cdot 10^{-5}$	$1.0\cdot 10^{-3}$	[158]
$90V_2O_5\cdot 10P_2O_5$	420 °C	$8–10\cdot 10^{-5}$	$2\cdot 10^{-3}$	[171]
$90V_2O_5\cdot 10P_2O_5$	550 °C	$2–8\cdot 10^{-5}$	0.07	[175]
$Li_3V_2(PO_4)_2F_3$	500 °C	$2\cdot 10^{-7}$	$1.7\cdot 10^{-3}$	[176]
$75V_2O_5\cdot 15B_2O_3\cdot 10P_2O_5$	275 °C for 5 h	$2.8\cdot 10^{-8}$	$7.5\cdot 10^{-7}$	[162]

glass transition temperature (T_g) to remove microstresses and prevent cracking. In most part of papers glass has been obtained in such way. Lithium carbonate (Li_2CO_3), boric acid (H_3BO_3) and vanadium oxide (V_2O_5) can be used as the starting reagents for lithium vanadate-borate glasses. The molten mixture is typically held at 900–1200 °C for several hours [79]. Platinum crucibles are recommended to use to avoid an interaction between the melt and crucible material. To obtain glass the melt is quenched between stainless steel plates.

Solid electrolyte can be obtained by traditional solid-state synthesis or sol–gel method. $Li_7La_3Zr_2O_{12}$ is usually obtained by sol–gel technique [92, 181]. Lithium carbonate, Li_2CO_3, lanthanum oxide, La_2O_3, and acidified solution of zirconyl nitrate, $ZrO(NO_3)_2$ are usually taken as initial materials for the LLZO synthesis. The most widely used stabilizing dopant is aluminum [87, 92, 97]. It can be introduced by hydrate of aluminum nitrate, $Al(NO_3)_3\cdot 9H_2O$. The sol–gel technique of LLZO synthesis is given in a number of works [92, 97, 181]. A comparison of LLZO synthesized by sol–gel and solid-state methods can be found in [181]. Composite electrolytes of LLZO with a glass are prepared by a mechanical mixing of components with a subsequent sintering [98–101]. Dense ceramic with garnet structure can also be obtained by self-propagating high-temperature syntheses [182].

The single electrochemical cell can be fabricated as follows: powdered glassy vanadate cathode material is poured onto one side of a sintered electrolyte pellet with subsequent annealing of the obtained "sandwich" [54]. The annealing temperature must be sufficiently high to soften vanadate glass for a short time, typically, higher than the melting temperature. Experiments show that the optimal temperature is located in the range of 600–850 °C for vanadate systems. The annealing time needs to be long enough for a tight contact formation but not so long to prevent full interaction between cathode and electrolyte. Thus, the cathode material fills the pores at the interface due to the thermal treatment. Then the obtained two-layered tablet is cooled and anode can be applied. In [54] gallium–silver alloy was used as anode and reference electrode due to its good adhesion properties and easy operation.

13.4.2 Characterization Methods

X-ray diffraction (XRD) analysis is the simplest way to confirm a glassy state of the obtained material. In the case of amorphous substance, an XRD pattern is halo which significantly differs from XRD patterns of crystalline substances where sharp reflexes are present. Typical XRD patterns for crystalline, glass-ceramic and glassy materials are given in Fig. 13.3. The XRD analysis can also be used to determine the phase composition of solids. Generally, it is possible to calculate the degree of crystallinity from XRD data, but it is much more complicated techniques [183].

Thermal behavior of the system is usually studied by thermal analysis. Most common method is differential scanning calorimetry (DSC). A typical DSC curve of glasses contains a bend corresponding to glass transition temperature and exothermic effects corresponding to glass crystallization (Fig. 13.4). When DSC is performed at temperatures higher than the melting temperature, a process of melting is observed as an endothermic effect. A combination of XRD and DSC can provide more detailed information about a mechanism of glass crystallization and help to choose conditions for glass ceramic preparation.

It was noted in Sect. 13.2 that in oxide glasses vanadium can be present in different oxidation degrees [79, 184]. The fraction of V^{4+} in vanadate systems is one of the key parameters of electrochemical properties. To estimate this parameter a redox titration can be used [131]. Another method of the determination of V^{4+} concentration is an electron paramagnetic resonance (EPR) spectroscopy. Since V^{4+} ion ($3p^6 3d^1$ configuration) contains one unpaired electron, it is EPR-active in contrast to V^{5+} ions. The quantity of V^{4+} ions can be estimated as an area restricted by EPR spectra. This area is obtained by the double integration of EPR spectrum after the subtraction of baseline [185, 186]. In the case when other EPR-active ions are absent, the value of resulting spectrum area is proportional to the number of V^{4+} ions. The EPR

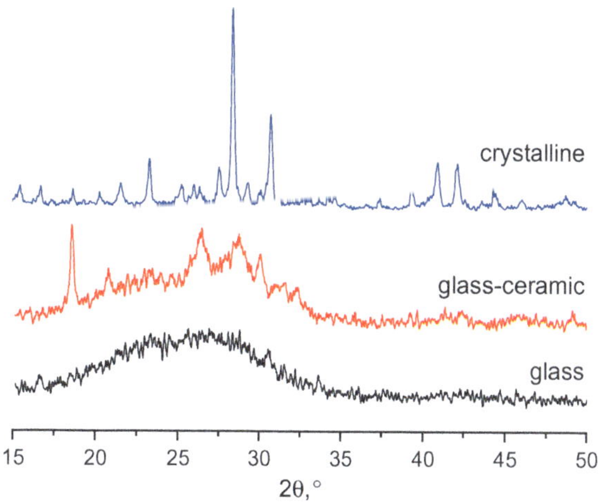

Fig. 13.3 Typical XRD patterns of glassy, glass-ceramic and crystalline materials

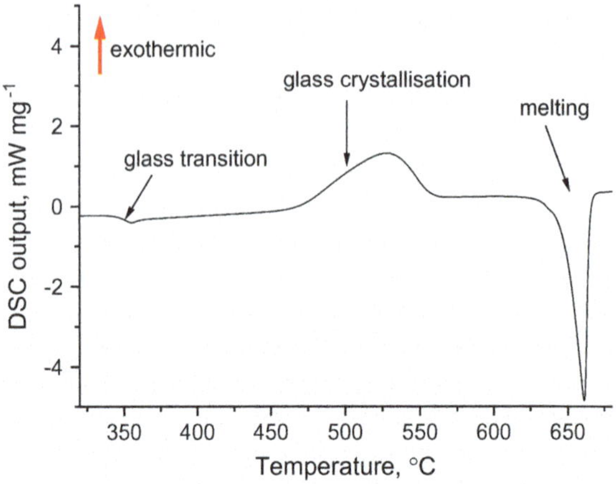

Fig. 13.4 A typical DSC curve of glassy material

spectrum of vanadate glasses typically has a well-resolved hyperfine structure at low vanadium concentration. This structure vanishes with increasing vanadium oxide concentration due to V^{4+}–O–V^{5+} super exchange mechanism [163]. Figure 13.5 shows some examples of EPR spectra of glassy vanadates.

13.4.3 Electrochemical Methods

Electrochemical impedance spectroscopy (EIS, IS) is the most common method of studying both ohmic and polarization resistance of the systems. EIS is used for determination of conductivity of individual components as well as whole power sources. To study the conductivity of solid electrolyte by EIS, blocking electrodes is usually applied. They can be made of gold, platinum or other metals sputtered in a different manner. An impedance spectra processing gives a value of electrical resistance. The conductivity can be calculated by the following equation:

$$\sigma = \frac{h}{R \cdot S}, \tag{13.2}$$

where R is the electrical resistance, h is the sample thickness and S is the sample area. The uncertainty of conductivity can be expressed via a differentiation of (13.2). In the case when h and S can be accurately measured, for example, two-electrode cell with plane-parallel electrodes, the conductivity uncertainty can be expressed as:

$$\mathrm{d}\sigma = \frac{\sigma \cdot \mathrm{dR}}{\mathrm{R}}, \tag{13.3}$$

where dR is an uncertainty of the resistance determination. For interpretation of impedance spectra of batteries, different equivalent circuits are constructed. Usually,

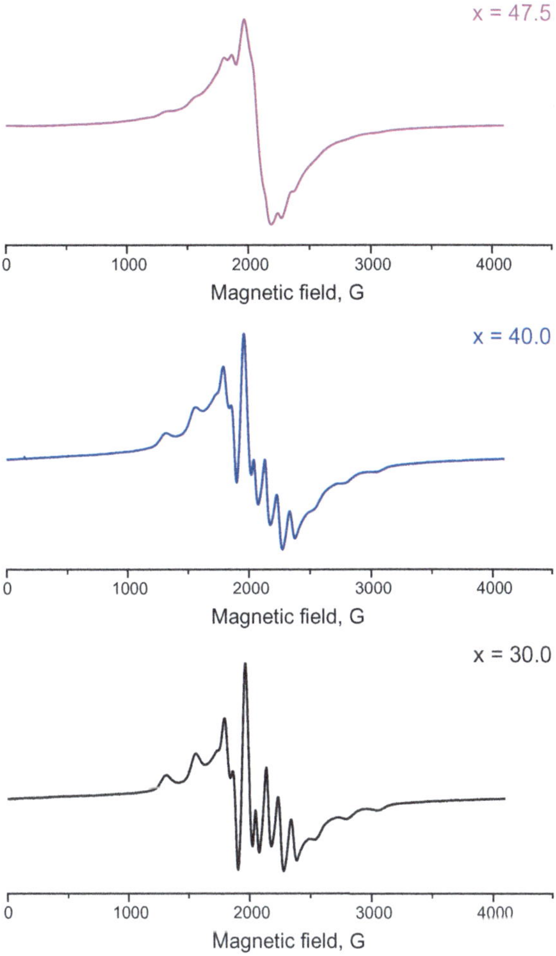

Fig. 13.5 EPR spectra of glasses in $30Li_2O–(70\text{-}x)B_2O_3–xV_2O_5$ system [79]

one relaxation process is fitted by the scheme, including a resistor and a capacitor (or a constant phase element, CPE) connected in parallel. Some examples of EIS equivalent circuits for the description of ASSB impedance spectra can be found in [26, 34, 44]. Typical impedance spectra of ASSB are presented in Fig. 13.6. An advantage of EIS technique is the simplicity of its realization. A disadvantage is an ambiguity of its interpretation, since the same spectra can be successfully fitted by absolutely different equivalent circuits. Conformity of every part of the spectrum to a certain process can be proved by the use of model cells. However, this is time-consuming and laborious work. Moreover, impedance spectra change during charge or discharge processes [44] by the reason of change in material properties due to potential variations.

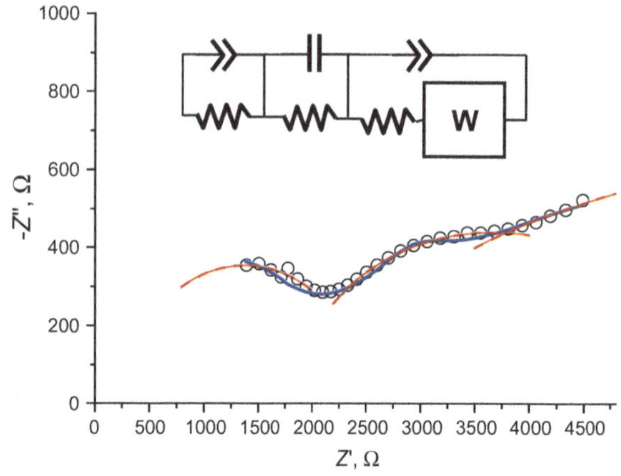

Fig. 13.6 An example of impedance spectrum of ASSB with an equivalent circuit

Chronopotentiometry is a method based on tracking of voltage value during charge or discharge at a constant current. This method is used rather for practical purposes than for fundamental ones. Chronopotentiometry shows how processes of charge or discharge are realized in the battery, at which voltage it can operate, what is a value of its polarization, can this device be used in practice and so on. For chronopotentiometry of ASSB very low current densities are typically applied, about several [30, 34, 38, 44] or several tens [27, 31, 36, 53, 187] $\mu A \cdot cm^{-2}$ because of high resistance.

Cyclic voltammetry (CV) is a method based on measurements of current while voltage changes cyclically with time. This method can help to estimate some dynamic characteristics of a system. The presence of irreversible peaks on CV curve can be attributed to some outside processes [44]. The position of CV peak is determined by an electroactive pair charged/discharged at this potential. Thus, the voltage difference between main cathodic and anodic peaks must correspond to electromotive force (EMF) of a battery (Fig. 13.7). It should be noted that the presence of incidental

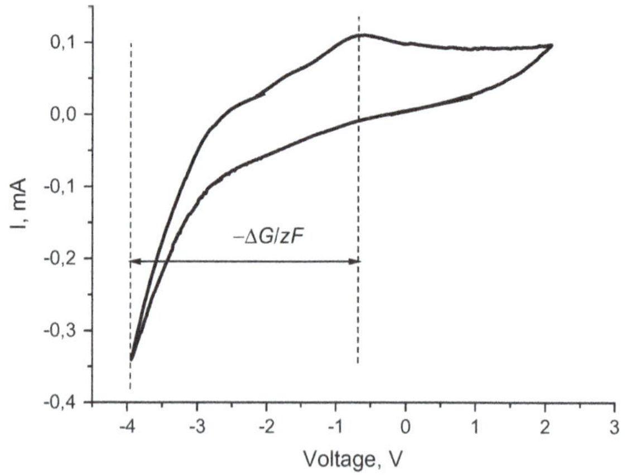

Fig. 13.7 An example of a cyclic voltammogram of vanadate-based ASSB

electroactive pairs, which are undesirable impurities, can be identified by the location and the number of peaks, see work [44] for an example.

Current interrupt (CI) method is based on the measurement of instant voltage drop after switching off the current. Since the ohmic (IR) losses immediately disappear after reducing the current to zero, the resistance can be calculated as the voltage drop divided by the current before the interrupt [188–190]. This method allows one to split voltage drop on an ohmic drop and a polarization. It should be noted that electrical lines of force disappear with the speed of light, but a potentiostat measures data with limited speed. Therefore, the value of IR determines by extrapolation of time dependency of voltage on time of the interrupt, and this is the main inaccuracy of the method.

13.4.4 Three-Electrode Cell

In practice, we need only two electrodes in power sources, namely, cathode and anode. To obtain all discharge and charge profiles and to demonstrate usability and characteristics of battery it is enough to make a two-electrode cell. But how can we determine electrode polarizations? Which electrode works better during charging or discharging? What is a reason of overvoltage? These and similar questions cannot be answered by performing experiments in two electrode cell because, in this case, a difference between two electrodes is always measured and contribution of electrodes to this difference cannot be distinguished. Thus, when it is necessary to receive more information about electrode processes at least three-electrode cell should be used. The third electrode is called "reference electrode" and its potential has to be constant during experiment. There are several ways to construct a measurement cell with three electrodes. Indeed, all electrodes have to be separated by electrolyte.

It is easy to apply the third electrode in liquid cells; it is enough to put a wire with the third electrode in the same electrolyte in which cathode and anode are embedded (Fig. 13.8a). The design of the three-electrode all-solid-state cell is much more complicated task. In fact, there are only a few works where a three-electrode all-solid-state cell is studied [28, 54]. At the moment, a combined search query of all-solid-state battery in Title and "three-electrode cell" in all fields returns only 18 papers on Sciencedirect.com website. Some of them do not deal with all-solid-state batteries or do not use three-electrode scheme for ASSB investigation. In [46], a Li–Al strip inserted in the electrolyte was used as a reference electrode. In 1996, Takada et al. [25] used a silver wire embedded in a glassy solid electrolyte. There are no details in this work and it can be only suggested that the wire was fused into the glass. In other works [28, 191, 192] a scheme with combined solid–liquid electrolyte was used. On one side of a solid electrolyte tablet something like a vessel with liquid electrolyte is constructed (Fig. 13.8b). Lithium reference and counter electrodes are embedded in this liquid electrolyte. In works [193, 194] carried out by the same authors, there is no description of a three-electrode cell. It is only mentioned that both reference and counter electrodes were lithium foil, but a mention of liquid electrolyte

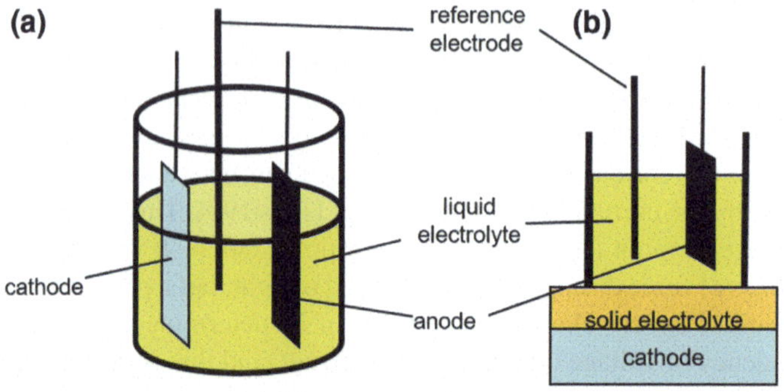

Fig. 13.8 A three-electrode cell for **a** a convenient liquid-electrolyte battery and **b** an all-solid-state battery (cross-section)

in the experimental part [193, 194] allows concluding that the same scheme with the vessel filled with electrolyte was used. Electrochemical processes in solids with the use of different electrodes were discussed in detail by Chebotin [195].

In [54] the next principal scheme of the design of a three-electrode cell for ASSB with ceramic or glassy electrolyte has been suggested. Electrolyte in the form of tablet or plate is covered by the positive electrode on the one side and, on the other side, two open-circuit counter electrodes are located (Fig. 13.9). For more accurate interpretation of measurement results, working and counter electrodes should be symmetrical (Fig. 13.9a), but one side of the electrolyte can be fully covered by an electrode to decrease current densities (Fig. 13.9b). Let us consider electric potentials

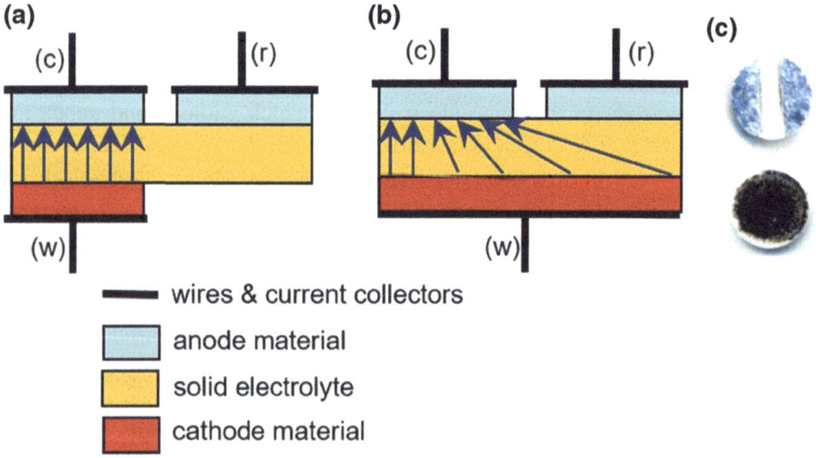

Fig. 13.9 The schematic representation of three-electrode cell for an all-solid-state electrochemical cell: **a** A principal scheme, **b** a real scheme applied in [54] and **c** photographs of anodic and cathodic side of the cell assembled in [54]. Arrows designate electric lines of force. Symbols (w), (c) and (r) mean working, counter and reference electrodes, respectively

observed in three-electrode cell in detail. The value of the voltage consists of electrode potentials, ohmic drop due to current flow and overvoltage. Let us observe in Fig. 13.9a and consider that we pass a constant current between working and counter electrodes (cathode and anode). The voltage drops between various electrode pairs are equal to:

$$U_{wc} = \varphi_w(q) - \varphi_c(q) + iR_{el} + iR_w(q) + iR_c(q) + |\eta_w(q,t)| + |\eta_c(q,t)|, \quad (13.4)$$

$$U_{wr} = \varphi_w(q) - \varphi_r + iR_{el} + iR_w(q) + |\eta_w(q,t)|, \quad (13.5)$$

$$U_{cr} = \varphi_c(q) - \varphi_r + iR_c(q) + |\eta_c(q,t)|, \quad (13.6)$$

where U is a voltage drop, i is the current, R is an electric resistance, φ is an electrode potential, η is an overvoltage, q is a state of charge (SOC) and t is time. The subscripts w, c, r and el mean working, counter and reference electrodes and electrolyte, respectively. Note that almost all terms in (4–6) are variables and they depend on the state of charge or/and time. The state of charge is, in turn, also the function of time. In general, we can only assume that R_{el} is constant (if we work in the region of its stability) and the potential of the reference electrode is constant. If we remember that one of the electrode is metallic (lithium or alloy) its resistance can be neglected. In our case it is the counter electrode (Fig. 13.9a). Furthermore, if the value of i is low enough all iR terms are vanishingly small. For example, when the resistance is equal to 1 kΩ and applied current is about 1 μA, the voltage drop is equal to 1 mV. This value is sufficiently lower than typical electrode potential differences which have an order of volts. Thus, (13.4–13.6) can be noticeably simplified:

$$U_{wc} = \varphi_w(q) - \varphi_c(q) + |\eta_w(q,t)| + |\eta_c(q,t)|, \quad (13.7)$$

$$U_{wr} = \varphi_w(q) - \varphi_r + |\eta_w(q,t)|, \quad (13.8)$$

$$U_{cr} = \varphi_c(q) - \varphi_r + |\eta_c(q,t)|. \quad (13.9)$$

According to (13.4) and (13.7), cathodic and anodic contributions to potential or overvoltage cannot be distinguished, as we mentioned above. U_{wc} is exactly that value we measure in a two-electrode cell. Since U_{wc} can be expressed as $U_{wc} = U_{wr} - U_{cr}$, it is enough to measure only two voltage drops instead of the three. Usually, a potentiostat can measure only one voltage difference simultaneously and one needs to perform at least two measurements to obtain enough information, as it was done in [54]. There are some potentiostats which allow measuring simultaneously two voltage differences in one cell by themselves (Bio-Logic) or via additional auxiliary electrodes (Solartron, Parstat, WonATech, Autolab with a pX1000 module).

13.5 All-Solid-State Battery with Vanadate Glass-Ceramic Cathode

A description of electrochemical tests of ASSB with vanadate glass-ceramic cathode is given in [54]. Values of electrode potentials for many redox pairs referenced to the standard hydrogen electrode (SHE) are tabulated at standard conditions. According to these data, potentials of V^{5+}/V^{4+} and Li^+/Li^0 pairs are about 1 and −3 V, correspondingly. Therefore, it can be expected that in the charged state the potential of such battery could reach a value about 4 V. The close value of voltage was obtained by Levy et al. [119] for electrochemical system Li-$60V_2O_5{\cdot}40TeO_2$ glass. In the case when lithium activity is decreased by the formation of alloys, the value of EMF is slightly lower. For example, in case of Li–In alloy this difference is equal to ~0.6 V [25, 31]. The value of EMF can be expressed via thermodynamic characteristics of the system:

$$E = -\frac{\Delta G_r}{zF}, \tag{13.10}$$

where ΔG_r is Gibbs energy difference of the reaction between cathode and anode materials, z is a number of electrons participated in electrochemical reaction and F is the Faraday constant, ~96484 $C{\cdot}mol^{-1}$. From (10) it is evident that EMF of a power source depends on temperature.

The initial potential of the (Ga-Ag | Li^+ | vanadate glass-ceramics) cell is equal to ~1 V at 280 °C [54]. This value corresponds to the fully discharged state. After passing of ~0.36 °C of electricity through the system, the potential reaches a value of ~3.3 V [54]. As one can see, the reached potential is very close to the expected one. CV curve (Fig. 13.7) demonstrates that voltage difference between anodic and cathodic peaks is also about 3.3 V. All these facts confirm that vanadium switches between 4+ and 5+ states on the cathode, and metallic lithium (or its alloy) liberates on the anode. The electrochemical half-reactions can be written in the next way:

(Cathode): $Li_2O–B_2O_3–V_2O_{5\text{-}\delta} \rightarrow Li_{2\text{-}x}O–B_2O_3–V_2O_{5\text{-}\delta} + xLi^+ + e^-$,

(Anode): $Ga–Ag + Li^+ + e^- \rightarrow Li–Ga–Ag$.

The reactions are written for the battery charging; for discharging they pass in opposite direction. As we mentioned above, the amount of V^{4+} in cathode material must be higher than the amount of deintercaleted lithium. Otherwise, the cathode half-reaction is impossible because maximal oxidation degree of vanadium is 5+. It is preferable to assemble the electrochemical cells in the discharged state, because in this case anode and cathode materials remain in more stable and, consequently, less reactive state.

In such ASSB only very low current densities are achievable because of high resistance and significant thickness of the solid electrolyte. The charge/discharge process are carried out in a galvanostatic mode with current densities of tens microamperes per square centimeter. An example of such charging curve is shown in Fig. 13.10. The voltage of the charging curve does not correspond to open-circuit voltage (OCV);

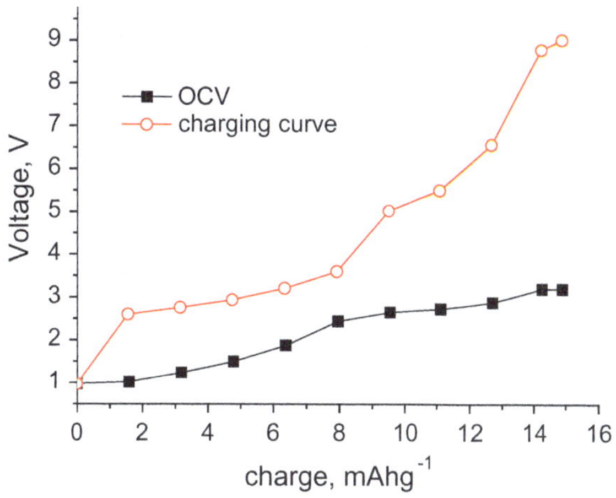

Fig. 13.10 Charging curve and OCV of (Li–Ga–Ag | Li^+ | $25Li_2O{\cdot}30B_2O_3{\cdot}45V_2O_5$) cell [54]. Current density on the cathode was about 19.8 $\mu A{\cdot}cm^{-2}$

the voltage includes all terms from (13.4), see Sect. 4.4. One can estimate the value of OCV during charge (discharge) by a stepwise charging (discharging). For this purpose, the current should be turned off at regular intervals of time and then we need to wait for the relaxation of a potential value. The "OCV" curve in Fig. 13.10 was obtained in such way, and it can be seen that the main contribution to the battery voltage is given by such useless components as IR and/or overvoltage.

As we discussed above, more detailed information can be obtained using a reference electrode. In [54] reference electrode is made of the same material as anode, namely, gallium–silver alloy. At first, the cathode potential referred to the reference electrode was measured during a following cycle: charge-relaxation–discharge-relaxation. Then, the anode potential was measured in the same manner (Fig. 13.11). A step change in potential immediately after the turning on (off) current can be attributed to the ohmic drop, but this estimation is very rough due to the low registration frequency. Figure 13.11 shows that the cathode potential grows sufficiently during the charge while anode potential after the starting step remains approximately constant. The plateau on the anodic curve is attributed, more likely, to the formation of two-phase system. The second phase can be metallic lithium or, rather, intermetallic Li_3Ga_{14} compound. Thermodynamics of Li–Ga systems can be found in [196–198].

At the discharge the situation changes to the opposite. The cathode potential changes smoothly and slightly while the anode potential grows extremely. The reason of such anode behavior at the discharge can be lithium loss, low diffusion of lithium to the interface or breaking strong covalent bonds between lithium and gallium. It is interesting that in other ASSB, a discharge is also worse realized than a charge, for example, Trong et al. [199] and Peng et al. [49].

Let us consider one more contribution to the battery voltage which was not discussed early. This situation can appear in the case of planar construction, like TFB in Figs. 13.1b, 13.9, when a current collector is located on only one side of the electrode. During charging/discharging, electrons are removed from (or inserted to) electrode/current collector interface and ions are removed from (or inserted to) to

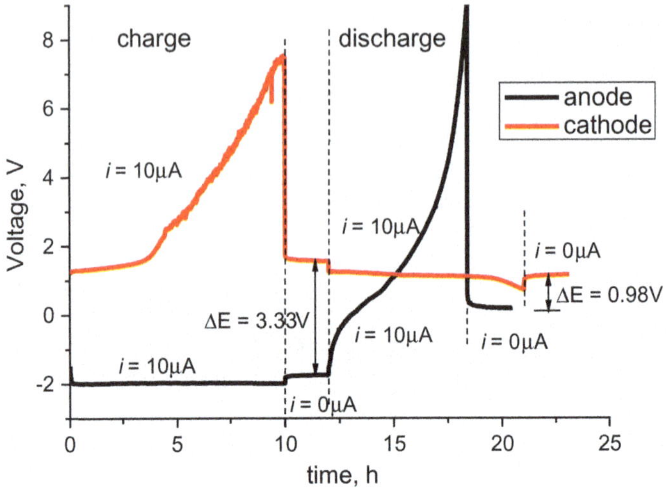

Fig. 13.11 Cathode and anode voltage referred to Ga–Ag electrode for a (Li–Ga–Ag | Li^+|$25Li_2O{\cdot}30B_2O_3{\cdot}45V_2O_5$) cell [54]

electrode/electrolyte interface. When the resistance of electrode material is high enough, charge separation can occur and, consequently, a capacitor can be obtained. The capacitance (C) of a plane parallel capacitor is determined by the formula:

$$C = \frac{\varepsilon S}{d}, \tag{13.11}$$

where ε is the absolute permittivity of dielectric located between conductive plates, S is an area of conductive plates and d is a distance between them. On the other hand, the capacitance provides relations between charge (q) and voltage (U) of the capacitor

$$U = \frac{q}{C}. \tag{13.12}$$

Combining (11) and (12) one can obtain:

$$U = q\frac{d}{\varepsilon S} \tag{13.13}$$

or for galvanostatic mode:

$$U = It\frac{d}{\varepsilon S}, \tag{13.14}$$

where I is the current and t is the time. One can see that in this case voltage grows linearly and can give an additional term(s) in (13.4–13.9). The system will relax to an equilibrium potential due to ionic and electronic motion and a rate of this relaxation depends on mobility of charge carriers in the electrode material. This phenomenon is especially noticeable when a current is switched off and other sources of the potential variation (growing overvoltage and changing OCV of the cell) disappear with time, see (13.4). Note that the effect of such charge accumulation is undesirable because it hinders a battery charging. A sufficient fraction of voltage in Fig. 13.10 can be associated with this effect. Electron conducting additions to cathode material are required to neutralize this phenomenon.

13.6 Conclusion

In this chapter, reasons of growing interest to all-solid-state batteries are considered. A brief review of challenges of ASSB development is presented and approaches to their overcoming are discussed. Advantages of glassy and glass-ceramic materials are given. Among glassy cathode materials, vanadate systems are the most interesting since they have the highest conductivity. According to the literature review, the conductivity of vanadium glasses can reach the value of order about 10^{-5} S·cm^{-1} at room temperature. Values of conductivity can be extremely increased by the heat treatment followed by glass-ceramic formation; the conductivity is obtained after the heat treatment glass-ceramics achieves 10^{-1} S·cm^{-1}. In all these cases conductivity is realized through polaron hopping mechanism. Further, conventional methods and approaches to the work with ASSB are described and the original design of all-solid-state cell with Ga-Ag reference electrode. Results of the electrochemical study of the behavior of the all-solid-state cell model with glass-ceramic vanadate cathode are presented. It is shown that OCV of the charged cell is in a full compliance with thermodynamic properties of the studied system. Reasons of a high overvoltage are suggested; overvoltages can be caused by lithium loss, the difficulty of the destruction of covalent Li–Ga bonds, and rather high resistance of vanadate electrode and originating in it charge separation.

Acknowledgements The authors are grateful to Prof. R. Murugan for kindly invitation to write this chapter.

The research has been carried out with the equipment of the Shared Access Center "Composition of Compounds" of the Institute of High-Temperature Electrochemistry of Ural Branch of RAS, Yekaterinburg, Russian Federation.

References

1. Sudworth JL, Tilley AR (1985) The Sodium sulfur battery. Chapman & Hall, New York
2. Koenig AA, Rasmussen JR (1990) Development of a high specific power sodium sulfur cell. In: Proceedings of the 34th international power sources symposium, Cherry Hill, NJ, USA, June 1990. https://doi.org/10.1109/ipss.1990.145783

3. Sudworth JL (1994) Zebra batteries. J Power Sourc 51:105–114. https://doi.org/10.1016/0378-7753(94)01967-3
4. Sudworth JL (2001) Sodium/nickel chloride (ZEBRA) battery. J Power Sour 100:149–163. https://doi.org/10.1016/S0378-7753(01)00891-6
5. Parthasarathy G, Weber N, Virkar AV (2007) High temperature sodium–zinc chloride batteries with sodium Beta-alumina solid electrolyte. ECS Trans 6:67–76. https://doi.org/10.1149/1.2811944
6. Ellis BL, Nazar FN (2012) Sodium and sodium-ion energy storage batteries. Curr Opin Solid State Mat Sci 16:168–177. https://doi.org/10.1016/j.cossms.2012.04.002
7. Li GS, Lu XC, Kim JY, Lemmon JP, Sprenkle VL (2013) Cell degradation of a Na–$NiCl_2$ (ZEBRA) battery. J Mater Chem A 1:14935–14942. https://doi.org/10.1039/C3TA13644B
8. Li GS, Lu XC, Kim JY, Meinhardt KD, Chang HJ, Canfield NL, Sprenkle VL (2016) Advanced intermediate temperature sodium–nickel chloride batteries with ultra-high energy density. Nature Comm 7:10683. https://doi.org/10.1038/ncomms10683
9. http://www.ngk.co.jp/english/announce/111031_nas.html. Accessed 26 Jun 2014
10. Bushkova OV, Andreev OL, Batalov NN, Shkerin SN, Kuznetsov MV, Tyutyunnik AP, Koryakova OV, Song EH, Chung HJ (2006) Chemical interactions in the cathode half-cell of lithium-ion batteries. Part I. Thermodyn Simul J Power Sour 157:477–482. https://doi.org/10.1016/j.jpowsour.2005.07.078
11. Tabuchi H (2013) New Problem for Boeing 787 Battery Maker. In: The New York Times. https://www.nytimes.com/2013/03/28/business/gs-yuasa-discovers-problems-with-its-car-battery.html. Accessed 27 March 2013
12. Mannstein (2013) Mitsubishi reports fire in i-MiEV battery pack, melting in Outlander PHEV pack. In: Green car congress. http://www.greencarcongress.com/2013/03/mmc-20130327.html. Accessed 27 March 2013
13. Jensen C (2013) Tesla Says Car Fire Started in Battery. In: The New York Times. https://wheels.blogs.nytimes.com/2013/10/02/highway-fire-of-tesla-model-s-included-its-lithium-battery/. Accessed 2 Oct 2013
14. Vlasic B, Trop J (2013) After 3 Fires, Safety Agency Opens Inquiry Into Tesla Model S. In: The New York Times. https://www.nytimes.com/2013/11/20/business/us-safety-agency-opens-inquiry-into-tesla-fires.html. Accessed 19 Nov 2013
15. Hradecky S (2010) The risks of lithium batteries in aircraft cargo. In: The Aviation Herald. http://avherald.com/h?article=431f0863. Accessed 8 Oct 2010
16. Jana A, García RE (2017) Lithium dendrite growth mechanisms in liquid electrolytes. Nano Energy 41:552–565. https://doi.org/10.1016/j.nanoen.2017.08.056
17. Kim JG, Son B, Mukherjee S, Schuppert N, Bates A, Kwon O, Choi MJ, Chung HYe, Park S (2015) A review of lithium and non-lithium based solid state batteries. J Power Sour 282:299–322. https://doi.org/10.1016/j.jpowsour.2015.02.054
18. Tan R, Yang J, Zheng J, Wang K, Lin L, Ji Sh, Liu J, Pan F (2015) Fast rechargeable all-solid-state lithium ion batteries with high capacity based on nano-sized Li_2FeSiO_4 cathode by tuning temperature. Nano Energy 16:112–121. https://doi.org/10.1016/j.nanoen.2015.06.016
19. Kamaya N, Homma K, Yamakawa Yu, Hirayama M, Kanno R, Yonemura M, Kamiyama T, Kato Yu, Hama Sh, Kawamoto K, Mitsuiet A (2011) A lithium superionic conductor. Nature Mat 10:682–686. https://doi.org/10.1038/nmat3066
20. Kuhn A, Duppel V, Lotsch BV (2013) Tetragonal $Li_{10}GeP_2S_{12}$ and Li_7GePS_8—exploring the Li ion dynamics in LGPS Li electrolytes. Energy Environ Sci 6:3548–3552. https://doi.org/10.1039/C3EE41728J
21. Yo Seino, Ts Ota, Takada K, Hayashi A, Tatsumisago M (2014) A sulphide lithium super ion conductor is superior to liquid ion conductors for use in rechargeable batteries. Energy Environ Sci 7:627–631. https://doi.org/10.1039/C3EE41655K
22. Shkerin SN, Profatilova IA, Lee JH (2009) Frequency dependence of conductivity of ethylene carbonate based electrolyte for Li-ion battery. Ionics 15:35–42. https://doi.org/10.1007/s11581-008-0235-y

23. Shkerin SN, Profatilova IA, Roh S (2009) Effect of ethylene carbonate concentration on the conductivity of carbonate-based electrolytes with $LiPF_6$ for Li-ion batteries. Ionics 15:761–764. https://doi.org/10.1007/s11581-009-0376-7
24. Tukamoto H, West AR (1997) Electronic Conductivity of $LiCoO_2$ and Its Enhancement by Magnesium Doping. J Electrochem Soc 144:3164–3168. https://doi.org/10.1149/1.1837976
25. Takada K, Aotani N, Iwamoto K, Kondo Sh (1996) Solid state lithium battery with oxysulfide glass. Solid State Ionics 86–88:877–882. https://doi.org/10.1016/0167-2738(96)00199-3
26. Ya Iriyama, Kako T, Yada Ch, Abe T, Ogumi Z (2005) Charge transfer reaction at the lithium phosphorus oxynitride glass electrolyte/lithium cobalt oxide thin film interface. Solid State Ionics 176:2371–2376. https://doi.org/10.1016/j.ssi.2005.02.025
27. Hayashi M, Takahashi M, Sakurai Y (2007) Preparation of positive $LiCoO_2$ films by electron cyclotron resonance (ECR) plasma sputtering method and its application to all-solid-state thin-film lithium batteries. J Power Sources 174:990–995. https://doi.org/10.1016/j.jpowsour.2007.06.081
28. Kim KH, Ya Iriyama, Yamamoto K, Kumazaki Sh, Asaka T, Tanabe K, Fisher CAJ, Ts Hirayama, Murugan R, Ogumi Z (2011) Characterization of the interface between $LiCoO_2$ and $Li_7La_3Zr_2O_{12}$ in an all-solid-state rechargeable lithium battery. J Power Sour 196:764–767. https://doi.org/10.1016/j.jpowsour.2010.07.073
29. Kotobuki M, Kanamura K, Sato Y, Yoshida T (2011) Fabrication of all-solid-state lithium battery with lithium metal anode using Al_2O_3-added $Li_7La_3Zr_2O_{12}$ solid electrolyte. J Power Sour 196:7750–7754. https://doi.org/10.1016/j.jpowsour.2011.04.047
30. Kotobuki M, Kanamura K (2013) Fabrication of all-solid-state battery using $Li_5La_3Ta_2O_{12}$ ceramic electrolyte. Ceram Int 39:6481–6487. https://doi.org/10.1016/j.ceramint.2013.01.079
31. Yubuchi S, Teragawa Sh, Aso K, Tadanaga K, Hayashi A, Tatsumisago M (2015) Preparation of high lithium-ion conducting Li_6PS_5Cl solid electrolyte from ethanol solution for all-solid-state lithium batteries. J Power Sour 293:941–945. https://doi.org/10.1016/j.jpowsour.2015.05.093
32. Huang B, Yao X, Huang Z, Guan Y, Yi J, Xu X (2015) Li_3PO_4-doped $Li_7P_3S_{11}$ glass-ceramic electrolytes with enhanced lithium ion conductivities and application in all-solid-state batteries. J Power Sour 284:206–211. https://doi.org/10.1016/j.jpowsour.2015.02.160
33. Okumura T, Takeuchi T, Kobayashi H (2016) All-solid-state lithium-ion battery using $Li_{2.2}C_{0.8}B_{0.2}O_3$ electrolyte. Solid State Ionics 288:248–252. https://doi.org/10.1016/j.ssi.2016.01.045
34. Schichtel P, Geiß M, Leichtweiß T, Sann J, Weber DA, Janek J (2017) On the impedance and phase transition of thin film all-solid-state batteries based on the $Li_4Ti_5O_{12}$ system. J Power Sour 360:593–604. https://doi.org/10.1016/j.jpowsour.2017.06.044
35. Woo SP, Lee SH, Soo Yoon YS (2017) Characterization of $LiCoO_2$/multiwall carbon nanotubes with garnet-type electrolyte fabricated by spark plasma sintering for bulk-type all-solid-state batteries. Compos Part B: Eng 124:242–249. https://doi.org/10.1016/j.compositesb.2017.05.025
36. Ito Y, Otoyama M, Hayashi A, Ohtomo T, Tatsumisago M (2017) Electrochemical and structural evaluation for bulk-type all solid-state batteries using Li_4GeS_4-Li_3PS_4 electrolyte coating on $LiCoO_2$ particles. J Power Sour 360:328–335. https://doi.org/10.1016/j.jpowsour.2017.05.112
37. Yoon M, Lee S, Lee D, Kim J, Moon J (2017) All-solid-state thin film battery based on well-aligned slanted $LiCoO_2$ nanowires fabricated by glancing angle deposition. J Power Sour 412:537–544. https://doi.org/10.1016/j.apsusc.2017.03.268
38. Lin J, Wu Y, Bi R, Guo H (2017) All-solid-state microscale lithium-ion battery fabricated by a simple process with graphene as anode. Sensor Actuat A-Phys 253:218–222. https://doi.org/10.1016/j.sna.2016.10.029
39. Glenneberg J, Andre F, Bardenhagen I, Langer F, Schwenzel Ju, Kun R (2016) A concept for direct deposition of thin film batteries on flexible polymer substrate. J Power Sour 324:722–728. https://doi.org/10.1016/j.jpowsour.2016.06.007

40. Ahmed NA, Hammache H, Eyraud M, Chassigneux C, Knauth P, lahreche A, Makhloufi L, Gabouzedoi N (2018) Morphological and optical properties of ZnO thin films grown on Si and ITO glass substrates. Ionics 24:277–284. https://doi.org/10.1007/s11581-017-2194-7
41. Bates JB, Dudney NJ, Neudecker B, Ueda A, Evans CD (2000) Thin-film lithium and lithium-ion batteries. Solid State Ionics 135:33–45. https://doi.org/10.1016/S0167-2738(00)00327-1
42. Pelé V, Flamary F, Bourgeois L, Pecquenard B, Le Cras F (2015) Perfect reversibility of the lithium insertion in FeS_2: the combined effects of all-solid-state and thin film cell configurations. Electrochem Commun 51:81–84. https://doi.org/10.1016/j.elecom.2014.12.009
43. Pat S, Özen S, Yudar H, Korkmaz Ş, Pat Z (2017) The transparent all-solid-state rechargeable micro-battery manufacturing by RF magnetron sputtering. J Alloy Compd 713:64–68. https://doi.org/10.1016/j.jallcom.2017.04.169
44. Yu R, Bao JJ, Chen TT, Zou BK, Wen ZY, Guo XX, Chen ChH (2017) Solid polymer electrolyte based on thermoplastic polyurethane and its application in all-solid-state lithium ion batteries. Solid State Ionics 309:15–21. https://doi.org/10.1016/j.ssi.2017.06.013
45. Julien C (1996) Electrochemical properties of disordered cathode materials. Ionics 2:169–178. https://doi.org/10.1007/BF02376017
46. Hammou A, Hammouche A (1988) All solid state Li-$Li_{1+x}V_3O_8$ secondary batteries. Electrochim Acta 33:1719–1720. https://doi.org/10.1016/0013-4686(88)85005-9
47. Zhu Y, Li J, Liu J (2017) A bifunctional ion-electron conducting interlayer for high energy density all-solid-state lithium-sulfur battery. J Power Sour 351:17–25. https://doi.org/10.1016/j.jpowsour.2017.03.072
48. Chen S, Zhao Y, Yang J, Yao L, Xu X (2017) Hybrid solid electrolytes with excellent electrochemical properties and their applications in all-solid-state cells. Ionics 23:2603–2611. https://doi.org/10.1007/s11581-016-1905-9
49. Peng G, Yao X, Wan H, Huang B, Yin J, Ding F, Xu X (2016) Insights on the fundamental lithium storage behavior of all-solid-state lithium batteries containing the $LiNi_{0.8}Co_{0.15}Al_{0.05}O_2$ cathode and sulfide electrolyte. J Power Sour 307:724–730. https://doi.org/10.1016/j.jpowsour.2016.01.039
50. Wakasugi J, Munakata H, Kanamura K (2017) Effect of Gold Layer on Interface Resistance between Lithium Metal Anode and $Li_{6.25}Al_{0.25}La_3Zr_2O_{12}$ Solid Electrolyte. J Electrochem Soc 164:A1022–A1025. https://doi.org/10.1149/2.0471706jes
51. Han X, Gong Y, Fu K, He X, Hitz GT, Dai J, Pearse A, Liu B, Wang H, Rubloff G, Mo Y, Thangadurai V, Wachsman ED, Hu L (2017) Negating interfacial impedance in garnet-based solid-state Li metal batteries. Nat Mater 16:572–579. https://doi.org/10.1038/nmat4821
52. Pershina SV, Druzhinin KV, Shevelin PYu, Raskovalov AA (2016) Cathode half-cell of all-solid-state battery, modified with $LiPO_3$ glass. Russ J Appl Chem 89:71–75. https://doi.org/10.1134/S1070427216060094
53. Suzuki Sh, Kawaji J (2017) Development of complex hydride-based all-solid-state lithium ion battery applying low melting point electrolyte. J Power Sour 359:97–103. https://doi.org/10.1016/j.jpowsour.2017.05.019
54. Il'ina EA, Saetova NS, Raskovalov AA (2016) All-solid-state battery Li–Ga–Ag | $Li_7La_3Zr_2O_{12}$ + Li_2O–Y_2O_3–SiO_2 | Li_2O–V_2O_5–B_2O_3. Russ J Appl Chem 89:1434–1439. https://doi.org/10.1134/s1070427216090081
55. Kercher AK, Chapel AS, Kolopus JA, Boatner LA (2017) Mixed polyanion glass cathodes: mixed alkali effect. J Electrochem Soc 164:A2777–A2782. https://doi.org/10.1149/2.1881712jes
56. Machida N, Fuchida R, Minami T (1989) Electrochemical insertion of lithium ions into V_2O_5 glasses containing transition-metal oxides. Solid State Ionics 35:295–298. https://doi.org/10.1016/0167-2738(89)90311-1
57. Machida N, Fuchida R, Minami T (1990) Cyclic voltammetric study on vanadium oxide based glasses as cathode in rechargeable lithium batteries. Solid State Ionics 37:299–302. https://doi.org/10.1016/0167-2738(90)90191-S

58. Machida N, Fuchida R, Minami T (1990) Preparation and lithium intercalation properties of rapidly quenched glasses in the system Fe_2O_3-V_2O_5. Solid State Ionics 40(41):589–592. https://doi.org/10.1016/0167-2738(90)90077-5
59. Denis S, Baudrin E, Touboul M, Tarascon JM (1997) Synthesis and electrochemical properties of amorphous vanadates of general formula RVO_4 (R = In, Cr, Fe, Al, Y) vs. Li. J Electrochem Soc 144:4099–4109. https://doi.org/10.1149/1.1838150
60. Piffard Y, Leroux F, Guyomard D, Mansot JL, Tournoux M (1997) The amorphous oxides $MnV_2O_{6+\delta}$ ($0 < \delta < 1$) as high capacity negative electrode materials for lithium batteries. J Power Sour 68:698–703. https://doi.org/10.1016/S0378-7753(96)02576-1
61. Sabi Yu, Sato S, Hayashi S, Furuya T, Kusanagi S (2014) A new class of amorphous cathode active material $Li_xM_yPO_z$ (M = Ni, Cu Co, Mn, Au, Ag, Pd). J Power Sour 258:54–60. https://doi.org/10.1016/j.jpowsour.2014.02.021
62. Kerchera AK, Ramey JO, Carroll KJ, Kiggans JO, Dudney NJ, Meisner RA, Boatner LA, Veith GM (2014) Mixed polyanion glass cathodes: iron phosphate vanadate glasses. J Electrochem Soc 161:A2210–A2215. https://doi.org/10.1149/2.0881414jes
63. Kerchera AK, Kolopus JA, Carroll KJ, Unocic RR, Kirklin S, Wolverton C, Stooksbury SL, Boatner LA, Dudney NJ (2016) Mixed polyanion glass cathodes: glass-state conversion reactions. J Electrochem Soc 163:A131–A137. https://doi.org/10.1149/2.0381602jes
64. Kercher AK, Kolopus JA, Sacci RL, Ruther RE, Gallego NC, Stooksbury SL, Boatner LA, Dudney NJ (2017) Mixed polyanion glass cathodes: effect of polyanion content. J Electrochem Soc 164:A804–A809. https://doi.org/10.1149/2.1341704jes
65. Aoyagi T, Fujieda T, Toyama T, Kono K, Takamatsu D, Hirano T, Naito T, Hayashi Y, Takizawa H (2016) Electrochemical properties and in-situ XAFS observation of Li_2O–V_2O_5–P_2O_5–Fe_2O_3 quaternary-glass and crystallized-glass cathodes. J Non-Cryst Solids 453:28–35. https://doi.org/10.1016/j.jnoncrysol.2016.09.016
66. Yamauchi H, Park G, Nagakane T, Honma T, Komatsu T, Sakai T, Sakamoto A (2013) Performance of lithium-ion battery with tin-phosphate glass anode and its characteristics. J Electrochem Soc 160:A1725–A1730. https://doi.org/10.1149/2.049310jes
67. Bates JB, Dudney NJ, Lubben DC, Gruzalski GR, Kwak BS, Yu X, Zuhr RA (1995) Thin-film rechargeable lithium batteries. J Power Sour 54:58–62. https://doi.org/10.1016/0378-7753(94)02040-A
68. Afyon S, Krumeich F, Mensing C, Borgschulte A, Nesper R (2014) New high capacity cathode materials for rechargeable Li-ion batteries: vanadate-borate glasses. Sci Rep 4:7113–7117. https://doi.org/10.1038/srep07113
69. Whittingham MS (2004) Lithium batteries and cathode materials. Chem Rev 10:4271–4302. https://doi.org/10.1021/cr020731c
70. Murphy DW, Christian PA, DiSalvo FJ, Carides JN (1979) Vanadium oxide cathode materials for secondary lithium cells. J Electrochem Soc 126:497–499. https://doi.org/10.1149/1.2129070
71. Cocciantelli JM, Doumer JP, Pouchard M, Broussely M, Labat J (1991) Crystal chemistry of electrochemically inserted $Li_xV_2O_5$. J Power Sour 34:103–111. https://doi.org/10.1016/0378-7753(91)85029-V
72. Delmas C, Cognac-Auradou H, Cocciantelli JM, Ménétrier M, Doumerc JP (1994) The $Li_xV_2O_5$ system: An overview of the structure modifications induced by the lithium intercalation. Solid State Ionics 69:257–264. https://doi.org/10.1016/0167-2738(94)90414-6
73. Miyaji F, Sakka S (1991) Structure of PbO-Bi_2O_3-Ga_2O_3 glasses. J Non-Cryst Solids 134:77–85. https://doi.org/10.1016/0022-3093(91)90013-V
74. Mekki A, Khattak GD, Wenger LE (2003) Structure and magnetic properties of lead vanadate glasses. J Non-Cryst Solids 330:156–167. https://doi.org/10.1016/j.jnoncrysol.2003.08.039
75. Milanova M, Iordanova R, Kostov KL (2008) Glass formation in the MoO_3-CuO-PbO system. J Non-Cryst Solids 355:379–385. https://doi.org/10.1016/j.jnoncrysol.2008.12.004
76. Doweidar H, El-Damrawi G, Mansour E, Fetouh RE (2012) Structural role of MgO and PbO in MgO–PbO–B_2O_3 glasses as revealed by FTIR; a new approach. J Non-Cryst Solids 358:941–946. https://doi.org/10.1016/j.jnoncrysol.2012.01.004

77. Cozar O, Ardelean I, Bratu I, Simon S, Craciun C, David L, Cefan C (2001) IR and EPR studies on some lithium-borate glasses with vanadium ions. J Mol Struct 563:421–425. https://doi.org/10.1016/S0022-2860(01)00442-2
78. Barczyński RJ, Król P, Murawski L (2010) AC and DC conductivities in V_2O_5–P_2O_5 glasses containing alkaline ions. J Non-Cryst Solids 356:1965–1967. https://doi.org/10.1016/j.jnoncrysol.2010.07.001
79. Saetova NS, Raskovalov AA, Antonov BD, Yaroslavtseva TV, Reznitskikh OG, Zabolotskaya EV, Kadyrova NI, Telyatnikova AA (2017) Conductivity and spectroscopic studies of Li_2O-V_2O_5-B_2O_3 glasses. Ionics Press. https://doi.org/10.1007/s11581-018-2452-3
80. Inaguma Y, Liquan C, Itoh M, Nakamura T, Uchida T, Ikuta H, Wakihara M (1993) High ionic conductivity in lithium lanthanum titanate. Solid State Commun 86:689–693. https://doi.org/10.1016/0038-1098(93)90841-A
81. Aono H, Sugimoto E, Sadaoka Y, Imanaka N, Adachi G (1989) Ionic conductivity of the lithium titanium phosphate ($Li_{1+X}M_XTi_{2-X}(PO_4)_3$, M = Al, Sc, Y, and La) systems. J Electrochem Soc 136:590–591. https://doi.org/10.1149/1.2096693
82. Leo CJ, Chowdari BVR, Rao GVS, Souquet JL (2002) Lithium conducting glass ceramic with Nasicon structure. Mater Res Bull 37:1419–1430. https://doi.org/10.1016/S0025-5408(02)00793-6
83. Thokchom JS, Gupta N, Kumar B (2008) Superionic conductivity in a lithium aluminum germanium phosphate glass-ceramic. J Electrochem Soc 155:A915–A920. https://doi.org/10.1149/1.2988731
84. Murugan R, Thangadurai V, Weppner W (2007) Fast lithium ion conduction in garnet-type $Li_7La_3Zr_2O_{12}$. Angew Chem Int Ed 46:7778–7781. https://doi.org/10.1002/anie.200701144
85. Murugan R, Ramakumar S, Janani N (2011) High conductive yttrium doped $Li_7La_3Zr_2O_{12}$ cubic lithium garnet. Electrochem Commun 13:1373–1375. https://doi.org/10.1016/j.elecom.2011.08.014
86. Ohta S, Kobayashi T, Asaoka T (2011) High lithium ionic conductivity in the garnet-type oxide $Li_{7-X}La_3(Zr_{2-X}, Nb_X)O_{12}$ (X = 0-2). J Power Sources 196:3342–3345. https://doi.org/10.1016/j.jpowsour.2010.11.089
87. Kumazaki S, Iriyama Y, Kim KH, Murugan R, Tanabe K, Yamamoto K, Hirayama T, Ogumi Z (2011) High lithium ion conductive $Li_7La_3Zr_2O_{12}$ by inclusion of both Al and Si. Electrochem Commun 13:509–512. https://doi.org/10.1016/j.elecom.2011.02.035
88. Wolfenstine J, Ratchford J, Rangasamy E, Sakamoto J, Allen JL (2012) Synthesis and high Li–ion conductivity of Ga-stabilized cubic $Li_7La_3Zr_2O_{12}$. Mater Chem Phys 134:571–575. https://doi.org/10.1016/j.matchemphys.2012.03.054
89. Li Y, Han JT, Wang CA, Xie H, Goodenough JB (2012) Optimizing. Li^+ conductivity in a garnet framework. J Mater Chem 22:15357–15361. https://doi.org/10.1039/C2JM31413D
90. Huang M, Dumon A, Nan CW (2012) Effect of Si, In and Ge doping on high ionic conductivity of $Li_7La_3Zr_2O_{12}$. Electrochem Commun 21:62–64. https://doi.org/10.1016/j.elecom.2012.04.032
91. Yang C, Yi-Qiu L, Xiang-Xin G (2013) Densification and lithium ion conductivity of garnet-type $Li_{7-x}La_3Zr_{2-x}Ta_xO_{12}$ (x = 0.25) solid electrolytes. Chin Phys B 22:0782011–0782015. https://doi.org/10.1088/1674-1056/22/7/078201
92. Raskovalov AA, Il'ina EA, Antonov BD (2013) Structure and transport properties of $Li_7La_3Zr_{2-0.75x}Al_xO_{12}$ superionic solid electrolytes. J Power Sources 238:48–52. https://doi.org/10.1016/j.jpowsour.2013.03.049
93. El Shinawi H, Janek J (2013) Stabilization of cubic lithium-stuffed garnets of the type "$Li_7La_3Zr_2O_{12}$" by addition of gallium. J Power Sour 225:13–19. https://doi.org/10.1016/j.jpowsour.2012.09.111
94. Deviannapoorani C, Dhivya L, Ramakumar S, Murugan R (2013) Lithium ion transport properties of high conductive tellurium substituted $Li_7La_3Zr_2O_{12}$ cubic lithium garnets. J Power Sour 240:18–25. https://doi.org/10.1016/j.jpowsour.2013.03.166
95. Song S, Yan B, Zheng F, Duong HM, Lu L (2014) Crystal structure, migration mechanism and electrochemical performance of Cr-stabilized garnet. Solid State Ionics 268:135–139. https://doi.org/10.1016/j.ssi.2014.10.009

96. Janani N, Deviannapoorani C, Dhivya L, Murugan R (2014) Influence of sintering additives on densification and Li^+ conductivity of Al doped $Li_7La_3Zr_2O_{12}$ lithium garnet. RSC Adv 4:51228–51238. https://doi.org/10.1039/C4RA08674K
97. Takano R, Tadanaga K, Hayashi A, Tatsumisago M (2014) Low temperature synthesis of Al-doped $Li_7La_3Zr_2O_{12}$ solid electrolyte by a sol-gel process. Solid State Ionics 255:104–107. https://doi.org/10.1016/j.ssi.2013.12.006
98. Tadanaga K, Takano R, Ichinose T, Mori S (2013) Low temperature synthesis of highly ion conductive $Li_7La_3Zr_2O_{12}$-Li_3BO_3 composites. Electrochem Commun 33:51–54. https://doi.org/10.1016/j.elecom.2013.04.004
99. Il'ina EA, Raskovalov AA, Saetova NS, Antonov BD, Reznitskikh OG (2016) Composite electrolytes $Li_7La_3Zr_2O_{12}$-glassy Li_2O-B_2O_3-SiO_2. Solid State Ionics 296:26–30. https://doi.org/10.1016/j.ssi.2016.09.003
100. Il'ina EA, Raskovalov AA, Antonov BD, Pankratov AA, Reznitskikh OG (2017) Composite electrolytes ceramic $Li_7La_3Zr_2O_{12}$/glassy Li_2O-Y_2O_3-SiO_2. Mater Res Bull 93:157–161. https://doi.org/10.1016/j.materresbull.2017.04.050
101. Pershina SV, Il'Ina EA, Reznitskikh OG (2017) Phase composition, density, and ionic conductivity of the $Li_7La_3Zr_2O_{12}$-based composites with $LiPO_3$ glass addition. Inorg Chem 56:9880–9891. https://doi.org/10.1021/acs.inorgchem.7b01379
102. Tammann G, Jenckel E (1929) Die zunahme der dichte von gläsern nach erstarrung unter erhöhtem Druck und die Wiederkehr der natürlichen dichte durch temperatursteigerung. Z Anorg allge Chem 184:416–420. https://doi.org/10.1002/zaac.19291840134
103. Stanworth JE (1952) Tellurite glasses. J Soc Glass Tech 36:217–241
104. Denton EP, Rawson H, Stanworth JE (1954) Vanadate glasses. Nature 173:1030–1032. https://doi.org/10.1038/1731030b0
105. Baynton PL, Rawson H, Stanworth JE (1957) Semiconducting properties of some vanadate glasses. J Electrochem Soc 104:237–240. https://doi.org/10.1149/1.2428544
106. Kitaigorodskii II, Karpechenko VG (1958) Synthesis and investigation of certain vanadium glasses. Glass and Ceram 15:294–297. https://doi.org/10.1007/BF00669325
107. Munakata M (1960) Electrical conductivity of high vanadium phosphate glass. Solid-State Electron 1:159–163. https://doi.org/10.1016/0038-1101(60)90001-0
108. Mott NF, Twose WD (1961) The theory of impurity conduction. Adv Phys 10:107–163. https://doi.org/10.1080/00018736100101271
109. Mott NF (1961) Electrons in disordered structure. Adv Phys 16:49–144. https://doi.org/10.1080/00018736700101265
110. Mott NF (1968) Conduction in glasses containing transition metal ions. J Non-Cryst Solids 1:1–17. https://doi.org/10.1016/0022-3093(68)90002-1
111. Lindsey GS (1970) Electron conduction in vanadium phosphate glasses. J Non-Cryst Solids 4:208–219. https://doi.org/10.1016/0022-3093(70)90043-8
112. Frazier LL, France PW (1977) Compositional dependence of the electrical conductivity of vanadium phosphate glass. J Phys Chem Solids 38:801–808. https://doi.org/10.1016/0022-3697(77)90075-0
113. Feltz A, Wollenhaupt R (1980) Electrical conductivity of VO_2 containing glasses in dependence on the polarizability of the oxide ions. J Non-Cryst Solids 35:1015–1020. https://doi.org/10.1016/0022-3093(80)90334-8
114. Chung CH, Mackenzie JD (1980) Electrical properties of binary semiconducting oxide glasses containing 55 mole % V_2O_5.J. Non-Cryst Solids 42:357–370. https://doi.org/10.1016/0022-3093(80)90036-8
115. Nassau K, Murphy DW (1981) The quenching and electrochemical behavior of Li_2O–V_2O_5glasses. J Non-Cryst Solids 44:297–304. https://doi.org/10.1016/0022-3093(81)90032-6
116. Dhawan VK, Mansingh A, Sayer M (1982) DC conductivity of V_2O_5–TeO_2 glasses. J Non-Cryst Solids 51:87–103. https://doi.org/10.1016/0022-3093(82)90190-9
117. Calestani G, Marghignani L, Montenero A, Bettinelli M (1986) DC conductivity of ZnO-V_2O_5 glasses. J Non-Cryst Solids 86:285–292. https://doi.org/10.1016/0022-3093(86)90016-5

118. Ghosh A, Chaudhuri BK (1986) DC conductivity of V_2O_5-Bi_2O_3 glasses. J Non-Cryst Solids 83:151–161. https://doi.org/10.1016/0022-3093(86)90065-7
119. Levy M, Rousseau F, Duclot MJ (1988) Electrochemical properties of glasses in the TeO_2–V_2O_5 system. Solid State Ionics 2:376–378. https://doi.org/10.1016/S0167-2738(88)80136-X
120. Levy M, Duclot MJ, Rousseau F (1989) V_2O_5-based glasses as cathodes for lithium batteries. J Power Sour 26:381–388. https://doi.org/10.1016/0378-7753(89)80150-8
121. Ghosh A, Chaudhuri BK (1988) Anomalous conductivity and other properties of V_2O_5-P_2O_5 glasses with Bi_2O_3 or Sb_2O_3 glasses. J Non-Cryst Solids 103:83–92. https://doi.org/10.1016/0022-3093(88)90419-X
122. Khawaja EE, Salim MA, Khan MA, Al-Adel FF, Khattak GD, Hussain Z (1989) XPS, auger, electrical and optical studies of vanadium phosphate glasses doped with nickel oxide. J Non-Cryst Solids 110:33–43. https://doi.org/10.1016/0022-3093(89)90179-8
123. Lebrun N, Lévy M, Souquet JL (1990) Electronic conductivity in glasses of the TeO_2-V_2O_5-MoO_3system. Solid State Ionics 40–41:718–722. https://doi.org/10.1016/0167-2738(90)90107-3
124. Miura T, Kishi T (1993) Inhomogeneous electrochemical lithiation of V_2O_5–TeO_2 binary glasses in a propylene carbonate solution. J Power Sour 44:645–649. https://doi.org/10.1016/0378-7753(93)80214-A
125. Mori H, Kitami T, Sakata H (1994) Electrical conductivity of V_2O_5–Sb_2O_3–TeO_2 glasses. J Non-Cryst Solids 168:157–166. https://doi.org/10.1016/0022-3093(94)90132-5
126. Mori H, Gotoh K, Sakata H (1995) Low-temperature dc conductivity of V_2O_5–SnO–TeO_2, glasses. J Non-Cryste Solids 183:122–125. https://doi.org/10.1016/0022-3093(94)00585-0
127. Sakata H, Amano M, Yagi T (1996) DC conductivity of V_2O_5–PbO–TeO_2 glasses and the effect of pressure. J Non-Cryst Solids 194:198–206. https://doi.org/10.1016/0022-3093(95)00478-5
128. Jayasinghe GDLK, Dissanayake MAKL, Careem MA, Souquet JL (1997) Electronic to ionic conductivity of glasses in the Na_2O–V_2O_5–TeO_2 system. Solid State Ionics 93:291–295. https://doi.org/10.1016/S0167-2738(96)00494-8
129. Jayasinghe GDLK, Dissanayake MAKL, Bandaranayake PWSK, Souquet JL, Foscallo D (1999) Electronic to ionic conductivity of glasses in the Li_2O–V_2O_5–TeO_2 system. Solid State Ionics 121:19–23. https://doi.org/10.1016/S0167-2738(98)00336-1
130. Ungureanu MC, Levy M, Souquet JL (1998) Mixed conductivity of glasses in the P_2O_5–V_2O_5–Na_2O system. Ionics 4:200–206. https://doi.org/10.1007/BF02375946
131. Mori H, Matsuno H, Sakata H (2000) Small polaron hopping conduction in V_2O_5–Sb–TeO_2 glasses. J Non-Cryst Solids 276:78–94. https://doi.org/10.1016/S0022-3093(00)00280-5
132. Salman FE, Snash NH, El-Haded HA, El-Mansy MK (2002) Electrical conduction and dielectric properties of vanadium phosphate glasses doped with lithium. J Phys Chem Solids 63:1957–1966. https://doi.org/10.1016/S0022-3697(02)00164-6
133. El-Desoky MM (2002) Potassium doping of semiconducting vanadium tellurate glasses. J Phys Chem Solids 73:259–262. https://doi.org/10.1016/S0254-0584(01)00390-X
134. El-Desoky MM (2005) Characterization and transport properties of V_2O_5–Fe_2O_3–TeO_2 glasses. J Non-Cryst Solids 351:3139–3146. https://doi.org/10.1016/j.jnoncrysol.2005.08.004
135. El-Desoky MM, Al-Assiri MS (2007) Structural and Polaronic transport properties of semiconducting CuO–V_2O_5–TeO_2 glasses. Mater Sci Eng B-Adv 137:237–246. https://doi.org/10.1016/j.mseb.2006.11.032
136. Montani RA, Lorente A, Vincenzo MA (2000) Effect of Ag_2O on the conductive behaviour of silver vanadium tellurite glasses. Solid State Ionics 130:91–95. https://doi.org/10.1016/S0167-2738(00)00280-0
137. Montani RA, Lorente A, Frechero MA (2002) Effect of Ag_2O on the conductive behaviour of silver vanadium tellurite glasses: Part II. Solid State Ionics 146:323–327. https://doi.org/10.1016/S0167-2738(01)01023-2

138. Montani RA, Frechero MA (2003) The conductive behaviour of silver vanadium–molybdenum tellurite glasses. Solid State Ionics 158:327–332. https://doi.org/10.1016/S0167-2738(02)00902-5
139. Montani RA, Frechero MA (2006) Mixed ion-polaron transport in lithium vanadium–molybdenum tellurite glasses. Solid State Ionics 177:2911–2915. https://doi.org/10.1016/j.ssi.2006.08.015
140. Turky G, Dawy M (2003) Spectral and electrical properties of ternary (TeO_2–V_2O_5–Sm_2O_3) glasses. Mat Chemy Phys 77:48–59. https://doi.org/10.1016/S0254-0584(01)00574-0
141. Dutta B, Fahmy NA, Pegg IL (2005) Effect of mixed transition-metal ions in glasses. I. The P_2O_5–V_2O_5–Fe_2O_3 system. J Non-Cryst Solids 351:1958–1966. https://doi.org/10.1016/j.jnoncrysol.2005.05.005
142. Dutta B, Fahmy NA, Pegg IL (2005) Effect of mixing transition ions in glasses. II. The P_2O_5–Fe_2O_3–MnO system. J Non-Cryst Solids 351:2552–2561. https://doi.org/10.1016/j.jnoncrysol.2005.06.044
143. Dutta B, Fahmy NA, Pegg IL (2006) Effect of mixed transition-metal ions in glasses. Part III: The P_2O_5–V_2O_5–MnO system. J Non-Cryst Solids 352:2100–2108. https://doi.org/10.1016/j.jnoncrysol.2006.02.043
144. Szu S, Chang F (2005) Impedance study of V_2O_5–TeO_2–BaO glasses. Solid State Ionics 176:2695–2699. https://doi.org/10.1016/j.ssi.2005.09.004
145. Krins N, Rulmont A, Grandjean J, Gilbert B, Lepot L, Cloots R, Vertruyen B (2006) Structural and electrical properties of tellurovanadate glasses containing Li_2O. Solid State Ionics 177:3147–3150. https://doi.org/10.1016/j.ssi.2006.07.034
146. Behzad H, Hekmatshoar MH, Mirzayi M, Azmoonfar M (2009) Activation energy and conductivity of glasses in the P_2O_5–V_2O_5–Li_2O system. Ionics 15:547–650. https://doi.org/10.1007/s11581-009-0319-3
147. Pant M, Kanchan DK, Sharma P (2010) Characterization and transport properties of $10BaO\text{-}xAg_2O\text{-}(85\text{-}x)V_2O_5\text{-}5TeO_2$ glass system. Ionics 16:797–805. https://doi.org/10.1007/s11581-010-0464-8
148. Souri D (2010) Small polaron hopping conduction in tellurium based glasses containing vanadium and antimony. J Non-Cryst Solids 356:2181–2184. https://doi.org/10.1016/j.jnoncrysol.2010.08.020
149. Shapaan M (2010) Effect of heat treatment on the hyperfine structure and the dielectric properties of $40P_2O_5\text{-}40V_2O_5\text{-}20Fe_2O_3$ oxide glass. J Non-Cryst Solids 356:314–320. https://doi.org/10.1016/j.jnoncrysol.2009.11.042
150. Devidas GB, Sankarappa T, Sujatha T, Kumar MP, Sadashivaiah PJ, Naik PS (2011) Polaron transport studies in La_2O–V_2O–P_2O_5 glasses. Solid State Sci 13:1385–1390. https://doi.org/10.1016/j.solidstatesciences.2011.04.008
151. Khattak GD, Mekki A, Siddiqui MN (2012) Compositional dependence of DC electrical conductivity of SrO-vanadate glasses. Solid State Ionics 211:5–11. https://doi.org/10.1016/j.ssi.2012.01.012
152. Mirzayi M, Hekmatshoar MH (2013) Study of electrical conductivity and memory switching in the zinc vanadium-phosphate glasses. Phys B 420:70–73. https://doi.org/10.1016/j.physb.2013.03.026
153. Szreder NA, Kosiorek P, Karczewski J, Gazda M, Barczyński RJ (2014) Microstructure and dielectric properties of barium-vanadate glasses. Procedia Engineer 98:62–70. https://doi.org/10.1016/j.proeng.2014.12.489
154. Salehizadeh SA, Melo BMG, Freire FNA, Valente MA, Graça MPF (2016) Structural and electrical properties of TeO_2–V_2O_5–K_2O glassy systems. J Non-Cryst Solids 443:65–74. https://doi.org/10.1016/j.jnoncrysol.2016.03.012
155. Mori H (2016) Effect of annealing in an H_2 gas atmosphere on the physical properties for $70V_2O_5{\cdot}30TeO_2$ (mol%) glasses. J Non-Cryst Solids 463:163–168. https://doi.org/10.1016/j.jnoncrysol.2017.02.018
156. Dixit VG, Singh K (1983) Electrical conductivity of transition metal oxide glasses in the system V_2O_5–B_2O_3–TeO_2 at high temperatures. J Phys Chem Solids 44:859–864. https://doi.org/10.1016/0022-3697(83)90122-1

157. Sharma BK, Dube DC, Mansingh A (1984) Preparation and characterization of V_2O_5–B_2O_3 glasses. J Non-Cryst Solids 65:39–51. https://doi.org/10.1016/0022-3093(84)90353-3
158. Culea E, Nicula Al (1986) Electrical properties of V_2O_5-B_2O_3 glasses. Solid State Commun 58:545–549. https://doi.org/10.1016/0038-1098(86)90793-3
159. Ichinose N, Nakai Y (1996) Electrical conductivity of V_2O_5–SrO–B_2O_3 glasses. J Non-Cryst Solids 203:353–358. https://doi.org/10.1016/0022-3093(96)00369-9
160. El Mkami H, Deroide B, Backov R, Zanchetta JV (2000) dc and ac Conductivities of $(V_2O_5)_x(B_2O_3)_{1-x}$ oxide glasses. J Phys Chem Solids 61:819–826. https://doi.org/10.1016/S0022-3697(99)00306-6
161. Sindhu S, Sanghi S, Sonam AA, Seth VP Kishore N (2005) The role of V_2O_5 in the modification of structural, optical and electrical properties of vanadium barium borate glasses. Phys B Cond Mat 1:65–75. https://doi.org/10.1016/j.physb.2005.04.037
162. Choi S, Ryu B (2016) Effects of crystallization on the structural, electrical, and catalytic properties of $75V_2O_5$-$15B_2O_3$-$10P_2O_5$ glass. J Non-Cryst Solids 431:112–117. https://doi.org/10.1016/j.jnoncrysol.2015.05.009
163. Sujatha B, Viswanatha R, Nagabushana H, Reddy CN (2017) Electronic and ionic conductivity studies on microwave synthesized glasses containing transition metal ions. J Mat Res Tech 6:7–12. https://doi.org/10.1016/j.jmrt.2016.03.002
164. Rajashekara G, Sangamesh J, Arunkumar B, Nagaraja N, Kumar MP (2017) Anomalous DC electrical conductivity in mixed transition metal ions doped borate glasses. J Non-Cryst Solids 481:289–294. https://doi.org/10.1016/j.jnoncrysol.2017.10.056
165. Garbarczyk JE, Wasiucionek M, Jozwiak P, Tykarski L, Nowinski JL (2002) Studies of Li_2O–V_2O_5–P_2O_5 glasses by DSC, EPR and impedance spectroscopy. Solid State Ionics 154:367–373. https://doi.org/10.1016/S0167-2738(02)00574-X
166. Jozwiak P, Garbarczyk JE (2005) Studies of Li_2O–V_2O_5–P_2O_5 glasses by DSC, EPR and impedance spectroscopy. Solid State Ionics 176:2163–2169. https://doi.org/10.1016/j.ssi.2004.06.028
167. Garbarczyk JE, Jozwiak P, Wasiucionek M, Nowinski JL (2004) Enhancement of electrical conductivity in lithium vanadate glasses by nanocrystallization. Solid State Ionics 175:691–694. https://doi.org/10.1016/j.ssi.2004.08.025
168. Garbarczyk JE, Jozwiak P, Wasiucionek M, Nowinski JL (2007) Nanocrystallization as a method of improvement of electrical properties and thermal stability of V_2O_5-rich glasses. J Power Sources 173:743–747. https://doi.org/10.1016/j.jpowsour.2007.05.069
169. Garbarczyk JE, Jozwiak P, Wasiucionek M, Nowinski JL (2006) Effect of nanocrystallization on the electronic conductivity of vanadate–phosphate glasses. Solid State Ionics 177:2585–6258. https://doi.org/10.1016/j.ssi.2006.04.038
170. Garbarczyk JE, Wasiucionek M, Jozwiak P, Nowinski JL, Julien CM (2009) Novel nanomaterials based on electronic and mixed conductive glasses. Solid State Ionics 180:531–536. https://doi.org/10.1016/j.ssi.2008.09.021
171. Pietrzak TK, Garbarczyk JE, Gorzkowska I, Wasiucionek M, Nowinski JL, Gierlotka S, Jozwiak P (2009) Correlation between electrical properties and microstructure of nanocrystallized V_2O_5–P_2O_5 glasses. J Power Sources 173:743–747. https://doi.org/10.1016/j.jpowsour.2009.02.031
172. Takahashi H, Karasawa T, Sakuma T, Garbarczyk JE (2010) Electrical conduction in the vitreous and crystallized Li_2O–V_2O_5–P_2O_5 system. Solid State Ionics 181:27–32. https://doi.org/10.1016/j.ssi.2009.12.001
173. Pietrzak TK, Garbarczyk JE, Wasiucionek M, Gorzkowska I, Nowinski JL, Gierlotka S (2011) Electrical properties vs. microstructure of nanocrystallized V_2O_5–P_2O_5 glasses - An extended temperature range study. Solid State Ionics 192:210–214. https://doi.org/10.1016/j.ssi.2010.05.018
174. Pietrzak TK, Wasiucionek M, Nowiński JL, Garbarczyk JE (2013) Isothermal nanocrystallization of vanadate–phosphate glasses. Solid State Ionics 251:78–82. https://doi.org/10.1016/j.ssi.2013.01.004

175. Pietrzak TK, Pawliszak Ł, Michalski PP, Wasiucionek M, Garbarczyk JE (2013) Highly conductive $90V_2O_5 \cdot 10P_2O_5$ nanocrystalline cathode materials for lithium-ion batteries. Procedia Engineer 251:78–82. https://doi.org/10.1016/j.proeng.2014.12.483
176. Pietrzak TK, Michalski PP, Wasiucionek M, Garbarczyk JE (2016) Synthesis of nanostructured $Li_3Me_2(PO_4)_2F_3$ glass-ceramics (Me = V, Fe, Ti). Solid State Ionics 288:193–198. https://doi.org/10.1016/j.ssi.2015.11.021
177. Al-Syadi AM, Al-Assiri MS, Hassan HMA, El-Desoky MM (2016) Grain size effects on the transport properties of $Li_3V_2(PO_4)_3$ glass–ceramic nanocomposites for lithium cathode batteries. J Mater Sci-Mater El 27:4074–4083. https://doi.org/10.1007/s10854-015-4266-7
178. Al-Syadi AM, Al-Assiri MS, Hassan HMA, El-Desoky MM (2016) Grain size effects on dynamics of Li-ions in $Li_3V_2(PO_4)_3$ glass-ceramic nanocomposites. Ionics 22:2281–2290. https://doi.org/10.1007/s11581-016-1772-4
179. El-Desoky MM, Al-Syadi AM, Al-Assiri MS, Hassan HMA, El Enany G (2016) Electrochemical performance of novel $Li_3V_2(PO_4)_3$ glass-ceramic nanocomposites as electrodes for energy storage devices. J Solid State Electrochem 20:2663–2671. https://doi.org/10.1007/s10008-016-3267-7
180. Austin IG, Mott NF (1969) Polarons in crystalline and non-crystalline materials. Adv Phys 18:41–102. https://doi.org/10.1080/00018736900101267
181. Il'ina EA, Andreev OL, Antonov BD, Batalov NN (2012) Morphology and transport properties of the solid electrolyte $Li_7La_3Zr_2O_{12}$ prepared by the solid-state and citrate–nitrate methods. J Power Sources 201:167–173. https://doi.org/10.1016/j.jpowsour.2011.10.108
182. Tolkacheva AS, Shkerin SN, Plaksin SV, Vovkotrub EG, Bulanin KM, Kochedykov VA, Ordinartsev DP, Gyrdasova OI, Molchanova NG (2011) Synthesis of Dense Ceramics of Single-Phase Mayenite $(Ca_{12}Al_{14}O_{32})O$. Russ J Appl Chem 84:907–911. https://doi.org/10.1134/S1070427211060012
183. Pecharsky VK, Zavalij PY (2009) Fundamentals of Powder Diffraction and Structural Characterization of Materials. Springer, Boston, MA. https://doi.org/10.1007/978-0-387-09579-0
184. McKnight M, Whitmore KA, Bunton PH, Baker DB, Vennerberg DC, Feller SA (2010) EPR study of $RLi_2O \cdot V_2O_5$, $RNa_2O \cdot V_2O_5$, $RCaO \cdot V_2O_5$ and $RBaO \cdot V_2O_5$ modified vanadate glass system. J Non-Cryst Solids 356:2268–2272. https://doi.org/10.1016/j.jnoncrysol.2010.07.016
185. Svistunenko DA, Sharpe MA, Nicholls P, Wilson MT, Cooper CE (2000) A new method for quantitation of spin concentration by EPR spectroscopy: application to methemoglobin and metmyoglobin. J Magn Resonan 142:226–275. https://doi.org/10.1006/jmre.1999.1935
186. Eaton GR, Eaton SS, Barr DP, Weber RT (2010) Quantitative EPR. SpringerWienNewYork, Wien
187. Choi HU, Jin JS, Park J-Y, Lim H-T (2017) Performance improvement of all-solid-state Li-S batteries with optimizing morphology and structure of sulfur composite electrode. J Alloys and Comp 723:787–794. https://doi.org/10.1016/j.jallcom.2017.06.135
188. Hayes M, Kuhn AT, Patefield W (1977) Techniques for the determination of ohmic drop in half-cells and full cells: A review. J Power Sources 2:121–136. https://doi.org/10.1016/0378-7753(77)80013 X
189. Cooper KR, Smith M (2006) Electrical test methods for on-line fuel cell ohmic resistance measurement. J Power Sources 160:1088–1095. https://doi.org/10.1016/j.jpowsour.2006.02.086
190. Mench MM (2008) Experimental diagnostic and diagnosis. Mench MM. Fuel cell engines. Wiley, New Jersey, pp 453–485
191. Amiki Y, Sagane F, Yamamoto K, Hirayama T, Sudoh M, Motoyama M, Iriyama Y (2013) Electrochemical properties of an all-solid-state lithium-ion battery with an in-situ formed electrode material grown from a lithium conductive glass ceramics sheet. J Power Sources 241:583–588. https://doi.org/10.1016/j.jpowsour.2013.05.006
192. Kato T, Yoshida R, Yamamoto K, Hirayama T, Motoyama M, West WC, Iriyama Y (2016) Effects of sintering temperature on interfacial structure and interfacial resistance for all-solid-state rechargeable lithium batteries. J Power Sources 325:584–590. https://doi.org/10.1016/j.jpowsour.2016.06.068

193. Iriyama Y, Yada C, Abe T, Ogumi Z, Kikuchi K (2006) A new kind of all-solid-state thin-film-type lithium-ion battery developed by applying a D.C. high voltage. Electrochem Commun 8:1287–1291. https://doi.org/10.1016/j.elecom.2006.03.003
194. Yada C, Iriyama Y, Abe T, Kikuchi K, Ogumi Z (2009) A novel all-solid-state thin-film-type lithium-ion battery with in situ prepared positive and negative electrode materials. Electrochem Commun 11:413–416. https://doi.org/10.1016/j.elecom.2008.12.004
195. Chebotin VN, Perfil'ev MV (1978) Electrohimiya tverdyh electrolitov (Electrochemistry of solid electrolytes). Himiya, Moscow
196. Sangster J, Pelton AD (1991) The Ga-Li (Gallium-Lithium) System. J Phase Equilib 12:33–36. https://doi.org/10.1007/BF02663670
197. Saint J, Morcrette M, Larcher D, Tarascon JM (2005) Exploring the Li–Ga room temperature phase diagram and the electrochemical performances of the Li_xGa_y alloys vs. Li. Solid State Ionics 176:189–197. https://doi.org/10.1016/j.ssi.2004.05.021
198. Dębski A, Gąsior W, Dębski R (2016) Thermodynamic properties of liquid Ga-Li alloys: experiment vs. modeling. J. Chem. Thermodyn 97:348–353. https://doi.org/10.1016/j.jct.2016.02.015
199. Trong LD, Thao TT, Dinh NN (2015) Characterization of the Li-ionic conductivity of $La_{(2/3-x)}Li_{3x}TiO_3$ ceramics used for all-solid-state batteries. Solid State Ionics 278:228–232. https://doi.org/10.1016/j.ssi.2015.05.027

Chapter 14
Fabrication of All-Solid-State Lithium Batteries with an Aerosol Process

Takeshi Kimura and Kiyoshi Kanamura

Abstract All-solid-state lithium batteries with $LiCoO_2$-Li_3BO_3 composite cathode layer was formed on Al-doped $Li_7La_3Zr_2O_{12}$ pellets by an aerosol deposition method. The effects of heat treatment after the aerosol process on the electrochemical performance of the all-solid-sate batteries were investigated. The overpotentials of charge-discharge cycles decrease after the heat treatment, and the cycle performance improved. The interfacial resistance of between $LiCoO_2$-Li_3BO_3 composite cathode layer and Al-doped $Li_7La_3Zr_2O_{12}$ electrolyte with the heat treatment was smaller than without the heat treatment.

14.1 Introduction

All-solid-state lithium batteries (All-SS-LBs) have been extensively investigated as a next-generation battery due to their high performance, high safety and high energy density. Sulfide materials for solid electrolytes have been developed and applied to rechargeable laminated cells. The sulfide solid electrolyte has a high Li^+ ion conductivity and mechanical softness. These properties enable the fabrication of All-SS-LBs by using a fabrication process similar to that used for lithium-ion batteries. In particular, the formation of an interface between the cathode material and the solid electrolyte is not very difficult due to the soft nature of the sulfide solid electrolyte. However, it is possible for the sulfide solid electrolyte and the oxide cathode material to react during long cycling of All-SS-LBs [1]. This problem should be solved to realize a long cycle life for All-SS-LBs. Another important solid electrolyte material is oxide. There are many kinds of oxide solid electrolytes for Li^+ ion transport.

T. Kimura (✉) · K. Kanamura
Graduate School of Urban Environmental Sciences, Tokyo Metropolitan University, 1-1 Minami-Ohsawa Hachioji, Tokyo 192-0397, Japan
e-mail: kimura-t-ah@tmu.ac.jp

K. Kanamura
e-mail: kanamura@tmu.ac.jp

R. Murugan and W. Weppner (eds.), *Solid Electrolytes for Advanced Applications*,
https://doi.org/10.1007/978-3-030-31581-8_14

For example, N-doped Li_3PO_4 (LIPON) [2–4], $Li_{1.3}Al_{0.3}Ti_{1.7}(PO_4)_3$ (LATP) [5–8], perovskite $Li_{3x}Li_{2/3-x}TiO_3$ (LLT) [9, 10] and garnet $Li_7La_3Zr_2O_{12}$ (LLZ) [11, 12] are popular materials as Li^+ ion-conducting solid electrolytes. All these materials are mechanically harder than sulfide materials. Therefore, the interfacial formation between a cathode layer and a solid electrolyte is an important issue for the fabrication of All-SS-LBs. To date, various kinds of fabrication methods have been utilized to fabricate an electrochemically active interface between the cathode and solid electrolyte materials. For example, a sintering process, chemical vapor deposition (CVD) or physical vapor deposition (PVD) process, and the flux crystal growth method have been used to establish good contact between the oxide cathode and the solid electrolyte. Recently, an aerosol deposition (AD) process has been reported to form an interface between the cathode and electrolyte. The AD process is a kind of coating technology that uses room temperature impact consolidation (RTIC) to fabricate a thick ceramic layer. A fine particle is aerosolized by mixing with Ar carrier gas and is propelled at high velocity toward a substrate. This process provides a thick ceramic layer with a short deposition time. The AD process is considered suitable for the preparation of All-SS-LBs since the process is much easier and more rapid than other conventional processes, such as CVD and sputtering processes.

A composite cathode with a Li^+ ionic conductor may be used to achieve good contact between the cathode layer and the solid electrolyte. All-SS-LBs constructed by the screen-printing process using a composite of $LiCoO_2$ and Li_3BO_3 (LBO) as a Li^+ ion conductor with a low melting point have been reported by Ohta et al. [13]. $Li_{2.2}C_{0.8}B_{0.2}O_3$, a Li^+ ion conductor, has also shown the same effect as LBO [14]. In addition, these additives provide a sufficient contact area between the cathode layer and the solid electrolyte pellet (separator part) after the annealing process of the composite cathode layer.

In this study, All-SS-LBs were fabricated using an Al-doped $Li_7La_3Zr_2O_{12}$ (LLZ) pellet, a composite particle of $LiCoO_2$ and Li_3BO_3 and a Li metal anode. Al-doped LLZ has a cubic garnet structure providing a higher Li^+ ion conductivity than that of non-doped LLZ. The cathode layer was prepared using the AD process and annealing process.

14.2 Experimental

Al-LLZ powder (DLZ-3 Daiichi Kigenso Kagaku Kogyo Co., Ltd., Japan, D_{50} = 11 μm) was pulverized by planetary ball milling before pelletization with a Newton press. The pellets were sintered at 1040 °C for 20 h by the powder BET method. The diameter of the sintered pellets was 12 mm, and the thickness was adjusted to 0.5 mm by polishing with sandpaper. The Li^+ ion conductivity was determined by impedance measurements at various temperatures.

$LiCoO_2$ (LCO) powder was synthesized via a sol–gel method using lithium acetate ($C_2H_3LiO_2$, 98.0%, FUJIFILM Wako Pure Chemical Corp.) and cobalt acetate tetrahydrate ($C_4H_6CoO_4$–$4H_2O$, 99.0%, FUJIFILM Wako Pure Chemical Corp.).

The reactants were dissolved in pure water, and then, citric acid ($C_6H_8O_7$, 99.0%, Kanto Chemical Co. Inc.) as a chelating agent for gelation was added to the solution. The [Li]/[Co] ratio was 1.1:1 to compensate for possible lithium loss during the post calcination process, and the $C_4H_6CoO_4$–$4H_2O/C_6H_8O_7$ molar ratio was 2. The solution was stirred for 1 h, followed by heating at 80 °C. The solution was dried by heating at 120 °C until a solid mixture was formed. The solid mixture was calcined in air at 400 °C for 9 h before sintering in air at 800 °C for 8 h to obtain crystalline LCO fine particles.

Li_3BO_3 powder was synthesized via a solid-state reaction with lithium carbonate (Li_2CO_3, 99.0%, FUJIFILM Wako Pure Chemical Corp.) and boron trioxide (B_2O_3, 95.0%, Kanto Chemical Co. Inc.). Stoichiometric amounts of reactants were mixed before calcination in air at 600 °C for 10 h to obtain LBO powders. The LBO powders were pulverized at 400 rpm using a planetary ball mill until the particle size of LBO was reduced to <1 μm. The LCO–LBO composite particles were synthesized from the prepared LCO and pulverized LBO with a weight ratio of 7:3 through sintering in air at 800 °C for 2 h after mixing of both oxides for 15 min. The structures of the prepared LCO and LCO–LBO composites were identified by X-ray diffraction (XRD). The morphology and particle size were observed with scanning electron microscopy (SEM).

Figure 14.1 shows a schematic diagram of the AD apparatus used in this study. The LCO–LBO composite powder was set in a tube between a gas valve and a nozzle before evacuating a vacuum chamber. The Al-LLZ pellet was masked with a punched polystyrene film and set in the chamber. The distance between the pellet and nozzle was 40 mm. The diameter of the nozzle hole was 1 mm. The deposition area on the Al-LLZ pellet was ϕ 10 mm. The chamber pressure was controlled to be 80 Pa by evacuation with a rotary pump. The LCO–LBO composite powder was propelled

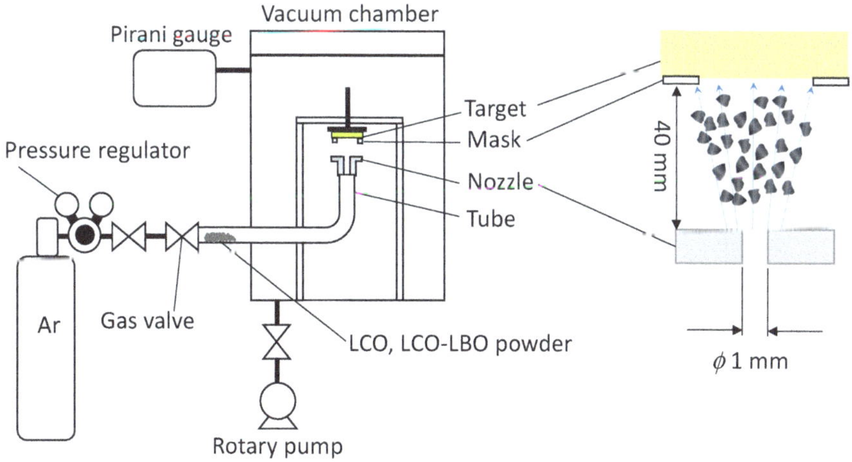

Fig. 14.1 Schematic diagram of the AD apparatus

onto the Al-LLZ target with Ar gas at 0.6 MPa. A 25 mg sample of the composite powder was used for one AD shot. A thick cathode layer was obtained after a number of AD shots. The amount of LCO was measured from the weight of pellets before and after AD.

All-SS-LBs were assembled by using the prepared LCO layer on an Al-LLZ pellet by AD. The Au layer as a current collector was sputtered on the cathode layer. For the preparation of the lithium metal anode side, the Au layer (50 nm) was first sputtered on the LLZ pellet. This Au layer is very important for connecting LLZ pellets and lithium metal. It is difficult for lithium metal layers to form directly on LLZ pellets with good contact [15]. The Au–Li alloy layer was formed by heat treatment at 100 °C for 6 h. In the course of this alloying process, good contact between lithium metal and LLZ pellets was obtained.

14.3 Results and Discussion

Figure 14.2 shows the XRD pattern of the synthesized LCO–LBO composite powder. Three peaks were observed at 18°, 36° and 44°, which were assigned to LCO. The main peaks corresponding to LBO were also observed. This peak intensity was very low, indicating lower crystallinity and a small amount of LBO in the composite. It is likely that LBO exists in an amorphous (glassy) state. Peaks corresponding to LCO and LBO were observed, and no other peaks are observed in Fig. 14.2. From this result, it can be said that LCO and LBO do not react during the preparation process of the composite particles.

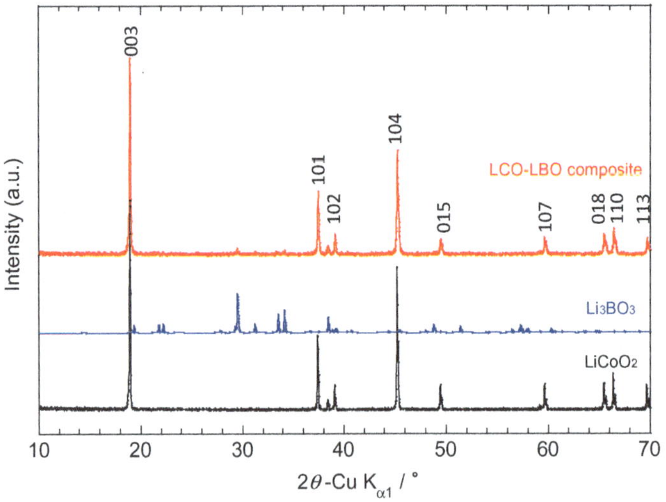

Fig. 14.2 XRD pattern of the synthesized LCO–LBO composite powder

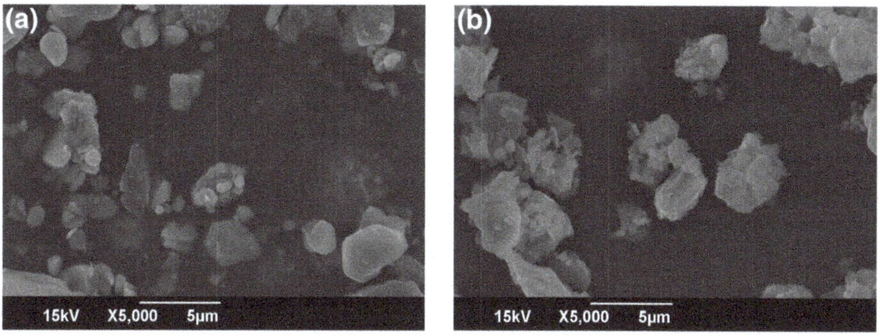

Fig. 14.3 SEM images of **a** LCO and **b** LCO–LBO composite powders prepared by sintering at 800 °C

Figure 14.3 shows the scanning electron microscopy (SEM) images for LCO and LCO–LBO composite powders prepared by sintering at 800 °C. The LCO–LBO composite grain size was smaller than 5 μm for both particles. The particle size of LCO was not so different from that of the LCO–LBO particles. The particle shape was also independent of the composite preparation process. In the SEM image of the LCO–LBO composite particles, a smaller particle was attached to a larger particle. The smaller particle may be LBO, and the larger particle may be LCO. Based on this observation, the LBO exists on the LCO particle surface.

Figure 14.4 a, b shows the photos of the Al-LLZ pellets with an LCO layer or LCO–LBO composite layer prepared by two AD shots of LCO powder or LCO–LBO powder. The color of the LCO cathode layer was different from that of LCO–LBO, indicating that the deposition of LCO–LBO was much better than that of LCO. From this result, it can be said that LBO works as a good additive to improve adhesion among particles and between the cathode layer and the Al-LLZ pellet. To obtain good contact, LBO is a very important additive in the AD process.

Figure 14.5a, e shows the cross-sectional SEM images of the LCO cathode layer

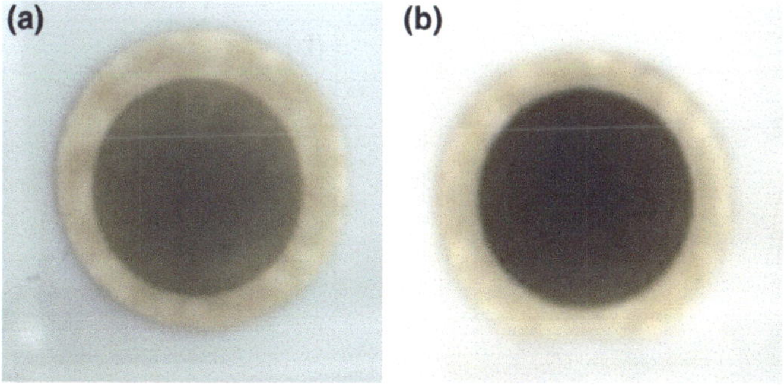

Fig. 14.4 Photos of the Al-LLZ pellets with **a** the LCO layer or **b** LCO–LBO composite layer prepared by two AD shots of LCO powder or LCO–LBO powder

on Al-LLZ (LCO/Al-LLZ) and the LCO–LBO composite cathode layer on Al-LLZ (LCO–LBO/Al-LLZ). Figure 14.5b–d, f–h shows the EDX mapping images for Fig. 14.5a, e. Co in the LCO and LCO–LBO composite cathode layers is indicated by yellow. La in Al-LLZ solid electrolytes is indicated by light blue. The thicknesses of the LCO and LCO–LBO composite layers were 0.5–1 and 1–2 μm, respectively. The deposited amount of LCO on LCO/Al-LLZ was 63 μg and that on LCO–LBO/Al-LLZ was 97 μg. LCO–LBO composite particles are useful for improving the efficiency (yield) of the AD process. The densities of the cathode layers were not so different from each other and were estimated to be 1.2–1.6 g cm^{-3}. The density was roughly estimated from the thickness and weight of the cathode layer.

Figure 14.6a, b shows the charge–discharge curves of All-SS-LB with LCO/Al-LLZ and LCO–LBO/Al-LLZ prepared by two AD shots. The charge–discharge tests were carried out in constant current mode at 0.3 C rate between 2.5 and 4.2 V in an Ar gas atmosphere at 30 °C. The first discharge capacity of the All-SS-LB with the

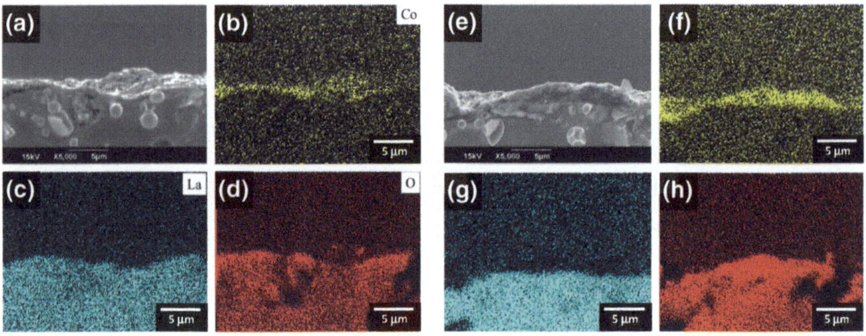

Fig. 14.5 Cross-sectional SEM images and EDX mapping images of LCO/Al-LLZ (**a**–**d**) and LCO–LBO/Al-LLZ (**e**–**h**) prepared by two AD shots

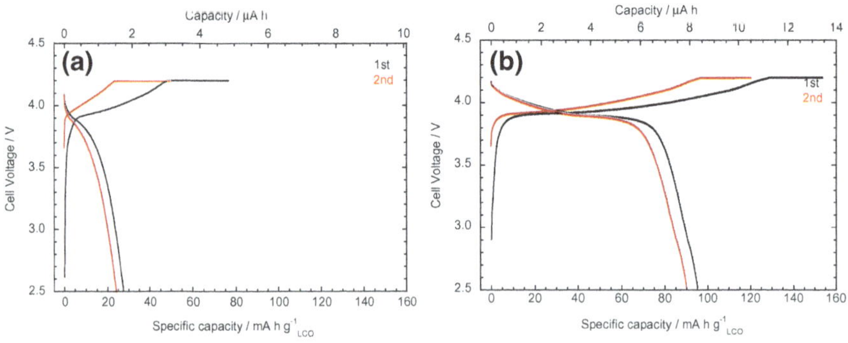

Fig. 14.6 Charge–discharge curves for the All-SS-LBs with **a** an LCO cathode layer and **b** LCO–LBO composite cathode layer prepared by two AD shots

LCO cathode was small. The utilization of LCO was only 20% (100% = 137 mA h g^{-1}; theoretical capacity of LCO). On the other hand, the utilization was higher than 70% when using an LCO–LBO composite cathode.

A thick cathode layer was obtained by a number of AD shots. A thick cathode layer was formed on the Al-LLZ pellet and was investigated for All-SS-LIB with higher capacity. Figure 14.7a, e shows the cross-sectional SEM images of LCO/Al-LLZ and LCO–LBO/Al-LLZ prepared by 20 AD shots. Figure 14.7b–d, f–h shows the EDX mapping images in Fig. 14.7a, e. LCO or LCO–LBO composite cathode layers are indicated by green. Al-LLZ solid electrolytes are pink. The thicknesses of the LCO and LCO–LBO composite layers were 5–7 and 9–11 μm, respectively. In this case, a thicker cathode layer was obtained using the composite particles.

Figure 14.8 shows the charge–discharge curves of All-SS-LB with LCO–LBO/Al-LLZ prepared by 20 AD shots. The charge–discharge test was carried out in constant current mode at 0.1 C rate between 2.5 and 4.2 V in an Ar gas atmosphere at 30 °C. The thick LCO–LBO composite cathode layer could be charged and discharged

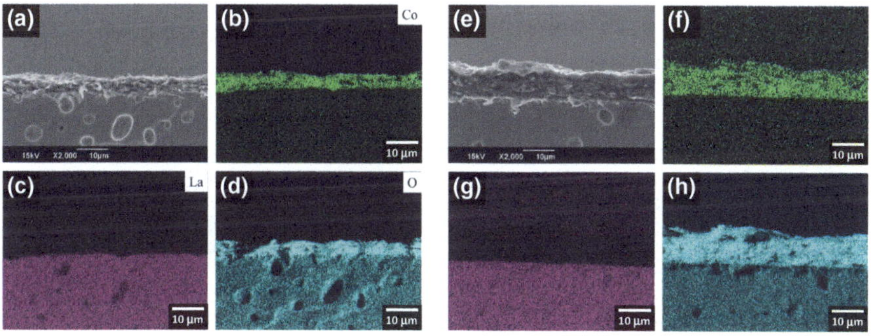

Fig. 14.7 Cross-sectional SEM images and EDX mapping images of LCO/Al-LLZ (**a**–**d**) and LCO–LBO/Al-LLZ (**e**–**h**) prepared by 20 AD shots

Fig. 14.8 Charge–discharge curves for the All-SS-LBs with an LCO–LBO composite cathode layer prepared by 20 AD shots

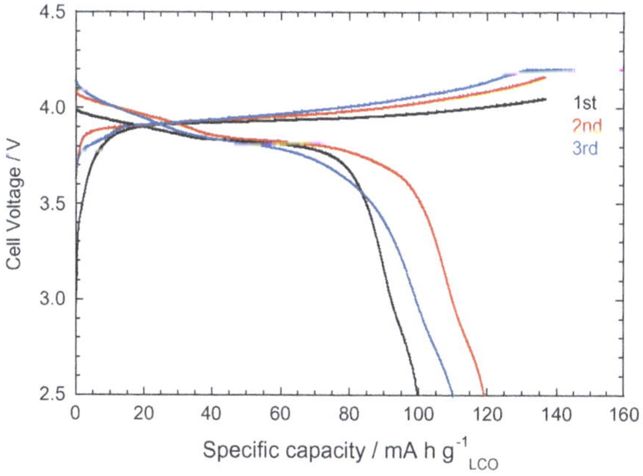

even at room temperature. The discharge capacity was less than 100 mA h g^{-1}, which was much better than that of the All-SS-LB with a thinner cathode layer. In addition, an irreversible capacity at the first charge and discharge cycle was observed. The irreversible capacity may be due to the electrochemical reaction of the amorphous impurity phase or degradation of the cathode layer, which provides no cathode reaction during the first charge process.

Figure 14.9a, b shows illustrations of the cathode layer and the solid electrolyte. When the LCO layer without LBO was deposited, few Li^+ ion conducting pathways were present in the LCO layer, and few ionically connected LCO/Al-LLZ interfaces were formed. The contact among the LCO particles in the cathode layer and between the cathode and Al-LLZ pellet was poor (Fig. 14.9a), resulting in the poor electrochemical performance of the All-SS-LB prepared in this study. On the other hand, when the LCO–LBO composite layer was deposited, a number of Li^+ ion conducting paths formed in the LCO layer and at the LCO cathode layer/Al-LLZ interface due to LBO co-deposition as a Li^+ ion conducting additive. Therefore, the utilization of LCO was increased by both types of good contact, as shown in Fig. 14.9b.

If LBO–LBO particles are connected to each other and the LCO and LBO connection is improved, Li^+ ion diffusion may be improved in the composite cathode layer and at the composite cathode/solid electrolyte interface. To realize a better connection of LBO–LBO and LCO–LBO/Al-LLZ, heat treatment at 750 °C for

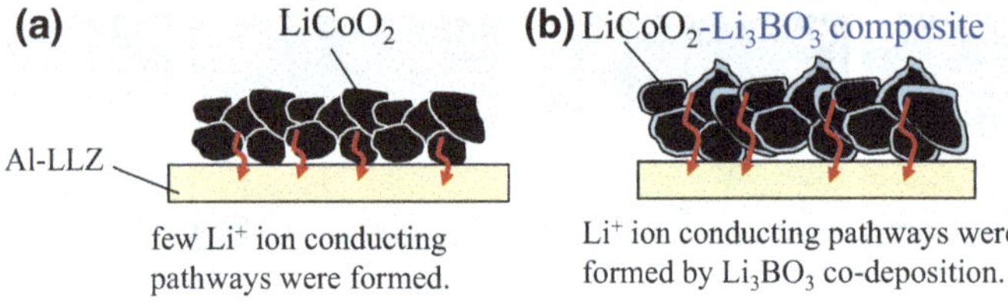

Fig. 14.9 Illustrations of the cathode layer (**a**) with/ (**b**) without LBO and the solid electrolyte

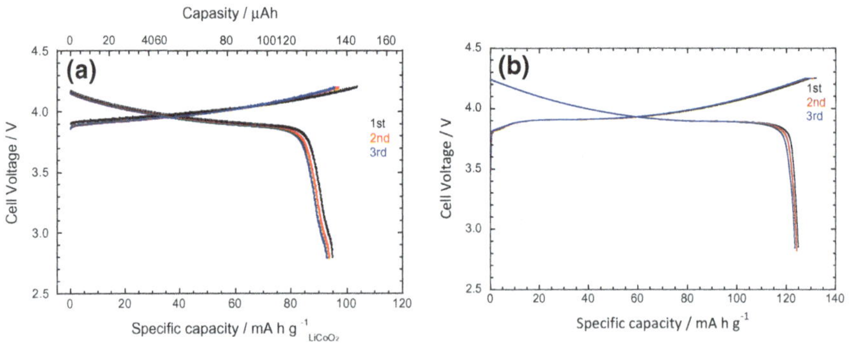

Fig. 14.10 Charge–discharge curves of **a** LCO–LBO/Al-LLZ with heat treatment after AD and **b** LCO with liquid-type electrolyte (1 mol dm^{-3} $LiPF_6$/EC:DEC = 1:1 in volume)

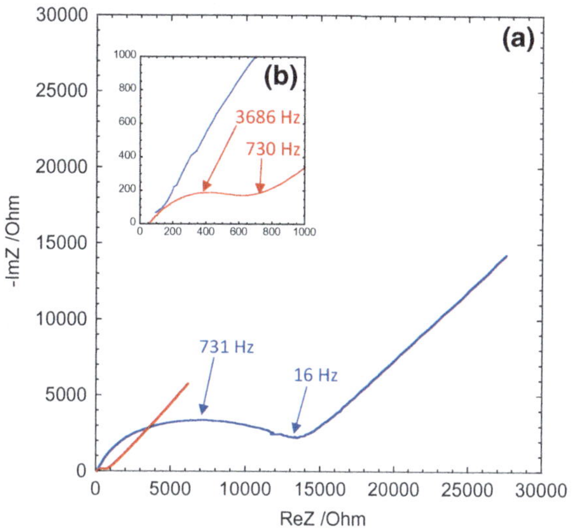

Fig. 14.11 AC impedance spectra of All-SS-LBs with and without heat treatment

1 h was given after the AD process. Figure 14.10a shows the charge–discharge curves of LCO–LBO/Al-LLZ with heat treatment after AD. The overpotentials of the charge and discharge cycles decreased after heat treatment, and the cycle performance improved. The charge–discharge curves are similar to those of the batteries with liquid-type electrolytes, such as 1 mol dm^{-3} $LiPF_6$/EC:DEC = 1:1 in volume. The capacity of the All-SS-LB was slightly smaller than that of the battery using the liquid-type electrolyte (Fig. 14.10b). To clarify the effect of the heat treatment after AD, the AC impedances of All-SS-LB with LCO–LBO/LLZ with and without heat treatment were measured.

Figure 14.11 shows the AC impedance spectra of All-SS-LBs with and without heat treatment. The impedance measurements were performed at 60 °C. The frequency range was from 3 MHz to 0.1 Hz. The AC amplitude was 100 mV. The interfacial resistance with the heat treatment was smaller than that without the heat treatment. During the heat treatment at 750 °C, the LBO layers were melted (the melting point of LBO is less than 750 °C). Therefore, the LBO particles were connected to each other, LCO particles and the Al-LLZ solid electrolyte pellet during this heat treatment process. The good connections among the solids result in a lower interfacial resistance for Li^+ ion transfer at the interfaces.

Figure 14.12 shows the cycle performance and the coulombic efficiency of All-SS-LB with the heat treatment. The charge–discharge test was performed in a constant current mode at 0.2 C rate between 2.8 and 4.3 V in an Ar gas atmosphere at 60 °C. Although the coulombic efficiencies were higher than 98.5%, the discharge capacities were slightly reduced. Figure 14.13 shows the rate capability of the all-solid-state battery with heat treatment. The test was performed at 60 °C. Discharge capacities of 130, 110, 95 and 30 mA h g^{-1} were obtained at 0.1 C, 0.5 C, 1 C and 10 C, respectively. After the rate capability test, the discharge capacity at 0.1 C was still

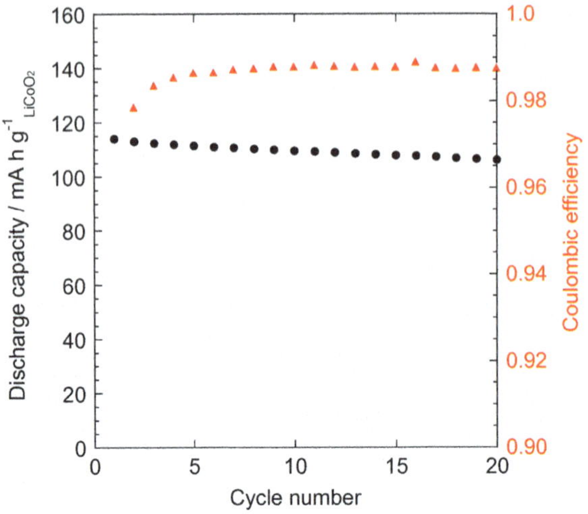

Fig. 14.12 Cycle performance and coulombic efficiency of All-SS-LB with heat treatment

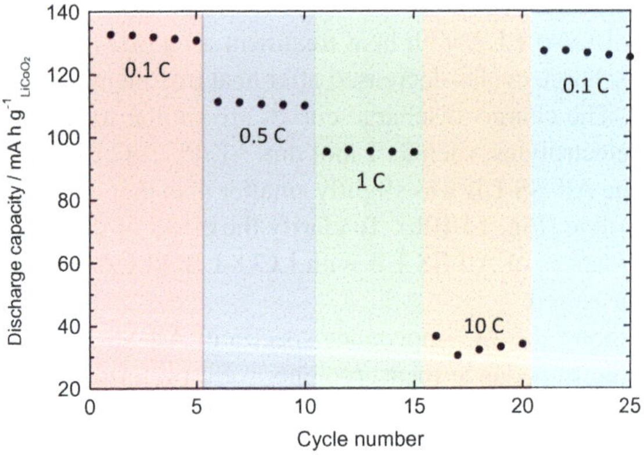

Fig. 14.13 Rate capability of All-SS-LB with heat treatment

120–130 mA h g^{-1}. The electrochemical performance of the prepared All-SS-LB was still lower than that of a Li-metal battery with liquid electrolyte. In the future, additional improvements will be made to prepare high-performance All-SS-LBs.

14.4 Conclusion

In this study, a $LiCoO_2$ cathode layer was formed on LLZ pellets by the aerosol deposition method. $LiCoO_2/Li_3BO_3$ composite particles were synthesized by a solid-state reaction to optimize the morphology and size of the particles for the AD method. Using this composite particle, the cathode layer was fabricated on an Al-LLZ pellet. All-SS-LB with a lithium metal anode was successfully prepared and showed good performance. In addition, the $LiCoO_2/Li_3BO_3$ cathode was improved by heat treatment at 750 °C. The discharge capacity was improved and higher than 70%. The interfacial resistance of $LiCoO_2$–Li_3BO_3/Al-LLZ decreased with the heat treatment.

Acknowledgments This study was financially supported by the ALCA-SPRING project of the Japan Science and Technology Agency. The authors would like to thank Sadahiro Yagishita and Tadahiro Niwa of Daiichi Kigenso Kagaku Kogyo Co., Ltd., Japan for providing the sample of Al-doped LLZ.

References

1. Sakuda A, Hayashi A, Tatsumisago M (2009) Interfacial observation between $LiCoO_2$ electrode and Li_2S-P_2S_5 solid electrolytes of all-solid-state lithium secondary batteries using transmission electron microscopy. Chem Mater 22:949–956. https://pubs.acs.org/doi/10.1021/cm901819c
2. Yu X, Bates JB, Jellison Jr GE, Hart FX (1997) A stable thin-film lithium electrolyte: lithium phosphorus oxynitride. J Electrochem Soc 144:524–532. http://jes.ecsdl.org/content/144/2/524.short
3. Hamon Y, Douard A, Sabary F, Marcel C, Vinatier P, Pecquenard B, Levasseur A (2006) Influence of sputtering conditions on ionic conductivity of LiPON thin films. Solid State Ionics 177:257–261. https://doi.org/10.1016/j.ssi.2005.10.021
4. Kim YG, Wadley HNG (2009) Plasma-assisted deposition of lithium phosphorus oxynitride films: substrate bias effects. J Power Sources 187:591–598. https://doi.org/10.1016/j.jpowsour.2008.11.070
5. Kunshima GB, Gromov OG, Lokshin EP, Kalinnikov VT (2014) Sol–gel synthesis of $Li_{1.3}Al_{0.3}Ti_{1.7}$ $(PO_4)_3$ solid electrolyte. Russ J Inorg Chem 59:424–430. https://link.springer.com/article/10.1134/S0036023614050118
6. Kim KM, Shin DO, Lee YG (2015) Effects of preparation conditions on the ionic conductivity of hydrothermally synthesized $Li_{1+x}Al_xTi_{2-x}$ $(PO_4)_3$ solid electrolytes. Electrochim Acta 176:1364–1373. https://doi.org/10.1016/j.electacta.2015.07.170
7. Ma F, Zhao E, Zhu S, Yan W, Sun D Jin Y, Nan C (2016) Preparation and evaluation of high lithium ion conductivity $Li_{1.3}Al_{0.3}Ti_{1.7}$ $(PO_4)_3$ solid electrolyte obtained using a new solution method. Solid State Ionics 295:7–12. https://doi.org/10.1016/j.ssi.2016.07.010
8. He S, Xu Y, Zhang B, Sun X, Chen Y, Jin Y (2018) Unique rhombus-like precursor for synthesis of $Li_{1.3}Al_{0.3}Ti_{1.7}$ $(PO_4)_3$ solid electrolyte with high ionic conductivity. Chem Eng J 345:483–491. https://doi.org/10.1016/j.cej.2018.03.151
9. Wolfenstine J, Allen JL, Read J, Sakamoto J, Donacel GG (2010) Hot-pressed $Li_{0.33}La_{0.57}TiO_3$. J Power Sources 195:4124–4128. https://doi.org/10.1016/j.jpowsour.2009.12.109
10. Trong LD, Thao TT, Dinh NN (2015) Characterization of the Li-ionic conductivity of $La_{(2/3-x)}Li_{3x}TiO_3$ ceramics used for all-solid-state batteries. Solid State Ionics 278:228–232. https://doi.org/10.1016/j.ssi.2015.05.027
11. Wolfenstine J, Rangasamy E, Allen JL, Sakamoto J (2012) High conductivity of dense tetragonal $Li_7La_3Zr_2O_{12}$. J Power Sources 208:193–196. https://doi.org/10.1016/j.jpowsour.2012.02.031
12. Li Y, Han JT, Wang CA, Vogel SC, Xie H, Xu M, Goodenough JB (2012) Ionic distribution and conductivity in lithium garnet $Li_7La_3Zr_2O_{12}$. J Power Sources 209:278–281. https://doi.org/10.1016/j.jpowsour.2012.02.100

13. Ohta S, Komagata S, Seki J, Saeki T, Morishita S, Asaoka T (2013) All-solid-state lithium ion battery using garnet-type oxide and Li_3BO_3 solid electrolytes fabricated by screen-printing. J Power Sources 238:53–56. https://doi.org/10.1016/j.jpowsour.2013.02.073
14. Okumura T, Takeuchi T, Kobayashi H (2016) All-solid-state lithium-ion battery using $Li_{2.2}C_{0.8}B_{0.2}O_3$ electrolyte. Solid State Ionics 288:248–252. https://doi.org/10.1016/j.ssi.2016.01.045
15. Wakasugi J, Munakata H, Kanamura K (2017) Effect of Gold Layer on Interface Resistance between Lithium Metal Anode and $Li_{6.25}Al_{0.25}La_3Zr_2O_{12}$ Solid Electrolyte.JElectrochemSoc164:A1022-A1025. http://jes.ecsdl.org/content/164/6/A1022.full

Chapter 15
Li Metal Polymer Batteries

Ismael Gracia, Michel Armand and Devaraj Shanmukaraj

Abstract Polymer electrolytes are of prime importance for advanced lithium-based batteries in terms of high-energy density, design flexibility and safety. In this chapter, we summarize the fundamental property requirements of materials for polymer electrolyte application. State of the art polymer hosts, salts and additives are reviewed along with the challenges faced in the current state of the art. Finally, the fabrication process (scaling-up) of Li metal polymer battery components, along with an overview of the current status of Li metal polymer batteries (Industrial), is presented. A brief conclusion and perspective for Li metal polymer batteries are also discussed.

15.1 Introduction to Global Challenges and Li-ion Batteries

Fossil fuels, including coal, oil and natural gas, are currently the world's primary energy sources, which are non-renewable, resulting in production of CO_2 and contributing to the climate change and global warming. If we want to meet the pledge of <2 °C increase in average temperature on earth (COP21), three-fourths of the known fossil fuels reserves should be kept in the ground. Yet, there is a projected doubling of world energy consumption within the next 50 years, and this has to be coupled with low or even zero emission sources. Thus, there is an increasing awareness of the need for efficient, clean and renewable energy sources. Currently, the global green-house gas emissions in the transportation sector account to around 14%, as projected in Fig. 15.1. One possible solution to minimize this global challenge (climate change) is to electrify transportation while generating electricity without CO_2 emission. Renewable energy sources input is estimated to grow consistently over the coming years. On the other hand, current lithium-ion technology provides the predominant battery technology to handle the issue of storing energy to allow electrification of transport, owing to their high energy, high voltage, cycle life and excellent storage characteristics.

I. Gracia · M. Armand · D. Shanmukaraj (✉)
CIC Energigune. Parque Tecnológico de Álava, Albert Einstein, 48, ED.CIC, 01510 Miñano, Álava, Spain
e-mail: dshanmukaraj@cicenergigune.com

R. Murugan and W. Weppner (eds.), *Solid Electrolytes for Advanced Applications*,
https://doi.org/10.1007/978-3-030-31581-8_15

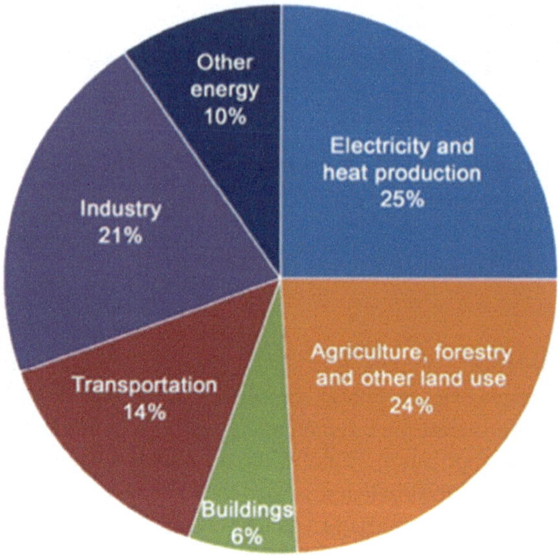

Fig. 15.1 Global green-house gas emissions by economic sector [1]

Lithium batteries have been highly sought after due to the unique property of Li metal, namely, lowest atomic number (high specific capacity 3884 mAh/g) and very low redox potential [−3.10 V vs. standard hydrogen electrode (SHE)]. The use of Li metal in batteries was crippled due to the formation of dendrites during cell cycling, thus posing high risk and safety concerns. Since the introduction of "rocking chair concept", that is, shuttling Li^+ ions between two intercalation hosts [2, 3], lithium-ion rechargeable batteries seemed a viable solution for circumventing the safety issue and poor rechargeability.

Most of the commercially available Li-ion batteries use liquid electrolytes, an important component that determines the safety and stability of a battery. Non-aqueous Li-ion batteries are typically composed of a lithium salt (lithium hexafluorophosphate) solubilized in an organic solvent (cyclic or linear) or mixture of solvents (binary or ternary). This leads to the formation of a protective solid electrolyte interface (SEI) on the negative electrode, mainly graphite and passivation of aluminum current collector on the positive electrode. However, these electrolytes generate PF5 and RO–($R = CH_3$, C_2H_5) which are very strong electrophile and base-nucleophile, respectively. Above the room temperature, the generation of such species is enhanced and becomes a major concern for maintaining the stability of the SEI and safety of the battery.

15.2 Introduction to Lithium Polymer Batteries

Lithium batteries using polymer electrolytes as separators fall under the category of Li polymer batteries. Polymer electrolytes are generally classified as dry polymer electrolyte and gel polymer electrolytes. Dry polymer electrolytes are composed of a polymer matrix and a salt, whereas gel polymer electrolytes are additionally composed of a polar solvent added as a plasticizer.

Ever since Armand et al. [4] proposed the first PEO/Li^+ dry solid polymer electrolyte (SPE) system (~10^{-4} S cm^{-1} at 40–60 °C) for lithium batteries in 1978, polymer electrolytes have attracted widespread attention. Solid-state batteries using lithium metal (SSLB) are of high interest due to their potential benefits in gravimetric and volumetric energy density and safety when compared to traditional liquid-electrolyte-based systems. The ever-growing need for design flexible energy storage devices for use in mobile electronics has kindled the growth of solid-state battery market.

From the point of view of mass production of dry polymer SSLB, the solid polymer electrolytes (SPEs) should comply with the following requirements [5, 6]:

(a) *High ionic conductivity*: SPEs should present high ionic conductivity ($\geq 10^{-4}$ S cm^{-1}), not only at high temperature but also at close to ambient temperatures, while being an electronic insulator to avoid shelf discharge. Fast charge/discharge rates in SSLBs are facilitated by high ionic conductivities at the working temperature. Besides, high Li^+ transference (t_{Li+}) number is another requirement in order to reduce concentration polarization of electrolyte components during charge/discharge cycles and abate dendrite formation [7, 8].

(b) *Good mechanical properties*: Unlike ceramic electrolytes that have a brittle nature leading to the generation of cracks, SPEs respond in an elastic manner to stress (Young's modulus $\geq$ 30 MPa), which makes them more suitable to accommodate volume changes. A polymer electrolyte membrane with high mechanical stability is a basic requirement for easy processability at an industrial scale. In addition, the polymer electrolytes are expected to be mechanically stable against the growth of dendrites. This could be usually tailored by the introduction of inorganic fillers [9], cross-linking including interpenetrated networks, block polymers, or by embedding SPEs into fibrous networks [10–12].

(c) *High electrochemical, chemical and thermal stability*: The electrochemical stability of SPEs has to be at least 4 V versus Li/Li^+, preferably $\geq$4.5 V versus Li/Li^+, such that the electrolyte is stable at the cut-off voltages of both electrodes [13]. This property has become more crucial with the industrial scale production of high-voltage cathode materials and lithium metal anodes [14]. Furthermore, the polymer membranes have to be stable toward any chemical species present in the battery at the open-circuit voltage as well as to any species generated upon cycling. Finally, a good thermal stability (Tdecomposition $\geq$ 150 °C) also presents a comparative advantage against liquid electrolytes in terms of safety against electrical or thermal abuse that can lead to fire or explosions [15].

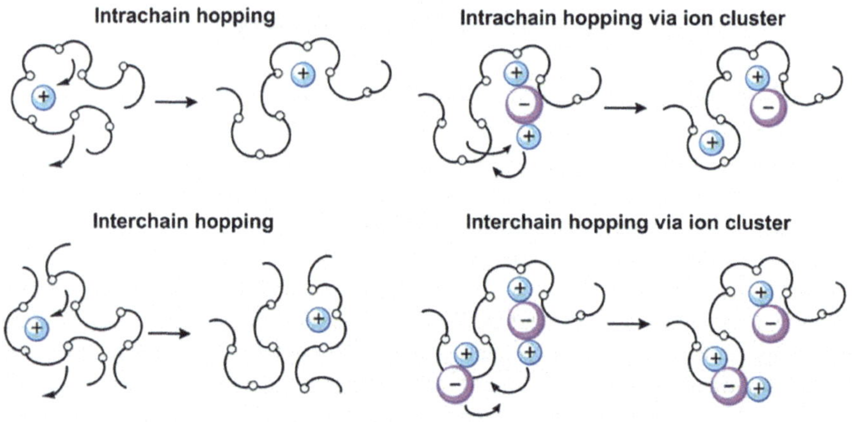

Fig. 15.2 Ion transport mechanism in PEO-based solid polymer electrolyte [21]

Despite the polymer acting as a host for lithium ions, it was promptly envisaged that its role is much more crucial than just behaving as a support. Ionic conductivity of SPEs highly depends on the nature of the polymer matrix in terms of its ability to solvate the lithium salt and the crystalline versus amorphous behavior of the solid solution.

The ions are transported within the polymer matrix through a commonly accepted ion transport mechanism [16–18] in which Li^+ ions that are dissociated from the counter ion and coordinated to the electron-donor groups present in the polymer host hop from one coordinating site (usually composed of more than three electron-donor groups) to another. Ion hopping is facilitated either by segmental motion of the polymer chain which is directly associated with the glassy/amorphous regions of the SPE, or by the ion cluster-assisted hopping mechanism, in which temporary re-association with the counter ion occurs before being re-solvated by electron-donor groups in the polymer (Fig. 15.2). This occurs in the amorphous part of the polymer and starts at the glass transition (T_g) of the macromolecule. The crystalline part of the polymer offers very limited or no ion motion [19]. The fraction of amorphous/crystalline region is determined for PEO/salts by a phase diagram. TFSI is the salt offering the widest domain of amorphicity in composition and temperature [20].

In general terms, it would be possible to state that a successful polymer matrix should be able to fulfill the following parameters in order to become a good host for a lithium salt:

(a) *Electron-donating nature of the functional groups present in the polymer backbone* or the functional groups pending from it. Cation solvation, as in organic liquid electrolytes, is a major requirement to achieve high ionic mobility in SPEs. Therefore, polymer–cation interactions need to be strong enough to favor the salt dissociation, but weak enough to favor the ionic hopping from one coordinating site to another. In this regard, functional polar groups, such as ether, amines, hydroxyl, carbonyl-based among others, have been extensively used in polymer matrices.

(b) *Dielectric constant* (ε) of the polymer host is also an important feature to tailor as the highest the constant is, the better the ionization of the salt that can be achieved in the physical mixture, leading to higher concentration of charge carriers. Note that when the pristine polymer is solvating but with a low ε, the addition of salt, introducing dipoles, increases the polarity (self-dissociation).

(c) *High backbone flexibility* is required in terms of lowering the energy barrier for bond rotation that facilitates the segmental motion of the polymer chain. Completely amorphous materials at their respective working temperatures are highly sought after as SPEs, thus low T_g values of the polymer host become a requisite.

(d) *High molecular weight* of the polymer host is desirable due to better mechanical properties and film-forming ability that is usually enhanced at high molecular weight (M_w) fractions.

15.3 Promising Candidates as Electrolytes for Solid-State Batteries

Research in polymer electrolytes for lithium-ion or lithium-based battery technologies is a well-developed field in which many progress has been made since the first steps back at the 1970s. Since the early studies on the feasibility of ionic conductivity in solid solutions or physical mixtures of a sodium or lithium salt with a high molecular weight polymer, the research on this topic has moved to the investigation of new materials in order to address the industrial needs. Some of the materials suitable for use as dry polymer electrolytes have been summarized as below.

15.3.1 Materials as Polymer Hosts

A huge effort has been devoted to the screening of multiple polymer matrices that range from polymer commodities, extending from PEO to PMMA or PVC [22–24], to complex synthetic architectures that include co-polymers and block-copolymers. Among all, the better performing polymer matrices for solid-state batteries, not considering gel electrolytes, usually are (Fig. 15.3) englobe polyethylene oxide (PEO) [25, 26], polyvinylalcohol (PVA) [27–29], and flexible aliphatic polycarbonates and polyesters [30, 31], such as polyethylene carbonate (PEC) or polycaprolactone (PCL). Besides, organic/inorganic polymers such as polysiloxanes and polyphosphazenes have also been envisaged as good candidates for SPEs due to their very low glass transition temperatures that lead to high ionic conductivities [32–35].

15.3.2 The Importance of Salt Anions

Beyond the polymer host, a highly dissociable lithium salt, as lithium ions source, is the other main component of SPE. Being soluble in the polymer host is of paramount importance as it will be directly linked to its ability to generate free charges, Li^+ ions.

The traditional, $LiClO_4$ or $LiPF_6$ inorganic salts used in common liquid organic electrolytes are not usually regarded as viable industrial alternatives as they will be possibly hazardous for users and other battery components. Perchlorate is plagued with safety issues and in $LiPF_6$ the equilibrium $LiPF_6 \Leftrightarrow PF_5 + LiF$ results in the cleavage of ether links, destroying the polymer. Therefore, sulfonate anions like $CF_3SO_3^-$ were envisaged in the first place as possible alternatives; however, their low dissociation constant and unfavorable phase diagram lead to low ionic conductivities, hampering their application to lithium batteries. To address these problems, series of bulkier anions have been developed over the years, being sulfonimides: $Li[CF_3SO_2)_2N]$ (LiTFSI), $Li[FSO_2)_2N]$ (LiFSI), $Li[C_2F_5SO_2)_2N]$ (LiBETI), etc.), borates: $Li[C_2O_4)_2B]$ (LiBOB), $Li[C_2O_4)(F_2)B]$ (LiDFOB), etc.) and phosphate-based $Li[F_3(C_2F_5)_3P]$ (LiFAP) anions, the most studied [36–39] as shown in Fig. 15.4. Negative charge delocalization through the structure of these anions, favored by the use of electron-withdrawing functional groups, allows higher ionic dissociation resulting in, typically, higher conductivities. Increasing the size of the anion is normally associated to a higher plasticizing capability of the Li salt, by increasing the amorphous/crystalline ratio.

Among all, lithium bis(trifluoromethane)sulfonimide (LiTFSI) shows the highest ionic conductivity in most polymer matrices and, therefore, usually acts as a control salt for new anion development [40–42]. However, it has been demonstrated that at potentials >3.8 V, LiTFSI and its decomposition products can corrode aluminum usually present as current collector in lithium ion battery (LIB) cathodes. Therefore, research in further alternatives capable of achieving high ionic conductivities and

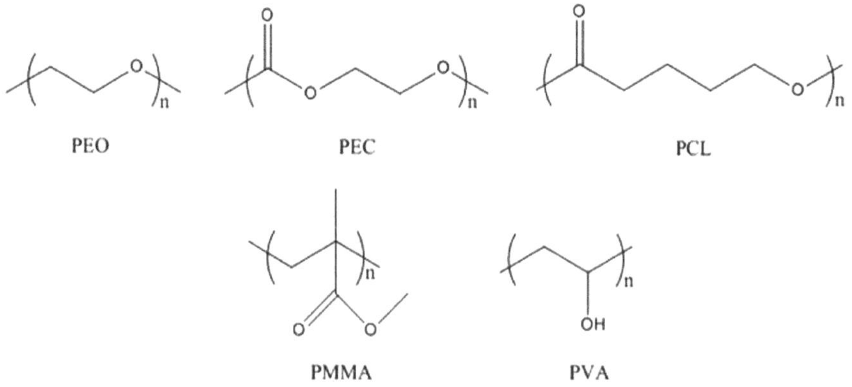

Fig. 15.3 Most common polymer matrices employed in dry SPEs

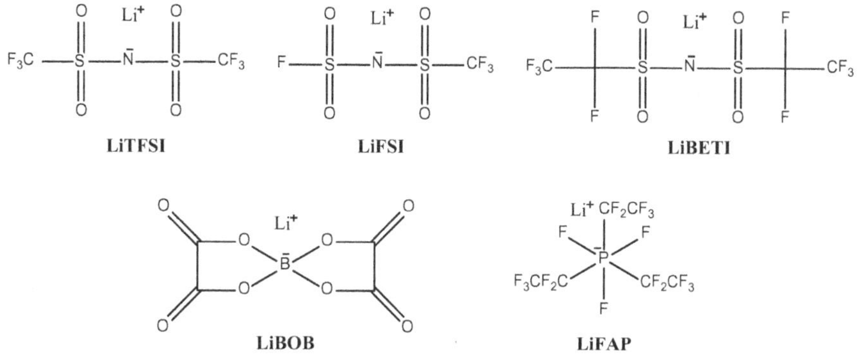

Fig. 15.4 Most commonly studied bulky anion based salts in SPEs

electrochemical stability against high-voltage cathode materials and aluminum is considered as an open field in solid-state batteries [43–45].

15.3.3 Polyether-Based SPEs

PEO (polyethylene oxide)-based SPEs have always been considered as the most promising candidate for application in actual LIBs since the pioneering research by Peter Wright in 1975, showing its suitability as a host for a number for Na salts [46]. A few years later, Armand and co-workers implemented PEO polymer matrices in rechargeable batteries [4, 47]. Li^+ transport in PEO typically responds to a coupled system in which Li^+ ions are coordinated to at least three EO (CH_2CH_2O) units [20]. In these systems ion mobility is directly linked to the segmental motion of PEO chain and can take place intra- and interchain in the amorphous domain.

While the easiest way to increase the amorphous behavior of SPE is to increase the temperature above the melting temperature, working at high temperatures can lead to compromised mechanical properties, reduction of the electrochemical window and reduction of the overall efficiency of the battery pack due to a higher energy consumption of the thermal management system. Several approaches have been described as possible means to reach the aforementioned characteristics in all-solid electrolytes, not taking into consideration gelation with organic solvents, which might not solve the safety issues concerning contemporary organic solvents used in Li-ion batteries:

(a) *Blend polymer electrolytes* consist of a physical mixture of two different polymer matrices. Commonly, the predominant polymer shows high ionic mobility, due to a rather low T_g, while the minor component of the solid solutions usually acts as a scaffolding polymer providing mechanical support to the SPE. PEO, PPO (polypropylene oxide) and other derivatives such as Jeffamines® have been

Table 15.1 Ionic conductivity of some common polymer, polymer blends and block co-polymer electrolytes

Polymer electrolyte system	Ionic conductivity (S cm^{-1})	Temperature (°C)	References
PEO/$LiClO_4$	$2.8\cdot10^{-7}$	27	[48]
PEO/LiTFSI	$3.8\cdot10^{-4}$	50	[20]
PPO/LiTFSI	$>10^{-4}$	50	[49]
PEO-PMMA/$LiCF_3SO_3$	$2\cdot10^{-5}$	30	[50]
PEO-P(VDF-TrFE)/$LiClO_4$	$7\cdot10^{-4}$	RT	[51]
PEO-PPC/$LiClO_4$	$6.8\cdot10^{-5}$	RT	[52]
PEO:PEI/$LiClO_4$	10^{-4}	30	[53]
PAN-PEO/$LiClO_4$	$6.8\cdot10^{-4}$	25	[54]
P(EO-PO)/LiTFSI	10^{-3}	80	[55]
PLMA-b-POEM/$LiCF_3SO_3$	10^{-5}	25	[56]
PEO-b-PS/LiTFSI	$>10^{-4}$	60	[57]

traditionally used as the main component of blends; however, their mechanical properties are not good enough to provide self-standing membranes that would allow industrial processing as well as provide enough hardness to avoid Li dendrite growth in Li-metal batteries. On the other side, PMMA (poly methyl methacrylate), PVC (polyvinyl chloride) or PAN (polyacrylonitrile) among others is considered as good supporting polymer matrix in order to achieve good mechanical properties and high voltage stability.

(b) *Block co-polymers*, which include strands of different nature being covalently linked in the same polymer chain, constitute another possibility to achieve both high ionic conductivity and sufficient mechanical robustness. The use of alternated sections of conductive and supporting polymers can have different effect depending on the solubility of the blocks in each other. If the blocks are miscible, final properties of the polymer matrix will be a combination of the original properties. Meanwhile, if they are not soluble the materials usually show phase segregation which implies that one phase will be acting as a supporting scaffold, as in polymer blends, while the other take the role of ion conductive media. The ionic conductivity of some common polymers, blend polymer and block co-polymer electrolytes with Li salts are tabulated in Table 15.1.

15.3.4 Polycarbonates and Polyesters

Owing to the rather low dielectric constant of the C–O bond in PEO, salt dissociation is not always fully achieved, resulting in the formation of ion aggregates, thus decreasing the ionic conductivity of the SPE. Aliphatic or cyclic carbonates (–O–(C = O)–O–) are common organic solvents in liquid electrolytes due to the high dielectric constant of the latter, which allows high salt dissociation even in mixtures [58]. The introduction of such structures in the polymer backbone or as side-chain functional groups in aliphatic polymers has also proven to be a very efficient method to reach high ionically conductive and electrochemically stable SPEs.

Poly(ethylene carbonate) (PEC), poly(trimethylene carbonate) (PTMC) and poly(propylene carbonate) (PPC) are the most interesting linear carbonate structures, of which the former shows higher ionic conductivities due to high concentration of donor groups that can favor ionic mobility through the polymer matrix.

Besides, aliphatic polyesters that tend to show low T_g values and high dielectric constant have also been used as polymer hosts. Polycaprolactone and polylactic acid are the most studied structures. The random copolymer of trimethylene carbonate (TMC) and caprolactone (CL) shows high conductivities [59]. The low degree of crystallinity favors the existence of large amorphous regions in the SPE, thus favoring ionic conduction. As shown in Fig. 15.5 (left), for an optimized ratio of TMC:CL (60:40), a high ionic conductivity of the order of 10^{-5} S/cm was achieved at 60 °C. Although compared with the conventional PEO:LiTFSI (Fig. 15.5 right) electrolytes, the conductivity seems to be lower. The basic advantage of using such polyester-based electrolytes is their high electrochemical stability suitable for use with high voltage cathodes. Polyesters have also been usually used as the major component of blends or block copolymers in which another supporting polymer structure is introduced to enhance the mechanical robustness of the SPE.

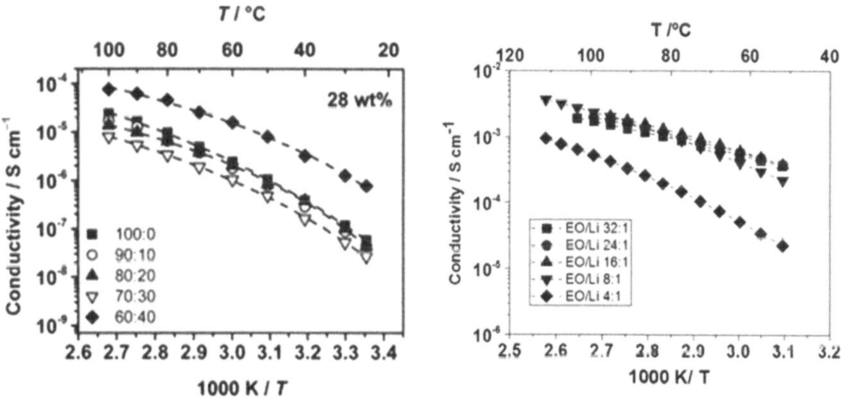

Fig. 15.5 (Left) Ionic conductivity of copolymers of TMC:CL with 28 wt% LiTFSI salt [59] and (right) ionic conductivity of PEO with LiTFSI salt for different EO:Li ratios (data extracted from reference 20)

Table 15.2 Electrochemical stability and ionic conductivity of polyester, polycarbonate and polysiloxane-based polymer electrolytes

Polymer electrolyte system	Ionic conductivity ($S\ cm^{-1}$)	Temperature (°C)	Electrochemical stability (Volts)	Battery configuration	References
PTMC/LiTFSI	10^{-7}	60	5.0	$LiFePO_4/Li$	[61]
PEC/LiFSI	$2.2 \cdot 10^{-4}$	60	–	–	[62]
PEC/LiTFSI	$4.7 \cdot 10^{-4}$	20	4.5	–	[63]
PCL-PTMC/LiTFSI	$4.1 \cdot 10^{-5}$	50	4.6	$LiFePO_4/Li$	[59]
Cellulose-PPC/LiTFSI	$3 \cdot 10^{-4}$	20	4.6	$LiFePO_4/Li$	[64]
PSi/LiTFSI	$4.5 \cdot 10^{-4}$	25	–	–	[65]
P(DMS-EO)/$LiClO_4$	$2.6 \cdot 10^{-4}$	25	5.0	–	[66]
P(Si-EO)/LiTFSI	10^{-4}	25	4.5	$LiNi_{0.8}Co_{0.2}O_2/Li$	[67]
PSi-PVDF/LiTFSI	$5.7 \cdot 10^{-5}$	25	5.1	$LiFePO_4/Li$	[68]

15.3.5 Polysiloxanes

Polysiloxanes are inorganic polymers that belong to the silicone family. One of their most appealing features is that they present very low T_g values due to the flexibility of the Si–O–Si bond, which allow them to behave as completely amorphous materials and thus reach high ionic conductivity values. Short pendant EO-based side chains ensure that the cation solvation requirements are met [60]. Owing to the relatively strong Si–O bond, they also present good electrochemical, chemical and thermal stabilities. On the contrary, such low glass transition results in poor mechanical properties that are often mitigated by cross-linking or embedding into networks of polymer fibers.

The electrochemical stability and ionic conductivity of polycarbonates, polyesters and polysiloxane-based electrolytes (Table 15.2) suggest the use of such electrolytes for high-voltage cathode materials, though most tests have been with $LiFePO_4$.

15.3.6 Polymer-in-Salt Systems

When discussing about SPEs, polymer usually acts as the solvent (host), the main ingredient of a solid solution for the Li–salt. This is called the "salt in polymer" regime. However, in order to increase ionic conductivity, a high percentage of lithium salt (>50 wt%) can also be used, such SPEs being termed as "polymer-in-salt systems" [69, 70].

These SPEs can partially retain the mechanical stability of the polymer matrix, while they show a complete rubbery behavior similar to those of glasses in which fast ion conduction, preferably cationic, is possible. One can consider that the mixture is plasticized by the salt (LiTFSI, LiFSI) which have extrapolated T_gs below ambient. In these systems ion transport is usually considered as decoupled from the segmental

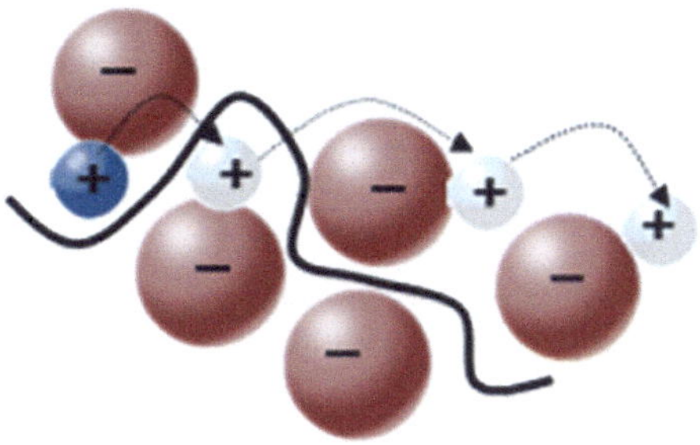

Fig. 15.6 Ion transport in polymer-in-salt systems

Table 15.3 Ionic conductivity of polymer-in-salt electrolytes

Polymer	Polymer to salt (Mole/Wt.) ratio	Li salt	Ionic conductivity (S cm-1)	Temperature (°C)	Electrochemical window (Volts)	References
PAN	1:1	LiTFSI	$1.1{\cdot}10^{-4}$	30	5.0	[78]
PEO	9–91	$LiClO_4$ + $LiNO_3$ + LiOAc	10^{-3}	RT	>4.5	[79]
PAM	1:1	LiTFSI	10^{-4}	RT	–	[76]
P(AN-co-BuA)	35–65	LiTFSI + LiI	10^{-5}	50	–	[80]

motion of polymer chains (Fig. 15.6). The high concentration of cations leads to the formation of clusters and percolation networks where cation mobility is highly favored [71–73].

Poly(acrylonitrile) (PAN) is a widely studied polymer matrix for polymer-in-salt systems as it allows complete solvation and stabilization of Li^+ cations due to the high concentration of the very polar nitrile groups [74–77]. The ionic conductivity achieved at room temperature or close with some widely used polymer matrices is tabulated in Table 15.3. However, Li–salts being the most expensive component of SPE force polymer-in-salt systems to be often regarded only from the scientific point of view.

15.3.7 *Single-Ion Conductors*

As explained above, ionic conduction in SPEs is usually ambipolar, leading to a low (<0.3) Li^+ transference number, lower than that of the anion. This fact causes salt concentration gradients upon charge/discharge cycles that could be avoided if the anion moieties were not mobile; for example, by anchoring them to the polymer matrix. Single-ion conductors or polymeric lithium salts are synthetic polymers in which the

Table 15.4 Ionic conductivity correlation with anionic centers for single-ion conducting polymer electrolytes

Anionic center	Conductivity (S cm^{-1})	Temperature (°C)	References
$-CO_2^-$	$1.5 \cdot 10^{-7}$	25	[84]
$-SO_3^-$	$1.8 \cdot 10^{-7}$	25	[85]
$-SO_2(CF_2)_4N(-)-$	10^{-6}	25	[86]
$-SO_2N(-)SO_2F$	10^{-5}	60	[87]
$-SO_2N(-)SO_2CF_3$	10^{-4}	60	[88]
$-OB(-)(O)(C_2O_4)$	$7 \cdot 10^{-6}$	25	[89]
$-OP(-)(C_2O_4)_2$	$1.6 \cdot 10^{-6}$	25	[90]

anionic moiety of the Li–salt is covalently linked to the polymer backbone, usually as a side-functional group [81, 82]. The poly(salt) can be alloyed with a solvating polymer or incorporated into the solvating chains. In principle, this should lead to close to unity t_{Li+} and therefore better C-rate performances for a given ionic conductivity. It is also expected that the formation of dendrites triggered by salt depletion at the Li side could be controlled. To date, one of the main challenges affecting single-ion conductors is the fact that highly dissociable anions such as those derived from $TFSI^-$ are difficult to be grafted into a polymer backbone and involves several steps, which also usually results in polymers with higher glass transition temperatures and higher crystalline degrees than ambipolar SPE systems. Although ionic mobility in these polymers is usually lower than for the already presented polymer and Li–salt mixtures (Table 15.4), a close to unity t_{Li+} can ensure a sufficiently high ionic conductivity for use as electrolytes in LIBs or Li-metal batteries [83]. A successful approach for single-ion conducting polymer electrolytes was designed based on a polyanionic block co-polymer (Fig. 15.7) using Styrene-TFSI as anion. The use of TFSI anion away from the solvating block enabled a high dissociation level leading to weak Li^+ interactions.

15.3.8 Use of Additives

The use of additives in solid polymer electrolyte can account for many different reasons. Improving ionic conductivity without inducing safety hazards, enhancing mechanical properties or improving electrochemical stability of the systems are some of the driving factors for the introduction of additives in polymer electrolytes.

Plasticizing additives have been widely explored as a means to increase ionic mobility in dry-polymer electrolytes. In contrast to gel polymer electrolytes wherein the additive is a low boiling point organic solvent, plasticizing additives includes a low melting point additive, with a plastic-crystalline phase below T_m, and able to solvate Li-salt, thereby decreasing the T_g of the whole SPE to a considerable extent. While ionic conductivity can be easily increased in this way, mechanical

Fig. 15.7 Chemical structure of single-ion conducting tri-block copolymer P(STFSILi)-b-PEO-b-P(STFSILi) electrolytes [83]

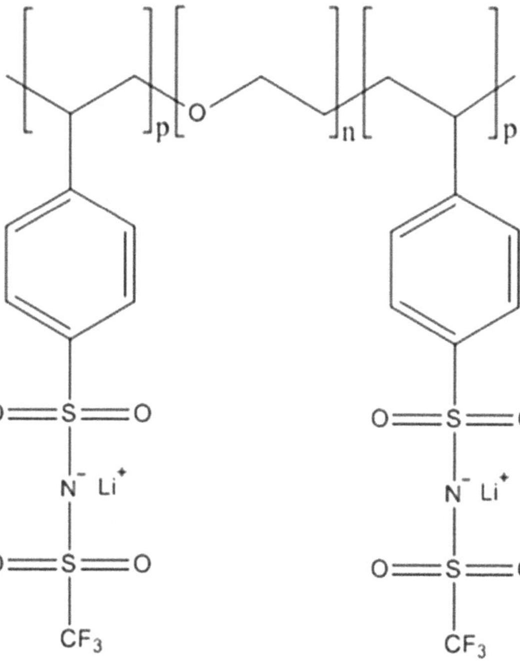

properties are usually compromised and require further reinforcement efforts. One of the most widely studied additives is succinonitrile (SN), which due to its high dielectric constant is able to easily dissociate ion pairs. SN plasticized SPEs tend to show very high ionic conductivities ($5{\cdot}10^{-4}$ S cm^{-1} at 30 °C) as well as high oxidation potential (>4.5 V) that makes it a very interesting additive toward high-voltage cathodes and high energy density batteries.

Composite polymer electrolytes are electrolytes in which an inorganic filler or additive has been dispersed in a polymer matrix [91]. Inorganic fillers can be inactive (Al_2O_3, SiO_2, TiO_2, etc.) or active (γ-$LiAlO_2$, Li_3N, etc.), depending on their contribution toward Li transport. Inactive or inert fillers indirectly modify the SPEs properties, by decreasing the crystallinity, increasing the segmental dynamics of the polymer chains or enhancing the mechanical or electrochemical stability of the electrolyte membranes. On the other hand, active fillers have a direct role in providing direct Li transport through the grains and being relayed by the polymer electrolyte, improving the ionic conductivity, increasing T_{Li+} or improving oxidation resistance. The improved ionic conductivity and electrochemical stability window of polymer electrolytes using different additives is shown in Table 15.5.

Table 15.5 Ionic conductivity of SPE systems with additives

Polymer electrolyte system	Additive	Ionic conductivity ($S\ cm^{-1}$)	Temperature (°C)	Electrochemical window (Volts)	References
PEO/LiTFSI	SN	$7\cdot10^{-4}$	RT	5	[92]
PVDF-HFP-ETPTA/LiTFSI	SN	10^{-3}	RT	5	[93]
PAN-PVA-CN/$LiPF_6$	SN	$4.5\cdot10^{-4}$	25	5	[94]
PEO/$LiClO_4$	Al_2O_3	10^{-5}	30	5.0	[91]
PEO/$LiClO_4$	TiO_2	$>10^{-5}$	30	5.0	[95]
PEO/$LiClO_4$	SiO_2	$3\cdot10^{-6}$	30	5.0	[96]
PEO/$LiBF_4$	γ-$LiAlO_2$	$3\cdot10^{-7}$	30	–	[97]
PEO/$LiCF_3SO_3$	Li_3N	$1.2\cdot10^{-4}$	30	–	[98]

15.3.9 Li Metal Polymer Solid-State Batteries

The weight and volume of the hosts into which the Li^+ intercalates poses a limitation for the energy content of the battery. Therefore, the requirement of alternative high-energy materials has become crucial. Employing lithium metal as an alternative to graphite anodes would result in a ~35% increase in volumetric specific energy and ~50% increase in gravimetric energy density at the cell level.

A comparison of the conventional Li-ion battery and Li metal solid-state battery at a cell level is shown in Fig. 15.8. Li metal solid-state batteries consist of a Li metal anode, polymer electrolyte and a composite cathode comprising a Li-intercalation host, conductive carbon and polymer electrolyte material as a binder in order to provide ionic pathways.

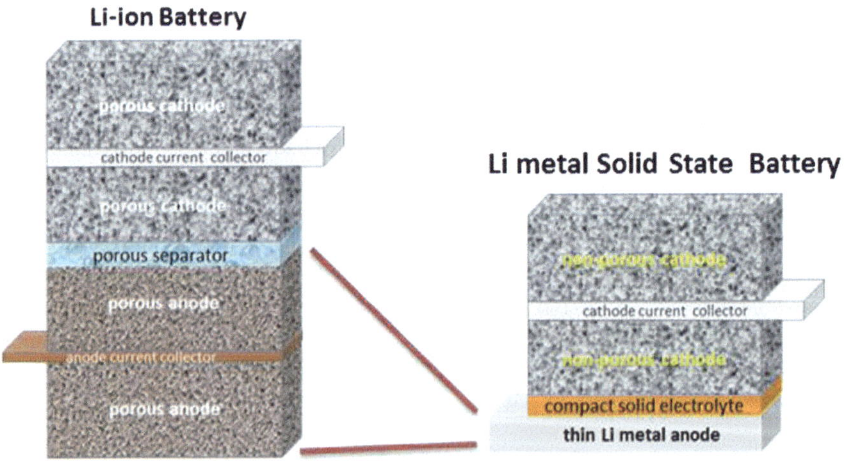

Fig. 15.8 Comparison of cell design—conventional Li-ion battery and Li metal solid-state battery [Courtesy: Hydro-Québec]

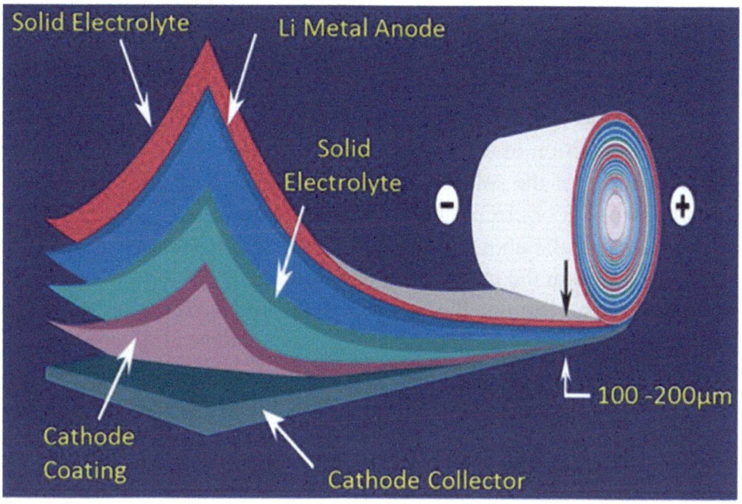

Fig. 15.9 Components of Li metal polymer battery stack (Courtesy: Hydro-Québec)

Bipolar stacking with cells connected in series is feasible in solid-state batteries that in turn increase the total cell voltage. Moreover, the thickness of the whole cell is reduced due to the thin Li metal anode and the thin polymer separator/electrolyte used. Safety concerns are diminished due to the absence of any flammable compounds and the need for a complex liquid cooling system could thus be avoided. At the pack level the solid-state cells are space-saving and can pack twice the energy density of conventional Li-ion batteries. The typical stacking of the electrodes in a Li metal polymer battery is shown in Fig. 15.9. The thin sheet of Li metal used in the cell reduces the requisite of an anode current collector, thus eliminating the use of any metal collector (Cu, etc.). This in turn reduces the total weight of the battery as well as improves the energy density, while reducing the overall production cost of the cell.

Foreseeing the advantages of using Li metal and with an aim to improve the energy density and specific capacity of the current Li-ion technology using liquid electrolytes, three types of battery systems are under vigorous research, namely Li metal battery with an intercalation cathode (high voltage), Li–S (sulfur) and Li–O_2 batteries. The primary goal of these systems was to harness the high specific capacity of sulfur in Li/S systems and the oxygen electrode in Li–O_2 batteries without being limited by the low specific capacity of graphite anode as in conventional Li-ion batteries.

Although these new battery systems are promising, their development has been hindered by the limitations of liquid electrolytes. In the case of high voltage cathodes the conventional carbonate-based electrolytes are not stable to oxidation, whereas Li–S batteries using liquid electrolytes undergo severe polysulfide shuttle effect leading to self-discharge and active material loss.

In the case of Li–O_2 batteries where the reaction involved in the formation of Li_2–O_2 involves the reaction from O_2 to $[O_2]^{\bullet-}$ superoxide anion, to –O–O– peroxide to form Li_2O_2, the peroxide oxidizes to form oxygen during recharge. The major issue is that $[O_2]^{\bullet-}$ is a powerful nucleophile, and O_2^{2-} is a powerful base. The oxidation of O_2^{2-} leads to a fraction of oxygen as singlet O_2, a powerful oxidizer, as a result of which the stability of the solvents is compromised leading to decomposition of electrolyte and hence poor electrochemical performance.

The issues faced in the above-mentioned three-battery systems are summarized in Fig. 15.10. Apart from the issues mentioned for the different battery systems, the most common problems limiting the use of Li metal are the formation of Li dendrites. One way to suppress the growth of these dendrites is mechanical suppression. A "holy grail" to the limitations mentioned in the advanced battery systems and the inhibition of dendrite formation could be the replacement of liquid electrolytes with polymer electrolytes.

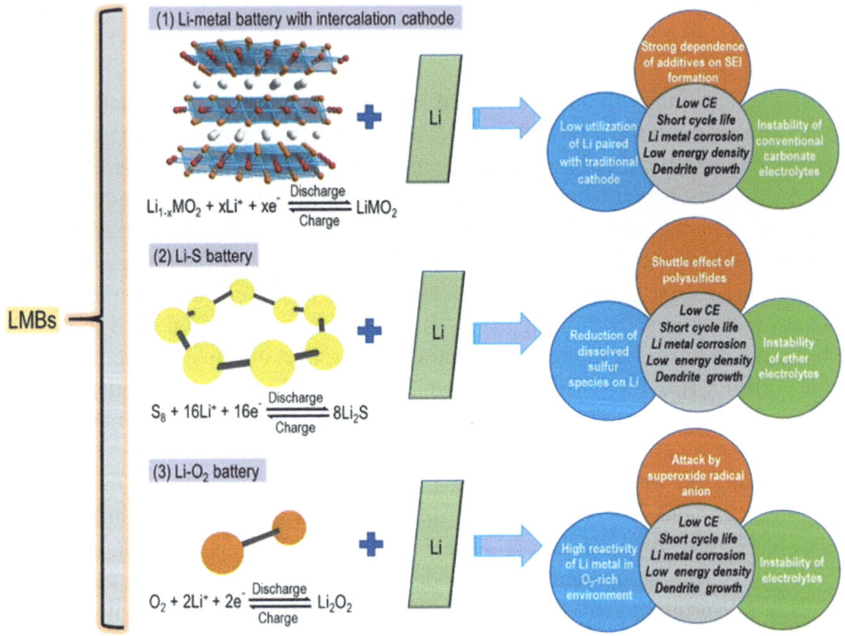

Fig. 15.10 Challenges in development of advanced Li metal batteries using liquid electrolytes [99]

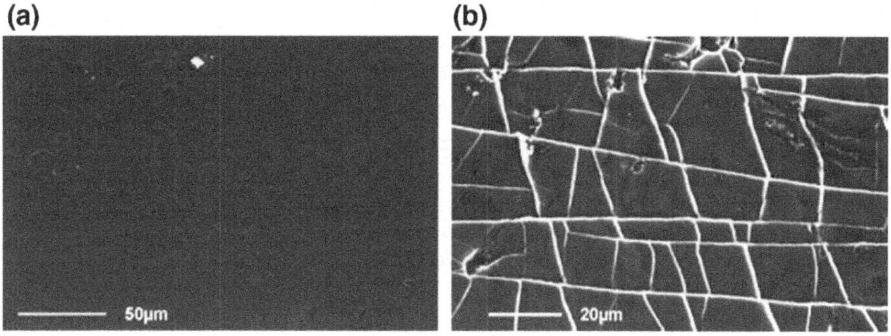

Fig. 15.11 SEM images of cracks on a cleaned Li metal surface [100]

15.4 Challenges of Li Metal Anodes/Polymer Electrolytes in Solid-State Polymer Batteries

Although using a solid polymer electrolyte and Li metal anode could be an alternative when compared to Li-ion batteries some issues exist when employing Li metal anodes, the prime concerns being the surface of Li metal, interface between Li metal and solid electrolyte (SEI) and the reaction of different cathodes with the polymer electrolytes.

15.4.1 Intercalation Host Cathodes

The Li metal surface due to its metallurgical origin is not regular and presents fine cracks along its surface. This leads to inhomogeneous current distribution on the surface of the Li metal where the cracks acts as nucleation center for Li growth. The growth of Li along the cracks leads to needle-like Li dendrites that affect the cycling capability of the cell. These dendrites although delayed by employing polymer electrolytes, due to a higher Young's modulus will eventually result in short circuit of the cell. A conventional cleaned Li metal surface (Fig. 15.11a) showing the fine cracks (Fig. 15.11b) on the surface is shown. Moreover, when high-voltage intercalation host materials are used as cathodes, the oxidative stability of the polymer electrolytes is questionable. Currently, there is a trade-off between highly electrochemically stable polymer electrolytes and the ionic conductivity of such electrolytes.

15.4.2 Sulfur Cathodes

In a Li/S cell the elemental sulfur is reduced to lithium polysulfides (Li_2S_x, $3 < x < 8$) that diffuse out from the cathode. The higher order polysulfides formed, which are

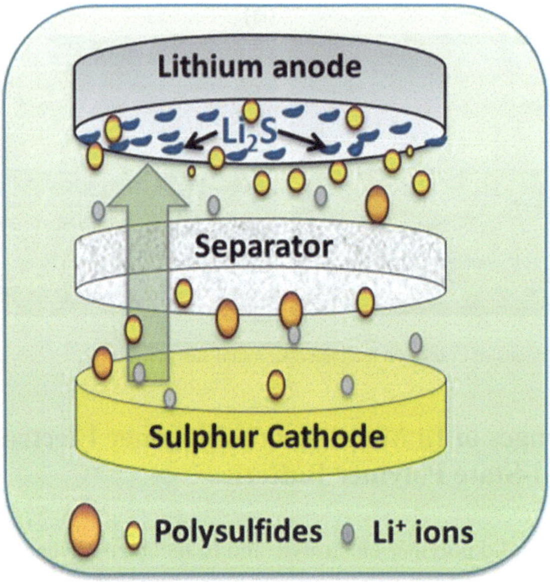

Fig. 15.12 Schematic of a typical Li/S cell with polysulfide shuttling onto the Li metal anode

appreciably soluble in most organic solvents, can easily diffuse out from the cathode side and migrate through the separator to the Li metal anode, a process called shuttling (Fig. 15.12). This shuttling effect passivates the Li metal anodes and results in a loss of active material. Even further the Li metal plating and stripping during cycling increases the surface area of the Li metal anode gradually. This increase causes an uneven current distribution along with other mechanical issues that result in low coulombic efficiency and capacity retention. The use of solid polymer electrolyte being employed can delay such undesired effects such as polysulfide solubility and Li metal dendrite formation, but has not been able to completely overcome this issue.

15.4.3 Oxygen Cathodes

PEO-based polymer electrolytes undergo degradation due to the formation of highly reactive O_2 species (O_2^-, LiO_2 and Li_2O_2) during oxygen reduction reaction (ORR) and oxygen evolution reaction (OER). These electrolytes are also expected to undergo oxidation in an oxygenated environment, as shown in Fig. 15.13. Another issue regarding the use of dry solid polymer electrolytes is the need to operate the cell at a higher temperature that is not considered feasible for Li–O_2 batteries. Apart from PEO polymer hosts, other polymer matrices that could undergo degradation from the formation of Li_2O_2 are poly(vinyl chloride) (PVC), poly(vinylidene fluoride) (PVDF), P(VDF–HFP) and poly(vinylpyrrolidone) (PVP).

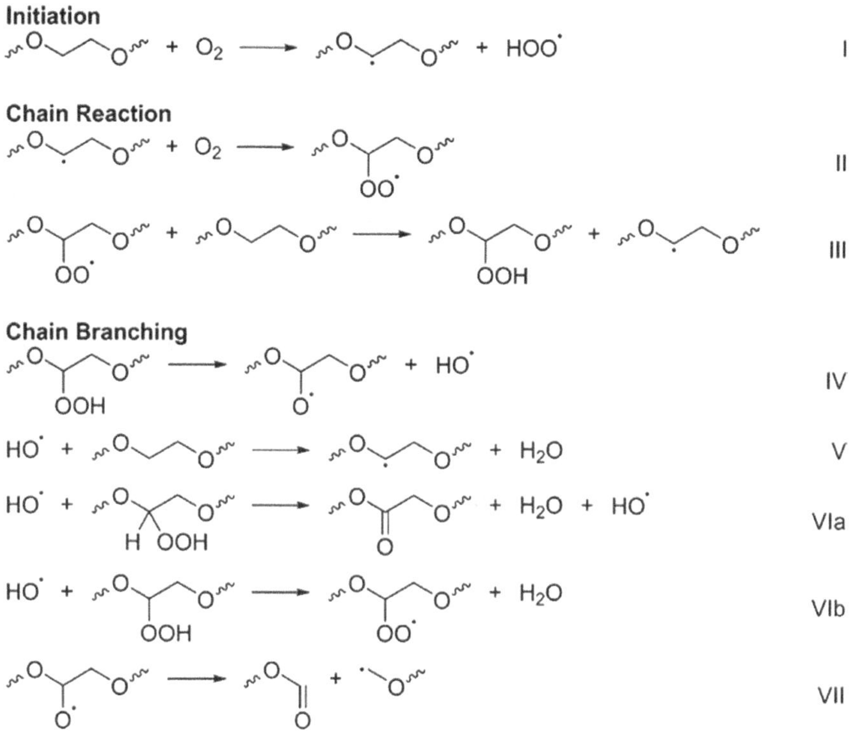

Fig. 15.13 Scheme of PEO oxidation in an oxygenated environment [101]

In order to overcome the issues faced with the Li metal polymer batteries, the processing of Li metal as well as suitable polymer matrices and Li salts are under vigorous research. The manufacturing process of Li metal batteries currently being used at an industrial scale is discussed in the following section.

15.5 Fabrication Process of Li Metal Polymer Battery

The industrial scale fabrication of Li metal polymer batteries is shown in Fig. 15.14. The polymer electrolyte slurry (lithium salts with suitable polymer matrix in solvent) is deposited on a polypropylene support and the solvents are evaporated. Simultaneously, the cathode slurry is prepared by mixing appropriate quantity of polymer, lithium salts, cathode active material and carbon black. The prepared slurry is deposited on the current collector followed by drying.

The dried cathode and electrolyte are then laminated and wound. Lithium metal anode is prepared by extrusion starting from the commercial lithium metal ingot. Once the lithium metal is extruded, thin films are formed by rolling/calendaring to

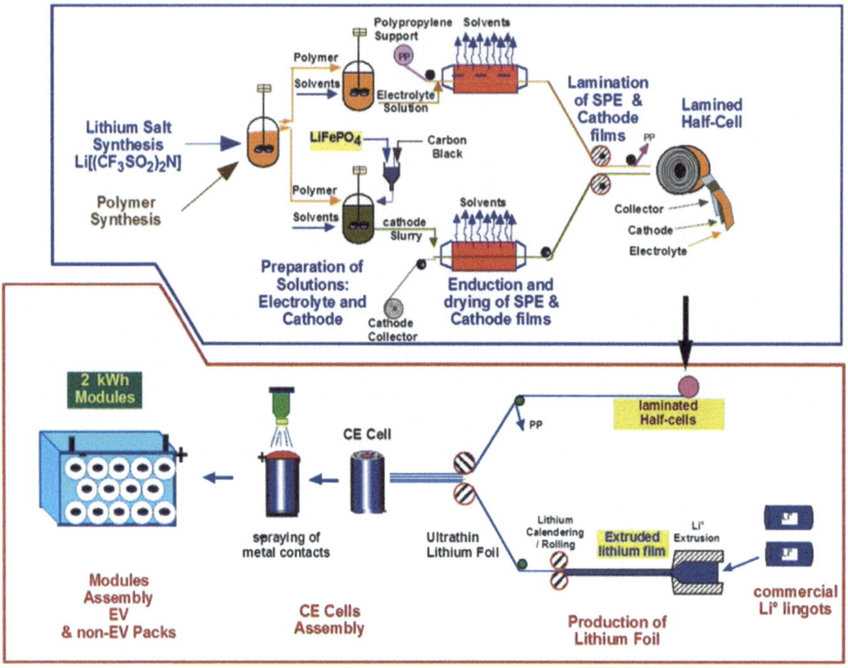

Fig. 15.14 Industrial-scale fabrication of Li metal polymer batteries

form ultra-thin foil (Fig. 15.15). The ultra-thin Li metal foil and the laminated half cell (cathode/electrolyte) are wound. Once the full cell is wound the metal contacts are sprayed and the cell modules are assembled.

Fig. 15.15 Ultra-thin Li metal foil obtained after extrusion and calendering (Courtesy: Hydro-Québec)

The lithium extrusion process followed by calendaring/rolling eliminates most of the crack defects that could be naturally found on a lithium metal surface, thus preventing the formation of nucleation centers for dendrite growth.

15.6 Current Status of Li Metal Polymer Batteries for EV Applications

Armand et al. pioneering work in the field of dry solid polymer electrolytes at Hydro-Québec led to more research efforts in this direction, foreseeing the safety and high energy density that these battery systems could deliver. Further developments were made in 1990s with cells based on lithium metal foil as anode, PEO $LiClO_4/LiN(CF_3SO_2)_2$ as electrolyte and VO_x (vanadium oxide) as cathode that was a collaborative work by Hydro-Québec (Canada) and 3 M (United States). Aiming at the transport sector (EVs) these cell modules displayed 155 KWh/kg specific energy with a cycle life of 600 cycles at 80% depth of discharge (DoD).

Based on this technology, the Bolloré group from France successfully developed the first commercial Li metal polymer battery (LMP) that is composed of $LiFePO_4$/PEO/Li (dry solid polymer electrolyte). These batteries have been claimed to combine high power and energy density with a driving range of 250 km for the cars in which these batteries are mounted. The batteries have been under practical application in cars through a car sharing program named as "blue car" by its subsidiary "Blue solutions" in 2011. The LMP battery has an energy density of 180 Wh kg^{-1} and 80% capacity retention after 1300 cycles. However, the operation temperature of the battery is between 60 and 80 °C. These batteries have also been incorporated in buses and trams that run locally. Currently, the Bolloré group has eight car-sharing services in Europe (Autolib in Paris, Bluely in Lyon, Bluecub Bordeaux and Bluetorino in Turin), in the United States (Blueindy in Indianapolis and BlueLA in Los Angeles) and in Asia (BlueSG in Singapore).

A typical electrochemical cell from Bolloré (Fig. 15.16) combines ultrathin cathode, anode and electrolyte films. The electrochemical cells are connected in series in a module, which in turn are connected in series to form the complete battery pack. In addition, the Blue solution team had been able to replace the coating of the positive electrode via a solvent by a solvent-free extrusion process, a considerable cost advantage. Some water is added to the extrudate that is evaporated in the dry room.

Another breakthrough in dry polymer electrolytes was from SEEO in USA, formerly owned by Bosch in Germany. The DryLyte™ polymer electrolyte developed exhibited a high energy density of 220 Wh kg^{-1} with an operating voltage of 3.42 V. Further developments were made by SEEO using $LiNi_{0.8}Co_{0.15}Al_{0.05}O_2$ (NCA) cathode material, a strategy involving the coating of cathode materials with a high voltage stable polymer binder. The LMP battery, which is still at the R&D level, was found to deliver a stable cycling at 4.2 V. At higher loadings this was further improved to 350 Wh kg^{-1} at 80% depth of discharge (DOD).

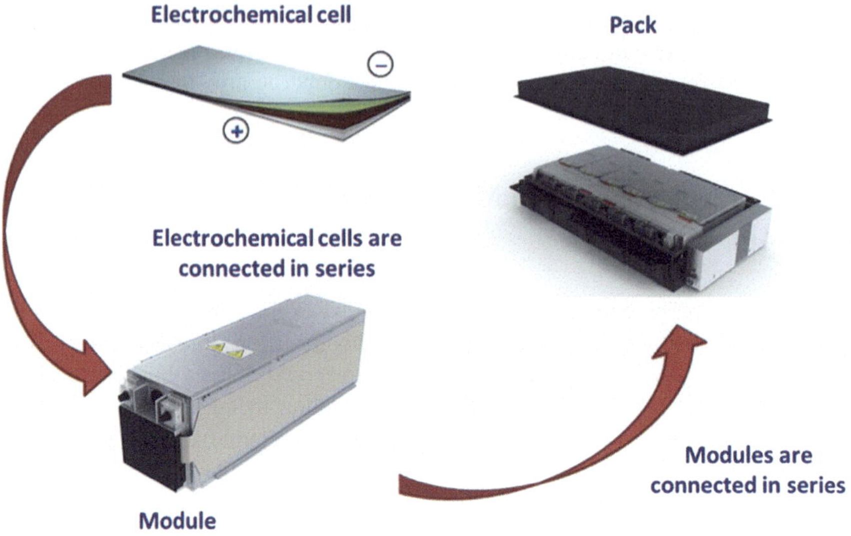

Fig. 15.16 A typical electrochemical cell, module and pack (Courtesy: Bolloré)

15.7 Conclusion and Future Perspectives

Lithium metal polymer batteries seems to be a promising system to enable the use of advanced Li-ion intercalating cathode materials as well as the development of Li/S and Li/O_2 batteries. More research initiatives toward these novel systems would be a plus point to rapidly reach our goals in terms of mobility, grid storage and so on. Li metal polymer batteries apart from the design flexibility for mobile electronic applications and space saver with high energy density for EV application are a safe alternative to liquid electrolyte Li-ion batteries. Moreover, the wide thermal stability of the polymer electrolytes in conjunction with an increased energy density makes them a promising technology. However, since the polymer electrolytes suffer from poor ionic conductivity and kinetics at room temperature, the best solution to bring down the operating temperature would be to move toward Li-ion conducting additives for composite polymer electrolytes or hybrid polymer electrolytes.

Selection of polymer electrolytes with single-ion conducting polymer (block co-polymers) or fillers that could anchor anion in the electrolytes (quasi single-ion conducting) could lead to electrolytes with high rate capability and avoid concentration gradients during cycling, thereby adding to the safety of battery by inhibiting Li dendrite formation. Achieving these above-mentioned conditions could make Li metal polymer electrolyte battery systems versatile without the need of any flame retardants as additive as in the case of liquid electrolytes. This in turn could create a safer bet to other Li-ion battery systems, thus making it a viable technology to reduce the percentage of CO_2 emission from the mobility sector and contribute to the deployment of renewable energy storage.

Acknowledgements I. Gracia, M. Armand and D. Shanmukaraj gratefully acknowledge the financial support from CIC Energigune Co-operative Center, Spain. The authors would further like to thank Dr. Karim Zhagib (Hydro-Québec) and Bolloré for providing the figures and authors/publishers for their generous permission to reproduce the figures.

References

1. Holmberg K, Erdemir A (2017) Influence of tribology on global energy consumption, costs and emissions. Friction 5:263–284
2. Armand MB (1980) Materials for advanced batteries, Springer US, Boston, MA pp 145–161
3. Murphy DW, Broadhead J, Steele BCH (1980) Materials for advanced batteries. Springer, US, Boston, MA, pp 331–342
4. Chabagno JM, Duclot M, Armand MB (1978) Proceedings of the Second International Meeting on Solid Electrolytes, St Andrews, Scotland, 20–22 September 1978
5. Long L, Wang S, Xiao M, Meng Y (2016) Polymer electrolytes for lithium polymer batteries. J Mater Chem A 4:10038–10069
6. Scrosati B, Vincent CA (2000) Polymer electrolytes: the key to lithium polymer batteries. MRS Bull 25:28–30
7. Matsuda Y, Morita M, Tsutsumi H (1993) Polarization behavior of lithium electrode in polymeric solid electrolytes. J Power Sour 44:439–443
8. Song JY, Wang YY, Wan CC (1999) Review of gel-type polymer electrolytes for lithium-ion batteries. J Power Sour 77:183–197
9. Manuel Stephan A, Nahm KS (2006) Review on composite polymer electrolytes for lithium batteries. Polymer (Guildf) 47:5952–5964
10. Youcef HB, Garcia O, Lago N, Shanmukaraj D, Armand MB (2016) Cross-linked solid polymer electrolyte for all-solid—state rechargeable lithium batteries. Electrochim Acta 220:587–594
11. Young WS, Kuan WF, Epps TH (2014) Block copolymer electrolytes for rechargeable lithium batteries. J Polym Sci, Part B: Polym Phys 52:1–16
12. Zhang Q, Liu K, Ding F, Liu X (2017) Recent advances in solid polymer electrolytes for lithium batteries. Nano Research 12:4139–4174
13. Armand MB, Duclot MJ, Rigaud PH (1981) Polymer solid electrolytes: Stability domain. Solid State Ionics 3–4.429–430
14. Goodenough JB, Park KS (2013) The Li–ion Rechargeable battery: a perspect. J Am Chem Soc 135:1167–1176
15. Tarascon JM, Armand M (2001) Issues and challenges facing rechargeable lithium batteries. Nature 414:359–367
16. Berthier C, Gorecki W, Minier M, Armand MB, Chabagno JM, Rigaud Ph (1983) Microscopic investigation of ionic conductivity in alkali metal salts-poly(ethylence oxide) adducts. Solid State Ionics 11:91–95
17. Armand MB (1994) The history of polymer electrolytes. Solid State Ionics 69:309–319
18. Lightfoot P, Mehta MA, Bruce PG (1993) Crystal structure of the polymer electrolytes poly(ethylene oxide)3: LiCF3SO3. Science 262:883–885
19. Roger F, Sangramithra C, Bruce PG (1999) Crystalline and amorphous phases in the poly(ethylene oxide)-$LiCF_3SO_3$ system. Macromolecules 32(3):808–813
20. Lascaud S, Perrier M, Vallée A, Besner S, Prud'homme J, Armand MB (1994) Phase diagrams and conductivity behavior of poly(ethylene oxide)- molten salt rubbery electrolytes. Macromolecules 27:7469–7477
21. Xue Z, He D, Xie X (2015) Poly(ethylene oxide)-based electrolytes for lithium-ion batteries. J Mater Chem A 3:19218–19253

22. Ali U, Karim KJA, Buang NA (2015) A review of the properties and applications of Poly(methyl methacrylate) (PMMA). Polym Rev 55:678–705
23. Ramesh S, Leen KH, Kumutha K, Arof AK (2007) FTIR studies of PVC/PMMA blend based polymer electrolytes. Spectrochim. Acta Part A Mol Biomol Spectrosc 66:1237–1242
24. Ramesh S, Arof AK (2001) Ionic conductivity studies of plasticized poly(vinyl chloride) polymer electrolytes. Mater Sci Eng, B 85:11–15
25. Quartarone E, Mustarelli P, Magistris A (1998) PEO-based composite polymer electrolytes. Solid State Ionics 110:1–14
26. Shin JH, Henderson WA, Passerini S (2005) PEO- based polymer electrolytes with ionic liquids and their use in lithium metal-polymer electrolyte batteries. J Electrochem Soc 152:A978–A983
27. Rajendran S, Sivakumar M, Subadevi R (2003) Effect of salt concentration in poly(vinyl alcohol)—based solid polymer electrolytes. J Power Sources 124:225–230
28. Hu XL, Hou GM, Zhang MQ, Rong MZ, Ruan WH, Giannelis EP (2012) A new nanocomposite polymer electrolyte based on poly(vinyl alcohol) incorporating hypergrafted nano-silica. J Mater Chem 22:18961
29. Xia M, Liu Q, Zhou Z, Tao Y, Li M, Liu K, Wu Z, Wang D (2014) A novel hierarchically structures and highly hydrophilic poly(vinyl alcohol-co- ethylene)/poly (ethylene terephthalate) nanoporous membrance for lithium-ion battery separator. J Power Sour 266:29–35
30. Tominaga Y (2017) Ion-conductive polymer electrolytes based on poly(ethylene carbonate) and its derivatives. Polym J 49:291–299
31. Zhang Q, Liu K, Ding F, Liu XJ (2017) Recent advances in solid polymer electrolytes for lithium batteries. Nano Res 10:4139–4174
32. Rossi NA, West R (2009) Silicon-containing liquid polymer electrolytes for application in lithium ion batteries. Polym Int 58:267–272
33. Karatas Y, Banhatti RD, Kaskhedikar N, Burjanadze M, Funke K, Wiemhöfer HD (2009) Synthesis and Modeling of polysiloxane-based salt-in-polymer electrolytes with various additives. J Phys Chem B 113:15473–15484
34. Burjanadze M, Karatas Y, Kaskhedikar N, Kogel LM, Kloss S, Gentschev AC, Hiller MM, Müller RA, Stolina R, Vettikuzha P et al (2010) Salt-in-Polymer Electrolytes for Lithium Ion Batteries Based on Organo-Functionalized Polyphosphazenes and Polysiloxanes. Zeitschrift für Phys. Chemie 224:1439–1473
35. Assary RS, Curtiss LA, Redfern PC, Zhang Z, Amine K (2011) Computational studies of polysiloxanes: oxidation potentials and decomposition reactions. J Phys Chem C 115:12216–12223
36. Xu W, Angell CA (2001) Weakly Cordinating anions and the exceptional conductivity of their nonaqueous solutions. Electrochem Solid-State Lett 4:E1–E4
37. Xu K, Zhang S, Jow TR, Xu W, Angell CA (2002) LiBOB as Salt for lithium-ion batteries. Electrochem Solid-State Lett 5:A26–A29
38. Aravindan V, Gnanaraj J, Madhavi S, Liu HK (2011) Lithium-ion conducting electrolyte salts for lithium batteries. Chem A Eur J 17:14326–14346
39. Marcinek M, Syzdek J, Marczewski M, Piszcz M, Niedzicki L, Kalita M, Plewa-Marczewska A, Bitner A, Wieczorek P, Trzeciak T et al (2015) Electrolytes for Li–ion transport- Review. Solid State Ionics 276:107–126
40. Ue M (1994) Mobility and Ionic association of lithium and quaternary ammonium salts in propylene carbonate and γ-butyrolactone. J Electrochem Soc 141:3336–3342
41. Dominey L, Koch V, Blakley T (1992) Thermally stable lithium salts for polymer electrolytes. Electrochim Acta 37:1551–1554
42. Younesi R, Veith GM, Johansson P, Edstrom K, Vegge T (2015) Lithium salts for advances lithium batteries: Li-metal, Li–O_2 and Li–S. Energy Environ Sci 8:1905–1922
43. Armand M, Tarascon JM (2008) Building better batteries. Nature 451:652–657
44. Tan S, Ji YJ, Zhang ZR, Yang Y (2014) Recent progress in research on high-voltage electrolytes for lithium-ion batteries. Chem Phys Chem 15:1956–1969

45. Yue L, Ma J, Zhang J, Zhao J, Dong S, Liu Z, Cui G, Chen L (2016) All solid-state polymer electrolytes for high-performance lithium ion batteries. Energy Storage Mater 5:139–164
46. Wright PV (1975) Electrical conductivity in ionic complexes of poly(ethylene oxide). Br. Polym. J.7: 319–327
47. Armand MB, JM Chabagno, Duclot MJ (1979) Fast Ion Transp. Solids, Electrodes Electrolytes (Ed: GKSP Vashitshta, JN Mundy), North Holland Publishers, Amsterdam pp 131–136
48. Choudhary S, Sengwa RJ (2012) Effect of different anions of lithium salt and MMT nanofiller on ion conduction in melt-compounded PEO-LiX-MMT electrolytes. Ionics 18:379–384
49. Vachon C, Labrèche C, Vallée A, Besner S, Dumont M, Prud'homme J (1995) Microphase separation and conductivity behavior of poly(ethylene oxide)—lithium salt electrolytes. Macromolecules 28:5585–5594
50. Sengwa RJ, Dhatarwal P, Choudhary S (2014) Role of preparation methods on the structural and dielectric properties of plasticized polymer blend electrolytes: Correlation between ionic conductivity and dielectric parameters. Electrochim Acta 142:359–370
51. Correia DM, Costa CM, Nunes-Pereira J, Silva MM, Botelho G, Ribelles JLG, Lanceros-Méndez S (2014) Physicochemical properties of poly(vinylidene fluoride-trifluoroethylene)/poly(ethylene oxide) blend membranes for lithium ion battery applications: Influence of poly(ethylene oxide) molecular weight. Solid State Ionics 268:54–67
52. Xiao-Yuan Y, Xiao M, Shuang-Jin W, Qi-Qiang Z, Yue-Zhong M (2010) Fabrication and characterization of PEO/PPC polymer electrolyte for lithium-ion battery. J Appl Polym Sci 115:2718–2722
53. Tanaka R, Sakurai M, Sekiguchi H, Mori H, Murayama T, Ooyama T (2001) Lithium ion conductivity in polyoxyethylene/ polyethylenimine blends. Electrochim Acta 46:1709–1715
54. Yuan F, Chen HZ, Yang HY, Li HY, Wang M (2005) PAN-PEO solid polymer electrolytes with high conductivity. Mater Chem Phys 89:390–394
55. Michiyuki K, Hayashi E, Masayoshi W (1998) Network polymer electrolytes with free chain ends as internal plasticizer. J Electrochem Soc 145:1521–1527
56. Soo PP, Huang B, Jang Y, Chiang YM, Sadoway DR, Mayes AM (1999) Rubbery block copolymer electrolytes for solid state rechargeable lithium batteries. J Electrochem Soc 146:32–37
57. Wang C, Sakai T, Watanabe O, Hirahara K, Nakanishi T (2003) All Solid State lithium- polymer battery using a sself-cross-linking polymer electrolyte. J Electrochem Soc 150:A1166–A1170
58. Xu K (2004) Nonaqueous liquid electrolytes for lithium-based rechargeable. Chem Rev 104:4303–4417
59. Mindemark J, Törmä E, Sun B, Brandell D (2015) Copolymers of trimethylene carbonate and ε-caprolactone as electrolytes for lithium-ion batteries. Polym 63:91–98
60. Hooper R, Lyon LJ, Moline DA, West R (1999) A highly conductive solid state polymer electrolyte based on a double comb polysiloxane polymer with oligo(ethylene oxide) side chains. Organometallics 18:3249–3251
61. Sun B, Mindemark J, Edström K, Brandell D (2014) Polycarbonate-based solid polymer electrolytes for Li–ion batteries. Solid State Ionics 262:738–742
62. Tominaga Y, Yamazaki K (2014) Fast Li–ion conduction in poly(ethylene carbonate)- based electrolytes and composites filled with TiO_2 nanoparticles. Chem Commun 50:4448–4450
63. Okumura T, Nishimura S (2014) Lithium ion conductive properties of aliphatic Polycarbonate. Solid State Ionics 267:68–73
64. Zhang J, Zhao J, Yue L, Wang Q, Chai J, Liu Z, Zhou X, Li H et al (2015) Safety- reinforced poly(propylene carbonate)-based all-solid-state polymer electrolyte for ambient-temperature solid polymer lithium batteries. Adv Energy Mater 5:1501082
65. Hooper R, Lyons LJ, Mapes MK, Schumacher D, Moline DA, West R (2001) Highly Conductive Siloxane Polymers. Macromolecules 34:931–936
66. Fonseca CP, Neves S (2002) Characterization of polymer electrolytes based on poly(dimethyl siloxane-co-ethylene oxide). J Power Sour 104:85–89

67. Oh B, Vissers D, Zhang Z, West R, Tsukamoto H, Amine K (2003) New interpenetrating network type poly(siloxane-*g*-ethylene oxide) polymer electrolyte for lithium battery. J Power Sour 119:442–447
68. Lin Y, Li J, Lai Yuan C, Cheng Y, Liu J (2013) A wider temperature range polymer electrolyte for all-solid-state lithium ion batteries. RSC Adv 3:10722–10730
69. Angell CA, Liu C, Sanchez E (1993) Rubbery solid electrolytes with dominant cationic transport and high ambient conductivity. Nature 362:137–139
70. Angell CA, Xu K, Zhang SS, Videa M (1996) Variations on the salt-polymer electrolyte theme for flexible solid electrolytes. Solid State Ionics 86–88:17–28
71. Forsyth M, Jiazeng S, MacFarlane DR (2000) Novel high salt content polymer electrolytes based on high T_g polymers. Electrochim Acta 45:1249–1254
72. Bushkova OV, Zhukovsky VM, Lirova BI, Kruglyashov AL (1999) Fast ionic transport in solid polymer electrolytes based on acrylonitrile copolymers. Solid State Ionics 119:217–222
73. Forsyth M, MacFarlane D, Hill AJ (2000) Glass transition and free volume behaviour of poly(acrylonitrile)/$LiCF_3SO_3$ polymer-in-salt electrolytes compared to poly(ether urethane)/$LiClO_4$ solid polymer electrolyte. Electrochim Acta 45:1243–1247
74. Ferry A, Edman L, Forsyth M, MacFarlane DR, Sun J (2000) NMR and Raman studies of a novel fast-ion conducting polymer-in-salt electrolyte based on $LiCF_3SO_3$ and PAN. Electrochim Acta 45:1237–1242
75. Wang Z, Gao W, Chen L, Mo Y, Huang X (2002) Study on roles of polyacrylonitrile in "salt-in-polymer" and "polymer-in-salt" electrolytes. Solid State Ionics 154–155:51–56
76. Zalewska A, Pruszczyk I, Sułek E, Wieczorek W (2003) New poly(acrylamide) based (polymer in salt) electrolytes: preparation and spectroscopic characterization. Solid State Ionics 157:233–239
77. Florjańczyk Z, Zygadło-Monikowska E, Wieczorek W, Ryszawy A, Tomaszewska A, Fredman K, Golodnitsky D, Peled E, Scrosati B (2004) Polymer-in-Salt Electrolytes Based on Acrylonitrile/Butyl Acrylate Copolymers and Lithium Salts. J. Phys. Chem. B 108:14907–14914
78. Wu B, Wang L, Li Z, Zhao M, Chen K, Liu S, Pu Y, Li J (2016) Performance of polymer-in-salt electrolyte PAN-LiTFSI enhances by graphene oxide filler. J Electrochem Soc 163:A2248–A2252
79. Fan J, Angell CA (1995) The preparation, conductivity, viscosity and mechanical properties of polymer electrolytes and new hybrid ionic rubber electrolytes. Electrochim Acta 40:2397–2400
80. Łasińska AK, Marzantowicz M, Dygas JR, Krok F, Florjańczyk Z, Tomaszewska A, Zygadło-Monikowska E, Zukowska Z, Lafont U (2015) Study of ageing effects in polymer-in-salt electrolytes based on poly(acrylonitrile-co-butyl acrylate) and lithium salts. Electrochim Acta 169:61–72
81. Zhang H, Li C, Piszcz M, Coya E, Rojo T, Rodriguez-Martinez LM, Armand MB, Zhou Z (2017) Single lithium-ion conducting solid polymer electrolytes: advances and perspectives. Chem Soc Rev 46:797–815
82. Porcarelli L, Shaplov AS, Bella F, Nair JR, Mecerreyes D, Gerbaldi C (2016) Single-ion conducting polymer electrolytes for lithium metal polymer batteries that operate at ambient temperature. ACS Energy Lett. 1:678–682
83. Bouchet R, Maria S, Meziane R, Aboulaich A, Lienafa L, Bonnet JP, Phan TNT, Bertin D, Gigmes D, Devaux D, Denoyel R, Armand MB (2013) Single-ion BAB triblock copolymers as highly efficient electrolytes for lithium-metal batteries. Nat Mater 12:452–457
84. Kim HT, Park JK (1997) Effects of cations on ionic states of poly(oligo-oxyethylene methacrylate-co-alkali metal acrylamidocaproate) single-ion conductor. Solid State Ionics 98:237–244
85. Zhang S, Deng Z, Wan G (1991) Cationic conductivity of blend complexes composed of poly[oligo(oxyethylene) methacrylate] and the alkali metal salts of poly(sulfoalkyl methacrylate) Polym J 23:73–78

86. Geiculescu OE, Yang J, ZhouS Shafer G, Xie Y, Albright J, Creager SE, Pennington WT, DesMarteau DD (2004) Solid Polymer electrolytes from polyanionin lithium salts based on the LiTFSI anion substrate. J Electrochem Soc 151:A1363–A1368
87. Ma Q, Xia Y, Feng W, Nie J, Hu YS, Li H, Huang X, Chen L, Armand M, Zhou Z (2016) Single lithium- ion conducting polymer electrolytes based on a super-delocalized polyanion. RSC Adv. 6:32454–32461
88. Feng S, Shi D, Liu F, Zheng L, Nie J, Feng W, Huang X, Armand MB, Zhou Z (2013) Single lithium-ion conducting polymer electrolytes based on poly[(4-styrenesulfonyl)(trifluoromethanesulfonyl) imide] anions. Electrochim Acta 93:254–263
89. Xu W, Williams MD, Angell CA (2002) Novel Polyanionic Solid electrolytes with weak coulomb traps and controllable caps and spacers. Chem Mater 14:401–409
90. Sun XG, Angell CA (2004) New Sulfone electrolytes part II. Cyclo alkyl group containing sulfones. Solid State Ionics 175:743–746
91. Croce F, Appetecchi GB, Persi L, Scroscati B (1998) Nanocomposite polymer electrolytes for lithium batteries. Nature 394:456–458
92. Yang R, Zhang S, Zhang L, Liu W (2013) Electrical properties of composite polymer electrolytes based on PEO-SN-$LiCF_3SO_3$. Int J Electrochem Sci 8:10163–10169
93. Ha HJ, Kil EH, Kwon YH, Kim JY, Lee CK, Lee SY (2012) UV-curable semi-interpenetrating polymer network- integrated, highly bendable plastic crystal composite electrolytes for shape-conformable all- solid-state lithium ion batteries. Energy Environ Sci 5:6491–6499
94. Zhou D, He YB, Liu R, Liu M, Du H et.al. (2015) In-situ synthesis of a hierarchical all-solid state electrolyte based on nitrile materials for high performance Lithium ion batteries. Adv Energy Mater.5: https://doi.org/10.1002/aenm.201500353
95. Croce F, Curini R, Martinelli A, Persi L, Ronci F, Scrosati B, Caminiti R (1999) Physical and Chemical properties of nanocomposite polymer electrolytes. J. Phys. Chem. B 103:10632–10638
96. Fan L, Nan CW, Zhao S (2003) Effect of modified SiO_2 on the properties of PEO-based polymer electrolytes. Solid State Ionics 164:81–86
97. Croce F, Persi L, Ronci F, Scrosati B (2000) nanocomposite polymer electrolytes and their impact on the lithium battery technology. Solid State Ionics 135:47–52
98. Skaarup S, West K, Zachau Christiansen B (1988) Mixed Phase solid electrolytes. Solid State Ionics 28–30:975–978
99. Liu B, Zhang JG, Xu W (2018) Advancing Lithium Metal batteries. Joule 2:833–845
100. Gireaud L, Grugeon S, Laruelle S, Yrieix B, Tarascon JM (2006) Lithium metal stripping/plating mechanisms studies: A metallurgical approach. Electrochem Commun 8:1639–1649
101. Harding JR, Amanchukwu CV, Hammond PT, Shao-Horn Y (2015) Instability of Poly(ethylene oxide) upon oxidation in Lithium—Air batteries. J Phys Chem C 119:6947–6955

The manufacturer's authorised representative in the EU is Springer Nature Customer Service Centre GmbH, Europaplatz 3, 69115 Heidelberg, Germany. If you have any concerns regarding our products, please contact ProductSafety@springernature.com

Printed and bound by CPI Group (UK) Ltd, Croydon, CR0 4YY
30/09/2026
02225652-0014